MOLECULAR
BIOLOGY
INTELLIGENCE
UNIT

Transcription Factors in Eukaryotes

Athanasios G. Papavassiliou, M.D., Ph.D.
Department of Biochemistry
School of Medicine
University of Patras
Patras, Greece

CHAPMAN & HALL
I(T)P An International Thomson Publishing Company

New York • Albany • Bonn • Boston • Cincinnati • Detroit • London • Madrid • Melbourne •
Mexico City • Pacific Grove • Paris • San Francisco • Singapore • Tokyo • Toronto • Washington

LANDES BIOSCIENCE
AUSTIN, TEXAS
U.S.A.

MOLECULAR BIOLOGY INTELLIGENCE UNIT

Transcription factors in eukaryotes

LANDES BIOSCIENCE
Austin, Texas, U.S.A.

Please address all inquiries to the Publishers:
Landes Bioscience, 810 South Church Street, Georgetown, Texas, U.S.A. 78626
Phone: 512/ 863 7762; FAX: 512/ 863 0081

North American distributor:

Chapman & Hall, 115 Fifth Avenue, New York, New York, U.S.A. 10003

CHAPMAN & HALL

U.S. and Canada ISBN: 0-412-12501-3

Library of Congress Cataloging-in-Publication Data

Transcription factors in eukaryotes/ [edited by] Athanasios Papavassiliou.
p. cm. — (Molecular biology intelligence unit)
Includes bibliographical references and index.
ISBN 1-57059-503-8 (alk. paper)
1. Transcription factors. I. Papavassiliou, Athanasios, 1961-. II. Series.
[DNLM: 1. Transcription Factors. 2. Eukaryotic Cells. 3. Transcription, Genetic. 4. Gene Expression Regulation. 5. Signal Transduction. 6. Cell Cycle—genetics. QH 450.2 T7718]
QP552.T68T744 1997S
572.8'845—dc21
DNLM/DLC
for Library of Congress

97-28283
CIP

Publisher's Note

Landes Bioscience produces books in six Intelligence Unit series: *Medical, Molecular Biology, Neuroscience, Tissue Engineering, Biotechnology* and *Environmental.* The authors of our books are acknowledged leaders in their fields. Topics are unique; almost without exception, no similar books exist on these topics.

Our goal is to publish books in important and rapidly changing areas of bioscience for sophisticated researchers and clinicians. To achieve this goal, we have accelerated our publishing program to conform to the fast pace at which information grows in bioscience. Most of our books are published within 90 to 120 days of receipt of the manuscript. We would like to thank our readers for their continuing interest and welcome any comments or suggestions they may have for future books.

Shyamali Ghosh
Publications Director
Landes Bioscience

CONTENTS

EDITOR

Athanasios G. Papavassiliou, M.D., Ph.D.
Department of Biochemistry
School of Medicine
University of Patras
Patras, Greece
Chapter 1

CONTRIBUTORS

Matthias Austen, Ph.D.
Institut für Molekularbiologie
Medizinische Hochschule Hannover
Hannover, Germany
Chapter 13

Aria Baniahmad, Ph.D.
Genetisches Institut
Justus-Liebig-Universität
Giessen, Germany
Chapter 6

Peter B. Becker, Ph.D.
Gene Expression Programme
European Molecular Biology Laboratory
Heidelberg, Germany
Chapter 18

Rachel C. Davies, B.Sc., Ph.D.
MRC Human Genetics Unit
Western General Hospital
Edinburgh, U.K.
Chapter 16

Martin Eggert, Ph.D.
Genetisches Institut
Justus-Liebig-Universität
Giessen, Germany
Chapter 6

Jonathan Frampton, Ph.D.
Institute of Molecular Medicine
John Radcliffe Hospital
University of Oxford
Oxford, U.K.
Chapter 15

Michael Z. Gilman, Ph.D.
ARIAD Pharmaceuticals, Inc.
Cambridge, Massachusetts, U.S.A.
Chapter 7

Martin Göttlicher, M.D.
Institut für Genetik
Forschungszentrum Karlsruhe
Karlsruhe, Germany
Chapter 5

Thomas Graf, Ph.D.
Cell Regulation Programme
European Molecular Biology Laboratory
Heidelberg, Germany
Chapter 15

Michael R. Green, M.D., Ph.D.
Howard Hughes Medical Institute
Program in Molecular Medicine
University of Massachusetts Medical Center
Worcester, Massachusetts, U.S.A.
Chapter 17

Nicholas D. Hastie, B.Sc., Ph.D.
MRC Human Genetics Unit
Western General Hospital
Edinburgh, U.K.
Chapter 16

Marie Henriksson, Ph.D.
Institut für Molekularbiologie
Medizinische Hochschule Hannover
Hannover, Germany
Chapter 13

Peter Herrlich, M.D., Ph.D.
Institut für Genetik
Forschungszentrum Karlsruhe
Karlsruhe, Germany
Chapter 5

Stefanie Hilfenhaus, B.Sc.
Institut für Molekularbiologie
Medizinische Hochschule Hannover
Hannover, Germany
Chapter 13

Barbara Kistler, M.Sc.
Institut für Medizinische Strahlenkunde und Zellforschung
Universität Würzburg
Würzburg, Germany
Chapter 9

Monica Lamas, Ph.D.
Institut de Génétique et de Biologie Moléculaire et Cellulaire
CNRS - INSERM - Université Louis Pasteur
Illkirch-Strasbourg, France
Chapter 4

David S. Latchman, M.A., Ph.D., D.Sc., M.R.C.Path.
Department of Molecular Pathology
University College London Medical School
London, U.K.
Chapter 11

Xiao-Yong Li, Ph.D.
Howard Hughes Medical Institute
Program in Molecular Medicine
University of Massachusetts Medical Center
Worcester, Massachusetts, U.S.A.
Chapter 17

Bernhard Lüscher, Ph.D.
Institut für Molekularbiologie
Medizinische Hochschule Hannover
Hannover, Germany
Chapter 13

Alfred Nordheim, Ph.D.
Institut für Molekularbiologie
Medizinische Hochschule Hannover
Hannover, Germany
Chapter 14

Christos A. Ouzounis, Ph.D.
The European Bioinformatics Institute
European Molecular Biology Laboratory
Cambridge, U.K.
Chapter 1

Giovanni Perini, Ph.D.
Howard Hughes Medical Institute
Program in Molecular Medicine
University of Massachusetts Medical Center
Worcester, Massachusetts, U.S.A.
Chapter 17

B. Franklin Pugh, Ph.D.
Center for Gene Regulation
Department of Biochemistry and Molecular Biology
The Pennsylvania State University
University Park, Pennsylvania, U.S.A.
Chapter 3

Hans J. Rahmsdorf, Ph.D.
Institut für Genetik
Forschungszentrum Karlsruhe
Karlsruhe, Germany
Chapter 5

Rainer Renkawitz, Ph.D.
Genetisches Institut
Justus-Liebig-Universität
Giessen, Germany
Chapter 6

Paolo Sassone-Corsi, Ph.D.
Institut de Génétique et de Biologie Moléculaire et Cellulaire
CNRS - INSERM - Université Louis Pasteur
Illkirch-Strasbourg, France
Chapter 4

Michèle Sawadogo, Ph.D.
Department of Molecular Genetics
M.D. Anderson Cancer Center
University of Texas
Houston, Texas, U.S.A.
Chapter 2

Ralf Schreck, Ph.D.
Institut für Medizinische Strahlenkunde und Zellforschung
Universität Würzburg
Würzburg, Germany
Chapter 9

Erika C. Shugart, B.Sc.
University of Virginia
Charlottesville, Virginia, U.S.A.
Chapter 8

Anette Sommer, B.Sc.
Institut für Molekularbiologie
Medizinische Hochschule Hannover
Hannover, Germany
Chapter 13

David J.D. Sourdive, Ph.D.
Unité des Virus Oncogènes
Département des Biotechnologies
Institut Pasteur
Paris, France
Chapter 10

Marilyn N. Szentirmay, Ph.D.
Department of Molecular Genetics
M.D. Anderson Cancer Center
University of Texas
Houston, Texas, U.S.A.
Chapter 2

Robert M. Umek, Ph.D.
University of Virginia
Charlottesville, Virginia, U.S.A.
Chapter 8

Marie-Luce Vignais, Ph.D.
Institut de Génétique Moléculaire de Montpellier
CNRS
Montpellier, France
Chapter 7

Bohdan Wasylyk, B.Sc., Ph.D.
Institut de Génétique et de Biologie Moléculaire et Cellulaire
CNRS - INSERM - Université Louis Pasteur
Illkirch-Strasbourg, France
Chapter 14

Thomas Wirth, Ph.D.
Institut für Medizinische Strahlenkunde und Zellforschung
Universität Würzburg
Würzburg, Germany
Chapter 9

Ronald M. Wisdom, M.D.
Department of Biochemistry
Vanderbilt University School of Medicine
Nashville, Tennessee, U.S.A.
Chapter 12

Moshe Yaniv, Ph.D.
Unité des Virus Oncogènes
Département des Biotechnologies
Institut Pasteur
Paris, France
Chapter 10

PREFACE

A painter one afternoon drew a train.
The last carriage cut away from the paper
and returned to the carbarn all by itself.

In precisely that carriage sat the artist.

Yánnis Rítsos, *Abstracted Painter*

Transcription of eukaryotic protein-coding genes is controlled by the interaction of diffusible proteins, called transcription factors, with specific regulatory DNA sequences, usually located in the upstream promoter region of the gene in question. Once bound to DNA, transcription factors alter expression of their target genes by interacting with other components of the transcriptional machinery (RNA polymerase complex and accessory factors). Thus, transcription factors contain at least two functional domains: (1) a DNA-binding domain, responsible for binding of the factor to specific recognition elements and (2) an activation domain, responsible for stimulating (or repressing) transcription through direct or indirect protein-protein interactions.

Regulating the expression and function of transcription factors clearly affords the cell a panoply of controls on tissue-specific and inducible gene expression. For instance, some transcription factors are expressed only in particular differentiated cell types and thus appear to control developmentally programmed patterns of gene expression. In other cases, the expression or activity of ubiquitous transcription factors is modulated either positively or negatively in response to growth factors, hormones, cytokines, antigens, oxidants, ultraviolet light etc.; therefore these transcription factors can be viewed as components in signal transduction pathways that regulate gene expression in response to a broad spectrum of extracellular stimuli. In addition, a variety of oncogenes and tumor suppressor genes have been found to encode transcription factors which are not only crucial to the development of many cancers, but also play pivotal roles in regulatory systems governing the proliferation and differentiation of normal cells. Consequently, understanding transcription factors is important not only from the standpoint of researchers working with them, but has become a key area for scientists interested in diverse aspects of biology and medicine.

Transcription Factors in Eukaryotes seeks to guide advanced students and postgraduates, physicians possessing strong background in the basic biological sciences and whose research interests relate to investigating molecular mechanisms of disease, and the specialist, through a survey of this fascinating field, by providing a state-of-the-art presentation of the nature of cellular and viral transcription factors and the manner in which they accomplish their role of affecting genetic programs in a wide range of situations. It is obviously impossible in a book of this size to cover the entire gamut of eukaryotic transcription factors comprehensively. Instead, a variety of well-defined systems were selected to serve as paradigms, together with a number of complex regulatory systems that are still in the burgeoning phase. The aim is to discuss in detail the available knowledge on several topical areas and to convey the broad scope of the theme and its relevance to other branches of molecular/cell biology and pathobiology. In doing this, and to give the book a personal flavor authors were encouraged to present the topics with their special perspectives, highlighting their individual research styles. Each chapter begins with an appreciable amount of background information, goes on to illuminate current understanding in the subject under review (this part varies in length among chapters) and, in most cases, concludes with an outlook for future directions. Emphasis has been given to the structure-function relationships of transcription factors in the context of cytoplasmic and nuclear milieux, their modes of action as well as their interplay in signal-dependent and tissue-specific gene regulation and the latest findings regarding their involvement in various aspects of cell growth, differentiation and oncogenesis. In editing the book, a conscious effort was made to limit redundancy among chapters. Where it was not done so, it is because it was felt that certain redundancies preserved the internal coherence of chapters and would serve the purposes of readers who might be using the book as an occasional reference.

Over the past ten years, striking progress has been made in elucidating the roles of transcription factors in a plethora of cellular processes and the development of multicellular organisms. None the less, the study of these vital proteins continues to be a flourishing area of biological science, our view of which is far from complete. A vast array of information continues to emerge at a rapid pace, and there remain many unresolved issues. Perhaps preeminent among them, posing a critical challenge for future exploration, is the problem of translating our advancements in understanding the intricate biochemistry and pathophysiology of transcription factors into practical progress in disease treatment and prevention. It is therefore hoped that this monograph will provide a conceptual framework to help the reader realize not only where we stand now, but also where future research may lead.

"....When the harvest was done and the hopes of crickets began."
Níkos Gátsos, *Amorgós*

Athanasios G. Papavassiliou
Patras, May 1997

Acknowledgments

I wish to express my deep appreciation to all of the contributing authors whose individual and research experience, as well as collective enthusiasm and interest, made this endeavor possible. I owe considerable debts to Renate Wise and Lynn O'Neill at Landes Bioscience for their most helpful advice and assistance. And, as ever, Effie has been wonderfully patient in the face of many late nights and early mornings.

CHAPTER 1

DNA-Binding Motifs of Eukaryotic Transcription Factors

Christos A. Ouzounis and Athanasios G. Papavassiliou

INTRODUCTION

Eukaryotic transcription factors are a very broad category of proteins that participate in the regulation of gene expression in higher organisms. This category is very heterogeneous, but many of its members are some of the most well-studied proteins to date. Although the subject has been reviewed extensively in some excellent publications[1] and it may seem redundant to write yet another review, we will attempt to show that there exist very many unsolved problems, even in such an area of intense study.

In this chapter, we will review the DNA-binding domains of a wide variety of transcription factors from a structural (not a historical) perspective. The classes of these domains will be defined using "molecular anatomy" criteria, if possible. Names associating the three-dimensional patterns with familiar objects will be adopted, according to tradition. Only in the absence of structural information will the functional associations of transcription factors will be used.

It is important to emphasize that many of the proteins (or indeed their DNA-binding domains) discussed here do not act in isolation but in concert with other molecules. Even so, an overview of their structural properties, and comparison with similar domains that may have different functions, can be instructive.

We will begin with a number of definitions, because even for such a familiar topic inaccuracies can creep in. Next, we will present an overview of the major DNA-binding domains of eukaryotic transcription factors, most of which will be encountered throughout the chapters of this book. Finally, we will raise a number of questions that may be used as guidelines for future research.

DEFINITIONS

What Is a Domain?

A structural domain can be defined as an independently folding unit within a protein that can be present in different contexts. A domain can be part of a protein, but a protein can

Transcription Factors in Eukaryotes, edited by Athanasios G. Papavassiliou.

also consist of a single domain. A protein in general and a transcription factor in particular may have multiple copies of the same or different domains. Similar domains do not necessarily share detectable sequence similarity and thus can be accurately defined only by using structural information.

What Is a Motif?

A sequence motif is a specific signature of a protein family that uniquely describes most of its members and can be used as a diagnostic pattern. A motif may belong to a domain or a protein. A single domain may have many motifs, but a single motif cannot be shared by unrelated domains. Motifs are assumed to be divergently related, all descending from a common ancestral protein. By definition, similar motifs should share sequence similarity.

What Is a Transcription Factor?

A transcription factor can be defined as a protein with a direct effect on transcription. The effect can be activation or repression of gene expression at the level of RNA transcription, and can be mediated either by direct binding (specific or not) to DNA, or by indirect interactions with other transcription factors. Thus, not all transcription factors are DNA-binding.

Traditionally, a *factor* was defined in a negative sense if the protein was not an enzyme (for which a clear nomenclature existed). In some cases, enzymes can also be transcription factors, or common domains can be shared between enzymes, or other proteins, and transcription factors.[2] In general, domains transcend the boundaries of a functionality class, and are present in proteins with different activities.

Thus, a more accurate definition of a transcription factor is: a regulatory molecule with a direct effect on transcription, whose substrate is another macromolecule and not a small molecule.

What Is a Eukaryotic Gene?

This is probably the most difficult question to answer. Very few eukaryotic organisms have been studied in molecular detail and the current databases contain extreme biases from medically or industrially important model species. In addition, it is not possible at the moment to predict the phylogenetic extent of a gene found in a given species. It can be assumed, however, that for any given gene, there is a large number of related species that contain a similar copy. Therefore, even if a unique gene is available, it is a safe assumption to state that this gene is the only known member of a gene family, with similar members existing within the same genome or across species. Most of the families discussed here have members in model eukaryotes such as yeast, *Drosophila* or human. It is important to bear in mind, however, that vast areas of the eukaryotic tree remain totally unknown: in plants or protista, there is very little information at present about various categories of transcription factors such as the ones discussed here.

A working definition is that a eukaryotic transcription factor is a protein family that has at least one known member in eukaryotes, while at the same time it is not known to be present in bacteria or archaea. This definition leaves open the issue of phylogenetic distribution: its members may be contained in a single species, a complete taxon (e.g., yeasts) or be present across the eukaryotic tree.

Some Persisting Difficulties

It follows from the previous definitions that the few transcription factor domains that are present both in eukaryotes and prokaryotes (bacteria and archaea) are not strictly eukaryotic, and will not be discussed here. Viral proteins will only be mentioned if they are members of the families discussed here (chapter 17 provides extensive coverage of this topic).

On the other hand, domains that are not DNA-binding, such as the leucine zipper domain responsible for dimerization in a number of eukaryotic transcription factors, will be briefly mentioned in conjunction with their associated DNA-binding domains (structural details on these domains can be found in the corresponding chapters).

Finally, certain regions in transcription factors cannot be strictly defined as domains, due to the unavailability of structural data: such examples are the glutamine- and proline-rich areas in a large number of proteins, assumed not to have a defined three-dimensional conformation. Such cases will not be mentioned here, but are discussed in certain chapters of the book.

The reader may have also noticed that in the definitions the issues of affinity and specificity were not included. The reason for this omission is that it is not generally known for all members of a certain class how these properties change with sequence similarity. Some transcription factors bind DNA with high specificity, while others have a broad specificity towards certain regulatory signals. A transcription factor may bind one or more promoters, enhancers, other elements yet undefined, or even other proteins.

Due to its pivotal importance and tremendous complexity, the process of transcription initiation has been much better studied than transcription elongation or termination. Although proteins involved in the latter two processes can also be defined as transcription factors, very little is known about them, compared with transcription initiation proteins (activators or repressors). It is anticipated that important steps in the regulation of gene expression at these levels will be discovered in the future, but until that moment we can only underline our ignorance.

Finally, if we followed our definitions strictly, many proteins (especially the general transcription initiation factors and also the core histones) would be excluded, since they have been discovered in archaea, organisms that preceded (and probably gave rise to) the higher forms of nucleated cells defined as eukaryotes. However, it is important to emphasize that many of the components of eukaryotic transcription seem to have been present before the emergence of the nucleus. In a very real sense, basic eukaryotic transcription is archaeal-like, and not vice versa.

All previous definitions and statements, followed in this chapter, serve a specific purpose: by providing a balanced view of the topic, we will be able to perceive the wide variety of different mechanisms that are available for the regulation of gene expression. Our views about DNA-binding domains of eukaryotic transcription factors can be less influenced by the classical cases, and enhanced by an equal treatment of natural diversity.

What Is the Organization into Classes?

Many transcription factors act synergistically within their functional continuum and usually interact with a number of associated proteins, some of which may also be transcription factors. Although such a classification can be illuminating, we will restrict the presentation to a structural perspective. Only in the absence of structural information will the functional viewpoint be adopted. We will see that it is not uncommon that these two schemes can in fact coexist, as many structurally related proteins are also functionally associated. What follows is a brief overview of the known families of DNA-binding domains of transcription factors in eukaryotes.

CLASSES OF DNA-BINDING DOMAINS OF EUKARYOTIC TRANSCRIPTION FACTORS

The Zinc Class

Numerous reviews have been written on zinc-containing domains in transcription factors and other proteins.[3-6] Since more than a decade ago, when the first zinc-'finger' domain was discovered (a historical note has been previously presented),[5] hundreds of proteins including transcription factors have been identified as zinc-binding.[7] In transcription factors domains, zinc (in the form of Zn^{++}) plays a structural role, stabilizing these short structural units by ionic interactions involving histidine or cysteine residues.[4] The variety and extent of zinc-binding domains in transcription can only be underestimated. At least four different types of zinc-binding domains have been described, which by no means are restricted to transcription.[6] The boundaries of these domains are not well-defined, since their conserved sequence patterns share elements that are suggestive of divergent evolution. In some cases, where no sequence similarity can be detected, these similar domains may have been discovered more than once during evolution.

Ten Types of Zinc-Binding Domains

There are at least ten zinc-finger types known to date,[7] that can be classified according to the type and arrangement of their amino acid ligands. Here, we will only discuss whose members are involved in transcription: the classical TFIIIA-like type, the hormone receptors, the binuclear cluster type, GATA-1, zinc ribbon, the LIM domain and finally the RING finger. The other types are the gag-like viral nucleocapsid proteins, a domain found in methionine-tRNA synthetase and the cysteine-rich domain in protein kinase C.[7]

The TFIIIA-Like Domain

A zinc-binding domain about 30 residues long was first discovered in transcription factor TFIIIA.[8] It contains two N-terminal cysteines and two C-terminal histidines that bind a zinc ion in tetrahedral coordination, and it is defined as a Cys2His2 (C2H2) zinc finger. The structure is a two-stranded ladder followed by a single helix: the cysteines are located on the β strands and the histidines on the α helix.[9] Usually this domain is found in multiple copies within a single protein.[10] Some degenerate copies can also be present. This domain has been found across eukaryotes, from yeast to human.

Hormone Receptors

The nuclear hormone receptors are a superfamily[11] of large proteins composed of three domains: the N-terminal modulation domain, a highly conserved central DNA-binding domain and a C-terminal hormone recognition domain.[12] The DNA-binding domain is about 65 residues long forming two subdomains, with four cysteines each that bind one zinc ion. It recognizes DNA sequences called hormone response elements (HREs)[13] (see also chapter 6). The first subdomain is responsible for specific DNA binding (P-box) while the second subdomain (D-box) facilitates dimerization of the domains.[14] Extensive structural studies have been performed on various nuclear hormone receptor DNA-binding domains,[15-18] with the main focus on the specificity of interactions.[19] Recent structural work underlines the importance of conformational changes on ligand binding (ref. 20 and references therein). Each subdomain contains four strands followed by a helix: the two helices are perpendicular to each other and resemble the GATA-1 and LIM domains (see below). The domain is present in animal phyla but has not been found in protista, yeasts or plants.

The Binuclear Zn2-Cys6 Cluster

A large number of fungal transcription factors contain a binuclear zinc cluster coordinated by six cysteine residues (Zn2-Cys6). The prototypical member of the family is GAL4. The domain is about 30 residues long and the cysteines are located at variable distances. It forms an irregular Y shape, with two short α helices (ref. 21 and references therein.) In general, this domain is found in the N-terminal section of these regulatory proteins. Fungal GAL4-like transcription factors are not found elsewhere in the eukaryotic tree.

The GATA-1 Domain

This domain is found in proteins that bind to a short consensus sequence (GATA) found in a number of genes. It is about 50 residues long, and has four cysteines that bind zinc. The structure contains two ladders, each having two β strands, and a C-terminal helix.[22] This type of zinc-containing DNA-binding domain is found in transcription factors from yeast to vertebrates.

Zinc Ribbon

A domain first found in transcription elongation factor TFIIS is about 50 residues long, was later discovered in an RNA polymerase 30 kDa subunit in vaccinia viruses, yeast, a small subunit (15 kDa) of RNA polymerase II in higher animals and also in archaea.[23] It contains four cysteines located at the top of the three-strand domain, and it binds DNA only in the presence of RNA polymerase II.[24] Recently, the same conformation (with no detectable sequence similarity) was found in the N-terminal domain of TFIIB[25] (see below). Although not exclusively eukaryotic, the domain is mentioned because of its unique all-β structure that extends the spectrum of zinc domains.

The LIM Domain

The LIM domain has been identified in a number of animal proteins, some of which are known to be transcription factors. This domain can occur in multiple copies within a single protein, and often in association with a homeodomain (see below). It is about 80 residues long, and contains two zinc-binding clusters, with a C2HC/C4 arrangement. The structure is reminiscent of a duplication.[26] The second subdomain is most similar to that of the GATA-1 family and to the first of the two subdomains in the hormone receptor family,[26] thus suggesting a role in DNA binding. While this domain is found in various proteins across eukaryotes, its function remains unknown; however, there is mounting evidence that this domain is a protein-protein interaction module.[27]

The RING Finger

A motif about 50 residues long has been identified in a large number of proteins with very diverse roles and important disease phenotypes. It forms a single domain and it contains eight ligands in a C3HC4 arrangement that forms two binding sites. The domain has a strand-strand-helix-strand topology, with the three strands forming a small sheet.[28] The RING finger is spread throughout eukaryotes and it is expected to have an important function (yet to be discovered). The RING finger domain has been reviewed recently.[29]

The PIC class

It is virtually impossible to summarize the complexity of the transcription preinitiation complex (PIC) in a single paragraph. To tag eukaryotic promoters, a large number of proteins associate with the TATA box (in type II promoters) or the corresponding elements in types I and III. Structural information exists for only two molecules associated with type II promoters, namely the TATA-binding protein TBP (a subunit of the general transcription factor TFIID) and TFIIB.

The Saddle Domain

Fine structural work has revealed the structure of the TBP,[30-33] one of the most important proteins in transcription. This protein recognizes the TATA box of type II promoters and is responsible for the formation of preinitiation complex where DNA-dependent RNA polymerase II starts transcription[34] (see also chapters 2 and 3). The DNA-binding C-terminal domain of TBP is approximately 180 residues, consisting of two highly conserved repeats of 90 residues. The structure of each repeat is a saddle-like domain with two helices and five strands, symmetrically arranged to form a ten-strand sheet concave surface which binds to DNA.[34] It has been found in all eukaryotic organisms, and also in archaea. Interestingly, pterin-4α-carbinolamine dehydratase (PCD) (dimerization cofactor of hepatic nuclear factor 1 (HNF1) or DCoH) has a similar saddle-like structure.[35] It is puzzling that PCD/DCoH is a transcriptional activator, associating with HNF1 for the regulation of liver-specific gene expression[35] (see also chapter 10).

A Cyclin Domain in TFIIB

TFIIB is the second protein that enters the PIC, forming a TFIIB-TFIID (TBP)-DNA complex that is subsequently recognized by RNA polymerase II and TFIIF[34] (see also chapters 2 and 3). TFIIB contains two domains, a N-terminal zinc-binding domain forming a zinc ribbon (see above) and a C-terminal core domain whose structure was recently solved.[36] The DNA-binding properties of TFIIB are probably mediated through the zinc ribbon[25] but binding with the C-terminal domain cannot be excluded.[36] The C-terminal core domain is formed by five helices per repeat, and is similar to the structure of cyclin A,[36] which contains the same structural elements.[37] Interestingly, the relationship is detectable at the sequence level (not shown), therefore strongly suggesting divergent evolution and cyclin origins from the general transcription machinery.[38] Members of this family have been found in archaea.[39] Transcription factor TFIIIB of the RNA polymerase III system is also similar to TFIIB.[40]

The Boot Domain of TFIIA

TFIIA is another general transcription factor that is required for transcription initiation of type II promoters (see also chapters 2 and 3). The structure of the TFIIA-TBP-DNA complex has been solved recently.[41] TFIIA is a large heterodimeric protein, composed of a large subunit (TOA1) and a small subunit (TOA2). The C-terminal domains of TOA1 and TOA2 each contribute six strands to form a 12-strand mixed β barrel, preceded by a four-helix bundle, formed by the N-terminus of TOA1.[41] The whole structure has a boot-like appearance, with the large β barrel forming the leg and the helix bundle forming the foot of the boot.[41] The protein binds the promoter mostly with the large subunit, through residues belonging to the edge of the barrel. This structure remains unique, bearing no resemblance to any other known fold. Although members of this family have not been found in archaea (as for TBP or TFIIB), TFIIA is present in all eukaryotic phyla examined so far.

Chromatin-associated folds

Chromatin structure, mediated by histones and nonhistone proteins, and nucleosomal displacement are key elements that relate to transcription in eukaryotes[42,43](see also chapter 18). However, chromatin-associated proteins such as histones were never considered as transcription factors in the classical sense. Recent discoveries relate core histones (H2A/B, H3 and H4) to TAF subunits and CBF-A and -C, and the globular domain of histone H5 to the HNF3-γ factor (see below), thus bringing histone structure into the issue of transcription.

TAF Subunits

A number of transcription factors, collectively known as TBP-associated factors (TAFs), are implicated in transcription activation in eukaryotes[44] (see also chapters 2 and 3). Recently, two subunits of TAFs, $dTAF_{II}42$ and $dTAF_{II}62$ from *Drosophila*, were shown to contain the core histone fold.[45] In addition, biochemical evidence suggests that other TAFs such as hTAF80, hTAF41 and hTAF20/15 from human also have a similar structure, thus indicating that the TFIID complex may contain a histone octamer-like TAF complex.[46]

CCAAT-Binding Subunits

The CCAAT-binding factor (CBF) is a family of conserved transcription factors (also known as NF-Y or CP1) that specifically interact with CCAAT motifs in a number of eukaryotic promoters,[47] in a heterotrimeric form composed of three subunits.[48] Subunits A and C are approximately 200 residues long, and their central region (55 residues long, positions 64-118 of CBF-A) has been shown to be homologous to the core histone fold.[49] It is not clear how CBF-B is involved in DNA binding. Members of the CBF family have been identified throughout eukaryotic phyla, and must have originated before the emergence of eukaryotes.[49]

The HMG-Box

High mobility group (HMG) proteins comprise the nonhistone elements in chromatin. One such group comprises the HMG-1/-2/-D proteins and the product of the testis-determining gene SRY (sex-determining region Y protein). These proteins contain a region called the HMG-box, approximately 90 residues long, which is the DNA-binding domain. Transcription factors LEF-1 and UBF also belong to this family and contain the HMG-box.[50] Members of this family have a variety of roles, including DNA repair and recombination and DNA packing, in addition to transcriptional activation. The structure of the DNA-binding domain from the HMG-1 proteins has been solved,[51,52] and it contains an extended N-terminal region followed by three helices forming a distorted L shape, also found in HMG-D from *Drosophila*.[53] Although the global fold remains the same in SRY,[54] detailed comparison shows differences that are important for its functional specificity.[55] This domain is widely spread in eukaryotes, including yeasts.

The winged-helix domain

A functionally diverse family of transcription factors in eukaryotes was found to encompass yet another variation of the helix-turn-helix domain (first described in prokaryotic regulators), defined as the winged-helix DNA-binding domain.[56] This 3-helix, 2/4-strand variant is present in the globular domain of noncore histones H5 and H1, the liver-specific transcription activator HNF-3γ, the products of the *ets* and *elk* oncogene families, and the proteins encoded by the forkhead family of *Drosophila* genes.[56] The domain is also present in proteins outside the realm of transcription, for example in the DNA-binding domain of the bacteriophage Mu transposase.[57]

The GH5 Fold

Histone H5 is the linker histone in the nucleosome, and the centrally located globular domain (GH5) is its most conserved region. The structure of GH5 has been determined, revealing a striking similarity to the catabolite gene activator protein CAP from bacteria.[58] The secondary structure elements are three helices followed by a single β hairpin,[58] also subsequently identified in the globular domain of histone H1.[59] Histones H1 and H5 are present in all eukaryotes.

The Ets Factors

Viral and cellular oncogenes of the *ets* family are known to encode nuclear DNA-binding proteins, involved in transcription.[60] The family encompasses various other proteins,[61] such as Sap1 (SRF accessory protein-1), the purine-rich sequence binding protein PU.1, and developmental regulators from *Drosophila* (see also chapter 14). The ETS domain shared by these proteins is 85 residues long, rich in aromatic and basic amino acids, and is usually localized at the C-terminal part of their molecule.[62] The structure of this domain from various representative members of the family has been solved, for example the PU.1 protein[63] and the Fli1 transcriptional activator,[64] revealing a similar arrangement of the helix-turn-helix domain to the CAP protein[65] and the other members of this class (see also chapter 14).

The HNF-3γ/Forkhead Family

A highly conserved domain about 100 residues long is shared between a number of eukaryotic transcription factors.[66] Some of the best known members of the family are the *Drosophila* protein forkhead and its mammalian homologue HNF-3γ. The structure determination of the HNF-3γ revealed its similarity to histone H5.[67] The sequence similarities with the Ets, H5 and particularly the CAP proteins are not detectable, and the proteins are considered similar only after the solution of the structure of some representative members. This family is present in animals and fungi.

Heat-Shock Factor

Another important protein family is the heat-shock factor, a DNA-binding protein that recognizes heat-shock promoter elements and contributes to the stress response in eukaryotes. A region about 90 residues long has been shown to be responsible for DNA binding.[68] The structure determination revealed a similarity to the winged-helix class, with the 3-helix, 4-strand structural domain recurrent throughout its members.[69] Sequence motif analysis has detected a similarity between heat-shock factors and Ets proteins,[70] thus unifying two structurally similar families at the sequence level, implying common origin from an ancestral form. It is puzzling why these families have diverged so significantly after their separation. This family is also present in all eukaryotic phyla.

IRF

Interferon regulatory factor (IRF) is responsible for the direct activation or indirect suppression of type I interferon genes.[71] Members of this family share a conserved N-terminal motif 110 residues long, involved in DNA binding.[72] These proteins are implicated in complex transcription regulation and affect cell growth and apoptosis.[71] Recently, structural studies revealed that the DNA-binding domain of IRF-2 is a member of the winged-helix class.[73]

Basic region-containing transcription factors

A large class of transcription factors contain a basic region (b) required for DNA binding of the regulatory sequences AP-1 and ATF,[74] usually followed by either a helix-loop-helix (HLH) structure or a leucine zipper (ZIP),[75] responsible for dimerization, or both these regions. These factors can thus be called the basic region-containing class.[76] The patterns of DNA binding[77] and transcription factor interactions,[78] homo- and heterodimeric, are very diverse. We will briefly discuss some representative members of the class.

Fos / Fra and Jun

Two related b/ZIP families represented by Fos/Fra- and Jun-related molecules share 40% sequence identity in the basic region, while the leucine zipper is more variable.[79] The basic region is at least 15 residues long, while the leucine zipper is about 30 residues long and immediately follows the basic region. The basic region is responsible for DNA binding, while the leucine zipper forms a coiled-coil structure and allows dimerization with other similar molecules[79] (see also chapter 5). The Fos and Jun proteins frequently heterodimerize and recognize DNA directly,[80] as has also been shown by the heteromeric complex bound to DNA.[81] This superfamily is widespread in all eukaryotes, from yeast to humans.

Myc-Like Factors

Members of another family of b/HLH/ZIP transcription factors, the Myc family, contain the basic region, followed by a helix-loop-helix domain and a leucine zipper.[82] Thus, the modular architecture of this family is different from the Fos/Jun family in that it contains a 30-residue long helix-loop-helix domain between the basic and the leucine-zipper regions. This battery of domains is variously located towards the C-terminus of the Myc subfamily, in the middle of the protein as in the Mad subfamily, or forms most of the protein as in the case of Max.[82] Max binds DNA as a dimer through its basic region.[83] The Myc family is involved in a number of fundamental cellular processes[84] (see also chapter 13). Members of the family have been found in all animal phyla and appear to have coevolved with their partners.[85]

Other b/ZIP and b/HLH Factors

A large number of transcription factors containing the leucine-zipper region represent variations containing this domain. Such factors are the CCATT-box and enhancer binding proteins (C/EBP)[86] (see also chapter 8), the cAMP response element (CRE) binding proteins[87] (see also chapter 4) and the fungal general transcription factor GCN4. The leucine zipper from GCN4 was the first whose structure was determined.[88] Subsequently, the structure of the full bZIP region bound to DNA was solved, showing that the complex network of contacts with DNA is mediated by residues that are conserved in this class.[89] In addition to the variety of b/ZIP factors, there are a number of proteins such as E47 and MyoD that only contain the helix-loop-helix region without a leucine zipper,[90] and are thus called b/HLH (see also chapter 13).

Helix-turn-helix domains in homeobox and Myb proteins

Developmental regulators in animals contain a small number of DNA-binding domains that are responsible for differential gene expression during development. Homeodomains are one of the most prominent families with hundreds of members[91] and some of the most well-studied eukaryotic DNA-binding domains,[92] present in a variety of regulatory proteins.[93] Homeodomain proteins recognize short DNA sequence elements and to achieve this they associate in complex formations with similar proteins.[92] The impact of structural work on the understanding of function in this family has been extensively reviewed elsewhere.[94] Curiously, the structure of the Myb-like DNA-binding domain resembles homeodomain elements. Another member of this class, not mentioned further, is the yeast RAP1 multifunctional protein, which binds telomeric DNA and is also a transcriptional regulator.[95]

Homeodomain

The homeodomain is a highly conserved DNA-binding domain 60 residues long, first identified in the *Drosophila* homeobox gene family, involved in the genetic control of development. Numerous structural studies[91] impossible to summarize here have been conducted on various members of this family.[96] The structure of representative members consists of three helices, forming a tightly packed helix-turn-helix domain,[97] similar to the bacterial regulatory proteins.[98] Helix 3 is the 'recognition helix' which recognizes the major groove of DNA.[99] A large number of homeodomain-containing proteins is known. Homeoboxes have extensively duplicated and diversified,[100] and have been found throughout eukaryotes. Three major subclasses can be identified: antennapedia,[101] engrailed[102] and paired (from the corresponding names of *Drosophila* genes).

Pax and Paired Box

Another domain, 125 residues long, found in some *Drosophila* genes[103] and the PAX family of proteins in vertebrates, is also known to bind DNA.[104,105] Generally, the domain is located at the N-terminus of these proteins. Some of these proteins contain a homeobox domain (see above) upstream of the paired box. The paired-box domain is very similar to the homeodomain structures already determined;[106] it induces a conformational change in the otherwise very short recognition sites, thus increasing the specificity.

The POU Family

The POU domain is a DNA-binding domain 75 residues long, found upstream of the so-called POU-type homeobox domain in a number of eukaryotic transcription factors (see also chapter 11). The structure of at least one member of the family, Oct-1, has been solved with a variety of approaches. Oct-1 consists of four α helices connected by short loops, with helices I and IV in a parallel coiled-coil arrangement.[107] It displays a remarkable similarity to the phage λ and 434 bacterial-type repressors.[107] It appears that the Oct-1 POU domain contacts the 5' half of the binding site, while the 3' half is recognized by the downstream POU-specific homeodomain.[108] POU domains are widespread in the animal kingdom.

Myb-Like Factors

Members of the *myb* oncogene family contain a region about 50 residues long in three tandem repeats, known to be involved in DNA binding,[109] usually located at the N-terminus of these proteins[110] (see also chapter 15). The domain adopts a helix-turn-helix structure, reminiscent of that of bacterial regulators.[98] It contains three conserved tryptophan residues,[111] which form a hydrophobic core.[112] The minimal DNA-binding domain is formed by repeats 2 and 3,[113] which undergo conformational changes upon binding to DNA.[114] In addition, repeat 2, which contains a cavity and is thermally unstable, seems to play a central role in DNA binding, especially recognition and activation.[115] This domain is widespread in eukaryotes, including viruses, plants and fungi.

The β sandwich immunoglobulin-like fold

Here, we describe two structural domains which presented yet another challenge to the idea of a limited structural 'vocabulary' of transcription factors in eukaryotes. These large, complex DNA-binding domains are found in some central proteins in cellular regulation, and are a contrast to the short, mobile domains that were thought to be the architectural components of eukaryotic transcription factors.

The Rel Family

Members of the Rel family of transcription factors are large proteins which contain a region about 300 residues long called the Rel homology region (RHR).[116] The most prominent member of this family is NF-κB, a heterodimeric (p50-p65) complex that is involved in a large number of stress-related transcriptional responses[116] (see also chapter 9). Both subunits contain a RHR and bind DNA in homodimeric or heterodimeric form. The structure determination of the p50 dimer[117,118] has shown that each RHR contains two domains resembling the immunoglobulin (Ig) fold, one of which is responsible for DNA binding and the other one for dimerization.[117,118] The structure of the Ig-like domains is reminiscent of the DNA-binding domain of tumor suppressor protein p53[119] (see below). Contacts with DNA are mediated through a complex network of electrostatic interactions,[116] displaying remarkable structural flexibility.[120] Members of the Rel family are known to be present in various animal phyla.[116]

The p53 Family

The tumor suppressor protein p53 is an important nuclear phosphoprotein that is implicated in a large number of human cancers[121] (see also chapter 16). The p53 protein is about 400 residues long, and the central 200 residues are involved in DNA binding.[121] The structure of this domain resembles an Ig antiparallel β sandwich,[122] reminiscent of the one found in NF-κB (see above). It is interesting that a totally different structural domain such as the Ig fold participates in DNA recognition, thus extending the examples of β-sheet interactions with DNA[123] (previously known only for members in the prokaryotic domain). For completeness, we mention that the structure of the C-terminal, 60-residue long domain of p53 involved in oligomerization has also been solved.[124] This family has so far been found only in vertebrates.

Unknown folds

There is a plethora of other known DNA-binding motifs in eukaryotic transcription factors, for which structural information is not available at present (spring of 1997). Ongoing structural work will surely reveal a wider variety of domains, extending the spectrum by which transcription factors are able to recognize, activate and regulate genetic elements.

The C-Terminal Domain of STATs

A protein family whose members have dual functions in *s*ignal *t*ransduction in the cytoplasm and *a*ctivation of *t*ranscription in the nucleus (hence the acronym STAT) has been recently discovered.[125] STAT proteins are over 700 residues long, and contain various other motifs towards their N-terminus, while their DNA-binding domain is located at the highly conserved C-terminus.[125] They interact with a variety of molecules and play pivotal roles in cellular function, in particular in response to activation by cytokines.[126] A complex pathway of interactions is being discovered and connections to other important pathways have been proposed[127] (see also chapter 7). The structure solution of one of the members of this family is anticipated with great interest.

The HMG-I / -Y AT-Hook

Another group of HMG proteins are the HMG-I/-Y and -C family. These are short proteins of about 100 residues. HMG-I/-Y are produced by alternative splicing of the same gene. These proteins bind to AT-rich DNA elements, and have been suggested to be implicated in nucleosome phasing and transcription regulation of AT-rich regions (see also chapter 18). They contain a short region of 11 residues, defined as the AT-hook, which can be repeated within their sequence,[128] and has been found to act in a cooperative manner.[129] The structure of this domain is not yet known.

Chromo and Chromo-Shadow Domains

The chromo (chromatin organization modifier) domain is a 60-residue long sequence pattern discovered in chromatin-modifying proteins.[130] These proteins are responsible for remodeling chromatin to its inactive state, heterochromatin, thus suppressing gene expression (see also chapter 18). They are not regarded as transcription factors, but because of their silencing function, they may be considered as such. A related domain located downstream of the chromo domain, called the chromo-shadow domain,[131] has been also defined more recently. The structure of these domains remains unknown. The domain is present in various eukaryotes, from yeasts to mammals.

The N-Terminal Domain of CTF/NF1

The CCAAT box-binding transcription factor (CTF) or nuclear factor 1 (NF1) composes a family of large CCAAT box-binding proteins that activate both transcription and DNA replication.[132] The 200-residue long N-terminal segment of CTF1 is sufficient for site-specific DNA recognition and protein dimerization. Full transcriptional activation requires an additional 100-residue long proline-rich C-terminal domain.[133] Structural information is not available for this family.

The N-Terminal Domain of GAGA

The GAGA factor is a transcriptional activator in *Drosophila* over 500 residues long, which binds the proximal region of the *engrailed* promoter and GA-rich regions.[134] It contains a N-terminal region about 100 residues long, shared with a number of eukaryotic transcription and regulatory proteins, such as the Broad Complex proteins and tramtrack (which also contains a zinc finger, see above). Other proteins include viral proteins of unknown function, the lola protein from *Drosophila*, and a few zinc finger-containing proteins, also of unknown function (C.A. Ouzounis, unpublished observations). Although the structure of this domain remains unknown, it provides another excellent opportunity for experimental analysis.

The Zinc-Binding Region of TP2

A small family of zinc-binding proteins, called nuclear transition proteins 2 (TP2), may have an important role in initiation of chromatin condensation and suppression of transcriptional activity during mammalian spermatogenesis.[135] These proteins have a middle, cysteine-rich, 30-residue long domain, possibly involved in DNA binding.[135] No structural information is available.

The TEA Domain

Various large (over 400 residues) eukaryotic proteins, including enhancer-binding factors, transcriptional activators and other regulators, share a 65-residue long DNA-binding region, the TEA domain.[136] Tissue specificity[137] and organ differentiation[138] are effected from members of this family. Again, structural information is not yet available.

The Copper-Fist Domain

A number of fungal transcription factors contain a short domain at their N-terminus, defined as the copper-fist domain.[139] The copper-fist domain is unusual in the sense that it is probably the only known DNA-binding eukaryotic domain that involves copper. Its structure remains unknown.

The MADS-Box

Proteins MCM1 and SRF[140] bind to the same DNA sequence and share a region of 60 amino acids.[141] The domain, known as the MADS-box,[142] is required for protein interactions and DNA binding (see also chapter 14), and it is also found in other proteins, such as ARG80, Deficiens and Agamous. These proteins are involved in a variety of cellular processes, including transcriptional activation, enhancer binding and metabolism-associated transcription regulation. Members of the family are present in fungi, plants and animals. The structure of this domain remains unknown.

FUTURE QUESTIONS

The structural diversity of known DNA-binding domains in the transcription factor repertoire of eukaryotes is staggering. From a handful of variants only a few years ago[143] and the hope that these would adequately generalize, in a situation analogous to the helix-turn-helix regulators in bacteria,[98] today we know the structure of at least thirty different types of transcription factor domains and the list is growing fast. Only the homeodomain and Myb-like proteins seem to be structurally similar to their bacterial counterparts,[144] and the hopes of finding general rules for DNA-protein interactions are challenged as never before.

Due to space limitations, we have only reviewed here these structural domains at 'high-resolution' without going into the details of the mechanisms of interaction, when known. Yet, is it possible, even at this level, to obtain a few general principles that are unique to this category of proteins? Below, we will attempt to list a number of key questions related to this issue, that will hopefully be used as guiding principles for future research.

First, why are DNA-binding domains of transcription factors so short? With the exception of the β barrel (TFIIA) and sandwich (NF-κB and p53) proteins, most domains described to date are short, composed of sequence segments of approximately 100 residues, and usually residing in the context of much larger macromolecules. This phenomenon of modular architecture, so pronounced in eukaryotic transcription factors, calls for an explanation. One trivial answer could be that this is an artifact of the limitations in the experimental methods used, particularly NMR spectroscopy, for the solution of the structures of large proteins. Another, more popular, explanation is that modular architecture represents an economical way to cope with the complexity and the requirements of elaborate regulation of gene expression.[145] A third, more attractive, explanation can be that the modules themselves have low specificity for their targets. Therefore, they need to be in a larger molecular context, where the functions of DNA binding and other macromolecular interactions are shared by the participating domains.[145] DNA-binding domains, in this cooperative process, undergo structural changes of induced fit, and are able to bind their targets with high specificity. In this view, the spectacular progress of X-ray crystallography to elucidate the structure of macromolecular complexes[41,146] is a most welcome step towards a deeper understanding of the role of short DNA-binding modules in context.

Second, are there any rules for DNA-protein recognition? Could we understand, for example, what makes a transcription factor activating or repressing? In the early days, when the repertoire of structural domains was small, and most of the known cases were probably intricate elaborations of a common theme, derived from an ancestral family, the issue was easier to answer, and repeated attempts have been made to generalize such observations. This 'recognition code'[147] has since suffered multiple criticisms.[148] Despite remarkable attempts for a simplified portrait of these complex interactions,[149,150] these generalizations will fall short of explanation if we do not reach closure in the description of the interactions under consideration. As a matter of fact, earlier concepts of DNA recognition based on helices alone were greatly challenged with the 'surprising' discoveries of β sheet-DNA interactions.[151] It seems that currently we are unable to generalize and find a common theme among all known DNA-binding domains of transcriptional regulators. It is even more remarkable that some enzymes are bifunctional, transcending their 'traditional' roles as passive converters of small molecules.[2] One particular example is nucleotide diphosphate kinase, whose structure has been solved, yet whose DNA-binding region remains unknown.[152] A positive attitude towards the broad spectrum of DNA-protein interaction modes may help us appreciate and understand their uniqueness and individuality.

We thus naturally come to the third, frequently asked question: is there a limited number of domains? We have seen a wide gamut of transcriptional regulators and transcription-associated proteins containing recurrent structural domains, presumably sharing a common ancestor (divergence) but in some cases possibly reinvented during evolution (convergence). Although convergence is only a hypothesis, and can never be disproved, it provides us with a single control experiment to test general theories of protein-nucleic acid interactions, free from accidents amplified during evolution. There is no general answer to the above question. Yet, we will always be amazed to discover structural similarities that we are unable to detect at the sequence level; we will also never cease to feel awe when faced with a yet another novel pattern of macromolecular beauty.[153]

The final question may be the most provocative of all: after all, is there such a thing as a *eukaryotic* transcription factor? In the cases we have seen, it is true that these domains are confined within eukaryotes, with some notable exceptions like core histones or TBP and TFIIB. On the other hand, very few of them are known to be present in representative species of fungi, plants and animals. One reason why we have provided the rather tedious information on phylogenetic distribution is now becoming apparent: for most of these proteins, we do not know whether there exist members present throughout the eukaryotic tree. Some of them will be found in representative eukaryotic taxa, but some will remain confined in a specific taxon, such as fungi, metazoa or vertebrates. What does this mean? Is there a correlation of the abundance and versatility of transcription factor families and their constituent domains with tissue structure, organ development and body plans? Have we defined the eukaryotic transcription factor in a negative sense, that is, a nonbacterial protein? There is no readily available answer to these questions, but it is worth considering redefining transcription factors according to their phylogenetic realm (and not beyond it).

A final comment, however, must be made. There is mounting evidence that the basal transcription machinery originated before the emergence of the eukaryotic cell, being at least as ancient as bacterial transcription.[154] In addition, nucleosomal packing, also a phenomenon older than eukaryotes,[49] can be regarded as an extreme manifestation of transcriptional repression and thus seen as a necessary condition for genome expansion during evolution. While prokaryotes need repressors to silence otherwise transcriptionally active genomes, eukaryotes may need more activators to locate and regulate transcription initiation sites, hidden in vast genomes.[49,154]

EPILOGUE

Only a continuous appreciation of our ignorance and the consciously active collaboration between structural, computational and in vivo and in vitro experimental biology will advance our understanding of the spectacular complexity of eukaryotic transcription and its components, until we will be able to visualize in full molecular detail the awesome genetic interactions that control the existence of all living organisms, including ourselves.

Note

Information about transcription factors, their structure, function and evolution, their functional associations, and pointers to public motif, sequence, structure and bibliographic databases will be maintained and continuously updated on the World Wide Web at the following URL: <http://www.ebi.ac.uk/research/transcription/>. The reader is also referred to the information provided in Table 1.1.

References

1. Pabo CO, Sauer RT. Transcription factors: structural families and principles of DNA recognition. Ann Rev Biochem 1992; 61:1053-1095.
2. Kyrpides NC, Ouzounis CA. Nucleic-acid binding metabolic enzymes: living fossils of stereochemical interactions? J Mol Evol 1995; 40:564-569.
3. Berg JM. Zinc finger domains: hypotheses and current knowledge. Ann Rev Biophys Biophys Chem 1990; 19:405-421.

4. Kaptein R. Zinc fingers. Curr Opin Struct Biol 1991; 1:63-70.
5. Coleman JE. Zinc proteins: enzymes, storage proteins, transcription factors and replication proteins. Ann Rev Biochem 1992; 61:897-946.
6. Berg JM, Shi Y. The galvanization of biology: a growing appreciation for the roles of zinc. Science 1996; 271:1081-1085.
7. Schwabe JWR, Klug A. Zinc mining for protein domains. Nature Struct Biol 1994; 1:345-349.
8. Miller J, McLachlan AD, Klug A. Repetitive zinc-binding domains in the protein transcription factor IIIA from *Xenopus* oocytes. EMBO J 1985; 4:1609-1614.
9. Lee MS, Gippert GP, Soman KV et al. Three-dimensional solution structure of a single zinc finger DNA-binding domain. Science 1989; 245:635-637.
10. Brown RS, Sander C, Argos P. The primary structure of transcription factor TFIIA has 12 consecutive repeats. FEBS Lett 1985; 186:271-274.
11. Evans RM. The steroid and thyroid hormone receptor superfamily. Science 1988; 240:889-895.
12. Ribeiro RC, Kushner PJ, Baxter JD. The nuclear hormone receptor gene superfamily. Ann Rev Med 1995; 46:443-453.
13. Lucas PC, Granner DK. Hormone response domains in gene transcription. Ann Rev Biochem 1992; 61:1131-1173.
14. Gewirth DT, Sigler PB. The basis for half-site specificity explored through a noncognate steroid receptor-DNA complex. Nature Struct Biol 1995; 2:386-394.
15. Schwabe JWR, Neuhaus D, Rhodes D. Solution structure of the DNA-binding domain of the oestrogen receptor. Nature 1990; 348:458-461.
16. Schwabe JWR, Chapman L, Finch JT et al. DNA recognition by the oestrogen receptor: from solution to the crystal. Structure 1993; 1:187-204.
17. Lee MS, Kliewer SA, Provencal J et al. Structure of the retinoid X receptor α DNA binding domain: a helix required for homodimeric DNA binding. Science 1993; 260:1117-1121.
18. Luisi BF, Xu WX, Otwinowski Z et al. Crystallographic analysis of the interaction of the glucocorticoid receptor with DNA. Nature 1991; 352:497-505.
19. Schwabe JWR, Chapman L, Finch JT et al. The crystal structure of the oestrogen receptor DNA-binding domain bound to DNA: how receptors discriminate between their response elements. Cell 1993; 75:1-12.
20. Schwabe JWR. How nuclear receptors get turned on. Curr Biol 1996; 6:372-374.
21. Gardner KH, Anderson SF, Coleman JE. Solution structure of the *Kluyveromyces lactis* LAC9 Cd_2Cys_6 DNA-binding domain. Nature Struct Biol 1995; 2: 898-905.
22. Clore GM, Bax A, Omichinski JG et al. Localization of bound water in the solution structure of a complex of the erythroid transcription factor GATA-1 with DNA. Structure 1994; 2:89-94.
23. Kaine BP, Mehr IJ, Woese CR. The sequence, and its evolutionary implications, of a *Thermococcus celer* protein associated with transcription. Proc Natl Acad Sci USA 1994; 91:3854-3856.
24. Qian X, Jeon C, Yoon H et al. Structure of a new nucleic-acid-binding motif in eukaryotic transcriptional elongation factor TFIIS. Nature 1993; 365:277-279.
25. Zhu W, Zeng Q, Colangelo CM et al. The N-terminal domain of TFIIB from *Pyrococcus furiosus* forms a zinc ribbon. Nature Struct Biol 1996; 3:122-124.
26. Pérez-Alvarado GC, Miles C, Michelsen JW et al. Structure of the carboxy-terminal LIM domain from the cysteine rich protein CRP. Nature Struct Biol 1994; 1:388-398.
27. Gill GN. The enigma of LIM domains. Structure 1995; 3:1285-1289.
28. Barlow PN, Luisi B, Milner A et al. Structure of the C3HC4 domain by 1H-nuclear magnetic resonance spectroscopy. A new structural class of zinc-finger. J Mol Biol 1994; 237:201-211.
29. Saurin AJ, Borden KLB, Boddy MN et al. Does this have a familiar RING? Trends Biochem Sci 1996; 21:208-214.
30. Nikolov DB, Hu S-H, Lin J et al. Crystal structure of TFIID TATA-box binding protein. Nature 1992; 360:40-46.
31. Kim Y, Geiger JH, Hahn S et al. Crystal structure of a yeast TBP/TATA-box complex. Nature 1993; 365:512-520.

32. Kim JL, Nikolov DB, Burley SK. Co-crystal structure of TBP recognizing the minor groove of a TATA element. Nature 1993; 365:520-527.
33. Nikolov D, Burley SK. 2.1 Å resolution refined structure of a TATA box-binding protein (TBP). Nature Struct Biol 1994; 1:621-637.
34. Burley SK. The TATA box binding protein. Curr Opin Struct Biol 1996; 6:69-75.
35. Kim JL, Burley SK. PCD/DcoH: more than a second molecular saddle. Structure 1995; 3:531-534.
36. Bagby S, Kim S, Maldonado E et al. Solution structure of the C-terminal core domain of human TFIIB: similarity to cyclin A and interaction with TATA-binding protein. Cell 1995; 82:857-867.
37. Brown NR, Noble ME, Endicott JA et al. The crystal structure of cyclin A. Structure 1995; 3:1235-1247.
38. Gibson TJ, Thompson JD, Blocker A et al. Evidence for a protein domain superfamliy shared by the cyclins, TFIIB and RB/p107. Nucleic Acids Res 1994; 22:946-952.
39. Ouzounis C, Sander C. TFIIB, an evolutionary link between the transcription machineries of archaebacteria and eukaryotes. Cell 1992; 71:189-190.
40. Buratowski S, Zhou H. A suppressor of TBP mutations encodes an RNA polymerase III transcription factor with homology to TFIIB. Cell 1992; 71:221-230.
41. Tan S, Hunziker Y, Sargent DF et al. Crystal structure of a yeast TFIIA/TBP/DNA complex. Nature 1996; 381:127-134.
42. Adams CC, Workman JL. Nucleosome displacement in transcription. Cell 1993; 72:305-308.
43. Grunstein M. Histone function in transcription. Ann Rev Cell Biol 1990; 6:643-678.
44. Goodrich JA, Tjian R. TBP-TAF complexes: selectivity factors for eukaryotic transcription. Curr Opin Cell Biol 1994; 6:403-409.
45. Xie X, Kokubo T, Cohen SL et al. Structural similarity between TAFs and the heterotetrameric core of the histone octamer. Nature 1996; 380:316-322.
46. Hoffmann A, Chiang C-M, Oelgeschläger T et al. A histone octamer-like structure within TFIID. Nature 1996; 380:356-359.
47. Maity SN, Golumbek PT, Karsenty G et al. Selective activation of transcription by a novel CCAAT binding factor. Science 1988; 241:582-585.
48. Maity SN, Sinha S, Ruteshouser CE et al. Three different polypeptides are necessary for DNA binding of the mammallian heteromeric CCAAT binding factor. J Biol Chem 1992; 267:16574-16580.
49. Ouzounis CA, Kyrpides NC. Parallel origins of the nucleosome core and eukaryotic transcription from Archaea. J Mol Evol 1996; 42:234-239.
50. Ner SS. HMGs everywhere. Curr Biol 1992; 2:208-211.
51. Read CM, Cary PD, Crane-Robinson C et al. Solution structure of a DNA-binding domain from HMG1. Nucleic Acids Res 1993; 21:3427-3436.
52. Weir HM, Kraulis PJ, Hill CS et al. Structure of the HMG box motif in the B-domain of HMG1. EMBO J 1993; 12:1311-1319.
53. Jones DNM, Searles MA, Shaw GL et al. The solution structure and dynamics of the DNA-binding domain of HMG-D from *Drosophila melanogaster*. Structure 1994; 2:609-627.
54. Werner MH, Huth JR, Gronenborn MA et al. Molecular basis of human 46X,Y sex reversal revealed from the three-dimensional solution structure of the human SRY-DNA complex. Cell 1995; 81:705-714.
55. Bramham J, Norman DG. hSRY: molecular gender bender. Structure 1995; 3:631-633.
56. Brennan RG. The winged-helix DNA-binding motif: another helix-turn-helix takeoff. Cell 1993; 74:773-776.
57. Clubb T, Omichinski JG, Savilahti H et al. A novel class of winged helix-turn-helix protein: the DNA-binding domain of Mu transposase. Structure 1994; 2:1041-1048.
58. Ramakrishnan V, Finch JT, Graziano V et al. Crystal structure of globular domain of histone H5 and its implications for nucleosome binding. Nature 1993; 362:219-223.

59. Cerf C, Lippens G, Ramakrishnan V et al. Homo- and heteronuclear two-dimensional NMR studies of the globular domain of histone H1: full assignment, tertiary structure, and comparison with the globular domain of histone H5. Biochemistry 1994; 33:11079-11086.
60. McLeod K, Leprince D, Stehelin D. The *ets* gene family. Trends Biochem Sci 1992; 17:251-256.
61. Laudet V, Niel C, Duterque-Coquillaud M et al. Evolution of the ets gene family. Biochem Biophys Res Comm 1993; 190:8-14.
62. Wasylyk B, Hahn SL, Giovane A. The Ets family of transcription factors. Eur J Biochem 1993; 211:7-18.
63. Kodandapani R, Pio F, Ni C-Z et al. A new pattern for helix-turn-helix recognition revealed by the PU.1 ETS-domain-DNA complex. Nature 1996; 380: 456-460.
64. Liang H, Mao X, Olejniczak ET et al. Solution structure of the *ets* domain of Fli-1 when bound to DNA. Nature Struct Biol 1994; 1:871-875.
65. Schultz SC, Shields GC, Steitz TA. Crystal structure and CAP-DNA complex: the DNA is bent by 90°. Science 1991; 253: 1001-1007.
66. Weigel D, Jaeckle H. The fork head domain: a novel DNA-binding motif of eukaryotic transcription factors? Cell 1990; 63:455-456.
67. Clark KL, Halay ED, Lai E et al. Cocrystal structure of the HNF-3/*forkhead* DNA-recognition motif resembles histone H5. Nature 1993; 364:412-420.
68. Vuister GW, Kim S-J, Orosz A et al. Solution structure of the DNA-binding domain of *Drosophila* heat shock transcription factor. Nature Struct Biol 1994; 1:605-613.
69. Harrison CJ, Bohm AA, Nelson HCM. Crystal structure of DNA binding domain of the heat shock transcription factor. Science 1994; 263:224-227.
70. Landsmann D, Wolffe AP. Common sequence and structural features in the heat-shock factor and Ets families of DNA-binding domain. Trends Biochem Sci 1995; 20:225-226.
71. Taniguchi T, Harada H, Lamphier M. Regulation of the interferon system and cell growth by the IRF transcription factors. J Cancer Res Clin Oncol 1995; 121:516-520.
72. Harada H, Fujita T, Miyamoto M et al. Structurally similar but functionally distinct factors, IRF-1 and IRF-2, bind to the same regulatory elements of IFN and IFN-inducible genes. Cell 1989; 58: 729-739.
73. Uegaki K, Shirakawa M, Harada H et al. Secondary structure and folding topology of the DNA binding domain of interferon regulatory factor 2, as revealed by NMR spectroscopy. FEBS Lett 1995; 359:184-188.
74. Hai T, Curran T. Cross-family dimerization of transcription factors Fos/Jun and ATF/CREB alters DNA binding specificity. Proc Natl Acad Sci USA 1991; 88: 3720-3724.
75. Landschulz WH, Johnson PF, McKnight SL. The leucine binding zipper: a hypothetical structure common to a new class of DNA binding proteins. Science 1988; 240:1759-1764.
76. Ellenberger TE. Getting a grip on DNA recognition: structure of the bZIP and bHLH DNA-binding domains. Curr Opin Struct Biol 1994; 4:12-18.
77. Kim J, Tzamarias D, Ellenberger T et al. Adaptability at the protein-DNA interface is an important aspect of sequence recognition by bZIP proteins. Proc Natl Acad Sci USA 1993; 90:4513-4517.
78. Phillips SEV. Built by association: structure and function of helix-loop-helix DNA-binding proteins. Structure 1994; 2:1-4.
79. Kerppola TK, Curran T. Transcription factor interactions: basics on zippers. Curr Opin Struct Biol 1991; 1:71-79.
80. Abate C, Luk D, Gentz R et al. Expression and purification of the leucine zipper and DNA binding domains of Fos and Jun: both Fos and Jun contact DNA directly. Proc Natl Acad Sci USA 1990; 87:1032-1036.
81. Glover JN, Harrison SC. Crystal structure of the heterodimeric bZIP transcription factor c-Fos-c-Jun bound to DNA. Nature 1995; 373:257-261.

82. Ryan KM, Birnie GD. *Myc* oncogenes: the enigmatic family. Biochem J 1996; 314:713-721.
83. Ferré-D' Amaré AR, Prendergast GC, Ziff EB et al. Recognition by Max of its cognate DNA through a dimeric b/HLH/Z domain. Nature 1993; 363:38-45.
84. Marcu KB, Bossone SA, Patel AJ. *myc* function and regulation. Ann Rev Biochem 1992; 61:809-860.
85. Atchley WR, Fitch WM. Myc and Max: molecular evolution of a family of protooncogene products and their dimerization partner. Proc Natl Acad Sci USA 1995; 92:10217-10221.
86. Wedel A, Ziegler-Heitbrock HW. The C/EBP family of transcription factors. Immunobiology 1995; 193:171-185.
87. Laoide BM, Foulkes NS, Schlotter F et al. The functional versatility of CREM is determined by its modular structure. EMBO J 1993; 12:1179-1191.
88. O' Shea EK, Klemm JD, Kim PS et al. X-ray structure of the GCN4 leucine zipper, a two-stranded, parallel coiled coil. Science 1991; 254:539-544.
89. Ellenberger TE, Brandl CJ, Struhl K et al. The GCN4 basic region leucine zipper binds DNA as a dimer of uninterrupted α helices: crystal structure of the protein-DNA complex. Cell 1992; 71:1223-1237.
90. Wolberger C. b/HLH without the zip. Nature Struct Biol 1994; 1:413-416.
91. Kornberg TB. Understanding the homeodomain. J Biol Chem 1993; 268:26813-26816.
92. Wolberger C. Homeodomain interactions. Curr Opin Struct Biol 1996; 6:62-68.
93. Gehring WJ, Affolter M, Bürglin T. Homeodomain proteins. Ann Rev Biochem 1994; 63:487-526.
94. Gehring WJ, Müller M, Affolter M et al. The structure of the homeodomain and its functional implications. Trends Genet 1990; 6:323-329.
95. Konig P, Giraldo R, Chapman L et al. The crystal structure of the DNA-binding domain of yeast RAP1 in complex with telomeric DNA. Cell 1996; 85:125-136.
96. Gehring WJ, Qian YQ, Billeter M et al. Homeodomain-DNA recognition. Cell 1994; 78:211-223.
97. Treisman J, Harris E, Wilson D et al. The homeodomain: a new face for the helix-turn-helix? BioEssays 1992; 14:145-150.
98. Harrison SC, Aggarwal AK. DNA recognition by proteins with the helix-turn-helix motif. Ann Rev Biochem 1990; 59:933-969.
99. Tullius T. Homeodomains: together again for the first time. Structure 1995; 3:1143-1145.
100. Wilson DS, Desplan C. Homeodomain proteins: cooperating to be different. Curr Biol 1995; 5:32-34.
101. Qian YQ, Billeter M, Otting G et al. The structure of the *Antennapedia* homeodomain determined by NMR spectroscopy in solution: comparison with prokaryotic repressors. Cell 1989; 59:573-580.
102. Kissinger CR, Liu B, Martin-Blanco E et al. Crystal structure of an engrailed homeodomain/DNA complex at 2.8 Å resolution: a framework for understanding homeodomain/DNA interactions. Cell 1990; 63:579-590.
103. Bopp D, Burri M, Baumgartner S et al. Conservation of a large protein domain in the segmentation gene *paired* and in functionally related genes in *Drosophila.* Cell 1986; 47:1033-1049.
104. Epstein J, Cai J, Glaser T et al. Identification of a Pax paired domain recognition sequence and evidence for DNA-dependent conformational changes. J Biol Chem 1994; 269:8355-8361.
105. Treisman J, Harris E, Desplan C. The paired box encodes a second DNA-binding domain in the Paired homeo domain protein. Genes Devel 1991; 5:594-604.
106. Wilson DS, Guenther B, Desplan C et al. High resolution crystal structure of a paired (Pax) class cooperative homeodomain dimer on DNA. Cell 1995; 82:709-719.
107. Dekker N, Cox M, Boelens R et al. Solution structure of the POU-specific DNA-binding domain of Oct-1. Nature 1993; 362:852-855.
108. Klemm JD, Rould MA, Aurora R et al. Crystal structure of the Oct-1 POU domain bound to an octamer site: DNA

recognition with tethered DNA-binding modules. Cell 1994; 77:21-32.
109. Biedenkapp H, Borgmeyer U, Sippel AE et al. Viral *myb* oncogene encodes a sequence-specific DNA binding activity. Nature 1988; 335:835-837.
110. Klempnauer K-H, Sippel AE. The highly conserved amino-terminal region of the protein encoded by the v-*myb* oncogene functions as a DNA-binding domain. EMBO J 1987; 6:2719-2725.
111. Anton IA, Frampton J. Tryptophans in *myb* proteins. Nature 1988; 336:719-719.
112. Ogata K, Hojo H, Aimoto S et al. Solution structure of a DNA-binding unit of Myb: a helix-turn-helix-related motif with conserved tryptophans forming a hydrophobic core. Proc Natl Acad Sci USA 1992; 89:6428-6432.
113. Ogata K, Morikawa S, Nakamura H et al. Solution structure of a specific DNA complex of the Myb DNA-binding domain with cooperative recognition helices. Cell 1994; 79:639-648.
114. Ogata K, Morikawa S, Nakamura H et al. Comparison of the free and DNA-complexed forms of the DNA-binding domain from c-Myb. Nature Struct Biol 1995; 2:309-320.
115. Ogata K, Kanei-Ishii C, Sasaki M et al. The cavity in the hydrophobic core of Myb DNA-binding domain is reserved for DNA recognition and *trans*-activation. Nature Struct Biol 1996; 3:178-187.
116. Chytil M, Verdine GL. The Rel family of eukaryotic transcription factors. Curr Opin Struct Biol 1996; 6:91-100.
117. Ghosh G, Van Duyne GD, Ghosh S et al. Structure of the NF-κB p50 homodimer bound to a κB site. Nature 1995; 373:303-310.
118. Müller CW, Rey FA, Sodeoka M et al. Structure of the NF-κB p50 homodimer bound to DNA. Nature 1995; 373: 311-317.
119. Kuriyan J, Thanos D. Structure of the NF-κB transcription factor: a holistic interaction with DNA. Structure 1995; 3:135-141.
120. Müller CW, Rey FA, Harrison SC. Comparison of two different DNA-binding modes of the NF-κB p50 homodimer. Nature Struct Biol 1996; 3:224-227.
121. Burley SK. p53: a cellular Achilles' heel revealed. Structure 1994; 2:789-792.
122. Cho Y, Gorina S, Jeffrey P et al. Crystal structure of a p53 tumor suppressor-DNA complex: understanding tumorigenic mutations. Science 1994; 265: 346-355.
123. Phillips SEV. The β-ribbon DNA recognition motif. Ann Rev Biophys Biomol Struct 1994; 23:671-701.
124. Lee W, Harvey TS, Yin Y et al. Solution structure of the tetrameric minimum transforming domain of p53. Nature Struct Biol 1994; 1:877-890.
125. Schindler C, Darnell JEJ. Transcriptional responses to polypeptide ligands: the JAK-STAT pathway. Ann Rev Biochem 1995; 64:621-651.
126. Darnell JEJ, Kerr IM, Stark GR. Jak-STAT pathways and transcriptional activation in response to IFNs and other extracellular signaling proteins. Science 1994; 264:1415-1421.
127. Ihle JN. STATs and MAPKs: obligate or opportunistic partners in signaling. BioEssays 1996; 18:95-98.
128. Reeves R, Nissen MS. The A.T-DNA-binding domain of mammalian high mobility group I chromosomal proteins. A novel peptide motif for recognizing DNA structure. J Biol Chem 1990; 265: 8573-8582.
129. Maher JF, Nathans D. Multivalent DNA-binding properties of the HMG-I proteins. Proc Natl Acad Sci USA 1996; 93:6716-6720.
130. Paro R. Imprinting a determined state into the chromatin of *Drosophila*. Trends Genet 1990; 6:416-421.
131. Aasland R, Stewart AF. The chromo shadow domain, a second chromo domain in heterochromatin-binding protein1, HP1. Nucleic Acids Res 1995; 23:3163-3173.
132. Santoro C, Mermod N, Andrews PC et al. A family of human CCAAT-box-binding proteins active in transcription and DNA replication: cloning and expression of multiple cDNAs. Nature 1988; 334: 218-224.
133. Mermod N, O' Neill EA, Kelly TJ et al. The proline-rich transcriptional activator of CTF/NF-I is distinct from the repli-

cation and DNA binding domain. Cell 1989; 58:741-753.
134. Soeller WC, Oh CE, Kornberg TB. Isolation of cDNAs encoding the *Drosophila* GAGA transcription factor. Mol Cell Biol 1993; 13:7961-7970.
135. Baskaran R, Rao MRS. Mammalian spermatid specific protein, TP2, is a zinc metalloprotein with two finger motifs. Biochem Biophys Res Comm 1991; 179: 1491-1499.
136. Bürglin TR. The TEA domain: a novel, highly conserved DNA-binding motif. Cell 1991; 66:11-12.
137. Yasunami M, Suzuki K, Houtani T et al. Molecular characterization of cDNA encoding a novel protein related to transcriptional enhancer factor-1 from neural precursor cells. J Biol Chem 1995; 270:18649-18654.
138. Campbell SD, Inamdar M, Rodrigues V et al. The *scalloped* gene encodes a novel, evolutionarily conserved transcription factor required for sensory organ differentiation in *Drosophila*. Genes Devel 1992; 6:367-379.
139. Jungmann J, Reins HA, Lee J et al. MAC1, a nuclear regulatory protein related to Cu-dependent transcription factors is involved in Cu/Fe utilization and stress resistance in yeast. EMBO J 1993; 12:5051-5056.
140. Treisman R, Ammerer G. The SRF and MCM1 transcription factors. Curr Opin Genet Devel 1992; 2:221-226.
141. Müller CG, Nordheim A. A protein domain conserved between yeast MCM1 and human SRF directs ternary complex formation. EMBO J 1991; 10:4219-4229.
142. Shore P, Sharrocks AD. The MADS-box family of transcription factors. EMBO J 1995; 229:1-13.
143. Struhl K. Helix-turn-helix, zinc-finger, and leucine-zipper motifs for eukaryotic transcriptional regulatory proteins. Trends Biochem Sci 1989; 14:137-140.
144. Brennan RG. Interactions of the helix-turn-helix binding domain. Curr Opin Struct Biol 1991; 1:80-88.
145. Frankel AD, Kim PS. Modular structure of transcription factors: implications for gene regulation. Cell 1991; 65:717-719.
146. Nikolov DM, Chen H, Halay ED et al. Crystal structure of a TFIIB-TBP-TATA-element ternary complex. Nature 1995; 377:119-128.
147. Pabo CO, Sauer RT. Protein-DNA interactions. Ann Rev Biochem 1984; 53: 293-321.
148. Matthews BW. Protein-DNA interaction: no code for recognition. Nature 1988; 335:294-295.
149. Suzuki M. A framework for the DNA-protein recognition code of the probe helix in transcription factors: the chemical and stereochemical rules. Structure 1994; 2:317-326.
150. Suzuki M. Common features in DNA recognition helices of eukaryotic transcription factors. EMBO J 1993; 12: 3221-3226.
151. Phillips SEV. Specific β-sheet interactions. Curr Opin Struct Biol 1991; 1: 89-98.
152. Chiadmi M, Moréra S, Lascu I et al. Crystal structure of the Awd nucleotide diphosphate kinase from *Drosophila*. Structure 1993; 1:283-293.
153. Janin J. Proteins with a ring. Structure 1994; 2:571-573.
154. Ouzounis C, Kyrpides N. The emergence of major cellular processes in evolution. FEBS Lett 1996; 390:119-123.
155. Holm L, Sander C. The FSSP database of structurally aligned protein fold families. Nucleic Acids Res 1996; 24:206-210.

Table 1.1. Domain families, their FSSP classification numbers,[155] the PDB structure database identifiers and their closest neighbors in structure space.

Domain	FSSP class	PDB code	Nearest neighbor
Ig-like			
NF-κB (rel)	10.2.2.4.	1svcP	*papD*
p53	10.2.4.1.	1tupA	*N-cadherin*
RING finger			
C3HC4 finger	11.1.1.1.	1chc	NA
Saddle domain			
DCoH	27.1.1.2.	1dchA	Arginine Repressor
TBP	27.2.1.1.	1ytbA	*Acylphosphatase*
Winged helix			
GH5	41.1.1.2.	1hstA	Diphteria Toxin Repressor
CAP	41.1.1.2.	1berA	BirA Biotin Operon Repressor
Homeodomain			
A1	42.2.1.1.	1yrnA	engrailed
engrailed	42.2.1.1.	1enh	A1
POU	42.2.1.1.	1octC	engrailed
Paired	42.2.1.1.	1fjlA	A1
LFB1	42.2.1.1.	1lfb	Paired
Paired C	42.2.2.2.	1pdnC	Trp Aporepressor
Myb	42.2.3.1.	1mseC	Hin Recombinase Dna-Binding
Winged helix			
HSF	63.1.1.1.	2hts	NA
Ets-1	64.1.1.1.	1stwA	NA
Leucine zipper			
MyoD	94.1.1.1.	1mdyA	Fos-Jun dimer
Fos-Jun dimer	94.2.2.1.	1fosE	GCN4
GCN4	94.2.2.2.	2dgcA	Fos-Jun dimer
Jun homodimer	94.2.2.5.	1junA	GCN4
GATA finger			
GATA-1	109.1.1.1.	1gatA	NA
Zinc ribbon			
TFIIS	113.1.1.1.	1tfi	NA

Table 1.1. (Con't) Domain families, their FSSP classification numbers,[155] the PDB structure database identifiers and their closest neighbors in structure space.

Domain	FSSP class	PDB code	Nearest neighbor
TFIIIA finger			
SWI5 C2H2 finger	128.1.1.1.	1ncs	NA
MBP-1	129.1.1.1.	1bbo	ZIF268
ZF268	129.1.1.2.	1zaa	Tramtrack finger
Tramtrack finger	129.1.1.2.	2drp	ZIF268
Core histone			
Archaean histone	135.1.1.1.	1bfmA	NA
HMG box			
SRY	139.1.1.1.	1hryA	HMG-D
HMG-D	139.1.1.2.	1hma	SRY
Fungal finger			
GAL4	222.1.1.1.	1d66A	CYP1
CYP1	222.1.2.1.	1pyc	GAL4

SMALL CAPITALS signify structural neighbors of prokaryotic DNA-binding domains, while *italics* signify neighbors not involved in transcription. A number of interesting points must be made. First, due to the strict criteria employed in this classification scheme (Liisa Holm, private communication), the first number of a class does not signify a distance in the numerical sense from other classes. Thus, two winged-helix subclasses appear to be distant while they are not. Second, there are five zinc-finger classes represented from the ones discussed here. Third, most domains have neighbors within their classes, with the exception of the Ig-like fold and the saddle domain, both fairly large proteins. Finally, some domains listed here remain unique at present.

CHAPTER 2

Specific Transcription Initiation by RNA Polymerase II and Associated Factors

Michèle Sawadogo and Marilyn N. Szentirmay

INTRODUCTION

During the past 20 years, investigations into the basic and regulatory mechanisms of RNA synthesis in eukaryotes have generated a rapid succession of unexpected results. First was the discovery of the three forms of eukaryotic polymerases with specialized functions in ribosomal (RNA polymerase I), protein-encoding (RNA polymerase II) and small RNA (RNA polymerase III) gene transcription. Next came the revelation of the extraordinary structural complexity of these enzymes, the variety of promoter elements that govern their function, and the multitude of general and specialized cofactors required for promoter recognition. More recently, key findings in class II gene expression have included the unusual structure and posttranslational modification of the largest subunit of RNA polymerase II, the novel DNA-binding properties of the promoter-recognizing protein TFIID, and the functional importance of the RNA polymerase II holoenzyme complex. Within the short format of this review, we have summarized our current understanding of the mechanisms implicated in promoter recognition and specific transcription initiation by RNA polymerase II, outlining also some of the features of the major proteins involved.

CORE PROMOTER ELEMENTS

Traditionally, the promoters of genes transcribed by RNA polymerase II have been described as composed of two distinct regions. The minimum promoter region consists of essential core elements that attach the transcription initiation complexes and fix the transcription start sites. The more distal promoter regions are highly variable and play a regulatory role by binding gene-specific transcription factors. Yet it is important to note that the composition and arrangement of the core elements also vary in different promoters, often in ways that are crucial to gene regulation.[1]

Transcription Factors in Eukaryotes, edited by Athanasios G. Papavassiliou.

Many class II promoters contain a TATAAA sequence, the TATA box, located 25-30 bp upstream of the transcription start point (reviewed in ref. 2). Another common core element in class II genes is the initiator (Inr),[3] which spans the transcription start site and is characterized by the rather loose consensus sequence PyPyANTPyPy, with the A located at position +1.[4] In vitro, either a TATA or an Inr element alone can direct basal transcription by RNA polymerase II from a specific start point.[3] However, promoters are often more active, both in vitro and in vivo, when both elements are present.[5] The lack of both TATA and Inr elements in class II promoters usually correlates with transcription initiating from multiple start sites over a region that can span up to hundreds of nucleotides.[6,7]

Although the TATA box and the Inr are at this time the best characterized of the RNA polymerase II core elements, they are by no means the only ones. For instance, two new elements were recently identified within the proximal transcribed region of a number of class II genes.[6,8] Other elements may be present in only a limited set of promoters, as exemplified by the proximal sequence element (PSE), which is found in the small nuclear RNA genes that are transcribed by RNA polymerase II as well as those transcribed by RNA polymerase III.[2]

CORE RNA POLYMERASE II

Eukaryotic RNA polymerases in general, and RNA polymerase II in particular, have been studied for more than 20 years, and their composition and general properties are now well established (reviewed in refs. 9 and 10). These are structurally complex enzymes, each composed of two large subunits and a variety of small subunits. In yeast, all 12 of the core subunits of RNA polymerase II have now been cloned. This groundbreaking work, in conjunction with biochemical studies, has helped define the function of some of these components. The two large subunits and one of the small subunits of RNA polymerase II display significant homologies to the β', β and α subunits of *E. coli* RNA polymerase and probably play similar roles to their prokoryotic counterparts in subunit assembly and catalysis. Five of the small subunits of RNA polymerase II (rather than three as was originally believed) are also intrinsic components of RNA polymerases I and III.[11] These common subunits are all encoded by unique and essential genes, but their functions remain elusive. Finally, two of the RNA polymerase II-specific small subunits form a subcomplex that is more loosely associated with the core enzyme than the other subunits and has been implicated in regulatory events rather than basic transcription.[10]

Perhaps the most interesting discovery that arose from cloning of the core RNA polymerase II genes is the very unusual composition of the C-terminal domain (CTD) of the largest subunit (reviewed in ref. 12). The CTD comprises tandem repeats of the heptapeptide Tyr-Ser-Pro-Thr-Ser-Pro-Ser, which are essential for cell viability and are highly conserved among different species but have no equivalent in any other RNA polymerase including the other two eukaryotic enzymes. The repetitive nature and biased amino acid composition of the CTD explain many of the known biochemical properties of the largest RNA polymerase II subunit, including its high immunogenicity, susceptibility to proteolysis, and variable phosphorylation states. Notably, the number of heptapeptide repeats found in RNA polymerase II from different species increases with the genomic complexity of the organism, with 26 copies in yeast, 43 in *Drosophila*, and 52 in mammals. The CTD is phosphorylated in vivo at multiple sites, mostly on Ser, but also on Thr and Tyr residues, probably resulting in a more extended structure. In the cell, the existence of both hypophosphorylated (form IIA) and hyperphosphorylated (form II0) populations of RNA polymerase II can be demonstrated, suggesting that phosphorylation of the CTD plays an important regulatory role. Although CTD phosphorylation is not absolutely required for basal transcription, it is clear that the highly phosphorylated form of RNA polymerase II is involved in transcription elongation in vivo. On the other hand, in vitro evidence indicates that RNA polymerase IIA associates more readily than RNA polymerase II0 into preinitiation complexes. These observations have led to the hypothesis that phosphorylation of the CTD may trigger dissociation of the elongation complex from some of the components involved in the initiation step.[12]

Despite its extraordinary structural complexity, it is clear that the core RNA polymerase II has by itself no intrinsic ability to recognize promoter elements. In that sense, it is analogous to the *E. coli* core RNA polymerase devoid of sigma factor, and it is interesting to see that this analogy also extends to the structural level. The three-dimensional structures of RNA polymerases are all characterized by a thumb-like projection surrounding a channel with a diameter of 25Å, an appropriate size for binding double-stranded helical DNA. However, the structure of yeast RNA polymerase II resembles that of *E. coli* core RNA polymerase in that the thumb forms part of a ring that surrounds the channel.[13] In contrast, the thumb of *E. coli* holoenzyme defines a deep but open groove. This structural difference might be a specific feature distinguishing promoter-binding from elongation conformations of RNA polymerases, with closure of the channel around the DNA responsible for the remarkable stability of the elongating ternary complex.

CLASS II GENERAL TRANSCRIPTION FACTORS

In contrast to the nonspecific transcription displayed by purified RNA polymerase, specific transcription initiation from class II gene promoters can be observed in crude whole-cell or nuclear extracts. Fractionation and chromatographic purification of active components from these crude extracts has led to the identification of a set of proteins, referred to as the general transcription initiation factors, that are implicated in promoter recognition, formation of a stable preinitiation complex, and specific transcription initiation by RNA polymerase II. While there may be different pathways leading to the assembly of a complete class II preinitiation complex (see below), the different components are now well characterized, and most of them have already been cloned from both yeast and higher eukaryotes. Because the general properties of the RNA polymerase II initiation factors have been the topic of many reviews,[14,15] these will be briefly summarized here, with only the most recent results mentioned in greater detail.

TFIID

In the RNA polymerase II system, TFIID seems to be the only general transcription factor that can interact with the DNA in a sequence-specific manner. In recent years, the structure and properties of this protein, as well as its interaction with other transcription components, have been intensively scrutinized (see chapter 3). TFIID is composed of a TATA-binding protein (TBP), which is also involved in transcription by RNA polymerases I and III, and a variety of associated polypeptides termed TBP-associated factors or TAFs (Table 2.1). TBP is a highly conserved saddle-shaped protein that interacts with the minor groove of the consensus TATA sequence, sharply bending the DNA. In vitro, TBP alone can efficiently substitute for TFIID in promoting transcription by RNA polymerase II from TATA boxes, but the holo-TFIID complex is usually required for a response to transcriptional activators.

TFIIB

TFIIB is the only general transcription factor that consists of a single polypeptide. Yet, TFIIB can interact simultaneously with a TBP-DNA complex, RNA polymerase II, and TFIIF. Mutational analysis has outlined the existence of two distinct domains in TFIIB: an N-terminal zinc-binding domain required for interaction with RNA polymerase II and a C-terminal region composed of two imperfect direct repeats involved in the interaction with TFIID (reviewed in ref. 16). The crystal structure of a TFIIB/TBP/TATA element ternary complex[17] reveals that TFIIB contacts both TBP and the DNA, acting as a clamp to stabilize the TBP-DNA complex. The interaction of

Table 2.1. General transcription initiation factors of RNA polymerase II

Name	Composition	Enzymatic Activity	Assigned Function
TFIID	TBP (~30 kDa) and 7 TAFs (20 to 250 kDa)	TFIIF kinase ($TAF_{II}250$)	Recognition of basic promoter elements. Response to transcriptional activators.
TFIIB	Single polypeptide (~35 kDa)	None known	Interface between TBP and RNA polymerase/TFIIF. Start site selection.
TFIIF	2 or 3 subunits 30 to 100 kDa	None known	Decreases nonspecific DNA binding by RNA polymerase. Interface between RNA polymerase and TFIIB/TBP/DNA. Additional role in elongation.
TFIIE	2 subunits 30 to 70 kDa (tetramer)	None known	TFIIH recruitment. DNA unwinding. Promoter clearance.
TFIIH	9 subunits (30 to 100 kDa)	DNA helicase (ERCC2, ERCC3) CTD kinase (CDK7)	DNA unwinding. Promoter clearance. Link to cell cycle and DNA repair.

TFIIB with the DNA is not sequence-specific since the contacts involve only the DNA backbone, extending both upstream and downstream of the TATA box.[17]

TFIIF

TFIIF is unique among the general transcription factors because of its dual role in both the initiation and elongation steps of transcription by RNA polymerase II. This multimeric protein plays a key role in the assembly of the preinitiation complex by providing interaction surfaces between multiple components. For instance, TFIIF interacts with both TFIIB and RNA polymerase II, and can thus promote association of the core RNA polymerase II with a TBP/TFIIB/TATA box complex (refs. 18, 19 and citations therein). One of the subunits of TFIIF displays considerable homologies with the sigma subunit of *E. coli* RNA polymerase, and like sigma, TFIIF can prevent nonspecific interaction of RNA polymerase with DNA. In addition to its role in preinitiation complex assembly, TFIIF participates in the RNA elongation step by suppressing pausing of RNA polymerase II at certain sites on the DNA.[20]

In addition to two polypeptides that are homologous to the known subunits of human TFIIF, yeast TFIIF also contains a loosely associated subunit with no identified mammalian homologue. Interestingly, this particular polypeptide is also a component of the yeast TFIID complex and is encoded by a gene implicated in actin cytoskeleton function.[21] The significance of this observation remains to be determined.

TFIIE AND TFIIH

When preinitiation complexes are assembled stepwise in vitro with purified components, TFIIE and TFIIH are the last two transcription factors to enter the complex. While cloning of the two different subunits of TFIIE revealed little about their exact function in transcription initiation, identification of different components of TFIIH brought about unexpected results regarding the involvement of this protein in not only transcription, but other cellular processes as well (reviewed in ref. 22). First, TFIIH has a kinase activity capable of phosphorylating the RNA polymerase II CTD. This activity is due to the presence in TFIIH of a cyclin/kinase complex composed in human of the CDK7 (MO15) kinase and cyclin H. The corresponding cyclin/kinase pair in yeast TFIIH, KIN28/CCL1, can be chromatographically separated from the core TFIIH subunits and is sometimes referred to as TFIIK. The ability of CDK7 to phosphorylate other cyclin-dependent kinases such as CDC2 and CDK2 implies a link between the basic transcription machinery and cell cycle control.

TFIIH also contains polypeptides known to participate in the nucleotide excision repair process. These include ERCC3 and ERCC2, two ATP-dependent DNA helicases (termed respectively RAD25/SSL2 and RAD3 in yeast). These subunits are involved both in the general and in the transcription-coupled repair pathways. In humans, mutations in the corresponding genes cause genetic diseases associated with defective DNA repair (e.g., *xeroderma pigmentosum*). Thus TFIIH-containing complexes provide an unexpected link between different cellular processes including transcription, DNA repair, and cell cycle regulation.[22]

TFIIA

In yeast, transcription reactions reconstituted in vitro with highly purified components showed that the five general transcription factors TFIID (TBP), TFIIB, TFIIF, TFIIE and TFIIH are both necessary and sufficient for basal transcription by RNA polymerase II.[23] Therefore, the exact function of several other proteins, which were originally described as general transcription factors,[14,15] should probably be reexamined. For instance, TFIIA, an activity often tightly associated with the TFIID complex,[24] would be better classified as a specialized coactivator. TFIIA is sometimes required to prevent the interaction of negative components,[25] and its effect on basic transcription is promoter-dependent.[26] In vitro, TFIIA can associate directly with the TBP-DNA complex, extending the protein-DNA contacts upstream of the TATA box and stabilizing the complex independently of the TFIIB interaction.[24,27] However, there are many indications that, both in vivo[28] and in vitro,[29-32] TFIIA is mostly important for the response to transcriptional activators.

THE RNA POLYMERASE II HOLOENZYME

The holoenzyme is a stable form of RNA polymerase II that contains several general transcription factors in addition to all of the polypeptides previously defined as subunits of the core RNA polymerase II (Table 2.2). R.A. Young's group discovered the holoenzyme in yeast as a consequence of their analysis of the suppressors of RNA polymerase B(II) genes (*SRB* genes). These genes were isolated through their ability to suppress the cold-sensitive phenotype of cells containing mutant RNA polymerase II with CTD truncations.[33] The protein products of the *SRB* genes copurified as a 1.2 MDa multisubunit complex that also contained the 12-subunit core RNA polymerase II and, at least in the early stages of purification, TBP.[33] So far, nine SRB proteins, ranging in size from 121 (SRB6) to 1420 amino acids (SRB9), have been identified. All of them are associated with the holoenzyme.[34] The SRB10 and SRB11 proteins are a kinase- and cyclin-like pair of proteins involved in CTD phosphorylation and essential for a normal transcriptional response to galactose induction.[35] Genetic evidence based on the

Table 2.2. Components of the RNA polymerase II holoenzyme

Name		Composition
• Core RNA polymerase II		12 subunits, 10 to 220 kDa
• Mediator:	- Global transcriptional regulators	SRB proteins, Gal11, Sin4, Rgr1
	- Chromatin remodeling proteins	SWI/SNF gene products
	- Subset of general transcription factors	TFIIF, TFIIB, TFIIH (in yeast)

identity between some of the *SRB*s and previously known genes implicates the SRB proteins in mediating transcriptional repression as well as activation.[36]

In addition to the SRBs, the RNA polymerase II holoenzyme contains other global transcriptional regulators. Those include the *GAL11*, *SIN4*, and *RGR1* gene products, which are all known to play both positive and negative roles in the transcriptional regulation of many yeast genes.[37,38] Several general transcription factors are also tightly associated with the RNA polymerase II holoenzyme. In yeast, these include TFIIF, TFIIH, and TFIIB, while interaction of TBP with the holoenzyme in the absence of DNA is much less stable.[37,39] Finally, the *SWI/SNF* gene products are also integral components of the yeast holoenzyme.[40] These evolutionary conserved proteins are involved in the activation of many inducible genes by their ability to effect, in an ATP-dependent fashion, rearrangements of chromatin structure.[41]

Only a small fraction (5-20%) of the total RNA polymerase II pool in yeast cells seems to be present at any one time in the holoenzyme complex, which explains why this particular form of enzyme was not discovered earlier. However, the SRB proteins are found exclusively in this complex.[39] Thompson and Young used temperature-sensitive mutations in *SRB* genes to investigate the importance of the RNA polymerase II holoenzyme for transcription of different yeast genes.[42] Following transfer to the restrictive temperature, inactivation of SRB function caused a rapid and general shutdown of mRNA synthesis in these mutant cells. This effect on general transcription by RNA polymerase II was similar to that observed with a temperature-sensitive mutation in the RNA polymerase itself. This result therefore constitutes very strong evidence that the holoenzyme is the form of RNA polymerase II that is recruited to most of the promoters in vivo. Note that, at present, most of our knowledge of the holoenzyme is derived from studies performed in yeast. However, increasing evidence indicates that mammalian organisms also contain large RNA polymerase II complexes containing general transcription factors and SRB-like proteins.[43-45]

It must be emphasized that the discovery of the holoenzyme has helped reconcile a number of earlier biochemical and genetic observations. First, gene-specific transcriptional activators were known to require additional proteins in order to exert their effects on the general transcription machinery. These bridging transcription factors are generally referred to as coactivators or mediators of transcriptional regulation. On the other hand, mutations in the RNA polymerase II CTD and other components of the holoenzyme were known to reduce the response of many genes to transcriptional regulators. Kornberg and his collaborators demonstrated that the RNA polymerase II holoenzyme is formed by the interaction of the core RNA polymerase II with a multiprotein complex, the mediator, that can promote the response to transcriptional activators.[37] The core RNA polymerase and the mediator can be separated by chromatography on a column of immobilized CTD-specific antibodies, where the mediator elutes free of RNA polymerase in the column flowthrough. Such observations demonstrate the essential role played by the CTD in the interaction between the mediator and the core RNA polymerase II. In contrast to the results observed in transcription reactions reconstituted in vitro with general transcription factors and purified core RNA polymerase II, reactions containing the mediator are responsive to activator proteins such as GAL4-VP16.[37,39] Together with other in vivo data,[46,47] this observation leads to the idea that gene-specific transcription factors might function by facilitating positioning of the RNA polymerase holoenzyme near the transcription start site.

TRANSCRIPTION INITIATION PATHWAY

Now that most of the components of the general initiation machinery have been identified and cloned, a large number of protein-protein and protein-DNA interactions are being defined and the structure of the RNA polymerase II preinitiation complex is emerging.[17] Concurrently with these structural studies, functional assays, often based on similar investigations in prokaryotic systems, are being applied to explore the mechanism of transcription initiation by RNA polymerase II. The initiation pathway consists of a number of successive steps, some of which are presently better defined than others. Each of these steps is a potential target for regulation by gene-specific transcription factors.

Preinitiation Complex Assembly

Promoter recognition by components of the general transcription machinery leads to formation of a preinitiation complex. The TATA element is a binding site for TFIID, an interaction that is mediated by the TBP subunit. This specific binding commits the gene to the transcriptional process. Footprinting and functional analysis of Inr-containing promoters revealed that TFIID can also function as the primary promoter recognition factor in the absence of a TATA box.[48-51] In *Drosophila*, the Inr-recognizing component of TFIID is one the TBP-associated factors, $TAF_{II}150$, which was shown to bind directly to this particular core element.[52,53] Thus, the TAFs appear to function in core promoter recognition not only to discriminate among the three classes of RNA polymerases, but also to assist TFIID binding to various types of promoters.

In vitro, various subcomplexes of promoter DNA and general transcription initiation factors are stable, which led to the original belief that assembly of the preinitiation complex on the promoter DNA may also take place in a stepwise fashion in vivo.[14,15] However, evidence linked to the existence of the holoenzyme now suggests that the assembly of a complete preinitiation complex in cells may instead occur in a single step through recognition of the DNA-TFIID complex by the RNA polymerase II holoenzyme.[42]

By several criteria, TATA and Inr elements seem to direct transcription through similar pathways.[50,54,55] Either element can position the initiation complex to give site-specific transcription initiation,[55] and a similar set of general transcription factors is used in each case.[48,50] However, recent results demonstrate that the core promoter structure plays an important role in transcriptional regulation.[1,56] Certain activators, for example, have a preference for the Inr or for TATA elements. The core promoter preferences of various activators or activator domains may be linked to their mechanisms of activation. For example, various upstream DNA-binding regulators can participate in the promoter recognition step of initiation by stabilizing the preinitiation complexes.[2] On the other hand, the alignment of upstream factor sites and the TATA box is important for specifying the direction of transcription, since TATA elements are not by themselves directional.[57,58] Thus, the overall architecture of promoters—both core and regulatory sequences—may program the transcriptional mechanisms of genes, for example by specifying which steps in the initiation pathway will be rate limiting.[59]

Start Site Selection

From isolation of suppressers of start site aberrations and other genetic studies in yeast, it is clear that TFIIB and at least two of the subunits of RNA polymerase II, RPB1 and RPB9, play key roles in the selection of transcription start sites.[60-62] In *Schizosaccharomyces pombe*, transcription starts 30 bases downstream of the TATA box, while it starts 40-120 downstream in *Saccharomyces cerevisiae*. In vitro studies using reconstituted reactions and exchange of individual components purified from these two species also confirmed the essential role of the TFIIB-RNA polymerase II pair in start site selection.[61]

In footprinting experiments at TATA-containing promoters, TFIID protects regions as far as 30-40 bases past the transcription start site, suggesting that core promoter elements as yet uncharacterized reside downstream of the start site.[53,63] Recently, two conserved downstream elements in TATA-less genes have been identified, both of which show characteristics of core promoter elements. One, found along with the Inr in a number of *Drosophila* promoters, appears to help anchor TFIID.[8] The second element, termed MED-1, plays a role in start site selection in the case of promoters with multiple start sites.[6]

Transcription Initiation

RNA polymerase II preinitiation complexes initially assemble on closed DNA. In subsequent steps, the promoter DNA unwinds to make an open complex,[64] the first phosphodiester bond forms, several more nucleotides are added to make a short RNA chain, and the RNA polymerase II finally moves away from the promoter. A distinguishing feature of the RNA polymerase II system is that at least one of these early transcription steps requires ATP (or dATP) hydrolysis.[9,14] A similar energy requirement has not been found for initiation by any other RNA polymerase.

There is general agreement that ATP hydrolysis is needed for synthesis of the first phosphodiester bond by RNA polymerase II,[65-67] although dinucleotide priming experiments initially suggested that ATP hydrolysis was not required until the promoter clearance stage.[68] A great deal of evidence points to the involvement of an ATP-dependent helicase activity in TFIIH for promoter opening prior to the actual transcription initiation event.[66] First, ATP and complete preinitiation complexes containing TFIIH are both required to observe unwound DNA during investigations of promoter opening.[67] Second, transcription initiation can occur in the absence of TFIIH on easily unwound (negatively supercoiled) promoters[69,70] or on heteroduplex promoters.[71,67] Initiation in these latter cases also bypasses the need for ATP and TFIIE, in agreement with the notion that TFIIE helps TFIIH bind to the preinitiation complex.[72] Finally, it is clearly the helicase rather than the CTD kinase function of TFIIH that is involved in promoter opening, since ATP-dependent formation of a first phosphodiester bond can take place in reactions reconstituted with TFIIH mutants defective in the kinase activity.[73,74]

Promoter Clearance and Early Elongation Events

The transition from the initiation to the elongation step of transcription requires that the RNA polymerase sever its ties to the promoter DNA and to some or all of the general initiation factors. This process of promoter clearance is ill defined, but it coincides in vivo with a modification of the RNA polymerase II CTD from a hypophosphorylated state (initiation complex) to a hyperphosphorylated state (elongation complex).[12] CTD phosphorylation is apparently not required for the initiation and early elongation steps, since RNA polymerase II molecules detected on cellular genes 20-30 nucleotides past the transcription start site are still hypophosphorylated.[75]

In vitro, the requirement for phosphorylation of the CTD is variable, depending on the promoter and the purity of the transcription components. It is dispensable for basal transcription in highly purified reconstitutions.[12,76,77] In fact, experiments with native versus proteolyzed RNA polymerase

II preparations revealed that the CTD is not required at all for transcription from some promoters (reviewed in ref. 12). In contrast, the CTD is essential for transcription of other genes such as dihydrofolate reductase, but even then phosphorylation of the CTD by TFIIH is not required for synthesis of abortive transcripts.[74] Thus, the evidence at this point would suggest that the CTD kinase activity associated with TFIIH is involved in the promoter clearance or early elongation steps, rather than transcription initiation per se.

Investigations initiated by the group of J.T. Lis have clearly demonstrated that early elongation events may be a common rate-limiting step for class II gene transcription in vivo (see ref.78 and citations therein). As best exemplified by the *Drosophila* heat-shock[78] and human c-*myc* genes,[79,80] the expression of many eukaryotic genes is restricted by the presence of an engaged but paused RNA polymerase at their 5' end. Activation of these genes must therefore involve factors that can release the elongation block. In agreement with this idea, several transcriptional activators have been reported to function by increasing the processivity of RNA polymerase II.[81]

Postinitiation Events

In order for a gene to be highly transcribed, multiple polymerases must initiate from the same promoter in rapid succession. This requires complete clearance of the promoter region at each round of initiation to allow access by the subsequent RNA polymerase. At present, little is known about the postinitiation events in class II gene transcription. Footprinting evidence suggests that minimally TFIID and perhaps another protein are present in the postinitiation complex,[82] which raises the possibility that an additional step may be required before the promoter can be reused. Successive rounds of transcription from the same gene have been observed in vitro in crude extracts and partially purified systems,[83] but the mechanisms that control the frequency of transcription initiation at different promoters remain obscure.

CELLULAR REGULATION OF RNA POLYMERASE II ACTIVITY

Although transcriptional regulation is most often mediated by gene-specific transcription factors, recent work from U. Schibler's laboratory suggests that transcription can also be regulated in a cell-type-specific manner by changes in the concentration of components of the basal RNA polymerase II machinery. In rodent testis, TBP, TFIIB, and RNA polymerase II accumulate to high levels in round spermatids, allowing the rapid synthesis of mRNAs that takes place during the final stages of spermiogenesis.[84] The RNA polymerase II basic machinery may also be involved in general events such as those associated with cell size regulation.[85]

On a somewhat different topic, one fascinating new concept set forward by P.R. Cook's laboratory relates to the relative movements of the eukaryotic RNA polymerase and its template (reviewed in ref. 86). Traditionally, the RNA polymerase is thought to be the moving component, tracking along the DNA during the RNA elongation step. However, it is also possible to consider the alternative possibility where it is instead the DNA that passes through a static RNA polymerase. This alternative model has the advantage of eliminating topological problems such as the RNA becoming entwined about the template. It is also entirely consistent with many other observations. First, RNA polymerases are surprisingly efficient at converting the free energy liberated during the RNA synthesis condensation reaction into mechanical work.[87] This mechanical force can be used by the polymerase to move along the DNA, but it can also be used by an immobilized polymerase to move the template. Accordingly, RNA polymerases that are artificially immobilized in vitro will carry out the elongation step of transcription as efficiently as soluble enzymes. Second, biochemical evidence indicates that RNA polymerases are attached to some structure within the nuclei. High-resolution immunofluorescence studies revealed that the sites of nascent mRNA synthesis and processing are not spread throughout the nuclei but are instead concentrated in some 300-500 foci. These "transcription factories" seem to be responsible for the immobilization of chromatin loops.[86,88]

Of course, many findings in transcriptional regulation may have to be reinterpreted in the context of an attached RNA polymerase II. For instance, one may have to think in terms of regulatory DNA elements and their cognate transcription factors as ultimately responsible for bringing a promoter into the vicinity of the RNA polymerases rather than serving as entry sites for soluble RNA polymerases. On the other hand, seemingly surprising results, e.g., the fact that different genes that are dependent on the same distant regulatory elements are activated alternately rather than at the same time,[89] may be much simpler to interpret if the RNA polymerases remain at fixed locations within the nucleus.

Although these mechanisms are clearly beyond the scope of this review, it is important to also keep in mind that transcriptional regulation in eukaryotes is not restricted to a control of the initiation and promoter clearance steps. For instance, the size of transcription units, which in mammals spans from less than 1 kb to as much as 2000 kb, plays an important role by determining the duration of transcription (from 30 s to 11 h). Mitosis is associated with a generalized repression of transcription. At that time, many nonhistone proteins, including several transcription factors, become phosphorylated,[90] and the nascent transcripts are aborted.[91] Consequently, transcription of each gene has to begin anew in the next cell cycle, which probably contributes to regulating the expression of very large genes. Finally, the efficiency of transcription termination and its link to poly(A) site function defines an additional level of control in the case of closely spaced transcription units.[92]

CONCLUDING REMARKS

Now that most of the general components of the RNA polymerase II machinery have been identified, their involvement in various transcription steps can be explored. With the discovery of the RNA polymerase II holoenzyme, the mechanism of preinitiation complex assembly is better understood. However, it is still unclear how the mediator and other initiation components are released and recycled after transcription initiation, and how these events relate to phosphorylation of the RNA polymerase II CTD. It should now be feasible to determine the exact mechanism of action of various regulatory proteins, which would give meaning to the vast amount of information accumulated in recent years on the physical interaction of different activators with general initiation components. It would also be interesting to investigate the absolute and relative concentrations of transcriptional components in different cells and under different physiological conditions, and whether these levels are individually or globally controlled. Finally, it seems likely that some as-yet-unassigned protein domains of RNA polymerase II and associated factors will eventually be assigned structural roles within the context of the nuclear matrix, while others may couple transcription to other cellular processes such as RNA processing and transport.

ACKNOWLEDGMENTS

We are grateful to Michael W. Van Dyke for critical reading of the manuscript. Work in our laboratory on transcription mechanisms is supported by Grants ARP 000015056 from the Texas Higher Education and G-1195 from the Robert A Welch Foundation.

REFERENCES

1. Emami KH, Navarre WW, Smale ST. Core promoter specificities of the Sp1 and VP16 transcriptional activation domains. Mol Cell Biol 1995; 15:5906-5916.
2. Kollmar R, Farnham PJ. Site-specific initiation of transcription by RNA polymerase II. Proc Soc Exp Biol Med 1993; 203:127-139.
3. Smale ST, Baltimore D. The "initiator" as a transcription control element. Cell 1989; 57:103-113.
4. Javahery R, Khachi A, Lo K et al. DNA sequence requirements for transcriptional initiator activity in mammalian cells. Mol Cell Biol 1994; 14:116-127.

5. Colgan J, Manley JL. Cooperation between core promoter elements influences transcriptional activity in vivo. Proc Natl Acad Sci USA 1995; 92:1955-1959.
6. Ince T, Scotto KW. A conserved downstream element defines a new class of RNA polymerase II promoters. J Biol Chem 1995; 270:30249-30252.
7. Liao W-C, Geng Y, Johnson LF. In vitro transcription of the TATAA-less mouse thymidylate synthase promoter: multiple transcription start points and evidence for bidirectionality. Gene 1994; 146:183-189.
8. Burke TW, Kadonaga JT. *Drosophila* TFIID binds to a conserved downstream basal promoter element that is present in many TATA-box-deficient promoters. Genes Dev 1996; 10:711-724.
9. Sawadogo M, Sentenac A. RNA polymerase B(II) and general transcription factors. Annu Rev Biochem 1990; 59:711-754.
10. Young RA. RNA polymerase II. Annu Rev Biochem 1991; 60:689-715.
11. Treich I, Carles C, Riva M et al. RPC10 encodes a new mini subunit shared by yeast nuclear RNA polymerases. Gene Expr 1992; 2:31-37.
12. Dahmus ME. The role of multisite phosphorylation in the regulation of RNA polymerase II activity. Progr Nucleic Acids Res Mol Biol 1994; 48:143-179.
13. Polyakov A, Severinova E, Darst SA. Three-dimensional structure of *E. coli* core RNA polymerase: promoter binding and elongation conformations of the enzyme. Cell 1995; 83:365-373.
14. Conaway RC, Conaway JW. General initiation factors for RNA polymerase II. Annu Rev Biochem 1993; 62:161-190.
15. Zawel L, Reinberg D. Common themes in assembly and function of eukaryotic transcription complexes. Annu Rev Biochem 1995; 64:533-561.
16. Bagby S, Kim S. Maldonado E et al. Solution structure of the C-terminal core domain of human TFIIB: similarity to cyclin A and interaction with TATA-binding protein. Cell 1995; 82:857-867.
17. Nikolov DB, Chen H, Halay ED et al. Crystal structure of a TFIIB-TBP-TATA element ternary complex. Nature 1995 377:119-128.
18. Wang BQ, Burton ZF. Functional domains of human RAP74 including a masked polymerase binding domain. J Biol Chem 1995; 270:27035-27044.
19. Sun ZW, Hampsey M. Identification of the gene (SSU71/TFG1) encoding the largest subunit of transcription factor TFIIF as a suppressor of a TFIIB mutation in Saccharomyces cerevisiae. Proc Natl Acad Sci USA 1995; 92:3127-3131.
20. Tan S, Conaway RC, Conaway JW. Dissection of transcription factor TFIIF functional domains required for initiation and elongation. Proc Natl Acad Sci USA 1995; 92:6042-6046.
21. Henry NL, Campbell AM, Feaver WJ et al. TFIIF-TAF-RNA polymerase II connection. Genes Dev 1994; 8:2868-2878.
22. Seroz T, Hwang JR, Moncollin V et al. TFIIH: a link between transcription, DNA repair and cell cycle regulation. Curr Opin Gen Dev 1995; 5:217-221.
23. Sayre MH, Tschochner H, Kornberg RD. Reconstitution of transcription with five purified initiation factors and RNA polymerase II from Saccharomyces cerevisiae. J Biol Chem 1992; 267:23376-23382.
24. Yokomori K, Admon A, Goodrich JA et al. *Drosophila* TFIIA-L is processed into two subunits that are associated with the TBP/TAF complex. Genes Dev 1993; 7:2235-2245.
25. Kim TK, Zhao Y, Ge H et al. TATA-binding protein residues implicated in a functional interplay between negative cofactor NC2 (DR1) and general factors TFIIA and TFIIB. J Biol Chem 1995; 270:10976-10981.
26. Aso T, Sezawa H, Conoway RC et al. A TATA sequence-dependent transcriptional repressor associated with mammalian transcription factor IIA. EMBO J 1994; 13:435-445.
27. Geiger JH, Hahn S, Lee S et al. Crystal structure of the yeast TFIIA/TBP/DNA complex. Nature 1996; 272:830-836.
28. Stargell LA, Struhl K. The TBP-TFIIA interaction in the response to acidic activators in vivo. Science 1995; 269:75-78.
29. Kaiser K, Stelzer G, Meisterernst M. The coactivator p15 (PC4) initiates transcriptional activation during TFIIA-TFIID-promoter complex formation. EMBO J 1995; 14:3520-3527.

30. Ozer J, Moore PA, Bolden AH et al. Molecular cloning of the small (gamma) subunit of human TFIIA reveals functions critical for activated transcription. Genes Dev 1994; 8:2324-2335.
31. Yokomori K, Zeidler MP, Chen JL et al. *Drosophila* TFIIA directs cooperative DNA binding with TBP and mediates transcriptional activation. Genes Dev 1994; 8:2313-2323.
32. Shykind BM, Kim J, Sharp PA. Activation of the TFIID-TFIIA complex with HMG-2. Genes Dev 1995; 9:1354-1365.
33. Thompson CM, Koleske AJ, Chao DM et al. A multisubunit complex associated with the RNA polymerase II CTD and TATA-binding protein in yeast. Cell 1993; 73;1361-1375.
34. Hentgartner CJ, Thompson CM, Zhang J et al. Association of an activator with an RNA polymerase II holoenzyme. Genes Dev 1995; 9:897-910.
35. Liao S-M, Zhang J, Jeffery DA et al. A kinase-cyclin pair in the RNA polymerase II holoenzyme. Nature 1995; 374:193-196.
36. Song W, Treich I, Qian N et al. *SSN* genes that affect transcriptional repression in *Saccharomyces cerevisiae* encode SIN4, ROX3 and SRB proteins associated with RNA polymerase II. Mol Cell Biol 1996; 16:115-120.
37. Kim Y-J, Björklund S, Li Y et al. A multiprotein mediator of transcriptional activation and its interaction with the C-terminal repeat domain of RNA polymerase II. Cell 1994: 77:599-608.
38. Li Y, Bjorklund S, Jiang YW et al. Yeast global transcriptional regulators Sin4 and Rgr1 are components of mediator complex/RNA polymerase II holoenzyme. Proc Natl Acad Sci USA 1995; 92:10864-10868.
39. Koleske AJ, Young RA. An RNA polymerase II holoenzyme responsive to activators. Nature 1994; 368:466-469.
40. Wilson CJ, Chao DM, Imbalzano AN et al. RNA polymerase II holoenzyme contains SWI/SNF regulators involved in chromatin remodeling. Cell 1996; 84: 235-244.
41. Peterson CL, Tamkun JW. The SWI-SNF complex: a chromatin remodeling machine? Trends Biochem Sci 1995; 20: 143-146.
42. Thompson CM, Young RA. General requirement for RNA polymerase II holoenzymes in vivo. Proc Natl Acad Sci USA 1995; 92:4587-4590.
43. Ossipow V, Tassan J-P, Nigg EA et al. A mammalian RNA polymerase II holoenzyme containing all components required for promoter-specific transcription initiation. Cell 1995; 83:137-146.
44. Chao DM, Gadbois EL, Murray PJ et al. A mamalian SRB protein associated with an RNA polymerase II holoenzyme. Nature 1996; 380:82-85.
45. Maldonado E, Shiekhattar, R, Sheldon M et al. A human RNA polymerase II complex associated with SRB and DNA-repair proteins. Nature 1996; 381:86-89.
46. Barberis A, Pearlberg J, Simkovich N et al. Contact with a component of the polymerase II holoenzyme suffices for gene activation. Cell 1995:81:359-368.
47. Gerber H-P, Hagmann M, Seipel K et al. RNA polymerase II C-terminal domain required for enhancer-driven transcription. Nature 1995; 374:660-662.
48. Pugh BF, Tjian R. Transcription from a TATA-less promoter requires a multisubunit TFIID complex. Genes Dev 1991; 5:1935-1945.
49. Wang JC, Van Dyke MW. Initiator sequences direct downstream promoter binding by human transcription factor IID. Biochim Biophys Acta 1993; 1216: 73-80.
50. Martinez E, Chiang C, Ge H et al. TATA-binding protein-associated factor(s) in TFIID function through the initiator to direct basal transcription from a TATA-less class II promoter. EMBO J 1994; 13:3115-3126.
51. Kaufmann J, Smale ST. Direct recognition of initiator elements by a component of the transcription factor TFIID complex. Genes Dev 1994; 8:821-829.
52. Verrijzer CP, Yokomori K, Chen J-L et al. *Drosophila* $TAF_{II}150$: similarity to yeast gene TSM-1 and specific binding to core promoter DNA. Science 1994; 264: 933-941.
53. Verrijzer CP, Chen J-L, Yokomori K et al. Binding of TAFs to core elements directs promoter selectivity by RNA polymerase II. Cell 1995; 81:1115-1125.

54. Jiang Y, Smale ST, Gralla JD. A common ATP requirement for open complex formation and transcription at promoters containing initiator or TATA elements. J Biol Chem 1993; 268:6535-6540.
55. Zenzie-Gregory B, O'Shea-Greenfield A, Smale ST. Similar mechanism for transcription initiation mediated through a TATA box or an initiator element. J Biol Chem 1992; 267:2823-2830.
56. Pengue G, Lania L. Kruppel-associated box-mediated repression of RNA polymerase II promoters is influenced by the arrangement of basal promoter elements. Proc Natl Acad Sci USA 1996; 93: 1015-1020.
57. Xu L, Thali M, Schaffner W. Upstream box/TATA order is the major determinant of the direction of transcription. Nucleic Acids Res 1991; 19:6699-6704.
58. Huang WB, Wong JM, Bateman E. TATA elements direct bi-directional transcription by RNA polymerases II and III. Nucleic Acids Res 1996; 24:1158-1163.
59. Jacob GA, Kitzmiller JA, Luse DS. RNA polymerase II promoter strength in vitro may be reduced by defects at initiation or promoter clearance. J Biol Chem 1994; 269:3655-3663.
60. Berroteran RW, Ware DE, Hampsey M. The sua8 suppressors of Saccharomyces cerevisiae encode replacements of conserved residues within the largest subunit of RNA polymerase II and affect transcription start site selection similarly to sua7 (TFIIB) mutations. Mol Cell Biol 1994; 14:226-237.
61. Li Y, Flanagan PM, Tschochner H et al. RNA polymerase II initiation factor interactions and transcription start site selection. Science 1994; 263:805-807.
62. Hull MW, McCune K, Woychik NA. RNA polymerase II subunit RPB9 is required for accurate start site selection. Genes Dev 1995; 9:481-490.
63. Emanuel PA, Gilmour DS. Transcription factor TFIID recognizes DNA sequences downstream of the TATA element in the Hsp70 heat shock gene. Proc Natl Acad Sci USA 1993 90:8449-8453.
64. Gralla JD. Opening promoters for m-RNA transcription. Cell Mol Biol Res 1993; 39:319-322.
65. Jiang Y, Yan M, Gralla JD. Abortive initiation and first bond formation at an activated adenovirus E4 promoter. J Biol Chem 1995; 270:27332-27338.
66. Dvir A, Garrett KP, Chalut C et al. A role for ATP and TFIIH in activation of the RNA polymerase II preinitiation complex prior to transcription initiation. J Biol Chem 1996; 271:7245-7248.
67. Holstege FCP, van der Vliet PC, Timmers HThM. Opening of an RNA polymerase II promoter occurs in two distinct steps and requires the basal transcription factors IIE and IIH. EMBO J 1996; 15:1666-1677.
68. Goodrich JA, Tjian R. Transcription factors IIE and IIH and ATP hydrolysis direct promoter clearance by RNA polymerase II. Cell 1994; 77:145-156.
69. Parvin JD, Sharp PA. DNA topology and a minimal set of basal factors for transcription by RNA polymerase II. Cell 1993;73:533-540.
70. Parvin JD, Shykind BM, Meyers RE et al. Multiple sets of basal factors initiate transcription by RNA polymerase II. J Biol Chem 1994; 269:18414-18421.
71. Pan G, Greenblatt J. Initiation of transcription by RNA polymerase II is limited by melting of the promoter DNA in the region immediately upstream of the initiation site. J Biol Chem 1994; 269: 30101-30104.
72. Maxon ME, Goodrich JA, Tjian R. Transcription factor IIE binds preferentially to RNA polymerase IIa and recruits TFIIH: a model for promoter clearance. Genes Dev 1994; 8:515-524.
73. Makela TP, Parvin, JD, Kim J et al. A kinase-deficient transcription factor TFIIH is functional in basal and activated transcription. Proc Natl Acad Sci USA 1995: 92:5174-5178.
74. Akoulitchev S, Makela TP, Weinberg RA et al. Requirement for TFIIH kinase activity in transcription by RNA polymerase II. Nature 1995; 377:557-560.
75. O'Brien T, Hardin S, Greenleaf A et al. Phosphorylation of RNA polymerase II C-terminal domain and transcriptional elongation. Nature 1994; 370:75-77.
76. Serizawa H, Conaway JW, Conaway RC. Phosphorylation of C-terminal domain

of RNA polymerase II is not required in basal transcription. Nature 1993; 363: 371-374.
77. Li Y, Kornberg RD. Interplay of positive and negative effectors in function of the C-terminal repeat domain of RNA polymerase II. Proc Natl Acad Sci USA 1994; 91:2362-2366.
78. Giardina C, Perez-Riba M, Lis JT. Promoter melting and TFIID complexes on *Drosophila* genes in vivo. Genes Dev 1992; 6:2190-2200.
79. Strobl LJ, Eick D. Hold back of RNA polymerase II at the transcription start site mediates down-regulation of c-myc in vivo. EMBO J 1992; 11:3307-3314.
80. Krumm A, Hickey LB, Groudine M. Promoter-proximal pausing of RNA polymerase II defines a general rate-limiting step after transcription initiation. Genes Dev 1995; 9:559-572.
81. Bentley DL. Regulation of transcriptional elongation by RNA polymerase II. Curr Opin Genetics Dev 1995; 5:210-216.
82. Van Dyke MW, Sawadogo M. DNA-binding and transcriptional properties of human transcription factor TFIID after mild proteolysis. Mol Cell Biol 1990; 10:3415-3420.
83. Szentirmay MN, Sawadogo M. Sarkosyl block of transcription reinitiation by RNA polymerase II as visualized by the colliding polymerases reinitiation assay. Nucleic Acids Res 1994; 22:5341-5346.
84. Schmidt EE, Schibler U. High accumulation of the RNA polymerase II transcription machinery in rodent spermatids. Development 1995; 121:2373-2383.
85. Schmidt EE, Schibler U. Cell size regulation, a mechanism that controls cellular RNA accumulation: consequences on regulation of the ubiquitous transcription factors Oct1 and NF-Y, and the liver-enriched transcription factor DBP. J Cell Biol 1995; 128:467-483.
86. Cook PR. RNA polymerase: structural determinant of the chromatin loop and the chromosome. BioEssays 1994; 6: 425-430.
87. Yin HY, Wang MD, Svoboda K et al. Transcription against an applied force. Science 1995; 270:1653-1657.
88. Jackson DA, Cook PR. Transcriptionally active minichromosomes are attached transiently in nuclei through transcription units. J Cell Sci 1993; 105: 1143-1150.
89. Wijgerde M, Grosveld F, Fraser P. Transcription complex stability and chromatin dynamics in vivo. Nature 1995; 377: 209-213.
90. White RJ, Gottlieb TM, Downes CS et al. Mitotic regulation of a TATA-binding-protein-containing complex. Mol Cell Biol 1995; 15:1983-1992.
91. Shermoen AW, O'Farrell PH. Progression of the cell cycle through mitosis leads to abortion of nascent transcripts. Cell 1991; 67:303-310.
92. Eggermont J, Proudfoot NJ. Poly(A) signals and transcriptional pause sites combine to prevent interference between RNA polymerase II promoters. EMBO J 1993; 12:2539-2548.

CHAPTER 3

Dynamics of Transcription Complex Assembly

B. Franklin Pugh

MAJOR EVENTS IN REGULATING GENE EXPRESSION

As is evident throughout this book, eukaryotes have developed numerous strategies to regulate the expression of each of their approximately 75,000 genes in response to changing intra- and extracellular environments. Our challenge is to decipher the molecular events that transform intra- and extracellular signals into changes in gene expression. One such molecular transformation is the release of latent cytoplasmic transcription regulators such as NF-κB/Rel, so that they are free to translocate into the nucleus and recognize specific promoters (Fig. 3.1, marker 1 and chapter 9). Other transformations include cascades of protein phosphorylation by kinases that lead to the activation of latent nuclear transcription regulators. What happens next is unclear.

The quiescent target gene and its promoter are presumably encased in chromatin and thus relatively inaccessible to the transcription machinery (Fig. 3.1, marker 2). Since the primary mechanism by which a protein binds a specific DNA site is through precise docking of spatially positioned amino acids with a complementary base pair surface (chapter 1), the relevant portion of the gene's control region must be accessible to transcription regulators. Accessibility may simply be manifested by having the DNA site constitutively exposed on the surface of the nucleosome. Alternatively, the site may be buried within chromatin, and the regulator might randomly and transiently reconfigure the chromatin as it searches for its cognate site (Fig. 3.1, marker 3). Perhaps due to its stable presence at a particular DNA sequence, the regulator, in conjunction with other proteins, might elicit a more permanent reorganization of the chromatin such that it is now more or less accessible to additional components of the transcription machinery. Chromatin reorganization might involve remodeling proteins such as the α2 repressor which translationally repositions nucleosomes, the Ada complex which contains histone acetylase, and SWI/SNF complexes which use the energy of ATP hydrolysis to alter nucleosome structure (Fig. 3.1, markers 4 and 5).[1-3] These steps in gene regulation are described in chapter 18.

Transcription Factors in Eukaryotes, edited by Athanasios G. Papavassiliou.

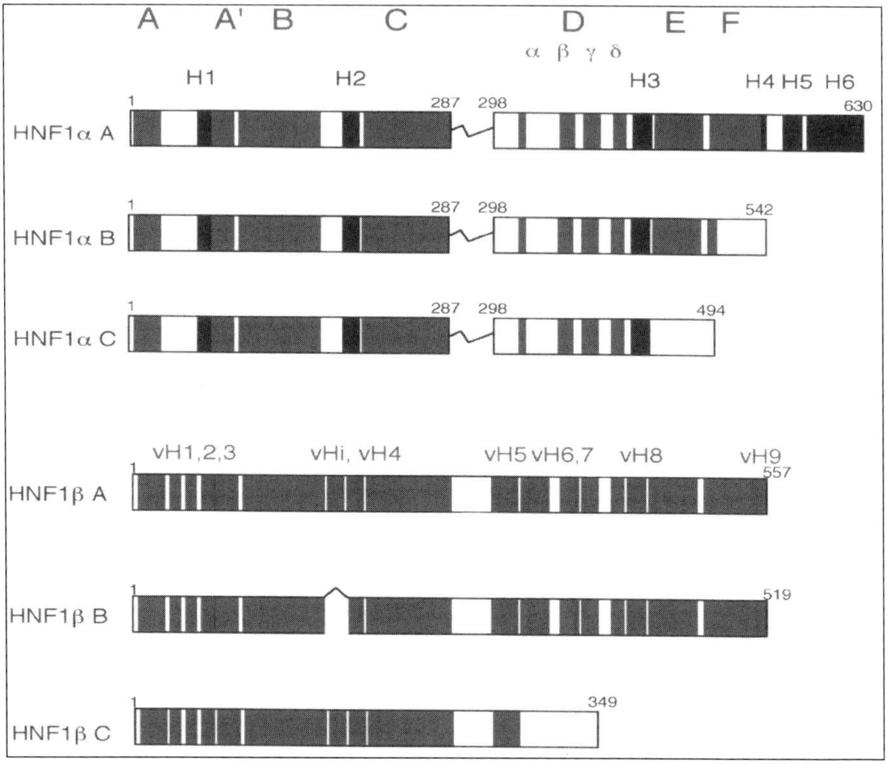

Fig. 3.1. Molecular events in the expression of a gene. 1. Transduction of signals from the cell surface to the nucleus. 2. 30 nm chromatin fiber. 3. Break-up of 30 nm fiber to 10 nm fiber by transcription factors. 4. Translational repositioning of nucleosomes by a sequence-specific regulator. 5. Nucleosome disassembly by SWI/SNF complexes and histone chaperones. 6. Spatial positioning of DNA by architectural transcription regulators. 7. A preinitiation complex (PIC) containing sequence-specific transcriptional activators (pink), a TBP-containing complex (yellow), basal initiation factors (gray), and pol II (black). 8. Gene transcription by RNA polymerase II (pol II). (Please see color insert.)

Once the chromatin is locally reconfigured at the target promoter, transcription complex assembly and transcription of the gene are not fait accompli. The promoter is apparently poised to accept additional transcription regulators and/or components of the transcription complex, each of which might be under the control of cellular signals. Efficient recruitment of components of the transcription machinery may require stereospecific alignment of sequence-specific activators at the promoter. Proper alignment may be the responsibility of architectural transcription factors such as HMG-I(Y), HMG1 and HMG2 (Fig. 3.1, marker 6).[4-6] Once the transcription complex has assembled at the promoter (Fig. 3.1, marker 7), regulators may govern the frequency of transcription initiation, clearance of RNA polymerase from the promoter, the rate of RNA polymerization (Fig. 3.1, marker 8), transcription termination, and reloading of RNA polymerase at the promoter (chapter 2). Once the RNA has been synthesized, additional factors that may also be under the control of cellular signals act on the nascent RNA to process and translocate it to its site of action.

THE ENZYMOLOGY OF GENE EXPRESSION

Enzymology is aimed in part at establishing the rate-limiting step in the transformation of a substrate into a product. Since the rate-limiting step is the slowest step in the process, it is this step which will be the subject of regulation by cellular signals. The transformation may involve a simple reaction such as phosphorylation of glucose, or may involve a complex pathway such as the conversion of glucose and ADP into ATP, CO_2, and water. In many respects, the control of gene expression can be viewed in the same way. If we are to understand the molecular basis of gene control, we must identify its rate-limiting step and elucidate how that step is regulated in response to cellular signals. In principle, any stage of gene activation described above (and in Fig. 3.1) may contain the rate-limiting step in the expression of a particular gene. As cellular signals change, the rate limiting step might also change within or between the various stages of gene activation. In effect, multiple levels of control provide a means of fine-tuning the expression of a gene, rather than simply switching it 'on' or 'off'.

Inasmuch as enzymology attempts to recapitulate enzyme function in vitro, understanding the control mechanism of gene expression requires the reconstruction of various stages of gene activation in vitro, where individual kinetic steps can be studied. The concern is whether the in vitro reaction faithfully mimics the in vivo reaction. Physiological relevance is particularly problematic for extremely complex reactions such as gene control, where in order to define individual steps unambiguously it is necessary to minimize the number of components in the reaction. As the number of components are reduced, the reaction under study becomes further removed from physiological context. While in vivo studies provide physiological context, they are unable to define molecular mechanisms. For example, in vivo studies alone cannot distinguish unequivocally whether a transcription regulator acts directly as an activator or indirectly as an anti-repressor. A solution to this quandary lies in an integrated approach where precise biochemical steps are defined in a minimal system but qualitatively verified in the context of the cell using genetic and cell biological techniques.

TRANSCRIPTION COMPLEX ASSEMBLY

Assembly of a transcription preinitiation complex (PIC) is thought to be a major point of gene regulation (recently reviewed in refs. 7 and 8). For the purposes of this chapter, a PIC is defined as all the stably integrated components present at a transcriptionally active promoter. What exactly comprises a PIC will vary from promoter to promoter. A particular PIC contains one or more sequence-specific regulatory proteins, the TATA binding protein (TBP), one of three nuclear RNA polymerases (pol I, II, or III), and a variety of other associated factors. In total, a particular PIC is likely to contain well over fifty polypeptides. Since many of these polypeptides are subunits of larger protein complexes, it is unlikely that each assembles step-wise into the developing PIC. On the other hand, the plasticity that is needed to assemble compositionally distinct transcription complexes at different promoters makes it implausible that PICs exist preassembled, simply poised to be loaded onto the appropriate promoter. While the exact steps in PIC assembly are unknown, substantial experimental evidence suggests that PIC assembly occurs through two major segments: The recruitment of a TBP-containing complex by activators bound to specific promoter sequences, followed by the recruitment of an RNA polymerase-containing complex (Fig. 3.2). In the absence of sequence-specific transcription regulators, genes are poorly transcribed even under conditions where the DNA is largely free of the repressive effects of chromatin. Speculation as to why transcription is inefficient in the absence of activators, and the mechanism by which activators enhance transcription complex assembly is the focus of the remainder of this chapter.

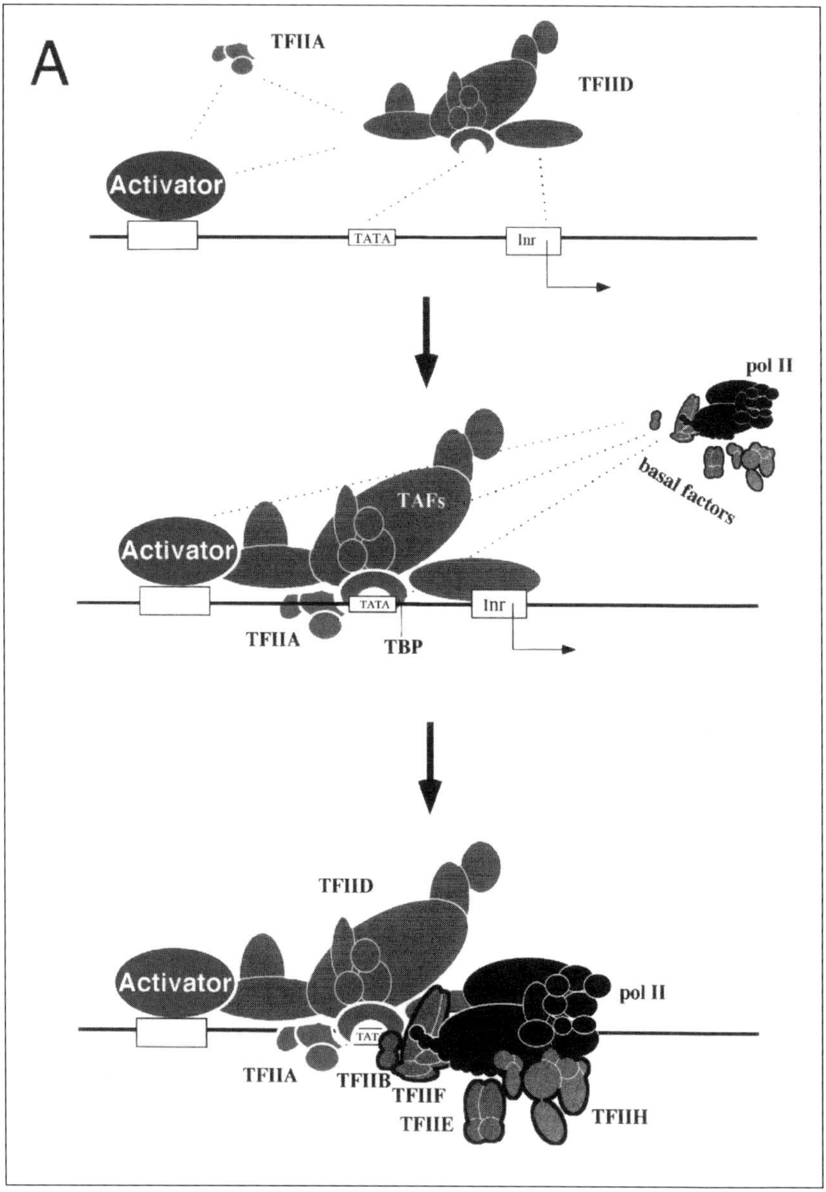

Fig. 3.2. Two related, but not exclusive views on transcription complex assembly. Each view has two major kinetic segments which differ only in detail. (A) Promoter bound activators (fuchsia), in conjunction with TFIIA (green) recruit the TBP-TAF TFIID complex (orange and blue, respectively) to TATA and/or initiator (Inr) elements. The activator-TFIIA-TFIID complex then incorporates additional basal initiation factors (TFIIB, TFIIE, TFIIF, and TFIIH shown in gray) and pol II (black) stepwise into the developing PIC. (B) (Opposite) The same as in A except that TAFs are replaced with the SRB/mediator complex (yellow), which interacts with TBP. The mediator, basal factors (gray), and the SWI/SNF complex (gray) are components of the pol II holoenzyme. The pol II holoenzyme is shown here to be preassembled and incorporated into the developing PIC in a single step. (Please see color insert.)

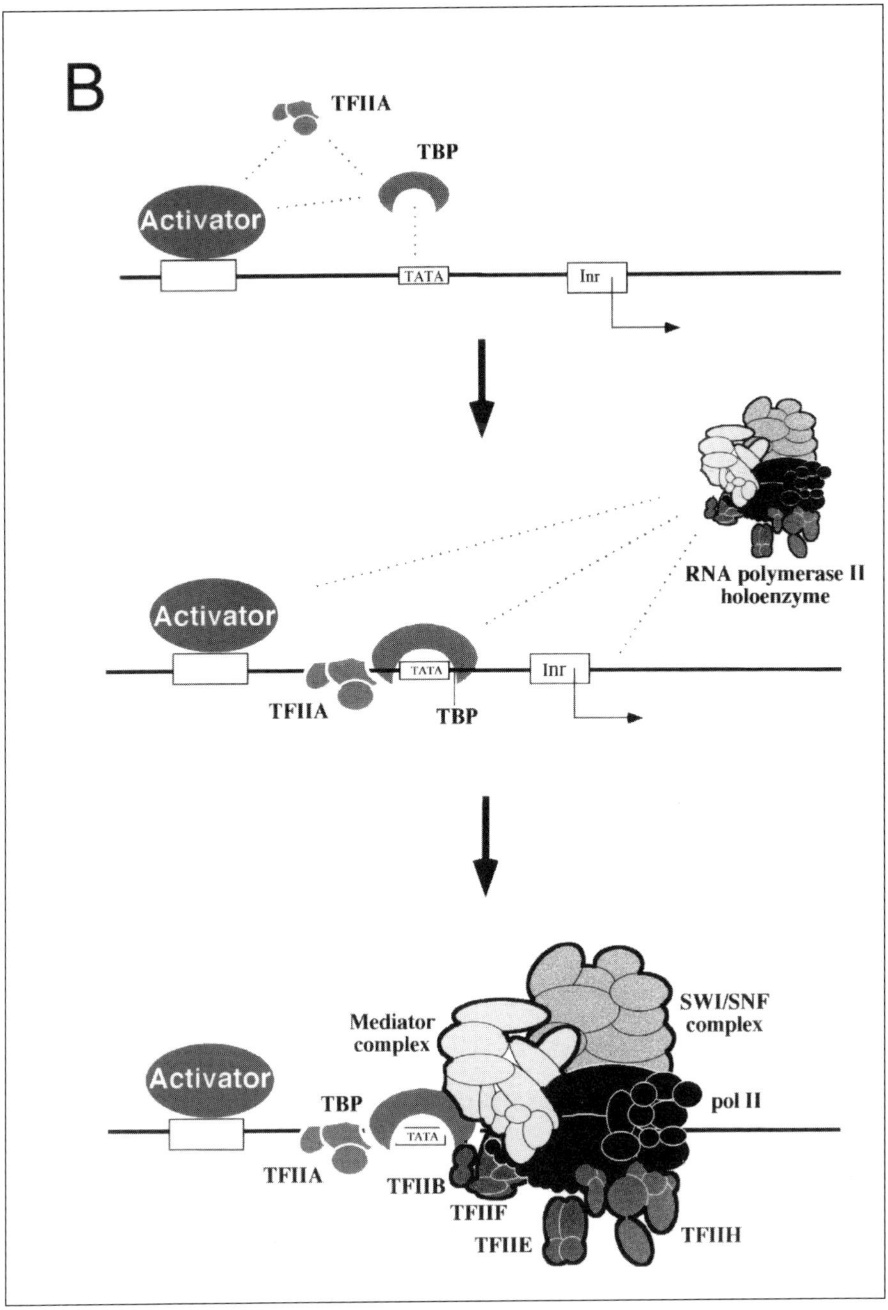
B
TFIIA
TBP
Activator
TATA
Inr
RNA polymerase II
holoenzyme
Activator
TATA
Inr
TFIIA
TBP
SWI/SNF
complex
Mediator
complex
Activator
TBP
TATA
pol II
TFIIA
TFIIB
TFIIF
TFIIE
TFIIH

For genes transcribed by pol II, the basal PIC was defined as containing TFIIA, TFIIB, TFIID, TFIIE, TFIIF, TFIIH, and pol II, but lacking sequence-specific transcription activators.[9,10] These basal factors were isolated from human tissue culture cells and were considered to be distinct entities since each could be separated from the other by standard chromatographic procedures.[11-13] Do these factors normally exist as distinct entities within the cell or are they part of a larger preassembled complex that disassembles during chromatography? More recently, portions of the total cellular TFIIB, TFIIE, TFIIF, TFIIH, pol II, and other factors have been isolated as an intact complex termed the pol II holoenzyme.[14] Others report the additional presence of TFIID and DNA repair enzymes in the holo-enzyme.[15,16] These preassembled complexes appear to be quite stable and are not dissociated upon standard chromatography. Do these preassembled complexes normally exist within the cell or are they products of an in vitro purification process? The existence of larger pol II-containing complexes and smaller subcomplexes may not be mutually exclusive. Both fully and partially preassembled complexes may exist in the cell and reflect the state of these components at different stages of the transcription cycle. What proportion of each factor resides in large and small complexes is unknown. From a regulatory perspective, it probably does not matter what the state of assembly is for those factors that enter into the PIC *after* the rate-limiting step. Since, by definition, the steps occurring after the rate-limiting step are relatively fast, it does not matter if they happen stepwise or in concert. However, since transcriptional activators must act on the rate-limiting step, an understanding of the molecular nature of this step is essential for defining the mechanism by which activators function.

THE RATE-LIMITING STEP IN BASAL PIC ASSEMBLY

If PIC assembly is a major point at which genes are regulated, then what rate-limiting step is targeted by activators? The first indication of the rate-limiting step in assembling a basal PIC came from order-of-addition experiments performed by Danny Reinberg and published in 1987.[13] He found that by preincubating a chromatin-free promoter with both TFIIA and TFIID, but not individually, the slow assembly of the basal transcription complex could be overcome. About ten years later, Kevin Struhl and colleagues provided in vivo evidence which is consistent with TFIID recruitment being rate limiting in transcription.[17] In vitro DNA-binding experiments from the labs of Berk, Carey, and Gralla provided further evidence that TFIID binding to a promoter is intrinsically slow.[18-20]

TFIID is a multisubunit complex consisting of TBP and at least eight TBP-associated factors (TAF_{II}s).[21,22] Distinct subunits of TFIID recognize core promoter elements and transcriptional activators (the latter discussed below). TBP binds to the TATA element located approximately 30 nucleotides upstream of the transcriptional start site, and $TAF_{II}150$ binds to DNA sequences located near the start site (Fig. 3.2A). Not all promoters contain recognition elements for TBP and $TAF_{II}150$, yet all pol II promoters tested thus far require TBP. Interestingly, $TAF_{II}150$ appears to be more readily dissociable from TFIID than the other TAFs.[23,24] $TAF_{II}150$ dissociation from TFIID might allow alternative promoter-specific factors to associate with TFIID.

It is not yet clear whether all pol II promoters utilize TAF_{II}s. Certainly, the pol II-transcribed snRNA promoters require a distinct TBP-TAF complex, and some promoters function in the absence of TAF_{II}s.[25-27] Promoters that appear to function in the absence of TAF_{II}s have been described primarily in the yeast system.[25] However, TAF_{II}s are present in yeast.[28-30] In vitro transcriptional activation in the absence of yeast TAF_{II}s does require the pol II-associated SRB/mediator complex which interacts with TBP (Fig. 3.2B).[31] It is possible that the SRB/mediator represents an alternative TAF-like complex.

If TFIID, TBP, or any other factor limits the rate of transcription, mechanistically there are two possibilities: the factor may be intrinsically slow in associating with the promoter (i.e., a bimolecular association rate constant, k_{on}, less than the diffusion limit of 10^8-10^9 $M^{-1}sec^{-1}$), and/or the factor may not remain stably associated with the promoter (i.e., rapid "off" rate). A number of studies have examined the 'on' and 'off' rates of TBP with the TATA box. TBP binds DNA quite stably ($k_{off} = 10^{-4}$ sec^{-1}, $t_{1/2}$ ~100 min.) and therefore dissociation of TBP is unlikely to limit transcription.[32,33]

Early studies from the Berk lab suggested that TBP was intrinsically slow in associating with TATA, which is consistent with this step being rate-limiting in transcription.[34] However, more quantitative studies by the labs of Hawley and Brenowitz revealed that TBP, at very high concentrations, can rapidly associate with DNA.[32,35] Nevertheless, the apparent k_{on} was still about a thousand fold lower than that expected of a diffusion limited reaction.

An explanation for the anomalously slow binding kinetics came from our findings that TBP, when not bound to DNA, self-associates to form dimers.[36,37] Consistent with this notion, the Burley and Kornberg laboratories had previously solved the crystal structure of TBP in the absence of DNA and found it to be a dimer.[37,38] Thus dimerization represents a competing reaction with DNA binding (Fig. 3.3). While TBP dimer dissociation is kinetically slow,[36] whether this step is the physiological rate-limiting step in transcription complex assembly remains to be determined. TBP dimerization does represent a physiological event in that TBP dimers have been detected in vivo.[39]

THE ROLE OF TFIIA

If TFIID is slow in binding to a promoter and thus potentially rate-limiting in transcription, then activators would be expected to accelerate the binding of TFIID. Indeed, activators such as Zta and GAL-AH increase the rate at which TFIID binds to a promoter.[18-20,40] Interestingly, the rate enhancement afforded by these activators requires TFIIA. As discussed above, TFIIA is necessary to bypass the rate limiting step in transcription by preincubating the promoter with TFIID. Thus, TFIIA appears to play an important role in recruiting TFIID to promoters (Fig. 3.2A).

TFIIA is the product of two genes, in which one of the products is proteolytically clipped in higher eukaryotes to form a three-subunit complex.[41,42] From crystal structure determinations, TFIIA appears to bind to the downstream stirrup of TBP (relative to the transcriptional start site) and project along the DNA upstream of TBP.[43,44] TFIIA is not required for basal transcription mediated by TBP (no TAFs), although TFIIA does stabilize TBP-TATA interactions as measured in mobility shift assays.[41,45]

TFIIA is important for basal transcription mediated by TFIID.[41] The Roeder and Reinberg labs have proposed that TFIIA counteracts negative regulators associated with TFIID (Fig. 3.3).[46-50] Dr_2/topo I binds TBP and inhibits the binding of TBP to TATA, and Dr_2/topo I is found to be associated with TFIID.[49] Inhibition of TBP-TATA binding by Dr_2/topo I is prevented by TFIIA, which suggests that at least one function of TFIIA is to overcome the repressive effects of Dr_2/topo I on TBP (or TFIID) binding to TATA. In addition, TFIIA appears to counteract other TFIID-associated negative cofactors such as Mot1 and Dr_1/NC2 (Fig. 3.3).[46,48,51] Mot1 is a transcriptional regulator that uses the energy of ATP hydrolysis to induce TBP to dissociate from DNA.[52] Mot1 is a member of the SWI/SNF family of conserved ATPases.[53] Dr_1/NC2 is a 19 kDa phosphoprotein that binds TBP but does not prevent TBP from binding to DNA.[46] Dr_1-TBP complexes cannot recruit TFIIB and are therefore inactive for PIC assembly. The requirement for TAF_{II}s to observe TFIIA activity might be related to the observation that TFIIA interacts with $TAF_{II}110$.[54] $TAF_{II}110$ may stabilize TFIIA's association with TFIID.

THE ROLE OF ACTIVATORS AND TAFS

Mechanistically, how might sequence-specific activators like Sp1 increase the rate of TFIID association with the promoter? The Tjian lab has shown quite convincingly

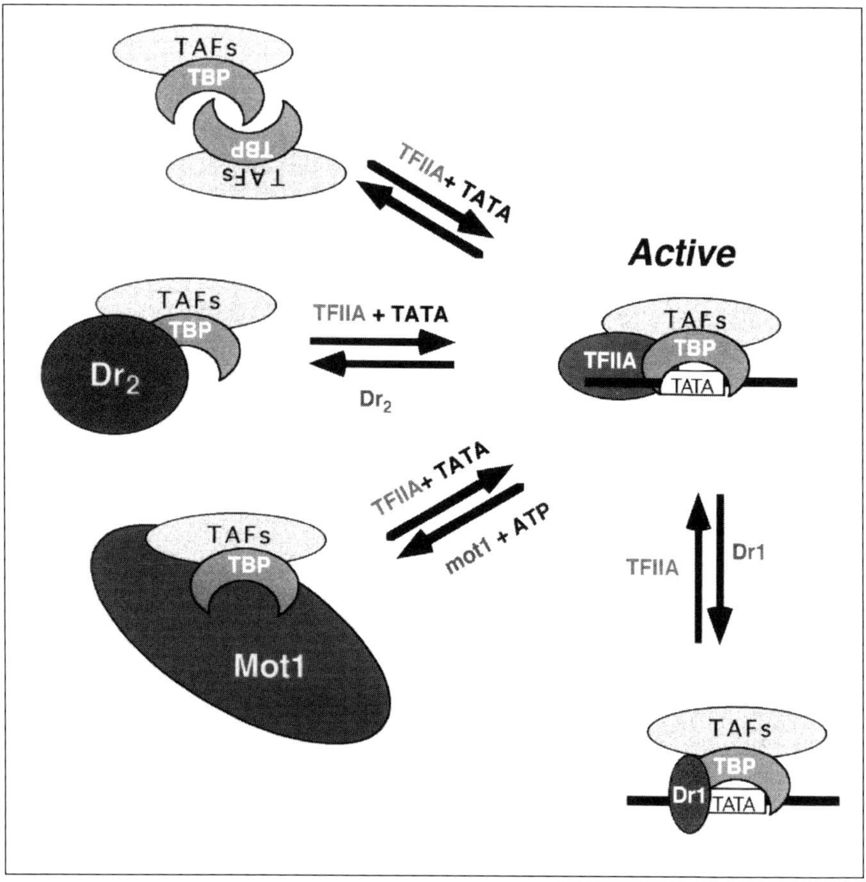

Fig. 3.3. Modes of negative regulation of TFIID. TFIID sequesters its DNA-binding surface through dimerization and by binding negative regulators such as Dr_2/topo I and Mot1. Mot1 function also requires ATP. It is not known whether the three types of interactions are mutually exclusive of each other. The negative regulator Dr_1/NC2 does not inhibit TATA binding by TBP (and presumably TFIID) but sequesters the complex in an inactive state by precluding the association of TFIIB. In all cases, TFIIA reverses the inhibitory effect and allows the TBP (or TFIID) complex to remain active for PIC assembly.

that distinct classes of activators interact directly with TAF components of TFIID (Fig. 3.2A).[21] Thus, Sp1 and Bicoid bind $TAF_{II}110$; Hunchback binds $TAF_{II}60$; GAL4-VP16 and p53 bind $TAF_{II}40$; and NTF1 binds $TAF_{II}150$.[55-59] By providing a larger "landing pad," activators may increase the frequency of productive collision of TFIID with the promoter. Alternatively, since activators typically require TFIIA, activator-TAF interactions might stabilize the binding of TFIIA to TFIID such that TFIIA can more productively remove negative regulators associated with TFIID.

Typical enhancement of TFIID binding to a promoter requires the presence of multiple activators, which can either be the same activator or distinct activators.[58,60-62] The mechanistic basis of multiple activators cooperating to recruit TFIID is unknown, although many possibilities exist. Multiple activators might cooperatively stabilize each others binding to the promoter, thereby providing a more stable "landing pad" for TFIID. The yeast GAL4 activator, through self-association, cooperatively stabilizes its binding to GAL4 DNA-binding sites.[63] In other cases, where experiments are per-

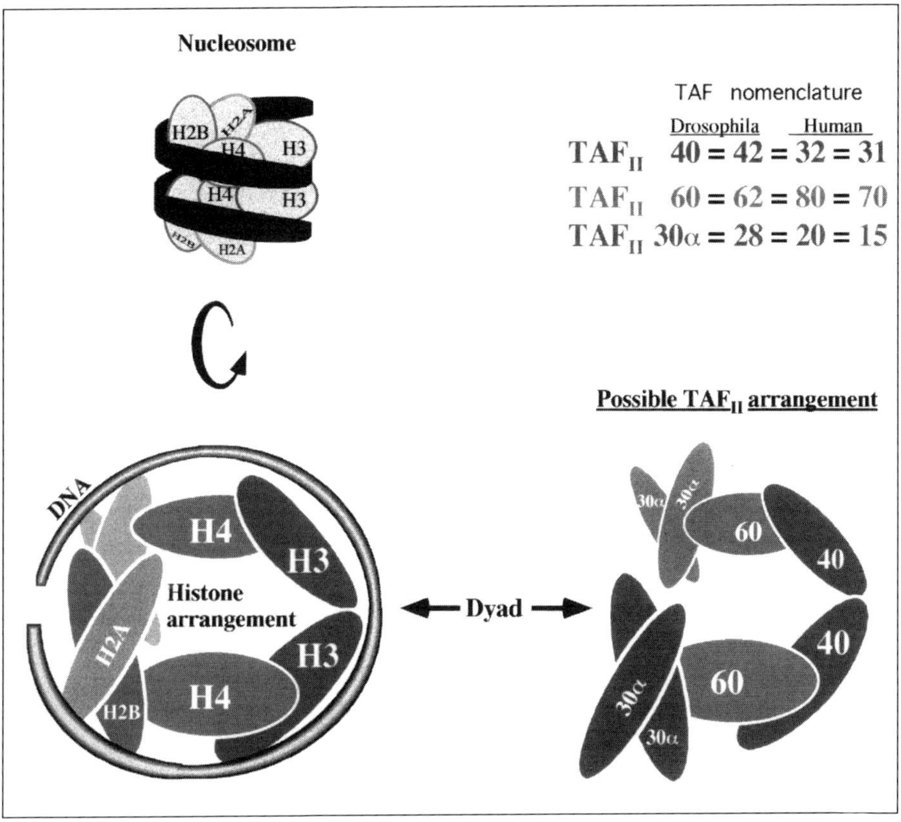

Fig. 3.4. Possible nucleosome-like structure of TFIID TAFs. A schematic of a nucleosome is shown in the upper left. A cross-sectional arrangement of histones within the nucleosome is shown in the lower left. Based on sequence and structural homology of $TAF_{II}40$, $TAF_{II}60$, and $TAF_{II}30\alpha$ to histones H3, H4, and H2B respectively, a homologous nucleosome-like structure has been proposed for the TAFs (lower right).[65-67] Equivalent TAF_{II} nomenclature for the *Drosophila* and human systems is presented in the upper right.

formed under binding site saturation such that cooperative stabilization of activators is not an issue, activators might provide a larger "landing pad" for TFIID (marker 7 in Fig. 3.1). For example, Sauer and Tjian have shown that Bicoid and Hunchback individually can recruit TFIID to their target promoters, and do so through interactions with their cognate TAFs.[58] Hence, Bicoid utilizes $TAF_{II}110$ but not $TAF_{II}60$, and Hunchback utilizes $TAF_{II}60$ but not $TAF_{II}110$. When bound at the same promoter, Bicoid and Hunchback provide a much greater enhancement of TFIID recruitment, as well as transcriptional activation, than the sum of the two regulator acting independently.

In cases where the same activator is bound to multiple sites on a promoter, synergism might arise if the activator contains multiple distinct activation domains that target different TAFs. Activators that contain one or more of a single type of activation domain might also synergize with itself by providing more of the same target for a particular TAF_{II}.[61] While the TAF_{II} might only interact with one activation domain, the presence of a large number of sites increases the probability that TFIID will collide with one of them.

$TAF_{II}40$ and $TAF_{II}60$ contain domains that are homologous to histones H3 and H4, respectively.[64] $TAF_{II}40$ and $TAF_{II}60$ interact to form dimers, and $TAF_{II}40$ self-associates to form $TAF_{II}40/60$ tetramers.[65,66] The crystal structure and circular dichroism spectra of the histone homology region of $TAF_{II}40/60$ reveals a structure that is very similar to that of the histone H3/H4 tetramer.[66,67] $TAF_{II}30\alpha$, which interacts with $TAF_{II}40/60$, has limited sequence homology to histone H2B.[65] The structural relationship of TFIID TAFs with the histone octamer suggests that TFIID may wrap DNA around its surface in a nucleosome-like manner (Fig. 3.4). However, at this point, such a nucleosome-like TFIID complex is speculative. A histone octamer forms a footprint on DNA that is approximately 140 bp, which corresponds to 1.75 wraps of the DNA around the histone octamer. TFIID appears to footprint no more than 70 bp of DNA. Thus, TFIID, which contains nearly seven times the mass of a histone octamer, binds to half as much DNA.

THE RATE-LIMITING STEP DURING ACTIVATED TRANSCRIPTION

While TFIID may be rate-limiting in transcription complex assembly in the absence of activators, is its recruitment still rate-limiting in the presence of an activator? The answer may vary depending on the type, position, and number of activators and negative regulators present. In the activated state, the rate-limiting step is still important in that it presents an opportunity for the cell to fine-tune the level of activated expression. Additional activators may further accelerate the recruitment of TFIID.

In the activated state, if TFIID recruitment is no longer rate-limiting, then what other step is limiting? Evidence suggests that once TFIID is bound to the promoter, the recruitment of TFIIB may be the next rate-limiting step.[68,69] TFIIB interacts with the upstream stirrup of TBP and directs the assembly of pol II and the remaining basal factors into the transcription complex.[70,71] If the recruitment of TFIIB is rate-limiting at any stage of the transcription cycle, then whether TFIIB exists as an isolated monomer or a component of the pol II holoenzyme, becomes a relevant issue. If TFIIB enters into the developing transcription complex as an isolated unit, then the activator-TFIID-TFIIA complex must target TFIIB directly. Indeed, a number of activators, including VP16, CREB/CBP, and others, have been shown to directly interact with TFIIB.[72,73] Other activators might exert their effect indirectly through TAFs. For example, p53, and VP16 interact with $TAF_{II}40$, and $TAF_{II}40$ interacts with TFIIB.[57,59,74] As with TFIID, activators might enhance TFIIB's interaction with the transcription complex by increasing its 'on' rate and/or decreasing its 'off' rate. The possibility that activators might function at two distinct steps of transcription complex assembly is supported by evidence that multiple activators synergistically enhance TFIID-TFIIA binding to a promoter as well as enhance TFIIB binding to the activator-TFIID-TFIIA complex.[62]

If the functional form of TFIIB lies within a pol II holoenzyme, then recruitment of TFIIB (and the remaining factors) may occur through activator interactions with other or additional components of the holoenzyme such as the SRB/mediator complex, TFIIF, or TFIIH (Fig. 3.2B).

TRANSCRIPTION COMPLEX DISASSEMBLY

Once the PIC has assembled, it is prepared to initiate transcription. DNA strand separation (open complex formation), transcription initiation, clearance of RNA polymerase from the promoter, and the rate of RNA polymerization are controlled by basal or general transcription factors TFIIE, TFIIF and TFIIH, and elongation factors such as TFIIS and others (chapter 2). Pol II holoenzyme disengages from TFIID (and presumably TFIIA) at an early stage, possibly during initiation or promoter clearance. TFIID appears to remain bound to the TATA box. Next, TFIIB, TFIIE, and TFIIH sequentially dissociate.[75] Only TFIIF and other elongation factors such as TFIIS remain stably bound to the transcribing polymerase.

Once pol II has cleared the promoter, promoter-bound TFIID is available to recruit TFIIB, which leads to the assembly of another PIC. Presumably reinitiation is rapid, inasmuch as the slow step of TFIID recruitment is bypassed. Transcriptional regulators might play an important role in transcription reinitiation by controlling TFIIB and/or pol II holoenzyme recruitment and controlling the stability of the TFIIA-TFIID-promoter complex.

CONCLUSIONS AND FUTURE PROSPECTS

Our understanding of how transcription complexes are assembled and regulated has followed the path of first defining important promoter DNA sequences. Second, the transcription regulatory factors that bind to these promoter elements have been identified and their genes cloned. Third, the proteins targets of these sequence-specific transcription factors have been identified. Fourth, the protein-protein and protein-DNA interfaces have been mapped to the amino acid and nucleotide level. Current analyses focus on determining the three dimensional structure of each component of the transcription complex, defining biochemical activities associated with each protein, and characterizing their in vivo properties. Information provided from structural, biochemical, and physiological studies will provide a foundation for the long term objective of understanding the dynamics of transcription complex assembly, and how cellular signals alter the dynamics in a manner that leads changes in gene expression.

ACKNOWLEDGMENTS

The author is supported by grants from the National Institutes of Health and the Searle/Chicago Community Trust.

REFERENCES

1. Roth SY, Shimizu M, Johnson L et al. Stable nucleosome positioning and complete repression by the yeast α 2 repressor are disrupted by amino-terminal mutations in histone H4. Genes Dev 1992; 6(3):411-425.
2. Peterson CL, Tamkun JW. The SWI-SNF complex: a chromatin remodeling machine? Trends Biochem Sci 1995; 20(4):143-146.
3. Brownell JE, Zhou J, Ranalli T et al. Tetrahymena histone acetyltransferase A: A homolog to yeast Gcn5p linking histone acetylation to gene activation. Cell 1996; 84:843-852.
4. Grosschedl R. Higher-order nucleoprotein complexes in transcription: Analogies with site-specific recombination. Curr Opin Cell Biol 1995; 7(3):362-370.
5. Shykind BM, Kim J, Sharp PA. Activation of the TFIID-TFIIA complex with HMG-2. Genes Dev 1995; 9(11): 1354-1365.
6. Tjian R, Maniatis T. Transcriptional activation: a complex puzzle with few easy pieces. Cell 1994; 77(1):5-8.
7. Pugh BF. Mechanisms of transcription complex assembly. Curr Opin Cell Biol 1996; 8(3):303-311.
8. Goodrich JA, Cutler G, Tjian R. Contacts in context: Promoter specificity and macromolecular interactions in transcription. Cell 1996; 84(6):825-830.
9. Zawel L, Reinberg D. Advances in RNA polymerase II transcription. Curr Opin Cell Biol 1992; 4(3):488-495.
10. Conaway RC, Conaway JW. General initiation factors for RNA polymerase II. Annu Rev Biochem 1993; 62:161-190.
11. Matsui T, Segall J, Weil PA et al. Multiple factors required for accurate initiation of transcription by purified RNA polymerase II. J Biol Chem 1980; 255(24):11992-11996.
12. Reinberg D, Roeder RG. Factors involved in specific transcription by mammalian RNA polymerase II. Purification and functional analysis of initiation factors IIB and IIE. J Biol Chem 1987; 262(7):3310-3321.
13. Reinberg D, Horikoshi M, Roeder RG. Factors involved in specific transcription in mammalian RNA polymerase II. Functional analysis of initiation factors IIA and IID and identification of a new factor operating at sequences downstream of the initiation site. J Biol Chem 1987; 262(7):3322-3330.
14. Koleske AJ, Young RA. The RNA polymerase II holoenzyme and its implications for gene regulation. Trends Biochem Sci 1995; 20(3):113-116.

15. Ossipow V, Tassan JP, Nigg EA et al. A mammalian RNA polymerase II holoenzyme containing all components required for promoter-specific transcription initiation. Cell 1995; 83(1):137-146.
16. Maldonado E, Shiekhattar R, Sheldon M et al. A human RNA polymerase II complex associated with SRB and DNA-repair proteins. Nature 1996; 381(6577): 86-89.
17. Klein C, Struhl K. Increased recruitment of TATA-binding protein to the promoter by transcriptional activation domains in vivo. Science 1994; 266(5183): 280-282.
18. Wang W, Gralla JD, Carey M. The acidic activator GAL4-AH can stimulate polymerase II transcription by promoting assembly of a closed complex requiring TFIID and TFIIA. Genes Dev 1992; 6(9):1716-1727.
19. Lieberman PM, Berk AJ. A mechanism for TAFs in transcriptional activation: activation domain enhancement of TFIID-TFIIA—promoter DNA complex formation. Genes Dev 1994; 8(9): 995-1006.
20. Kobayashi N, Boyer TG, Berk AJ. A class of activation domains interacts directly with TFIIA and stimulates TFIIA-TFIID-promoter complex assembly. Molec Cell Biol 1995; 15(11):6465-6473.
21. Goodrich JA, Tjian R. TBP-TAF complexes: selectivity factors for eukaryotic transcription. Curr Opin Cell Biol 1994; 6(3):403-409.
22. Tjian R. Molecular machines that control genes. Scientific American 1995; 272(2):54-61.
23. Hansen SK, Tjian R. TAFs and TFIIA mediate differential utilization of the tandem Adh promoters. Cell 1995; 82(4): 565-575.
24. Kaufmann J, Verrijzer CP, Shao J et al. CIF, an essential cofactor for TFIID-dependent initiator function. Genes Dev 1996; 10(7):873-886.
25. Kim YJ, Bjorklund S, Li Y et al. A multiprotein mediator of transcriptional activation and its interaction with the C-terminal repeat domain of RNA polymerase II. Cell 1994; 77(4):599-608.
26. Sadowski CL, Henry RW, Lobo SM et al. Targeting TBP to a non-TATA box cis-regulatory element: a TBP-containing complex activates transcription from snRNA promoters through the PSE. Genes Dev 1993; 7(8):1535-1548.
27. Henry RW, Sadowski CL, Kobayashi R et al. A TBP-TAF complex required for transcription of human snRNA genes by RNA polymerase II and III. Nature 1995; 374(6523):653-656.
28. Poon D, Weil PA. Immunopurification of yeast TATA-binding protein and associated factors. Presence of transcription factor IIIB transcriptional activity. J Biol Chem 1993; 268(21):15325-15328.
29. Poon D, Campbell AM, Bai Y et al. Yeast Taf170 is encoded by MOT1 and exists in a TATA box-binding protein (TBP)-TBP-associated factor complex distinct from transcription factor IID. J Biol Chem 1994; 269(37):23135-23140.
30. Reese JC, Apone L, Walker SS et al. Yeast TAFIIs in a multisubunit complex required for activated transcription. Nature 1994; 371(6497):523-527.
31. Hengartner CJ, Thompson CM, Zhang J et al. Association of an activator with an RNA polymerase II holoenzyme. Genes Dev 1995; 9(8):897-910.
32. Hoopes BC, LeBlanc JF, Hawley DK. Kinetic analysis of yeast TFIID-TATA box complex formation suggests a multi-step pathway. J Biol Chem 1992; 267(16): 11539-11547.
33. Coleman RA, Pugh BF. Evidence for functional binding and stable sliding of the TATA binding protein on nonspecific DNA. J Biol Chem 1995; 270(23): 13850-13859.
34. Schmidt MC, Zhou Q, Berk AJ. Sp1 activates transcription without enhancing DNA-binding activity of the TATA box factor. Molec Cell Biol 1989; 9(8): 3299-3307.
35. Petri V, Hsieh M, Brenowitz M. Thermodynamic and kinetic characterization of the binding of the TATA binding protein to the adenovirus E4 promoter. Biochemistry 1995; 34(31):9977-9984.
36. Coleman RA, Taggart AK, Benjamin LR et al. Dimerization of the TATA binding protein. J Biol Chem 1995; 270(23): 13842-13849.

37. Nikolov DB, Hu SH, Lin J et al. Crystal structure of TFIID TATA-box binding protein. Nature 1992; 360(6399):40-46.
38. Chasman DI, Flaherty KM, Sharp PA et al. Crystal structure of yeast TATA-binding protein and model for interaction with DNA. Proc Natl Acad Sci USA 1993; 90(17):8174-8178.
39. Taggart AKP, Pugh BF. Dimerization of TFIID when not bound to DNA. Science 1996; 272(5266):1331-1333.
40. Lieberman PM, Berk AJ. The Zta trans-activator protein stabilizes TFIID association with promoter DNA by direct protein-protein interaction. Genes Dev 1991; 5(12B):2441-2454.
41. Sun X, Ma D, Sheldon M et al. Reconstitution of human TFIIA activity from recombinant polypeptides: a role in TFIID-mediated transcription. Genes Dev 1994; 8(19):2336-2348.
42. Yokomori K, Zeidler MP, Chen JL et al. *Drosophila* TFIIA directs cooperative DNA binding with TBP and mediates transcriptional activation. Genes Dev 1994; 8(19):2313-2323.
43. Geiger JH, Hahn S, Lee S et al. Crystal structure of the yeast TFIIA/TBP/DNA complex. Science 1996; 272(5263): 830-836.
44. Tan S, Hunziker Y, Sargent DF et al. Crystal structure of a yeast TFIIA/TBP/DNA complex. Nature 1996; 381(6578): 127-134.
45. Imbalzano AN, Zaret KS, Kingston RE. Transcription factor (TF) IIB and TFIIA can independently increase the affinity of the TATA-binding protein for DNA. J Biol Chem 1994; 269(11):8280-8286.
46. Inostroza JA, Mermelstein FH, Ha I et al. Dr_1, a TATA-binding protein-associated phosphoprotein and inhibitor of class II gene transcription. Cell 1992; 70(3):477-489.
47. Meisterernst M, Roy AL, Lieu HM et al. Activation of class II gene transcription by regulatory factors is potentiated by a novel activity. Cell 1991; 66(5):981-993.
48. Meisterernst M, Roeder RG. Family of proteins that interact with TFIID and regulate promoter activity. Cell 1991; 67(3):557-567.
49. Merino A, Madden KR, Lane WS et al. DNA topoisomerase I is involved in both repression and activation of transcription. Nature 1993; 365(6443):227-232.
50. Cortes P, Flores O, Reinberg D. Factors involved in specific transcription by mammalian RNA polymerase II: purification and analysis of transcription factor IIA and identification of transcription factor IIJ. Molec Cell Biol 1992; 12(1): 413-421.
51. Auble DT, Hansen KE, Mueller CG et al. Motl, a global repressor of RNA polymerase II transcription, inhibits TBP binding to DNA by an ATP-dependent mechanism. Genes Dev 1994; 8(16): 1920-1934.
52. Auble DT, Hahn S. An ATP-dependent inhibitor of TBP binding to DNA. Genes Dev 1993; 7(5):844-856.
53. Davis JL, Kunisawa R, Thorner J. A presumptive helicase (MOT1 gene product) affects gene expression and is required for viability in the yeast Saccharomyces cerevisiae. Molec Cell Biol 1992; 12(4): 1879-1892.
54. Yokomori K, Admon A, Goodrich JA et al. *Drosophila* TFIIA-L is processed into two subunits that are associated with the TBP/TAF complex. Genes Dev 1993; 7(11):2235-2245.
55. Hoey T, Weinzierl RO, Gill G et al. Molecular cloning and functional analysis of *Drosophila* TAF110 reveal properties expected of coactivators. Cell 1993; 72(2): 247-260.
56. Chen JL, Attardi LD, Verrijzer CP et al. Assembly of recombinant TFIID reveals differential coactivator requirements for distinct transcriptional activators. Cell 1994; 79(1):93-105.
57. Thut CJ, Chen JL, Klemm R et al. p53 transcriptional activation mediated by coactivators $TAF_{II}40$ and $TAF_{II}60$. Science 1995; 267(5194):100-104.
58. Sauer F, Hansen SK, Tjian R. Multiple TAF(II)s directing synergistic activation of transcription. Science 1995; 270(5243): 1783-1788.
59. Goodrich JA, Hoey T, Thut CJ et al. *Drosophila* $TAF_{II}40$ interacts with both a VP16 activation domain and the basal transcription factor TFIIB. Cell 1993; 75(3):519-530.

60. Carey M. Mechanistic advances in eukaryotic gene activation. Curr Opin Cell Biol 1991; 3(3):452-460.
61. Emami KH, Carey M. A synergistic increase in potency of a multimerized VP16 transcriptional activation domain. EMBO J 1992; 11(13):5005-5012.
62. Chi TH, Lieberman P, Ellwood K et al. A general mechanism for transcriptional synergy by eukaryotic activators. Nature 1995; 377(6546):254-257.
63. Giniger E, Ptashne M. Cooperative DNA binding of the yeast transcriptional activator GAL4. Proc Natl Acad Sci USA 1988; 85(2):382-386.
64. Kokubo T, Gong DW, Wootton JC et al. Molecular cloning of *Drosophila* TFIID subunits. Nature 1994; 367(6462): 484-487.
65. Hoffmann A, Chiang CM, Oelgeschlager T et al. A histone octamer-like structure within TFIID. Nature 1996; 380(6572): 356-359.
66. Xie XL, Kokubo T, Cohen SL et al. Structural similarity between TAFs and the heterotetrameric core of the histone octamer. Nature 1996; 380(6572): 316-322.
67. Nakatani Y, Bagby S, Ikura M. The histone folds in transcription factor TFIID. J Biol Chem 1996; 271(12): 6575-6578.
68. Lin YS, Green MR. Mechanism of action of an acidic transcriptional activator in vitro. Cell 1991; 64(5):971-981.
69. Roberts SGE, Choy B, Walker SS et al. A role for activator-mediated TFIIB recruitment in diverse aspects of transcriptional regulation. Curr Biol 1995; 5(5) :508-516.
70. Nikolov DB, Chen H, Halay ED et al. Crystal structure of a TFIIB-TBP-TATA-element ternary complex. Nature 1995; 377(6545):119-128.
71. Zawel L, Reinberg D. Common themes in assembly and function of eukaryotic transcription complexes. Annu Rev Biochem 1995; 64:533-561.
72. Lin YS, Ha I, Maldonado E et al. Binding of general transcription factor TFIIB to an acidic activating region. Nature 1991; 353(6344):569-571.
73. Kwok RP, Lundblad JR, Chrivia JC et al. Nuclear protein CBP is a coactivator for the transcription factor CREB. Nature 1994; 370(6486):223-226.
74. Klemm RD, Goodrich JA, Zhou S et al. Molecular cloning and expression of the 32-kDa subunit of human TFIID reveals interactions with VP16 and TFIIB that mediate transcriptional activation. Proc Natl Acad Sci USA 1995; 92(13): 5788-5792.
75. Zawel L, Kumar KP, Reinberg D. Recycling of the general transcription factors during RNA polymerase II transcription. Genes Dev 1995; 9(12):1479-1490.

CHAPTER 4

Coupling Transcription to Signal Transduction: CREM and cAMP

Monica Lamas and Paolo Sassone-Corsi

INTRODUCTION

The control of gene expression is driven by transcription factors which integrate information from promoter sequences and signal transduction pathways. In eukaryotes the secondary messenger cAMP and a specific protein kinase (Protein Kinase A) characterize a major signal transduction system. Figure 4.1 shows schematically the cascade of events that occur upon activation of the cAMP signal transduction pathway. Intracellular levels of cAMP are regulated primarily by adenylate cyclase. This enzyme is modulated by various extracellular stimuli which mediate their action via receptors and their interaction with G proteins.[1] Thus, the binding of a specific ligand to a receptor results in the activation of the adenylate cyclase which converts ATP to cAMP. cAMP, in turn, binds cooperatively to two sites on the regulatory subunit of PKA, releasing the active catalytic subunit.[2,3] In the nucleus, PKA phosphorylates and thereby activates a number of transcription factors which bind as dimers to specific promoter sequences denominated cAMP-responsive elements (CREs).[3-5] Thus, PKA-mediated phosphorylation ultimately influences the transcriptional regulation of various genes through distinct cAMP-inducible promoter responsive sites.[4,5]

A consensus CRE site is constituted by an 8 bp palindromic sequence (TGACGTCA) with higher conservation in the 5' half of the palindrome with respect to the 3' sequence.[5] The CRE site is extremely similar to the consensus TPA-responsive element (TRE) (TGACTCA), the target DNA control element of a distinct signal transduction pathway (chapter 5). In fact, this similarity may provide a route for different intracellular signals to be integrated at the level of transcriptional control.[6,7]

CRE sites have been identified in a number of gene promoters, although not all of these cAMP effects have been clearly demonstrated to be transcriptional.[8-10] Many neuroendocrine-specific gene promoters present CRE sites and mutation of these elements causes a significant decline in promoter activity in the appropriate cell type. Therefore, CRE sites have been implicated in a tissue-specific regulation of gene expression.[11] However, CRE se-

Transcription Factors in Eukaryotes, edited by Athanasios G. Papavassiliou.

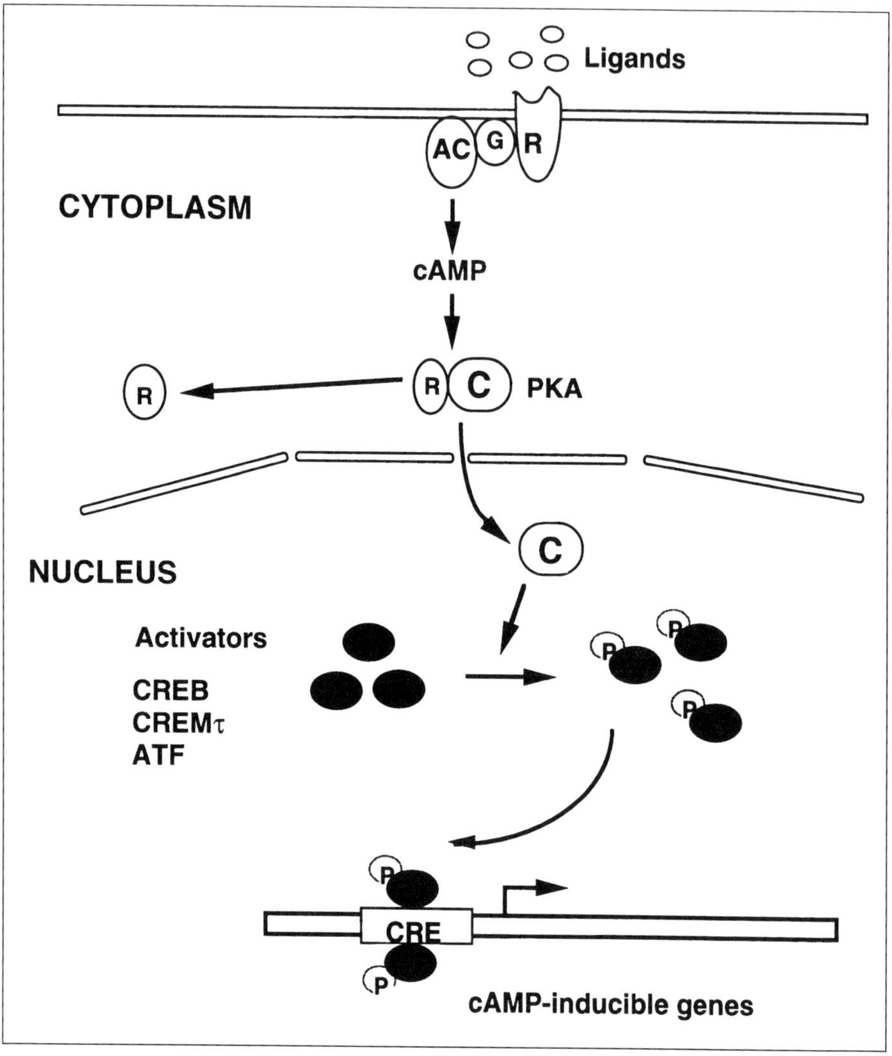

Fig. 4.1. The cAMP signal transduction pathway. Ligand-receptor (R) interaction activates coupled G proteins (G) which stimulate adenylyl cyclase activity (AC). As a consequence cAMP intracellular levels are increased. cAMP binds to the regulatory subunit of protein kinase A (PKA), releasing the active catalytic subunits which migrate into the nucleus and phosphorylate a series of transcriptional activators. Phosphorylated activators bind to cAMP response enhancer elements (CRE) in the promoter of responsive genes and modulate their expression.

quences have also been found in the context of regulatory regions of nonendocrine regulated genes such as the fibronectin promoter,[12] the MHC class II genes[13] and the enhancer of the HTLV-I LTR.[14] Thus the presence of a CRE is not entirely specific for neuroendocrine tissue-specific expression (although it may be necessary in the context of certain promoters for maintaining such expression).[11]

Subsequent to the identification of CREs in the promoters of several cAMP-responsive genes, it was shown that specific nuclear factors bind to these sites. These proteins

CREM /CREB subfamily	CREM →	s-CREM	Repressor
		τ	Activator
		τα	?
		τ2	Activator
		τ1	Activator
		α	Repressor
		β	Repressor
		γ	Repressor
		ICER	Repressor
	CREB		
	ATF1		
	ATF6		
	ATF4		
	ATF3		
	TREB5		
	E4BP4		
CRE-BP1 subfamily	CRE-BP1		
	ATFa		

Table 4.1. The family of cAMP-responsive nuclear factors. At least ten genes have been cloned that encode CRE-binding factors (CREBs) and activator transcription factors (ATFs). These transcriptional regulators exhibit a great functional diversity. Among them, CREM is a multiexonic gene that encodes a family of both activators and antagonists of cAMP-induced transcription.

bear different structural and functional characteristics and constitute a multigene family of transcription factors.

THE FAMILY OF cAMP-RESPONSIVE NUCLEAR FACTORS

The identification and the cloning of the first CRE-binding factor, CREB (CRE-binding protein), opened the way for the identification of a large number of cDNA clones encoding related but distinct CRE-binding factors. To date at least ten members of this family have been cloned (Table 4.1). All these proteins belong to the basic leucine zipper (bZIP) transcription factor class and outside of the bZIP region, homology between these factors is relatively poor. However, based upon regions of sequence similarity they can be divided into subfamilies. For example CREB, ATF1 and CREM share extensive homology.[15-17]

An added level of complexity arises from the fact that both CREB and CREM encode multiple, functionally different isoforms by alternative splicing.[18,19] Although the function of many CREB isoforms remains elusive the function and physiological role of many of the CREM isoforms has been determined.[17,19,20] Importantly, CRE-binding proteins act as both activators and repressors of transcription. CREB, CREMτ, and ATF1 mediate transcriptional activation.[17,21,22] The alternatively spliced CREM isoforms CREMα, β, and γ[19] and the products of the E4BP4 and CREB-2 gene members[23,24] act as antagonists of cAMP-induced transcription. The powerful repressor ICER (Inducible cAMP Early Repressor) deserves special mention since it is generated from an alternative cAMP inducible promoter of the CREM gene.[25,26] Indeed, the emerging picture is that the transcriptional regulation of cAMP-responsive genes results from a delicate balance between multiple activators and repressors of CRE containing promoters.

It ia noteworthy that the different factors are able to heterodimerize with each other but only in certain combinations. A "dimerization code" exists which seems to be a property of the leucine zipper structure of each factor. Some ATF/CREB factors are able to heterodimerize with the oncoproteins Fos and Jun, and this may change the specific affinity of binding to a CRE with respect to a Fos-Jun binding site.[6,27] In fact, part of the complexity of CRE-dependent regulation is accounted for by the large number of homo- and heterodimer combinations possible. This demonstrates the versatility of the transcriptional response to signal transduction.

CRE-binding proteins, like many other transcription factors, are intrinsically modular. Interestingly, the relative position of the different functional domains is well conserved among various members of the family such as CREB, CREM, ATF1 and ATF2.[8,16]

Figure 4.2 shows the general structure of the activator proteins CREB, CREMτ and ATF1. The transcriptional activation domain contains two independent regions.[3] The first, known as the phosphorylation box (P-box) or kinase inducible domain (KID), contains several consensus phosphorylation sites for various kinases, such as PKA, PKC, $p34^{cdc2}$, glycogen synthase kinase-3 and casein kinases (CK) I and II.[21,29-31]

The two domains flanking the P-box contain about three-times more glutamine residues than in the remainder of the protein in both CREB and CREM. Glutamine-rich domains have been characterized in other factors, such as AP-2 and Sp1, as transcriptional activation domains.[32,33] The current notion is that they constitute surfaces of the protein which can interact with other components of the transcriptional machinery. The Q2 domain appears to make a more significant contribution to the trans-activation function than Q1. This is demonstrated by the properties of the two naturally occurring CREM isoforms CREMτ1 and CREMτ2.[19] It is likely that the Q1 and Q2 regions function in an additive manner to generate the full activation potential of CREMτ. Furthermore, ATF1 lacks a counterpart of the Q1 domain, and still functions as an efficient transcription activator.[22] Thus, the P-box region and the Q2 domain are sufficient to mediate cAMP-induced transcription. Experimental evidence reinforces the notion that the activation domain is inherently a modular structure. Indeed, the P-box is able to confer PKA inducibility on a heterologous acidic activation domain (e.g., GAL4) both in trans and in cis, while the Q2 domain when fused to a heterologous DNA-binding domain, still retains a noninducible activation function.[19] It could be speculated that the P-box is involved in the regulation not only of the adjacent Q domains but also in controlling the activation function of other factors bound to separate promoter elements.[3]

The exonic structure of CREM and CREB has been determined. Although highly homologous there are several differences between the two genes.[17] Most striking is the presence of two bZIP domains in the CREM gene which are used alternatively by differential splicing (DBDI and DBDII) (Fig. 4.2).

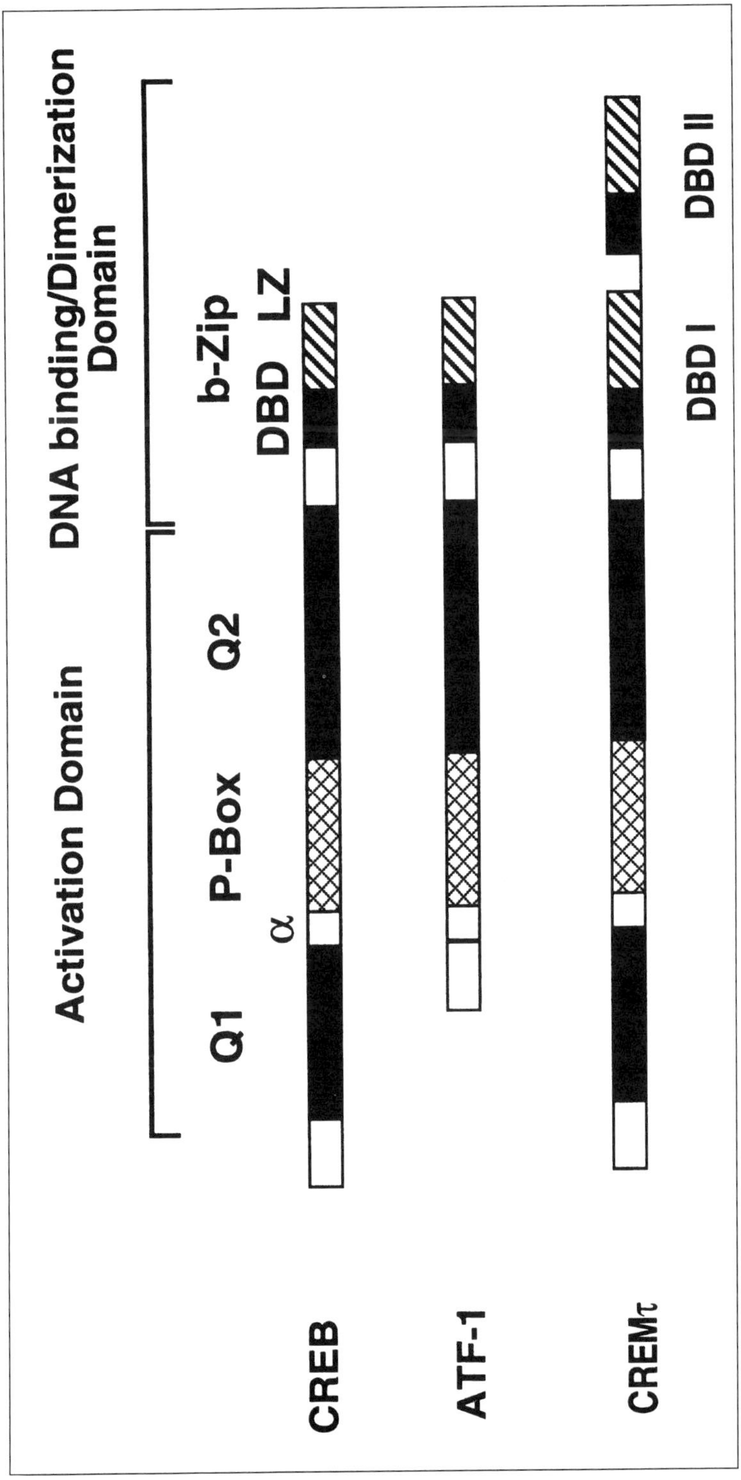

Fig. 4.2. General structure of the activator proteins CREB, ATF-1 and CREMτ. Schematic representation of the structural features of CREB, ATF-1 and CREMτ. These proteins belong to the bZIP transcription factor class and the bZIP domain is indicated (DBD and LZ). The activation domain consists of two independent regions: the phosphorylatable region (P-box) and the glutamine-rich activation domains (Q1 and Q2). CREM presents a remarkable characteristic: it contains two bZIP domains which are used alternatively by differential splicing (DBDI and DBDII).

COUPLING TRANSCRIPTION TO SIGNAL TRANSDUCTION: MOLECULAR ASPECTS

Phosphorylation-Dephosphorylation Events

Over the past years it has become evident that phosphorylation is one of the major mechanisms regulating the activity of transcription factors. Indeed, phosphorylation can regulate the activity of transcription factors at multiple levels including nuclear transport, dimerization, DNA binding and transcriptional activation.[34] In eukaryotes, upon activation of the adenylyl cyclase pathway, a serine residue at position 133 of CREB and at position 117 of CREMτ is phosphorylated by PKA.[21,30] The major effect of phosphorylation seems to occur at the level of the transactivation function of these proteins, transforming CREB and CREMτ into powerful transcriptional activators.

Interestingly, alternative signal transduction pathways can also induce phosphorylation of the PKA phosphoacceptor serine. It has been shown that Ser-117 in CREM is the target for phosphorylation by PKC, CamK, $p34^{cdc2}$ as well as PKA in vitro.[30,31] Within the P-box, the PKA phosphoacceptor site is located in a region of about 50 amino acids containing an abundance of phosphorylatable serines and acidic residues; this region was shown to be essential for transactivation by CREB and CREMτ.[29,30] Since CREM and other factors of the CREB/ATF family represent the final targets of the cAMP-pathway, these results show that they may also act as effectors of converging signaling systems and possibly as mediators of pathway cross-talk.[31] In fact, the emerging picture is that this particular phosphorylatable domain represents the convergence point for phosphorylation events stimulated by various signal transduction pathways.

The importance of phosphorylation in stimulating CRE-binding factors activity suggested that phosphatases could be conversely important in the attenuation of the transcriptional response. Indeed, it has been proposed that after the initial burst of phosphorylation in response to cAMP, CREB is dephosphorylated in vivo by protein phosphatase-1 (PP-1) and thereby its transactivation potential is drastically reduced.[35] Additional results show that both PP-1 and PP-2A can dephosphorylate CREB in vitro resulting in an apparent decreased binding to low affinity CRE sites.[36] Therefore, the precise role of PP-1 and PP-2A in the dephosphorylation of CREB remains controversial.

Interaction with Coactivators: CBP

The recent characterization of a coactivator that interacts specifically with the phosphorylated CREB represented a further step in the understanding of the mechanisms that mediate transcriptional responses to cAMP.[37] The coactivator CBP (CREB-binding protein) presents two zinc finger domains, a glutamine-rich domain at its C-terminus and a single consensus PKA recognition site. Phosphorylation of Ser-133 promotes binding to CBP and consequently the interaction with TFIIB, a general transcription factor involved in RNA polymerase II activity.[38] Thus, CBP may act as a link between CREB and the transcription preinitiation complex. This interaction may need some RNA polymerase II cofactors, such as TAF110. Some recent evidence suggests that although the binding of CREB to CBP is necessary, it is not sufficient for transcriptional activation.[39] Finally, the adenoviral E1A oncoprotein-associated p300, which is thought to play a role in preventing the cell cycle G0/G1 transition, is structurally very closely related to CBP.[40] Both CBP and p300 appear to have intrinsic activating properties which are inhibited by the E1A protein.[40] Thus, it is clear that further studies of the transcriptional activation domains of CRE-binding factors are likely to provide important insights into the function of transcription factors in general.

Alternative Polyadenylation

By the use of alternative polyadenylation sites, the CREM gene generates transcripts which present different stability.[20] Ten copies of the destabilizer element AUUUA are distributed throughout the 3' untranslated region of the CREM gene. Exclusion of these elements by alternative polyadenylation leads to a dramatic increase in the stability of the CREM message.[20] The choice of the polyadenylation site was shown to be physiologically regulated by the pituitary hormone FSH in germ cells.[20] Importantly, the implication of these findings is that hormones can regulate gene expression at the level of RNA processing and stability. This is likely to involve specific polyadenylation factors whose function may be modulated by signaling pathways activated by the hormone.

Negative Autoregulation: ICER

An important insight into the molecular mechanisms underlying the control of CRE-binding factors was given by the characterization of a new family of CREM isoforms, ICER (Inducible cAMP Early Repressor) whose expression is inducible by cAMP.[25,41] ICER represents the first and only case of a CRE-binding factor whose levels are shown to be inducible.

The CREB gene promoter and the CREM P1 promoter are GC-rich and direct constitutive expression of these proteins, a feature characteristic of housekeeping genes.[25,42] An alternative CREM promoter (P2), which is strongly inducible by cAMP, lies within an intron near the 3' end of the CREM gene and directs ICER transcription. The ICER open reading frame is constituted by the C-terminal segment of CREM and encodes a small protein of 120 amino acids with an expected molecular weight of 13.4 kDa. This protein, compared with the previously described CREM isoforms, essentially consists of only the DNA-binding domain. The intact DNA-binding domain directs specific ICER binding to a consensus CRE element. ICER functions as a powerful repressor of cAMP-induced transcription in transfection assays using an extensive range of reporter plasmids carrying individual CRE elements or cAMP-inducible promoter fragments.[25]

The cAMP inducibility of the CREM P2 promoter is driven by the presence of two pairs of closely-spaced CRE elements organized in tandem where the separation between each pair is only three nucleotides. Its characteristic kinetic of expression classifies CREM as an "early response gene".

In unstimulated cells the level of CREM transcript is below the threshold of detectability. However, upon treatment with forskolin (or other cAMP analogs), within 30 minutes there is a rapid increase in CREM transcript levels which peak after 2 hours and then progressively decline to basal levels by 5 hours.[25] Importantly CREM inducibility is enhanced and prolonged by cycloheximide, implying that a de novo synthesized factor could downregulate CREM transcription.[25] This observation, combined with the presence of CRE elements in the P2 promoter, suggested that the transient nature of ICER inducibility could be due to ICER itself. Consistently, the CRE elements in the P2 promoter have been shown to bind to the ICER proteins. In fact, the decline of CREM expression after induction is due to the newly synthesized ICER proteins which bind to the ICER promoter allowing negative regulation of transcription (Fig. 4.3).

The specific characteristics of CREM induction and the existence of a negative feedback loop allows the prediction of a refractory inducibility period in the gene's transcription[43] (Fig. 4.4). This aspect of CREM expression has been analyzed in detail both in the AtT20 corticotroph cell line and in the rat pineal gland (M. Lamas; in press).[44] Indeed in vivo studies have demonstrated that CREM is refractory to cAMP treatment under certain physiological conditions. In the rat pineal gland for example, the length of the refractory period is determined by the photoperiod.[44] This remarkable feature of gene expression is likely to be involved in the adaptation of the animals to environmental changes.

THE PHYSIOLOGICAL RESPONSE TO cAMP

A large number of genes regulated by endocrine stimuli contain CRE-like sequences in their regulatory regions. Thus,

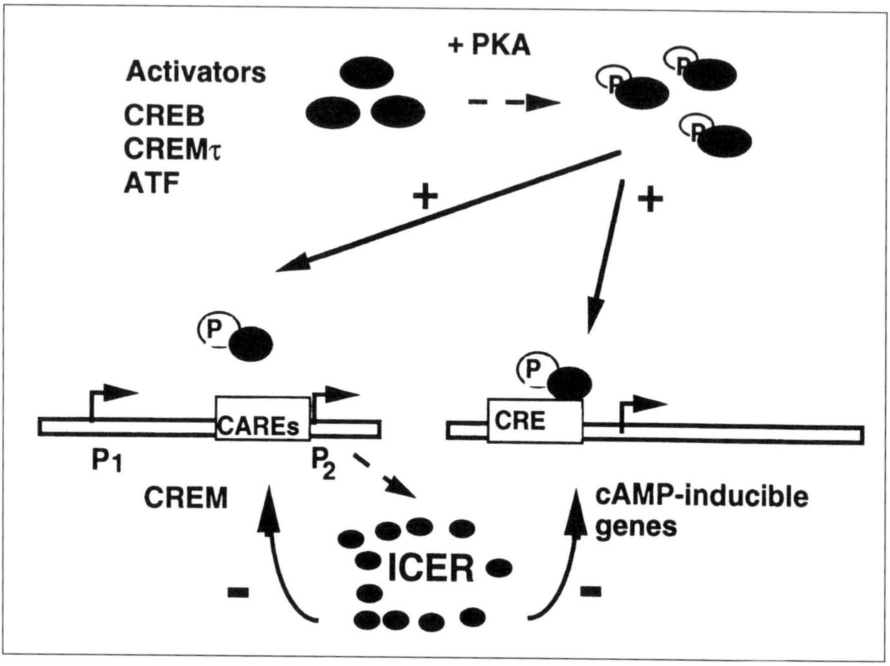

Fig. 4.3. The ICER negative autoregulatory feed-back loop. Upon activation of the cAMP signal transduction pathway, phosphorylated and thereby activated nuclear transcription factors stimulate transcription from cAMP-responsive promoters. Activation of transcription from the P2 CREM promoter leads to a dramatic increase of ICER protein levels. ICER is a powerful repressor of cAMP-induced transcription and exerts its effect both on heterologous promoters and on its own promoter.

CRE-binding proteins have been postulated to play an important role in the physiology of a variety neuroendocrine processes. The discovery of the CREM gene opened a new dimension in the study of the transcriptional response to cAMP.[45] The dynamic and versatile pattern of CREM expression combined with its tissue- and developmental-specific pattern, contrasts with that of the remaining members of the CRE-binding factor family which seem to be constant and ubiquitous.[5,16] This offered the first clue that CREM occupied a privileged position amongst this group of factors and pointed further to it fulfilling a pivotal role in the neuroendocrine response to cAMP.

Pituitary and Thyroid Glands

cAMP serves as mitogenic signal for the somatotroph cells of the anterior pituitary and CRE-binding proteins are likely to have pivotal functions in normal pituitary development. In fact, transgenic mice that express a CREB mutant which cannot be phosphorylated by PKA exhibit atrophied pituitary glands and a dwarf phenotype.[46] These effects might arise from repression of genes involved in proliferation and pituitary-specific gene expression, such as c-*fos* and GHF1/Pit-1, although the expression of these genes was not analyzed in these transgenic animals.[46] The transgenic phenotype obtained by blocking CREB function by a dominant repressor is equivalent to the one obtained by targeted cell death of the somatomammotrophs.[47] This could be an indication that CRE-binding proteins are likely to have pivotal functions in normal pituitary development.

CREM is inducible by activators of the cAMP-responsive signaling pathway in various endocrine tissues. In particular, the re-

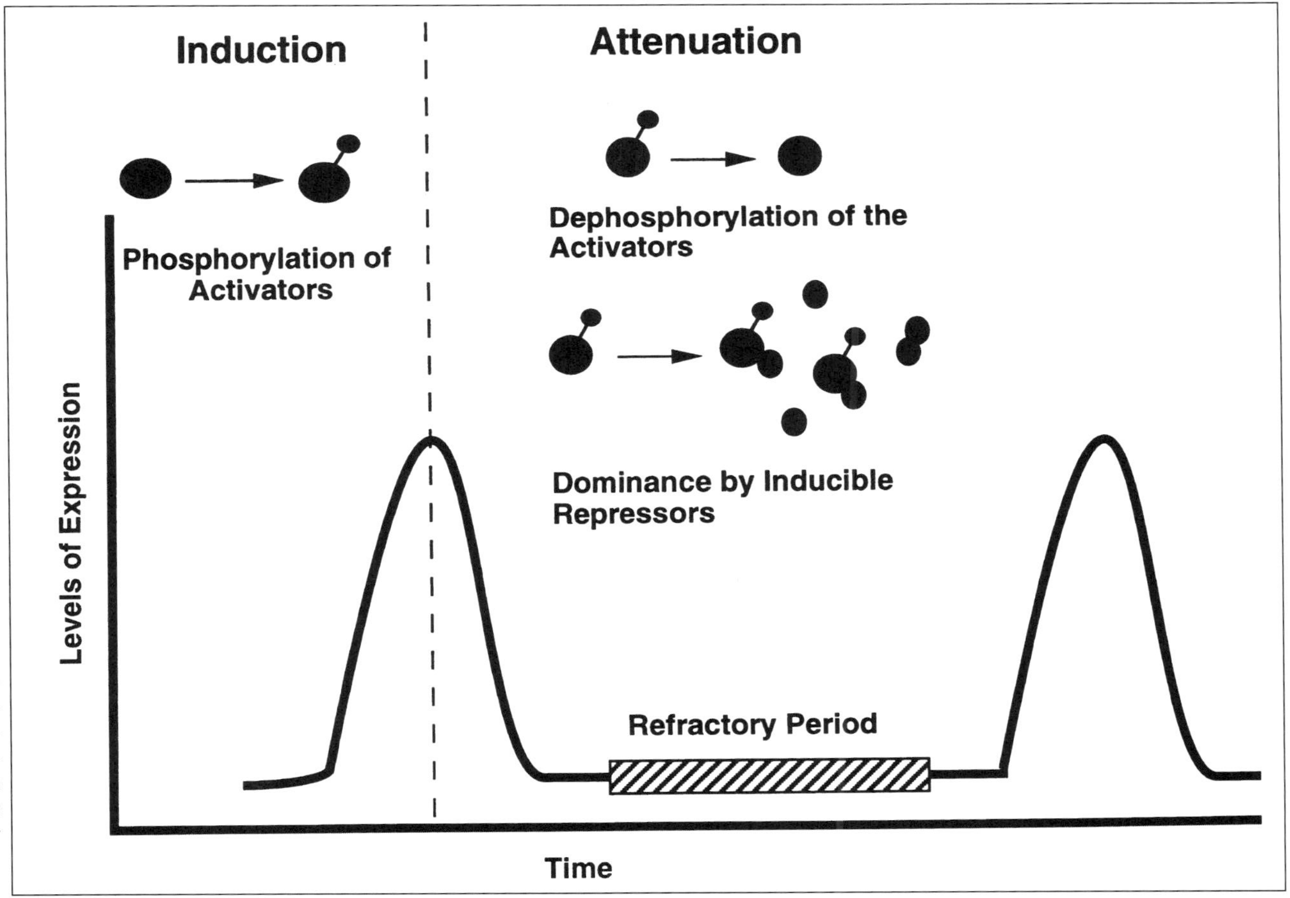

Fig. 4.4. CREM transcriptional refractory period. CREM is the only cAMP-inducible member of the CRE-binding factors family. Its kinetics of induction are those of an early response gene: the level of expression increases rapidly but transiently after stimulation. Different molecular mechanisms modulate both the induction and the attenuation of the gene's expression. A transcriptional refractory period precedes the second wave of CREM gene induction. The existence and the duration of this period are the consequences of a delicate balance between phosphorylation events and the dominance by the inducible repressor ICER.

pressor isoform ICER was shown to be inducible in both the rat thyroid gland and in the differentiated thyroid cell line FRTL-5.[48] The kinetics of ICER protein induction mirrors the down-regulation in the expression of the TSH receptor. It has been shown that ICER binds to a CRE-like sequence in the TSH-R promoter and represses its expression.[48] Thus, CREM induction by TSH in the thyroid gland represents a paradigm of the molecular mechanism by which pituitary hormones elicit homologous long-term desensitization of the receptor.

Spermatogenesis

The process of maturation of male germinal cells is regulated by the pituitary gonadotropins, luteinizing hormone (LH) and follicle-stimulating hormone (FSH). LH and FSH act on the Leydig and Sertoli cells in the seminiferous epithelium, respectively. The cAMP signal transduction pathway is responsible for converting these hormonal signals into long-term changes in gene expression.

CREM appears to play a privileged role in the regulation of the spermatogenic process. In testis, CREM is the subject of a developmental switch in expression.[20] It represents a highly abundant transcript in adult testis, while in prepubertal animals it is expressed at very low levels. Further characterization revealed that the abundant CREM transcript encodes exclusively the CREMτ activator, while in prepubertal testis only the repressor forms were detected at low levels. Thus, the developmental switch of CREM expression also constitutes a reversal of function.[17] Several lines of evidence demonstrate that CREMτ constitutes the first step of a transcriptional cascade which is responsible for the activation of several germ specific genes.[49-51]

CREM function in testis is not limited to the germ cells. Indeed, FSH induces ICER expression in Sertoli cells.[52] It has been shown that cAMP-induced ICER levels in Sertoli cells accompany the transcriptional down-regulation of the FSH receptor and have been implicated in its long-term desensitization. These results are analogous to the effect that ICER elicits on the expression of the TSH receptor in thyroid cells.[51]

Consistent with the previous data, gene knock-out studies have elegantly demonstrated the essential role of CREM in the complex process of spermatogenesis.[53,54] Disruption of the CREM gene by homologous recombination techniques leads to a severe impairment of spermatogenesis. CREM-deficient mice completely lack spermatozoa and are sterile (Fig. 4.5). Preparations from consecutive segments of the seminiferous epithelium demonstrated that spermatogenesis in these mice is interrupted at the very early spermatid stage. In addition, a remarkable increase in the apoptosis of the germ cells is observed.[53] Furthermore, the analysis of CREM potential downstream genes reveals the lack of postmeiotic cell-specific gene expression. Importantly, other members of the CRE binding proteins family (CREB, ATF-1, ATF-2) as well as the coactivator CREB-binding protein (CBP) appear unchanged in these mice.[53] This demonstrates that no redundant function can substitute for CREM in germ cells.

Pineal Gland: Circadian Rhythmicity

Circadian rhythms are day-night oscillations in an organism's biochemistry, physiology and behavior. Crucial elements for the synchronization of biological rhythms in mammals are the pineal gland[55] and the suprachiasmatic nucleus (SCN).[56] The pineal gland is considered to be a neuroendocrine transducer, as the neural input to this organ is converted into an endocrine output. Thus, environmental lighting conditions are transduced by the pineal gland from a neuronal to an endocrine message, the rhythmic secretion of melatonin.[55] This hormone synthesis is controlled by the SCN, being elevated at night and low during the day.[56] Importantly, the cAMP-dependent signal transduction pathway serves as a relay to stimulate melatonin synthesis.[57]

In the past few years a great advance has been achieved in the analysis of the molecular mechanisms which underlie rhythmic behaviors. A crucial step in the understand-

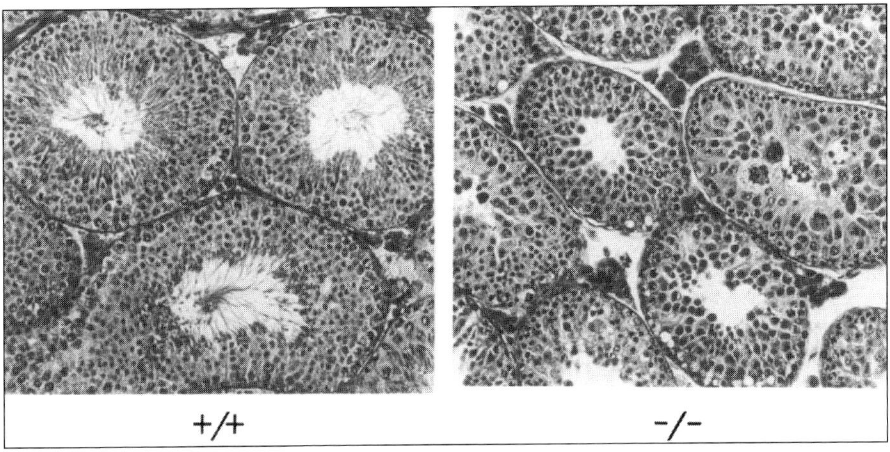

Fig. 4.5. Spermiogenesis deficiency in CREM-mutant mice. Histological analysis of testis sections from eight-week-old wild-type (+/+) and CREM-mutant (-/-) littermates. (Magnification x200). CREM-deficient mice show a reduced tubule diameter and lack developing spermatids and spermatozoa.

ing of the physiological role of CRE binding proteins in the establishment of circadian rhythms arises from the observation that the CREM gene is strongly induced at night in the pineal gland by clock-derived adrenergic signals.[26,58] The induced transcript encodes the repressor ICER and recent evidence demonstrates that the dynamics of ICER expression are determined by night length.[44] Therefore CREM is able to adapt to changes in photoperiod by modifying its mode of induction in response to adrenergic stimuli.[44] Thus, rats that have been adapted to different photoperiods show differential speed of ICER induction at the beginning of the night. In fact, the transcriptional response of the CREM gene appears to be determined by the memory of past photoperiods.

Additional evidence reinforces the notion that cAMP and CRE-binding proteins play a central role in the establishment and entrainment of circadian rhythms. The photoperiod-dependent changes in ICER correlate well with seasonal variations in melatonin. In the pineal gland, the circadian production of melatonin is controlled by a circadian fluctuation in the enzyme serotonin *N*-acetyl transferase (NAT). The fact that NAT activity correlates with increased levels of pineal cAMP suggested that NAT transcription could be modulated by CRE binding factors.[59] The very recent cloning of NAT both from rat and sheep[60,61] should greatly facilitate the analysis of the molecular means of controlling melatonin synthesis and the role of CRE-binding proteins in this process. Interestingly, it has been shown that the pineal expression of *Fra-2*, a member of the Fos family of transcription factors, is strongly modulated in a circadian fashion.[62] These studies demonstrate that the regulation of *Fra-2* and NAT is remarkably similar and suggest that their expression could be controlled by the same molecular mechanisms. Importantly, regulation of *Fra-2* expression appears to involve a cAMP-dependent mechanism probably via phosphorylation of a CRE-binding protein. The precise role of ICER in this process is under study.

Long-Term Memory Formation

Recently a combination of cellular and molecular approaches has made possible the identification of at least some of the key players in the complex process of long-term memory (LTM) formation. Importantly, most of the gene products identified so far to be involved in the LTM process belong to the cAMP signal transduction pathway. Thus, cAMP-dependent transcription appears to be required in the establishment of LTM. Experiments performed in organisms as different as the mollusk *Aplysia*, *Drosophila* and the mouse, indicate that CRE-binding proteins are likely to play a central role in the molecular processes leading to long-term memory.[63] The notion that such mechanisms may be conserved during evolution, from invertebrates to mammals, is supported by the fact that cAMP is amongst the oldest of signaling molecules. In *Aplysia*, direct injection of cAMP triggers short- and long-term facilitation, while injection of PKA inhibitors blocks both forms of facilitation.[64] Recent work in this organism describes the cloning of an inhibitory form of CREB (ApCREB2).[65] The injection of anti-ApCREB2 antibodies in the nucleus of sensory neurons blocks the repressor, allows the action of the activator ApCREB1 and enhances long-term synaptic facilitation. In *Drosophila*, construction of transgenic flies expressing a dominant negative form of CREB induces a specific and complete block of long-term memory to olfactory learning stimuli.[66] Finally, using the technique of homologous recombination in embryonic stem cells, mice containing a targeted mutation for one of the CREB isoforms show defects in long-term memory, while short-term memory is unaffected.[67]

Various CRE-binding factors may associate with each other and, in some cases, with other leucine zipper transcription factors,[28] using a specific dimerization code. Thus, it is likely that the opposing actions of activator and repressor isoforms could provide an important molecular mechanism of control in the process of long-term memory formation. Indeed, it has been proposed that CREB could act as a memory modulator.[68] However, many issues remain to be explored in delineating the means by which memory is established. Despite the fundamental role of cAMP and CRE binding proteins, it can not be excluded that other second messengers, acting in combination with the cAMP cascade, could have an important participation in this process.

CONCLUSIONS

In recent years great progress has been made in understanding transcriptional regulation by cAMP. Importantly, the study of CRE-binding proteins has provided a major advance in understanding the molecular mechanisms by which environmental stimuli can induce changes in the patterns of gene expression. Furthermore, increasing evidence demonstrates that the cAMP signal transduction pathway is crucial in regulating a large number of neuroendocrine processes. Indeed, it has been demonstrated that the CRE-binding proteins play a fulcral role in the efficient transduction of endocrine signals.

Finally, the use of homologous recombination to inactivate specific gene products has proven to be a powerful tool to reveal the molecular mechanisms that underlie hormonal regulation and metabolism. The generation of a mouse with an inactivated CREM gene represents a major step forward in our understanding since it has provided an informative phenotype.[53,54] Further studies in this model system should facilitate the much greater challenge that lies ahead: relating transcriptional control mechanisms to the physiology and biology of the organism.

REFERENCES

1. McKnight SG, Clegg CH, Uhler MD et al. Analysis of the cAMP-dependent protein kinase system using molecular genetic approaches. Rec Progr Horm Res 1988; 44:307-335.
2. Roesler WJ, Vanderbark GR, Hanson RW. Cyclic AMP and the induction of eukaryotic gene expression. J Biol Chem 1988; 263:9063-9066.
3. Lalli E, Sassone-Corsi P. Signal Transduction and Gene Regulation:The nuclear Response to cAMP. J Biol Chem 1994; 269:17359-17362.
4. Ziff EB. Transcription factors: a new family gathers at the cAMP response site. Trends Genet 1990; 6:69-72.

5. Borrelli E, Montmayeur JP, Foulkes NS, Sassone-Corsi P. Signal transduction and gene control:the cAMP pathway. Critical Rev Oncogenesis 1992; 3:321-338.
6. Hai TY, Curran T. Cross-family dimerization of transcription factors Fos:Jun and ATF/CREB alters DNA binding specificity. Proc Natl Acad Sci USA 1991; 88:3720-3724.
7. Masquilier D, Sassone-Corsi P. Transcriptional cross-talk: nuclear factors CREM and CREB bind to AP-1 sites and inhibit activation by Jun. J Biol Chem 1992; 267:22460-22466.
8. Ishiguro H, Kim KT, Johs TH et al. Neuron-specific expression of the human dopamine β-hydroxylase gene requires both the cAMP-response element and a silencer region. J Biol Chem 1993; 268: 24138-24144.
9. Steger DJ, Hecht JH, Mellon PL. GATA-binding proteins regulate the human gonadotropin α-subunit gene in the placenta and pituitary. Mol Cell Biol 1994; 14:5592-5602.
10. Horiuchi M, Nakamura N, Tang SS et al. Molecular mechanisms of tissue-specific regulation of mouse renin gene expression by cAMP. Identification of an inhibitory protein that binds nuclear transcriptional factor. J Biol Chem 1991; 266:16247-16254.
11. Wu H, Mahata SK, Mahata M et al. A functional cyclic AMP response element plays a crucial role in neuroendocrine cell type-specific expression of the secretory granule protein chromogranin A. J Clin Invest 1995; 96:568-578.
12. Dean DC, Mc Quillan JJ, Weintraub S. Serum stimulation of fibronectin gene expression appears to result from rapid serum-induced binding of nuclear proteins to cAMP response element. J Biol Chem 1990; 265:3522-3527.
13. Liou HC, Boothby MR, Glimcher LH. Distinct cloned class II MHC DNA-binding proteins recognize the X-box transcription element. Science 1988; 242: 69-71.
14. Wagner S, Green MR. HTLV-I *tax* protein stimulation of DNA-binding of bZIP proteins by enhancing dimerization. Science 1993; 262:395-399.
15. Hoeffler JP, Meyer TE, Yun Y et al. Cyclic AMP-responsive DNA-binding protein: structure based on a cloned placental cDNA. Science 1988; 242:1430-1433.
16. Hai TY, Liu F, Coukos WJ, Green MR. Transcription factor ATF cDNA clones: An extensive family of leucine zipper proteins able to selectively form DNA binding heterodimers. Genes & Dev 1989; 3:2083-2090.
17. Foulkes NS, Sassone-Corsi P. More is better: activators and repressors from the same gene. Cell 1992; 68:411-414.
18. Yamamoto KK, Gonzalez GA, Menzel P et al. Characterisation of a bipartite activator domain in transcription factor CREB. Cell 1990; 60:611-617.
19. Laoide BM, Foulkes NS, Schlotter F et al. The functional versatility of CREM is determined by its modular structure. EMBO J 1993; 12:1179-1191.
20. Foulkes NS, Mellström B, Benusiglio E, Sassone-Corsi P. Developmental switch of CREM function during spermatogenesis: from antagonist to transcriptional activator. Nature 1992; 355:80-84.
21. Gonzalez GA, Montminy MR. Cyclic AMP stimulates somatostatin gene transcription by phosphorylation of CREB at Ser 133. Cell 1989; 59:675-680.
22. Rehfuss RP, Walton KM, Loriaux MM et al. The cAMP-regulated enhancer-binding protein ATF-1 activates transcription in response to cAMP-dependent protein kinase A. J Biol Chem 1991; 266: 18431-18434.
23. Cowell IG, Skinner A, Hurst HC. Transcriptional repression by a novel member of the bZIP of transcription factors. Mol Cell Biol 1992; 12:3070-3076.
24. Karpinski BA, Morle GD, Huggenvik J et al. Molecular cloning of human CREB-2: An ATF/CREB transcription factor that can negatively regulate transcription from the cAMP response. Proc Natl Acad Sci USA 1992; 89:4820-4824.
25. Molina CA, Foulkes NS, Lalli E et al. Inducibility and negative autoregulation of CREM: An alternative promoter directs the expression of ICER, an early response repressor. Cell 1993; 75:875-886.
26. Stehle JH, Foulkes NS, Molina CA et al. Adrenergic signals direct rhythmic expression of transcriptional repressor

CREM in the pineal gland. Nature 1993; 365:314-320.
27. Kim J, Struhl K. Determinants of half-site spacing preferences that distinguish AP-1 and ATF/CREB bZIP domains. Nucleic Acid Research 1995; 23: 2531-2537.
28. Sassone-Corsi P. Transcription factors responsive to cAMP. Ann Rev Cell Dev Biol 1995; 11:355-377.
29. Lee CQ, Yun Y, Hoeffler JP et al. Cyclic-AMP-responsive transcriptional activation involves interdependent phosphorylated subdomains. EMBO J 1990; 9: 4455-4465.
30. de Groot RP, den Hertog J, Vandenheede et al. Multiple and cooperative phosphorylation events regulate the CREM activator function. EMBO J 1993; 12: 3903-3911.
31. de Groot RP, Derua R, Goris J et al. Phosphorylation and negative regulation of the transcriptional activator CREM by p34cdc2. Mol Endocrinol 1993; 7: 1495-1501.
32. Williams T, Admon A, Luscher B et al. Cloning and expression of AP-2, a cell-type-specific transcription factor that activates inducible enhancer elements. Genes Dev 1988; 2:1557-1569.
33. Courey AJ, Tjian R. Analysis of SP1 in vivo reveals multiple transcriptional domains, including a novel glutamine activation motif. Cell 1989; 55:887-898.
34. Hunter T, Karin M. The regulation of transcription by phosphorylation. Cell 1992; 70:375-387.
35. Hagiwara M, Alberts A, Brindle P et al. Transcriptional attenuation following cAMP induction requires PP-1-mediated dephosphorylation of CREB. Cell 1992; 70:105-113.
36. Nichols M, Weih F, Schmid W et al. Phosphorylation of CREB affects its binding to high and low affinity sites: implications for cAMP induced gene transcription. EMBO J 1992; 11: 3337-3346.
37. Chrivia JC, Kwok RPS, Lamb N et al. Phosphorylated CREB binds specifically to the nuclear protein CBP. Nature 1993; 365:855-859.
38. Kwok RP, Lundblad JR, Chrivia JC et al. Nuclear protein CBP is a coactivator for the transcription factor CREB. Nature 1994; 370:223-226.
39. Sun P, Maurer RA. An inactivating point mutation demonstrates that interaction of cAMP response element binding protein (CREB) with the CREB binding protein (CBP) is not sufficient for transcriptional activation. J Biol Chem 1995; 270:7041-7044.
40. Arany Z, Newsome D, Oldread E et al. A family of transcriptional adaptor proteins targeted by the E1A Oncoprotein. Nature 1995; 374:81-84.
41. Masquilier D, Foulkes NS, Mattei MG, Sassone-Corsi P. Human CREM gene: evolutionary conservation, chromosomal localization and inducibility of the transcript. Cell Growth & Differentiation 1993; 4:931-937.
42. Meyer TE, Waeber G, Lin J et al. The promoter of the gene encoding 3' 5'-cyclic adenosine monophosphate (cAMP) response element binding protein contains cAMP response elements: Evidence for positive autoregulation of gene transcription. Endocrinology 1993; 132: 770-777.
43. Sassone-Corsi P. Rhythmic transcription and autoregulatory loops: winding up the biological clock. Cell 1994; 78:361-364.
44. Foulkes NS, Duval G, Sassone-Corsi P. Adaptative inducibility of CREM as transcriptional memory of circadian rhythms. Nature 1996; 381:83-85.
45. Foulkes NS, Borrelli E, Sassone-Corsi P. CREM gene: Use of alternative DNA binding domains generates multiple antagonists of cAMP-induced transcription. Cell 1991; 64:739-749.
46. Struthers RS, Vale WW, Arias C et al. Somatotroph hypoplasia and dwarfism in transgenic mice expressing a nonphosphorylatable CREB mutant. Nature 1991; 350:622-624.
47. Borrelli E, Heyman R, Arias C et al. Transgenic mice with inducible dwarfism. Nature 1989; 339:538-541.
48. Lalli E, Sassone-Corsi P. Thyroid-stimulating hormone (TSH)-directed induction of the CREM gene in the thyroid gland participates in the long-term desensitization of the TSH receptor. Proc Natl Acad Sci USA 1995; 92:9633-9637.

49. Delmas V, van der Hoorn F, Mellström B et al. Induction of CREM activator proteins in spermatids: downstream targets and implications for haploid germ cell differentiation. Mol Endocrinol 1993; 7:1502-1514.
50. Kistler MK, Sassone-Corsi P, Kistler SW. Identification of a functional cAMP response element in the 5'-flanking region of the gene for transition protein 1 (TP1), a basic chromosomal protein of mammalian spermatids. Biol Reprod 1994; 51:1322-1329.
51. Sun Z, Sassone-Corsi P, Means A. Calspermin gene transcription is regulated by two cyclic AMP response elements contained in an alternative promoter in the calmodulin kinase IV gene. Mol Cell Biol 1995; 15:561-571.
52. Monaco L, Foulkes NS, Sassone-Corsi P: Pituitary follicle-stimulating hormone (FSH) induces CREM gene expression in Sertoli cells: involvement in long-term desensitization of the FSH receptor. Proc Natl Acad Sci USA 1995; 92: 10673-10677.
53. Nantel F, Monaco L, Foulkes NS et al. Spermiogenesis deficiency and germ-cell apoptosis in CREM-mutant mice. Nature 1996; 380:159-162.
54. Blendy JA, Kaestner KH, Weinbauer GF et al. Severe impairment of spermatogenesis in mice lacking the CREM gene. Nature 1996; 380:162-165.
55. Tamarkin L, Baird CJ, Almeida OFX. Melatonin: A coordinating signal for mammalian reproduction? Science 1985; 227:714-720.
56. Moore RY. Organization and function of the central nervous system circadian oscillator; the suprachiasmatic hypothalamic nucleus. Federation Proc 1988; 42:2783-2789.
57. Klein DC. Photoneural regulation of the mammalian pineal gland. In *Photoperiodism, melatonin and the pineal gland.* Pitman, London; 1985.
58. Stehle JH, Foulkes NS, Pevet P, Sassone-Corsi P. Developmental maturation of pineal gland function: synchronized CREM inducibility and adrenergic stimulation. Mol Endocrinol 1995; 9:706-716.
59. Takahashi JS. ICER is nicer at night (sir!). Curr Biol 1994; 4:165-168.
60. Borjigin J, Wang MM, Snyder SH. Diurnal variation in mRNA encoding serotonin *N*-acetyl transferase in pineal gland. Nature 1995; 378:783-785.
61. Coon SL, Rosenboom PH, Baler R et al. Pineal serotonin *N*-acetyltransferase: expression cloning and molecular analysis. Science 1995; 270:1681-1683.
62. Baler R, Klein DC. Circadian expression of transcription factor Fra-2 in the rat pineal gland. J Biol Chem 1995; 270: 27319-27325.
63. Frank DA, Greenberg ME. CREB: a mediator of long-term memory from mollusks to mammals. Cell 1994; 79:5-8.
64. Huang YY, Li XC, Kandel ER. cAMP contributes to Mossy Fiber LTP by initiating both a covalently mediated early phase and macromolecular synthesis-dependent late phase. Cell 1994; 79:69-79.
65. Bartsch D, Ghirardi M, Skehel PA et al. Aplysia CREB2 represses long-term facilitation: relief of repression converts transient facilitation into long-term functional and structural change. Cell 1995; 83:979-992.
66. Yin JCP, Wallach JS, Del Vecchio M et al. Induction of a dominant negative CREB transgene specifically blocks long-term memory in *Drosophila.* Cell 1994; 79:49-58.
67. Bourtchuladze R, Frenguelli B, Blendy J et al. Deficient long-term memory in mice with a targeted mutation of the cAMP-responsive element-binding protein. Cell 1994; 79:59-68.
68. Yin JCP, Del Vechio M, Zhou H, Tully T. CREB as a memory modulator: induced expression of a dCREB2 activator isoform enhances long-term memory in *Drosophila.* Cell 1995; 81:107-115.

CHAPTER 5

The AP-1 Family of Transcription Factors: Multi-Level Control of Activity

Martin Göttlicher, Hans J. Rahmsdorf and Peter Herrlich

INTRODUCTION

The AP-1 transcription factors are thought to play key roles in oncogenic cell transformation, cell proliferation, differentiation and apoptosis,[1-9] as well as in the cellular response to various microenvironmental stimuli including physical and chemical stress or cellular damage.[10-13] Although the exact functions of the AP-1 factors under these various cellular conditions are ill-defined, putative roles can be derived from the fact that numerous stimuli impinge on AP-1 activity, and from the analysis of the transforming oncogenes *v-fos* and *v-jun* . The oncogenic versions of *c-fos* and *c-jun*, e.g., *v-fos* [14] and *v-Jun* [15,16] had been cloned prior to biochemical evidence of the structure of AP-1 and prior to cloning of the cellular genes *c-fos* and *c-jun*.[17-20] AP-1 factors are dimers formed by members of the Fos, Jun or ATF families of ATF subunits. Only the dimers are active as transcription factors.[21-24] The more concise current understanding of molecular functions of AP-1 proteins discloses a number of structural differences between the proteins encoded by the viral oncogenic genes as compared to the cellular transcription factors. Most of these differences cause an increase in activity or uncouple the viral proteins from the normal regulatory cellular networks. Thus, the interest in the viral copies is more than just historical since the differences between the cellular and oncogenic versions of the AP-1 components have provided information on the regulation of AP-1 activity, which occurs on several levels. These include transactivating capacity,[25-27] protein stability,[28,29] and protein localization.[30] The viral oncogene products may be considered 'gain-of function' mutants compared to their cellular counterparts. These mutant forms play a pivotal role in the process of oncogenic transformation, but the analysis of the oncogenic versions can not fully explain exactly how c-Jun and c-Fos function in the context of normal cell physiology.

Transcription Factors in Eukaryotes, edited by Athanasios G. Papavassiliou.

Efforts over the last few years have delineated the role of AP-1 complexes as transcription factors in the immediate-early response of the cell to growth stimulatory signals[31-35] (reviewed in refs. 6, 36) or adverse extracellular agents[10-13] (reviewed in ref. 37). In these immediate early responses, AP-1 components play a dual role. The transcription of several AP-1 encoding genes is induced so that AP-1 proteins accumulate. Both preexisting and newly synthesized AP-1 subunits are subject to posttranslational activation and, interestingly, subunits of the AP-1 family participate in *c-fos* and *c-jun* promoter regulation. There is also a link to the cell cycle in that the synthesis of cell cycle regulating proteins such as cyclin D1[38] is controlled by AP-1.

A major problem in the study of AP-1 is the existence of multiple homo- and heterodimeric complexes. The subunit complexity of the AP-1 transcription factors has substantially increased during the last decade by the cloning of numerous members of the gene families which code for the AP-1 subunit proteins (see below). In an attempt to define subunit functions, *c-jun*, *c-fos*, and *fos B* have been inactivated by germline mutation in mice and the knock-outs of other genes, e.g., *junB*, *junD* and *fra1*, are currently pursued. The phenotypes of mice lacking these genes are, though in one published case lethal, confined to a surprisingly limited set of tissues considering the fundamental role which AP-1 is thought to play in cell proliferation and transformation. Thus, the function of some AP-1 proteins may be redundant but other functions in particular cells or states of physiology may be unique. Certainly, the increase in number of AP-1 family members makes the puzzle appear more complicated than the initial cloning of *c-jun* and *c-fos* may have suggested. Gain-of-function mutations such as the oncogenic versions of AP-1 proteins or loss-of-function strategies which address a whole group of AP-1 proteins such as dominant-negatively acting mutants expressed in cultured cells[39] may not discriminate sufficiently sharply between individual family members. More precisely defined mutants and genetic inactivation strategies are required to fully understand the complex biological functions of AP-1.[4,40,41]

Considering the immense literature on AP-1, this chapter is of necessity selective. After a basic description of individual subunit proteins of AP-1, the focus of this review will be on the different levels at which AP-1 activity is regulated. Mechanisms of tight control of AP-1 activity, and particularly the differences between different family members, will be described. The concluding section will summarize the recent approaches to validate the implications from AP-1 function seen in cultured cells, in whole animals and specifically in mice which carry defined genetic manipulations of the AP-1 genes.

THE PLAYERS: THE FOS, JUN AND ATF FAMILIES OF AP-1 SUBUNITS

DNA-Binding Homo- or Heterodimers

An AP-1 transcription factor binds as a dimer of two subunits to target DNA-sequences such as the consensus element TGANTCA, which had initially been identified in the promoters of the collagenase I and metallothionein IIA genes and the SV40 virus enhancer.[42-44] The activity which binds to this element has been termed activating protein 1 (AP-1). Similar binding sites have since been found in a large number of genes and a current compilation of binding sites reveals the idealized palindromic binding motif of TGA $N_{(1\ or\ 2)}$ TCA which is recognized by AP-1 complexes, including the more recently described AP-1-like complexes composed of Jun/ ATF/ CREB subunits.

Early experiments revealed heterogeneity of AP-1 affinity-purified on TGANTCA-DNA columns.[42] Dimer composition was first identified for c-Fos/c-Jun[19,20,23] which prefers DNA target elements in which the half-sites are spaced by one nucleotide (TGANTCA). c-Fos and related proteins are usually not able to bind to DNA as homodimers, whereas members of the c-Jun-family can bind target DNA sequences in vitro as homodimers also, but the affinity is lower compared to a heterodimer with Fos.[45-47] Since the identification of the first AP-1 factor as a heterodimer of c-Jun and c-Fos, the number of players in both families has substantially grown and numerous genes have been cloned. In the Fos family there are three additional proteins, *FosB*[48,49] and the Fos-related antigens Fra 1 and Fra 2.[50-54] The Jun family includes JunB and JunD in addition to c-Jun and v-Jun.[55-57] A related family of proteins binds to the cAMP-responsive element (CRE): the CRE-binding protein (CREB) and the ATF-proteins[58-65] also form heterodimers with AP-1 proteins (Jun) and bind to AP-1 binding sites.[66-69] In contrast to the Jun/Fos heterodimers, the ATF/CREB-containing AP-1 dimers prefer DNA-binding sites in which the two half-sites are separated by two nucleotides (TGANNTCA).[67,68,70,71]

The combination of all existing AP-1 subunits theoretically allows the formation of more than 50 different homo- or heterodimeric AP-1 proteins. They may serve specific functions or, at least in part, constitute safeguards against the failure of single components of an essential signaling system. Specificity for target genes and for incoming signals is found on several levels of AP-1 function. First, different target elements (e.g., TGA N TCA vs. TGA NN TCA) preferentially select different AP-1 complexes. Additional bases outside the core consensus element contribute to specific DNA-recognition. Further, the abundance of AP-1 subunits in different tissues or during embryonic development determines the selection of specific genetic programs.[57,72,73] The AP-1 complexes differ not only in target gene selection, but also their transcriptional activation potency once bound to DNA varies greatly as discussed below. Despite these specificities certain AP-1 complexes can substitute for others in most tissues, as suggested by gene disruption studies.

Domain Architecture: The DNA-Binding and Dimerization Domain

AP-1 dimers bind to specific DNA-target elements through a so-called basic region-leucine zipper domain (bZIP).[74-79] The position of the bZIP region within the protein varies amongst different AP-1 members (Fig. 5.1). The bZIP can be fused to a heterologous transactivation domain and functions as an independent DNA-binding domain (DBD).[80,81] The architecture of the bZIP region comprises an amino-terminal stretch of basic amino acids which directly contacts the DNA (refs. 82,83 and references therein). This basic region is followed by a carboxy-terminally adjacent coiled coil structure formed by a heptad repeat of leucines and hydrophobic amino acids interspersed by hydrophilic amino acids. This helix allows dimerization of two proteins through the hydrophobic surface and exposure of the hydrophilic surface to the surrounding aqueous medium.[78,84-89] Hydrophilic amino acids adjacent to the hydrophobic residues on either protein come into the vicinity of corresponding amino acids on the partner protein. Ionic interactions of these amino acids have been considered to stabilize or to disrupt certain dimers.[46,77,83,89,90] For instance, the formation of a Fos homodimer is made impossible by the juxtaposition of acidic amino acids which, due to electrostatic repulsion, prevent dimer formation. In the case of c-Jun homodimers or, even more pronouncedly, the c-Jun/c-Fos heterodimers, the electrostatic conditions are more favorable, which may explain the higher stability of these dimers (for a graphic representation see refs. 85, 91). These proposed dimerization rules have been supported experimentally by creating mutant AP-1 proteins by domain swaps[77,83,92,93] or replacement of defined amino acids in the yeast bZIP protein GCN4 by those of mammalian AP-1 proteins.[90] Point mutations in the c-Fos protein indicate that in addition to electrostatic interactions of the hydrophilic amino acids the hydrophobic interaction surface itself also affects dimerization properties and may account for the lack of Fos homodimerization.[94] Most predictions on the structure and dimerization properties of the coiled coil structure from mutagenesis studies have been confirmed by the resolution of the X-ray crystal structure of a DNA-bound heterodimer of the c-Jun and c-Fos DNA-

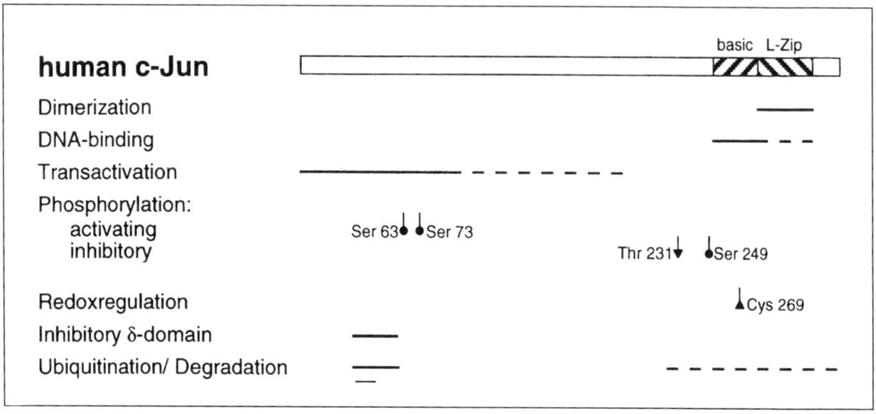

Fig. 5.1. Functional domains of c-Jun and sites of posttranslational modifications.

binding domains.[95] Dimerization is required for efficient binding of AP-1 factors to their target elements but does not require the presence of the basic region and of DNA contact. Mutant proteins lacking the basic region or containing additional amino acids between the leucine zipper and the basic region can still dimerize with other AP-1 proteins but the resulting dimer can not bind the AP-1 target element. Therefore these mutants, when overexpressed in a cell or tested in vitro, act as dominant-negative inhibitors of AP-1 activity presumably by sequestering the endogenous AP-1 proteins into heterodimers which are unable to bind to the DNA-target element.[39,46,96] Yet, DNA binding and coiled coil interaction may not happen completely independent of each other since analysis by circular dichroism and nuclear magnetic resonance spectroscopy suggest folding of the whole bZIP region into a helical structure and profound structural changes upon binding of DNA by the protein dimer.[97,98] Interestingly, dimerization rules in solution and dimerization at a DNA target seem to differ, also suggesting that the DNA element selects or influences dimer conformation (van Dam et al, in preparation).

The paradigm of the domain structure of transcription factors predicts that regulation of AP-1 activity occurs through the transactivating domain(s) of the constituting proteins (see below). Nevertheless at least in the case of c-Jun/c-Fos, the DNA-binding domain also seems susceptible to regulation. In vitro, the redox potential affects the oxidation state of a conserved cysteine residue in the basic regions of c-Jun and c-Fos and modulates the affinity of Jun and Fos to their DNA-target elements.[99,100] The cysteine residues in Jun and Fos react with gold (I) salts and this has been shown to inhibit AP-1 function in vitro and in vivo. Modulation of AP-1 function may mediate some of the anti-inflammatory activities of gold which is exploited in antirheumatic therapy.[101] Phosphorylation represents another modification affecting DNA-binding of AP-1: glycogen synthase kinase 3 (GSK-3) and casein kinase II phosphorylate serine and threonine residues just amino-terminally adjacent to the basic region of c-Jun, which reduces binding to the AP-1 recognition DNA motif.[102-105] Activators of protein kinase C and other inducers of AP-1 activity lead to decreased phosphorylation of these sites.[106-108] It seems that the DBD dephosphorylation of c-Jun is triggered by the complete activating phosphorylation of the amino-terminus (see below) presumably by intramolecular changes that lead to phosphatase accessibility of the DBD.[108] In vitro, mitogen activated protein kinases (MAP kinase, ERK 1 and 2, see below) can phosphorylate the same serine/threonines.[109]

This finding has still to be reconciled with the concept that MAP kinases are usually active in conditions which lead to AP-1 activation and not inactivation.

The Transactivation Domain(s)

The transactivation domain(s) are structurally less defined when compared between different AP-1 proteins. Mutational analysis of c-Jun has identified at least three individual motifs clustered in the amino terminus which contribute interdependently to transactivation.[25,110,111] A key role is played by the serine residues at position 63 and 73 which are phosphorylated by kinase cascades upon appropriate cellular stimulation (see below). Mutation of these residues impairs transactivation properties.[112] Full activation involves, in addition, threonines at positions 91 and 93.[108] Overexpression of the amino-terminal part of c-Jun, which lacks a DBD, interferes with transactivation by coexpressed full length protein, a phenomenon referred to as squelching.[113] This suggests that the amino terminus of c-Jun does not contact the basal transcription machinery directly but that interaction involves intermediary factor(s) which are required for transactivation but are available in limiting amounts only. Indeed, coprecipitation experiments from metabolically labeled cells support this proposal by showing two c-Jun interacting proteins with molecular weights of 52 and 54 kDa.[114] The c-Fos protein contains transactivation functions both in the amino-terminus, carboxy-terminal to the bZIP region (aa 226 to 236 and aa 267 to 276). The domains located carboxy-terminal to the bZIP region show similarities with the transactivating motifs in Jun and have therefore been referred to as homology boxes (HOBs).[25,111,115,116] Individual transactivating motifs of c-Fos act synergistically and synergism is maintained when HOBs are interchanged between Jun or Fos.[111]

Despite the similarities in DNA binding, different members of an AP-1 subunit family may differ substantially in their transactivating potential. JunD and JunB as well as Fra1, Fra2 and FosB are weaker transactivators than c-Jun or c-Fos. Therefore, when present in excess, they may displace c-Jun or c-Fos, respectively, and form AP-1 complexes with reduced transactivating potential.[57,117-121] This fine tuning of transactivation by differential composition of AP-1 complexes may be physiologically relevant, e.g., for cell growth, transformation, and susceptibility of cell proliferation to extracellular signals.[57,118,122,123]

Members of the ATF and CREB families of subunits which enforce preferred binding to palindromes with half sites spaced by two nucleotides, vary in their transactivation domains. ATF-2 resembles c-Jun in that serines 69 and 71 are phosphorylated in response to activating stimuli.[124] ATF-1 and CREB share a motif totally different from those of other AP-1 subunits. The critical serine to be modified is at position 133 (CREB) and position 63 (ATF-1) (see refs. 125,126 and references therein). For CREB a coactivator has been identified: CBP (CREB-binding protein).[127] CBP is apparently a member of a larger family of proteins. Another member of this family, p300, can also cooperate with CREB.[128] Specificity and function of coactivators are currently in debate and not yet well defined. CBP and p300 have been reported also to enhance transcription from Jun/Fos dependent promoters and from promoters regulated by nuclear receptors.[129-134] Possibly specificity is lost by the expression of unphysiologic levels. Alternatively, this class of coactivators serves a broader function, e.g., as an integrator of activities by more than one transcription factor or as a transcription factor-dependent modulator of chromatin.

FAST SIGNALING: AP-1 LINKS PROTEIN KINASE CASCADES TO A TRANSCRIPTIONAL RESPONSE

The members of the AP-1 family are switched within minutes into transcriptionally active conformations upon treatment of cells in culture with a variety of agents. Inducing agents include growth factors, interleukins, hormones, various types of tumor promoting agents, alkylating agents, organometals and electromagnetic radiation of different wavelength. The activating modifications of AP-1 subunits are by and large identical irrespective of the inducer, suggesting either that different inducers activate the same protein kinase cascade (e.g., the kinase cascade inducing the phosphorylation of c-Jun at serines 63 and 73) or that different protein kinases activate the transcription factors through phosphorylation of the same amino acids (as for instance CREB, which is phosphorylated at serine 133 by several nonrelated protein kinases). We will first describe the protein kinase cascades inducing the phosphorylation of c-Jun and ATF-2 and then consider the protein kinases which phosphorylate (and activate) CREB.

Both c-Jun and ATF-2 are activated through the phosphorylation of serines and threonines followed by prolines (c-Jun at serine residues 63 and 73 and at threonine residues 91 and 93 and ATF-2 at serine residue 90 and threonine residues 69 and 71).[107,110,124,135-138] This motif is recognized by proline directed protein kinases and evidence is overwhelming that these are the enzymes also relevant in vivo in the activation of AP-1. Since such activation depends on extracellular signals, it is clear that the proline directed protein kinases are in turn activated in response to a signaling event. Members of the linear signal transduction cascades involving several levels of protein kinases have been identified (Fig. 5.2). To date, three groups of proline directed protein kinases have been characterized. The extracellular signal regulated protein kinases (Erk 1,2), originally also named mitogen activated protein kinases (MAPK), with molecular weights of 42 and 44 kDa were discovered first. Erk 1,2 are activated through phosphorylation by the dual specificity kinases MEK 1,2 at one threonine and one tyrosine residue separated by a glutamic acid. MEK 1,2 are in turn activated through one of several protein kinases, including Raf-1, Raf B and MEKK 2, 3. The Jun N-terminal kinases (JNKs), also named stress activated protein kinases (SAPKs), were second to appear on the stage. The group has now grown to 10 isoforms which are encoded by three different genes.[139] The proteins with major molecular weights of 45 and 55 kDa are activated through phosphorylation by the JNK kinase (=MKK4=SEK1=SAPKK1) at both a threonine and a tyrosine residue separated by a proline. Like MEK 1,2, this dual specificity protein kinase is in turn phosphorylated by MEKK 2,3 and, in addition, by MEKK1. Recently, a new proline directed protein kinase, p38 (also named RK, reactivating kinase) has been found. This kinase is a relative of the HOG-1 kinase of yeast and is phosphorylated and activated through MKK3 (=SAPKK2), MKK4, and MKK6 (=SAPKK3) at a threonine and tyrosine residue separated by a glycine (Fig. 5.2, for reviews of proline-directed protein kinases see refs. 140,141). The term MAPK (and the upstream MAPK kinase and MAPKK kinase) had initially described the signaling cascade which we now call the Erk pathway. The terms MAPK, MAPKK and MAPKKK are now used to describe the three hierarchical levels in the Erk, the JNK and the p38 signaling cascades. Although very similar in hierarchical organization, the three well characterized pathways may serve partly distinct functions both with respect to upstream activating stimuli and downstream target molecules.

What is upstream of the MAP kinase kinase kinases? The answers are only partially known. EGF treatment of HeLa cells leads to the activation of all three groups of proline directed protein kinases, the Erk 1,2, the JNKs and p38/RK. It is possible but not yet proven that Ras is an obligatory intermediate in the activation of all these protein kinase cascades by EGF. EGF induces the autophosphorylation of its receptor at several cytoplasmic tyrosine residues at the cytoplasmic site of the protein. This then leads to the binding of several proteins with SH2 (src homology) domains possessing either enzymatic activity such as phospholipase Cγ or phosphoinositide-3-kinase or serving as bridging proteins (such as Grb2 and Shc) which recruit still other signaling proteins. The most important step seems to be the recruitment of the guanine nucleotide exchange factor "human son of sevenless, hSOS" which brings Ras into the active,

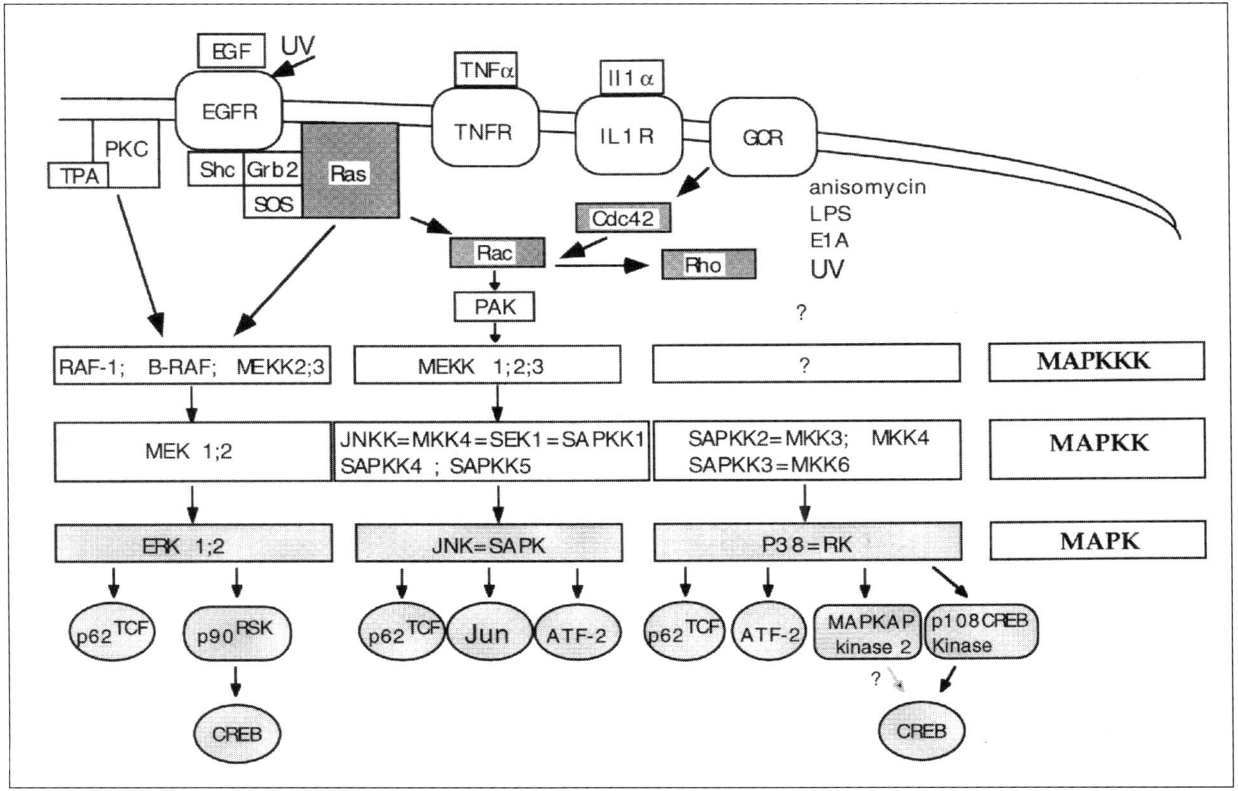

Fig. 5.2. Signal flow through proline directed kinases. For explanation see text and reviews by Symons, 1996, and Treisman, 1996. Signal flow is from top to bottom. GTP-binding proteins are shown in dark gray rectangles, the proline-directed protein kinases in light gray rectangles. Transcription factors phosphorylated by proline-directed protein kinases are presented in light gray ovals, protein kinases located downstream of proline-directed protein kinases are presented in light gray rounded rectangles. PAK = p21 associated protein kinase, TPA = 12-O-tetradecanoyl-phorbol-13-acetate, GCR = G-protein coupled receptor

GTP-binding form.[142] Ras, in turn, translocates c-Raf to the cytoplasmic membrane where it is activated by an as yet unknown protein kinase and in turn activates the MEK 1,2, Erk 1,2 pathway.

Besides activating Raf, Ras in its GTP-binding form activates other small GTP-binding proteins such as Rac and Rho (Ras → Rac → Rho), which had originally been thought to be exclusively involved in the regulation of the distinct structures making up the cytoskeleton. Their role in the control of cell proliferation is just emerging: Rac, for instance, is an upstream regulator of the protein kinase cascades leading to the activation of JNKs and p38. Rac acts on JNKs and p38 through as yet ill-defined steps, one of which seems to involve the p21-associated kinases (PAK) (Fig. 5.2, for a recent review see ref. 143).

Not only the receptor tyrosine kinases but also other receptors such as those for certain interleukins or tumor necrosis factor (TNF) as well as heterotrimeric G protein coupled receptors (GCR) activate all three protein kinase cascades. Chemical and physical stress appears to address all three pathways.[107,144,145] Certain agents, such as UV irradiation and TNFα, work through the activation of Ras.[136,146] GCRs possibly work through the activation of the GTP-binding protein Cdc 42 (Fig. 5.2, for a review see ref. 143), and/or the activation of receptor tyrosine kinases (for a review see ref. 146a). The primary targets of other agents leading to the activation of the protein kinase cascades, however, such as X-rays, mitomycin C, the mutagen N-methyl-N-nitro-N-nitrosoguanidine, anisomycin, lipopolysacharide or the viral protein E1A have not yet been delineated to any degree of certainty.

An interesting principle of action has recently been detected for several non-physiologic agents: the inactivation of phosphatases. That inactivation of a phosphatase can lead to a gain of function implies that there is a balance between protein kinase and protein phosphatase activities and that considerable signal flow must exist that is constantly downmodulated by the protein phosphatases. Okadaic acid, for instance, inhibits the serine-threonine protein phosphatases 1, 2A and 3 and thereby triggers events similarly to a tumor promoter acting on PKC (ref. 147 and references therein). UV irradiation, oxidants and organometals induce the phosphorylation of receptor tyrosine kinases and receptor-associated tyrosine kinases[11,148-150] through the inhibition of receptor-specific tyrosine phosphatases (ref. 151 and unpublished). As^{3+}, which is a potent tumor promoter and activator of AP-1, possibly induces the activation of JNKs through the inhibition of a dual specificity JNK phosphatase.[152]

Although the proline directed protein kinases Erk 1 and 2 were originally thought to phosphorylate and thus activate the c-Jun protein,[112] this proposition does not seem to hold (for discussion see ref. 36). In vitro, no detectable activity of Erk 1 or 2 for the transactivating domain of the c-Jun protein can be found, but rather an activity towards the DNA-binding domain (which may then inhibit DNA binding by c-Jun, see above). It appears that the only protein kinases which efficiently phosphorylate the transactivating domain of c-Jun are the JNKs,[144,153] also named SAPKs,[154] which are activated by all agents which activate c-Jun's transcriptional activity. The JNKs bind strongly to the δ-domain of the c-Jun protein, which can be exploited for affinity purification. They do not bind to v-Jun, which lacks the δ-domain, so that it is still an open question how the transactivating potential of v-Jun is regulated.[107] JNKs also bind to the Jun B protein, but without phosphorylating this target. In Jun B, the serines corresponding to serines 63 and 73 of c-Jun are not followed by proline residues[110] (for a review see ref. 36).

Also, the transcription factor ATF-2 is phosphorylated by JNKs and, as with c-Jun, binding of JNKs through a region distinct from the phosphorylated target sequence is obligatory for phosphorylation.[124,137-139] In contrast to c-Jun, ATF-2 is also phosphorylated and thus activated by the proline directed protein kinase p38 which also binds to the transactivating domain of ATF-2[145] (for a review see ref. 141). As discussed above, in cells stressed by chemical or physical means or treated with interleukins or TNFα, both proline directed protein kinases, JNKs and p38, are activated. Specific inhibitors may help in the future to decide which kinases are the predominant ATF-2 kinases in vivo.[155,156]

As detailed above, the transactivating domains of c-Fos and c-Jun are homologous, although they are located at different positions in the two proteins. The sequence surrounding the major JNK phosphorylation site in c-Jun is conserved in c-Fos, such that threonine 232 of c-Fos is homologous to serine 73 in c-Jun.[111] Threonine 232 is indeed instrumental in the activation of transcription by c-Fos upon overexpression of H-Ras or treatment of cells with EGF. Despite the apparent homology between threonine 232 in c-Fos and serine 73 in c-Jun, other Jun-activating stimuli such as UV-irradiation (which activates JNKs) or the phorbol ester 12-O-tetradecanoyl-phorbol-13-acetate (TPA, which activates Erk 1,2) do not lead to phosphorylation of threonine 232 in c-Fos. Thus, threonine 232 appears not to be a substrate of the known protein kinase signaling cascades (Erk 1,2/JNKs). Indeed, Fos threonine 232 is not phosphorylated by ERK 1,2 or JNKs in vitro. EGF or Ras may, therefore, act through a novel pathway to activate c-Fos. An EGF-inducible Fos-phosphorylating protein kinase with a molecular weight of 88 kDa has been found.[157] The relationship of this kinase to the other known signaling cascades is not yet understood. Other sites in c-Fos are phosphorylated by known protein kinases: serine 362 by p90^{RSK1} which is downstream of Erk 1,2 and serine 374 by Erk 2. These phosphorylation events do not affect transactivation ability but rather stabilize the protein and may potentiate the transforming and transcription repressing activities exerted by c-Fos.[158-160]

In contrast to the AP-1 members c-Jun, ATF-2 and c-Fos, the only relevant phosphorylation site in the CREB protein, serine 133,[161] is not followed by a proline and thus not a direct target of proline directed protein kinases. The classical activator of CREB is protein kinase A, which in the absence of cAMP is sequestered in the cytoplasm bound to a repressing regulatory subunit (for a review see ref. 162). Upon elevation of intracellular cAMP levels by stimulation of adenyl cyclase or by inhibition of cAMP phosphodiesterase, cAMP binds to the regulatory subunit releasing the active PKA. In addition to phosphorylating cytoplasmic substrates, PKA migrates into the nucleus and activates the CRE-binding protein CREB through phosphorylation of serine 133. In addition to PKA, several other serine/threonine specific protein kinases phosphorylate serine 133 of CREB. Among these are the calcium/calmodulin-dependent kinases type II and type IV.[163-166] However, while Ca^{2+}/calmodulin dependent protein kinase type IV is a potent activator of CREB transcriptional activity, type II protein kinase is not. This may relate to the fact that, as revealed in transient transfection studies, protein kinase type II does not enter the nucleus.[165] It was shown in a different study that type II kinase localized in the nucleus CREB phosphorylated at serine 142 in addition to serine 133. This second site of phosphorylation inhibited PKA or Ca^{2+}/calmodulin protein kinase type IV mediated activation of CREB. Replacement of serine 142 by alanine rendered the protein activatable by the type II kinase and enhanced the ability of Ca^{2+} to activate CREB.[166] Beside cAMP and Ca^{2+}, treatment of various cells with growth factors or UV induces CREB phosphorylation at serine 133. Since serine 133 is not followed by proline, growth factors and UV reach CREB either through a novel unknown pathway or CREB protein kinases are downstream of one of the proline directed kinases. The latter proved true: responding CREB kinases include p90^{RSK1} and p90^{RSK2} which are downstream of Erk 1,2,[167,168] MAPKAP 2 which is downstream of p38/RK,[169] and a p108 protein kinase, which is also downstream of p38/RK.[170] This latter enzyme may be identical to a p105 CREB kinase which is activated in PC12 cells by nerve growth factor.[171] Growth factor and UV-induced phosphorylation of CREB at serine 133 is functionally relevant for the induced transcription of the *c-fos* gene since a dominant negative version of CREB inhibits the activation of the *c-fos* promoter, presumably by blocking the CRE. Yet, a reporter driven by a single CRE is not efficiently activated by growth factors or UV.[170,171] The reason is presently not understood but may be due to additional, possibly inhibitory, phosphorylation (which would be distinct from the Ca^{2+}/calmodulin-dependent protein kinase II-dependent inhibitory phosphorylation of Ser-142, P. Sassone-Corsi, personal communication) or due to a lack of additionally required signaling events.

RECRUITMENT OF PLAYERS: REGULATED SYNTHESIS OF AP-1 COMPONENTS

Preexisting AP-1 links kinase-dependent signaling cascades to a transcriptional response of target genes. In addition, a largely overlapping set of stimuli which activates the protein kinase cascades described in the previous paragraph induces transcription of genes encoding various AP-1 subunits. This is because several of these genes are themselves controlled by AP-1 members, prime examples being the *c-fos* and *c-jun* genes (Fig. 5.3). Thus for example *c-fos* fulfills all the criteria of an immediate-early response gene,[172] e.g., rapidly increased expression upon stimulation of cells, followed by a rapid decrease in mRNA-levels to nearly basal levels, usually within less than an hour, and superinduction of mRNA-levels by protein synthesis inhibitors such as cycloheximide or anisomycin.

C-FOS PROMOTER REGULATION

The expression of *c-fos* is regulated by a wide variety of extracellular stimuli usually linked to cell activation, growth stimulation, or cell differentiation, such as stimulation of cells with serum or growth factors,[31-33] calcium,[32,163,173] activators of protein kinase C[32,174,175] or cAMP-inducing hormones.[176-178] Other, nonphysiological, adverse agents such as ultraviolet or ionizing radiation[177,179-181] (for further references see ref. 37) or certain mutagens also activate *c-fos* expression. The signaling pathways used by the latter stimuli are yet not fully understood, but UV irradiation activates several of the growth factor-induced signaling cascades.[11] Most of the *c-fos*-inducing signals trigger a typical immediate early response, e.g., a fast peak of gene activity followed by down-regulation to basal levels within one hour, but some stimuli such as UV irradiation also trigger in several cell types a second delayed response.[180,182]

The transcriptional response of *c-fos* is mediated by a number of cis-acting elements within the promoter of the gene as exemplified by the human *c-fos* promoter. The key cis acting element which integrates the response to serum and most of the growth factors is the so-called serum response element (SRE) located between positions -320 and -300.[183] The SRE binds a dimer of the serum response factor (SRF) but responsiveness to most stimuli involves a third protein, the ternary complex factor ($p62^{TCF}$).[184] Molecular cloning revealed that $p62^{TCF}$ is constituted by a group of proteins, Elk-1 and Sap1,2[185,186] (for a review see ref. 187) which belong to the family of the Ets proteins. Interestingly, occupancy of the SRE with transcription factors does not appear to change during serum stimulation of resting cells as analyzed by in vivo footprint analysis.[188] Thus, other mechanisms of regulation seem to prevail and indeed there is strong evidence that the SRE-bound ternary complex of SRF and $p62^{TCF}$ is converted to a transcriptionally active form by phosphorylation possibly while bound to DNA. The kinases phosphorylating $p62^{TCF}$ should thus be localized, at least in part, in the nucleus. All three groups of proline-directed protein kinases (Erk 1,2, JNKs and p38) have been shown to possess kinase activity both in vivo and in vitro towards $p62^{TCF}$ (for references see ref. 189). Thus, the protein kinase cascades discussed in the previous paragraph are the molecular targets by which a set of extracellular stimuli leads to both increased expression of *c-fos* (and *c-jun*, see below) and activation of preexisting as well as newly synthesized AP-1 complexes. These protein kinase cascades do not seem to be the only ones involved in the induction of the *c-fos* gene by lysophosphatidic acid (LPA) or heterotrimeric G proteins. LPA and trimeric G proteins can also target SRF without requiring stat factors or the TCF.[190]

A specific subset of growth factors, such as the platelet derived growth factor (PDGF) and the epidermal growth factor (EGF), use at least in part an additional signaling pathway leading to increased binding of a PDGF-inducible STAT factor (PIF) to the PDGF-inducible element (PIE) at position -346 of the human *c-fos* promoter.[191-193] This cis-acting element is unique in the *fos*-promoter since occupancy actually changes during promoter activation whereas all other elements analyzed so far seem preoccupied in the inactive state also and regulation occurs through post-translational modification of the protein factors. A third class of external signals converges on a binding site for the cAMP-responsive element binding protein. This cis-acting element, the CRE located around position -60, accounts for part of the cAMP-inducibility of *c-fos*.[194-196] Other CRE-like elements and a non-CRE-like element in the intragenic region of *c-fos* may contribute to cAMP inducibility of the gene.[195-197] The CRE at position -60 has also been implicated in Ca-inducibility due to CREB phosphorylation by calmodulin-dependent kinases.[163-166,198] The CRE also enhances the transcription of the *c-fos* gene induced by stimuli acting through the SRE.[167-171] Basal activity of the *c-fos* promoter is controlled by additional cis-regulatory elements such as a NF-1-like binding site around position -170, a direct repeat region between position -90 and -100 binding (an) as yet unknown factor(s) and possibly additional elements positioned upstream of the hitherto carefully analyzed promoter region (350 bp upstream of the transcriptional start site) (for references see ref. 199). Transgenic mouse experiments have revealed another interesting facet of the *c-fos* promoter. Mutation of individual elements had larger effects than expected from their role in the signaling events described above. It appears that the promoter represents an assembly whose function severely suffers upon loss of individual parts.[200]

The shut-off of transcription of the immediate early response genes involves common mechanisms used by most of the IE genes such as the protein-synthesis requiring accumulation of a transcriptional repressor (for a review see ref. 199). The down-regulation of *c-fos* gene expression appears to involve a negative feed-back loop of c-Fos onto its own promoter,[201-203] using a specific function in the carboxyl-terminus of c-Fos but not the DNA-binding properties of the bZIP domain. Site-directed mutagenesis of the c-Fos carboxy-terminus and of sequences different between v-Fos and c-Fos [204-206] impair the down-regulation of *c-fos* expression after the initial peak. c-Fos mutants in the bZIP region which cannot activate an AP-1 dependent promoter are however still capable of repressing the *c-fos* promoter in trans.[206,207] Fos transrepression appears to be effected through the SRE.[202,203,207-209]

c-jun Promoter Regulation

Similarly to *c-fos*, the *c-jun* gene is also both a target of AP-1-activating kinase cascades and codes itself for a gene product which converts the signals propagated by kinase cascades into a transcriptional response (Fig. 5.3). The major known trans-acting factors on the c-jun promoter are AP-1, SP-1, NF-1[20] and NF-Jun.[210] The binding sites for SP-1 and NF-1/CTF are found around positions -115 and -90 of the human, mouse or rat c-jun promoter. The cis-acting elements which respond to AP-1 are located at positions - 70 and - 190.[20,211] These AP-1 binding sites carry two nucleotides spacing the palindromic sequences and thus resemble ATF binding sites. The sites are preferentially occupied by Jun/ATF-2 heterodimers.[69] The AP-1 binding sites are the major cis-regulatory elements which mediate the response of the *c-jun* promoter to serum, TPA or 'cellular stress-inducing' agents such as UV irradiation, DNA-damage or oxidative stress.[20,69,211-213] As derived from genomic footprint analysis however, the occupancy of the AP-1 sites seems not to be changed by any of these stimuli.[212,214] This suggests that low levels of preexisting

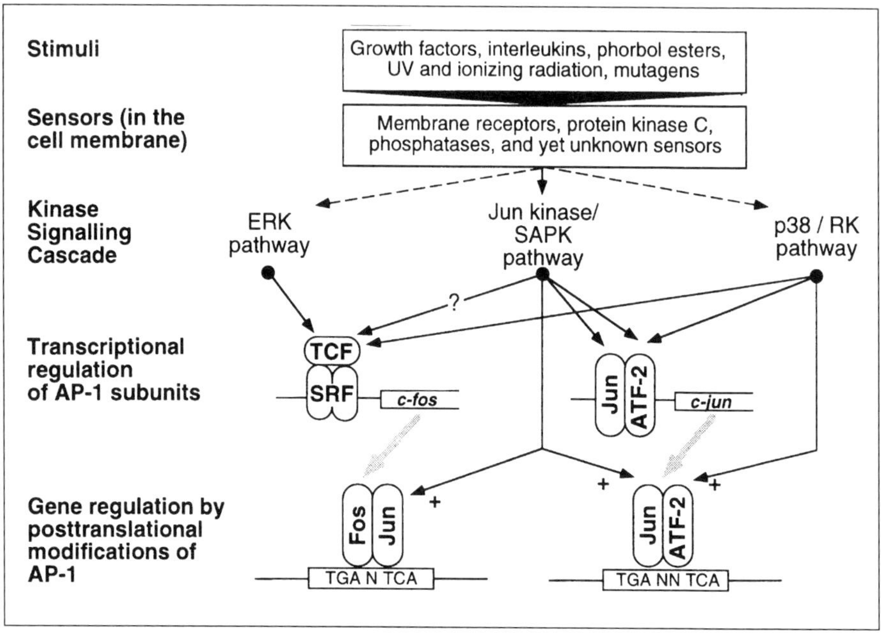

Fig. 5.3. Hierarchy and pathways of AP-1 regulation on the two levels of induced expression of the AP-1 subunits Fos and Jun as well as their posttranslational modifications. Gray arrows indicate the recruitment of newly synthesized proteins. Pathways, of which the relevance in the living cell is debatable, are marked by question marks. Some of the latter pathways are not shown in the figure for reasons of simplicity, but may be found in the main text. Dashed arrows show pathways which presumably involve intermediary signaling molecules.

Jun (and ATF-2) respond to signaling cascades by phosphorylation in the transactivation domains and thereby, in an autostimulatory loop, activate transcription of the *c-jun* gene. An additional element conferring responsiveness to growth factors around position -50 (overlapping a potential distal TATA-box) is related to the serum response element. Conditions can be found in which this element predominantly mediates the response of the *c-jun* promoter to growth factors such as EGF so that the proximal AP-1 site may be dispensable.[215]

In addition to the stimuli listed above, two other conditions trigger *c-jun* transcription: differentiation of F9 teratocarcinoma cells by retinoic acid and infection of cells with adenoviruses. On the side of the *c-jun* promoter, elements between -190 and -46 seem to be important together with putative contributions of further upstream sequences.[216-218] Since none of the elements of the *c-jun* promoter resembles a retinoic acid receptor binding site, the retinoic acid receptor seems not to act directly. In adenovirus-infected cells, the E1A protein and in particular the 12S E1A is responsible for *c-jun* transcription. The presence of both AP-1 sites is sufficient.[69,80]

In an example of cell-type specific regulation the *c-jun* promoter is activated by NF-Jun, a transcription factor which is specifically expressed in rapidly proliferating myeloid hematopoietic cells. NF-Jun resembles in several properties the transcription factor NF-κB and binds to a site between positions -141 and -131.[210] In summary, the promoter of *c-jun*, much like the *c-fos* promoter, integrates a number of signals related to cell growth. In contrast to the *c-fos* gene however, expression of the *c-jun* gene is subject to positive feed-back by the c-Jun protein.

TRANSCRIPTIONAL CROSS-TALK: REGULATION OF AP-1 ACTIVITY THROUGH OTHER TRANSCRIPTION FACTOR FAMILIES

In a cell the action of one given transcription factor family does not occur independently of others. Several transcription factors synergize or interfere with each other's function if a target promoter contains adjacent or overlapping binding sites for different factors. The clustering of AP-1 binding sites with binding sites for Ets proteins is found in many promoters, including the polyoma virus enhancer and numerous mammalian genes[219-222] (and references therein). The paradigm of the so-called transcriptional cross-talk, however, is the impact of nuclear receptors of the steroid hormone receptor superfamily on the AP-1 transcription factor (reviewed in refs. 223, 224). The predominant mechanism of cross-talk appears to involve protein-protein interactions. Interactions between different transcription factors apparently occur either when both transcription factors bind to DNA (e.g., on a composite response element as postulated in refs. 225-229) or when only one transcription factor directly acts on a promoter element.[230-233] Such protein-protein interactions may determine accessibility of the transcription factor complex for additional proteins by allosterically modulating each other's conformation. Interestingly, the nature of the AP-1 dimer determines the outcome of the interaction. As first found using the composite responsive element of the proliferin gene, the interaction can result in synergism (Jun homodimer and glucocorticoid receptor) or negative interference (Jun/Fos heterodimers and glucocorticoid receptor).[226] The authors postulate a specific role for the AP-1 subunit composition in the case of composite DNA-binding elements. Nevertheless, synergism vs. negative interference has also been observed on classical AP-1 sites.[81] This may, however, depend on cell-type specific additional factors.[234]

A strong case for protein-protein interaction as the key mechanism of cross-talk is generated by the finding of mutual interference. Steroid hormone receptors, such as the glucocorticoid receptor (GR) or the retinoic acid receptor, repress AP-1-dependent promoter activity without the presence of a DNA-binding site for the nuclear receptor and overexpression of AP-1 components represses the function of steroid hormone receptors in the absence of an AP-1 site.[230-234] It is not yet clear whether nuclear receptors directly interact with AP-1 and whether such an interaction would be sufficient for cross-talk. In vitro studies and mutational analysis in transfected cells have suggested that the zinc finger domain of the nuclear receptor and the bZIP DNA-binding domain of AP-1 participate in the negative interference, possibly by establishing contact between these proteins (reviewed in ref. 223). One in vivo footprint analysis indicates that the inhibited factor does not need to be kicked off its element.[235] This would be consistent with the idea that protein conformation and availability for other factors are modulated, while the transcription factor subject to cross-talk remains bound to the DNA target element. Mutations of the glucocorticoid receptor have been described which either retain only the transrepressing but not the transactivating activity of GR or specifically impair the transrepressing function of the steroid receptor without affecting transactivation.[236-238] Since some of the nonrepressing mutants still do interact with AP-1 in vitro, solely the formation of an AP-1 complex with a nuclear receptor does not appear to inevitably create a transcriptionally inactive complex (if one accepts that the protein interactions detected in vitro reflect the situation in vivo).[239] Rather, additional features, e.g., the structure of GR altered by the mutation, the interaction with coactivators or corepressors etc., may determine the transcriptional activity of the DNA-bound AP-1 together with the steroid hormone receptor.

The concept that a specific conformation of the nuclear hormone receptor is required for interference with AP-1 is further supported by the action of certain 'dissociating' receptor ligands which specifically and differentially affect either positive functions of the receptors or negative transcriptional cross-talk.[240,241,241a]

Transcription factor cross-talk is not limited to nuclear receptors and AP-1 but also occurs between nuclear hormone receptors and CREB, NF-κB or the Oct transcription factors (negative interference, reviewed in ref. 224), between AP-1 and NF-κB transcription factors (synergism)[242] and between AP-1 and MyoD (negative interference)[243-245] Physical interactions between these transcription factors have been documented in the cases of NF-κB and AP-1 as well as NF-κB and GR[242,246] and c-Jun and Myo D,[243] but it is not clear whether the interaction is solely responsible for the cross-talk. A currently debated alternative hypothesis proposes competition for common partners, e.g., CBP (as discussed earlier), as the mechanism of cross-talk.[132-134] Discussing the pro and contra arguments for such a mechanism, however, lies beyond the scope of this review.

INACTIVATION OF AP-1

A last step of AP-1 regulation to be discussed concerns the stability of the proteins. Both, c-Jun and c-Fos are proteins with a short half-life. Since increased activity of these proteins, for example in their oncogenic versions, leads to deregulated cell proliferation and cell transformation, appropriate inactivation of AP-1 proteins is essential. There is no evidence for regulated dephosphorylation. Rather, the AP-1 subunits seem to be removed by degradation. Proteins with a short half-life are often degraded by the ubiquitin/proteasome system,[247-249] and in fact, c-Jun is multi-ubiquitinated in vivo.[29] The δ-domain, which had been shown by previous studies to be the target of a cell specific repressor of c-Jun[26,27] (in addition to binding the JNKs as described above), is essential for c-Jun ubiquitination.[29] The δ-domain is not present in v-Jun and thus the oncogenic copy might escape ordered degradation and thereby, at least in part, acquire its transforming properties.[29] However, ubiquitination does not account for all the differences between c-Jun and v-Jun and not even for all the activities of the δ-domain, since the previously described titratable inhibitor of c-Jun activity requires the δ-domain but is apparently different from the ubiquitin-dependent degradation pathway.[26,27,29] The c-Jun ubiquitinating enzymes have not yet been clearly defined, but one of the putative E2 ubiquitin-conjugating enzymes, Ubc 9, interacts with c-Jun in a yeast two-hybrid screen.[239] The interaction does not require the δ-domain in c-Jun but occurs through the carboxyl-terminal part of the protein, which contains the second motif required for efficient c-Jun ubiquitination in vivo.[29] Other members of the ubiquitination machinery have not yet been defined.

Similarly to c-Jun, c-Fos, but not v-Fos, undergoes targeted degradation.[28] Direct ubiquitination in vitro and dependence of rapid c-Fos degradation on an intact ubiquitination system have been shown.[250] For c-Fos the ubiquitinating enzymes have also been identified, e.g., the UbcH5 E2-enzyme and a novel E3 enzyme.[250] c-Fos degradation adds an interesting twist to targeted protein degradation since a phosphorylation-dependent signal on its partner protein c-Jun[28] and the presence of c-Jun enhance ubiquitination in a cell-free system.[250] These observations give rise to the speculation that it is the activated AP-1 complex rather than the individual subunit which is subject to targeted degradation. Thus, degradation may preferentially eliminate the active forms of AP-1.

MUTATIONAL ANALYSIS IN VIVO

Biochemical and genetic work over the last decade has yielded a detailed understanding of AP-1 function, its modes of activation and inactivation, and how target genes are regulated. Despite this wealth of information, many questions have remained open, particularly with respect to the way that AP-1 acts in cell growth, differentiation and cell death. The emerging complexity of the AP-1 protein family has rather added to the puzzles than fulfilling the possibly premature expectations of simple answers. The unsolved problems are currently addressed by two types of complementary approaches. The knowledge on the structure and the dimerization properties of the bZIP DNA-binding domains allows creation of mutants of the AP-1 proteins with restricted dimerization properties (see above). Such mutants, when tested in cultured cells or in the animal, may help to assign defined AP-1 heterodimeric complexes to certain biological functions (van Dam et al, in preparation).

A second powerful tool to assess the function of different AP-1 subunits in the whole organism is provided by the analysis of mice carrying targeted mutations of the genes coding for AP-1 proteins. Although AP-1 proteins are expressed in many types of cells and tissues, the phenotypical consequences of disruption of the *c-fos* or *c-jun* gene by gene targeting in mice are confined to surprisingly few tissues. Thus, inactivation of *c-fos* primarily affects the osteoclast-macrophage cell lineage. The impaired function of bone-resorbing osteoclasts leads to osteopetrosis, reduction in bone marrow and presumably also to the observed defect in tooth development.[251-254] In addition to a reduced bone marrow space, other defects in the microenvironment appear to contribute to immune defects in *c-fos*$^{0/0}$ mice.[255] Furthermore all cells of the organism presumably suffer from the inability to rapidly increase AP-1 synthesis upon extracellular stimuli. For instance, fibroblasts from *c-fos*$^{0/0}$ mice are more sensitive to UV irradiation. This suggests that an UV-induced and c-Fos-dependent program serves stress-protective functions and is obviously defective in all cells despite the survival of mice under laboratory conditions.[13,256]

The inactivation of *c-jun* has much more dramatic consequences: the disruption of the gene is lethal during embryonic development but the lack of c-Jun still appears to affect only a limited set of tissues. The predominant phenotype is characterized by impaired development of the liver and by impaired erythropoiesis.[5,257] A limiting function of c-Jun in liver development is further supported by the finding that *c-jun*$^{0/0}$ cells may contribute to all somatic tissues in chimeric mice except for liver.[257] The knock-out of *fosB* leads to a behavioral phenotype, in that mice deleted homozygously for *FosB* do not nurture their offsprings.[258] Otherwise, embryonic development and adult physiology do not appear affected except for a slight decrease in body weight of the *fosB*$^{0/0}$ mice.

The knock-out of other genes coding for AP-1 subunits is currently in progress and at least one of theses genes, e.g., *junB*, appears essential for early embryonic survival (Schorpp-Kistner et al, manuscript in preparation).

In addition to the role in embryogenesis, AP-1 is part of numerous physiological and pathological processes and the contributions of individual AP-1 members may be deduced from knock-outs or from overexpressing the proteins in transgenic mice. c-Fos for example serves an important role in malignant progression of skin cancer[259] and it enhances, when ectopically expressed, the formation of chondrogenic tumors and osteosarcomas.[260-262]

Thus, any AP-1 family member which has been genetically inactivated so far is essential for a defined set of tissues or (patho)physiological processes. They serve functions which apparently can not be replaced by other family members while in other respects redundancy is prominent. Similarly all other family members will probably serve nonreplacable and distinct functions. There may also be tissues in which the inactivation of one AP-1 member, e.g., CREB, may be compensated by the upregulation of others (ref. 263 and references therein). Thus, despite a reasonably simple pattern of heterodimeric action of a transcription factor, the combinatorial possibilities demand further analyses, particularly in the organism. How a simple dimerizing factor can serve such key roles from embryogenesis to cell death, and from inflammation to carcinogenesis is not the only matter for future research. The possibilities of complex signaling network interactions which determine AP-1 activation have also been largely ignored.

Acknowledgments

The work was supported by the Deutsche Forschungsgemeinschaft (Go 473/2 to M.G. and Ra 247/6-1 to H.J.R.), the Dr. Mildred Scheel-Stiftung (W86/94 Ra I to H.J.R.) and by Fonds der Chemischen Industrie (to P.H.).

References

1. Bravo R, Burckhardt J, Curran T et al. Expression of c-fos in NIH3T3 cells is very low but inducible throughout the cell cycle. Embo J 1986; 5:695-700.
2. Angel P, Karin M. The role of Jun, Fos and the AP-1 complex in cell-proliferation and transformation. Biochim Biophys Acta 1991; 1072:129-157.
3. Kovary K, Bravo R. The jun and fos protein families are both required for cell cycle progression in fibroblasts. Mol Cell Biol 1991; 11:4466-4472.
4. Hilberg F, Wagner EF. Embryonic stem (ES) cells lacking functional c-jun: consequences for growth and differentiation, AP-1 activity and tumorigenicity. Oncogene 1992; 7:2371-2380.
5. Johnson RS, van Lingen B, Papaioannou VE et al. A null mutation at the c-jun locus causes embryonic lethality and retarded cell growth in culture. Genes Dev 1993; 7:1309-1317.
6. Sassone-Corsi P. Goals for signal transduction pathways: linking up with transcriptional regulation. Embo J 1994; 13:4717-4728.
7. Estus S, Zaks WJ, Freeman RS et al. Altered gene expression in neurons during programmed cell death: identification of c-jun as necessary for neuronal apoptosis. J Cell Biol 1994; 127: 1717-1727.
8. Pfarr CM, Mechta F, Spyrou G et al. Mouse JunD negatively regulates fibroblast growth and antagonizes transformation by ras. Cell 1994; 76:747-760.
9. Ham J, Babij C, Whitfield J et al. A c-Jun dominant negative mutant protects sympathetic neurons against programmed cell death. Neuron 1995; 14: 927-939.
10. Dévary Y, Gottlieb RA, Lau LF et al. Rapid and preferential activation of the c-jun gene during the mammalian UV response. Mol Cell Biol 1991; 11: 2804-2811.
11. Sachsenmaier C, Radler-Pohl A, Zinck R et al. Involvement of growth factor receptors in the mammalian UVC response. Cell 1994; 78:963-972.
12. Bandyopadhyay RS, Phelan M, Faller DV. Hypoxia induces AP-1-regulated genes and AP-1 transcription factor binding in human endothelial and other cell types. Biochim Biophys Acta 1995; 1264:72-78.
13. Schreiber M, Baumann B, Cotten M et al. Fos is an essential component of the mammalian UV response. Embo J 1995; 14:5338-5349.
14. Finkel MP, Biskis BO, Jinkins PB. Virus induction of osteosarcomas in mice. Science 1966; 151:698-701.
15. Maki Y, Bos TJ, Davis C et al. Avian sarcoma virus 17 carries the jun oncogene. Proc Natl Acad Sci USA 1987; 84: 2848-2852.
16. Bos TJ, Bohmann D, Tsuchie H et al. v-jun encodes a nuclear protein with enhancer binding properties of AP-1. Cell 1988; 52:705-712.
17. van Straaten F, Müller R, Curran T et al. Complete nucleotide sequence of a human c-onc gene: deduced amino acid sequence of the human c-fos protein. Proc Natl Acad Sci USA 1983; 80: 3183-3187.
18. Van Beveren C, van Straaten F, Curran T et al. Analysis of FBJ-MuSV provirus and c-fos (mouse) gene reveals that viral and cellular fos gene products have different carboxy termini. Cell 1983; 32:1241-1255.
19. Bohmann D, Bos TJ, Admon A et al. Human proto-oncogene c-jun encodes a DNA binding protein with structural and functional properties of transcription factor AP-1. Science 1987; 238:1386-1392.
20. Angel P, Allegretto EA, Okino ST et al. Oncogene jun encodes a sequence-specific trans-activator similar to AP-1. Nature 1988; 332:166-171.
21. Curran T, Van Beveren C, Verma IM. Viral and cellular fos proteins are complexed with a 39,000-dalton cellular protein. Mol Cell Biol 1985; 5:167-172.
22. Chiu R, Boyle WJ, Meek J et al. The c-Fos protein interacts with c-Jun/AP-1 to stimulate transcription of AP-1 responsive genes. Cell 1988; 54:541-552.

23. Rauscher FJ, Voulalas PJ, Franza BR Jr et al. Fos and Jun bind cooperatively to the AP-1 site: reconstitution in vitro. Genes Dev 1988; 2:1687-1699.
24. Sassone-Corsi P, Lamph WW, Kamps M et al. fos-associated cellular p39 is related to nuclear transcription factor AP-1. Cell 1988; 54:553-560.
25. Angel P, Smeal T, Meek J et al. Jun and v-jun contain multiple regions that participate in transcriptional activation in an interdependent manner. New Biol 1989; 1:35-43.
26. Bohmann D, Tjian R. Biochemical analysis of transcriptional activation by Jun: differential activity of c- and v-Jun. Cell 1989; 59:709-717.
27. Baichwal VR, Tjian R. Control of c-Jun activity by interaction of a cell-specific inhibitor with regulatory domain delta: differences between v- and c-Jun. Cell 1990; 63:815-825.
28. Papavassiliou AG, Treier M, Chavrier C et al. Targeted degradation of c-Fos, but not v-Fos, by a phosphorylation-dependent signal on c-Jun. Science 1992; 258:1941-1944.
29. Treier M, Staszewski LM, Bohmann D. Ubiquitin-dependent c-Jun degradation in vivo is mediated by the delta domain. Cell 1994; 78:787-798.
30. Roux P, Blanchard JM, Fernandez A et al. Nuclear localization of c-Fos, but not v-Fos proteins, is controlled by extracellular signals. Cell 1990; 63:341-351.
31. Müller R, Bravo R, Burckhardt J et al. Induction of c-fos gene and protein by growth factors precedes activation of c-myc. Nature 1984; 312:716-720.
32. Greenberg ME, Ziff EB. Stimulation of 3T3 cells induces transcription of the c-fos proto-oncogene. Nature 1984; 311: 433-438.
33. Kruijer W, Cooper JA, Hunter T et al. Platelet-derived growth factor induces rapid but transient expression of the c-fos gene and protein. Nature 1984; 312: 711-716.
34. Kruijer W, Schubert D, Verma IM. Induction of the proto-oncogene fos by nerve growth factor. Proc Natl Acad Sci USA 1985; 82:7330-7334.
35. Curran T, Bravo R, Müller R. Transient induction of c-fos and c-myc is an immediate consequence of growth factor stimulation. Cancer Surv 1985; 4: 655-681.
36. Karin M. The regulation of AP-1 activity by mitogen-activated protein kinases. J Biol Chem 1995; 270:16483-16486.
37. Rahmsdorf HJ. Modulation of Fos and Jun in response to adverse environmental agents. In: P. Angel and P. Herrlich, eds. The Fos and Jun families of transcription factors. Boca Raton: CRC-press, 1994:169-189
38. Albanese C, Johnson J, Watanabe G et al. Transforming p21ras mutants and c-Ets-2 activate the cyclin D1 promoter through distinguishable regions. J Biol Chem 1995; 270:23589-23597.
39. Ransone LJ, Visvader J, Wamsley P et al. Trans-dominant negative mutants of Fos and Jun. Proc Natl Acad Sci USA 1990; 87:3806-3810.
40. Field SJ, Johnson RS, Mortensen RM et al. Growth and differentiation of embryonic stem cells that lack an intact c-fos gene. Proc Natl Acad Sci USA 1992; 89:9306-9310.
41. Brusselbach S, Mohle-Steinlein U, Wang ZQ et al. Cell proliferation and cell cycle progression are not impaired in fibroblasts and ES cells lacking c-Fos. Oncogene 1995; 10:79-86.
42. Angel P, Imagawa M, Chiu R et al. Phorbol ester-inducible genes contain a common cis element recognized by a TPA-modulated trans-acting factor. Cell 1987; 49:729-739.
43. Lee W, Haslinger A, Karin M et al. Activation of transcription by two factors that bind promoter and enhancer sequences of the human metallothionein gene and SV40. Nature 1987; 325:368-372.
44. Lee W, Mitchell P, Tjian R. Purified transcription factor AP-1 interacts with TPA-inducible enhancer elements. Cell 1987; 49:741-752.
45. Halazonetis TD, Georgopoulos K, Greenberg ME et al. c-Jun dimerizes with itself and with c-Fos, forming complexes of different DNA binding affinities. Cell 1988; 55:917-924.

46. Smeal T, Angel P, Meek J et al. Different requirements for formation of Jun: Jun and Jun: Fos complexes. Genes Dev 1989; 3:2091-2100.
47. Ransone LJ, Wamsley P, Morley KL et al. Domain swapping reveals the modular nature of Fos, Jun, and CREB proteins. Mol Cell Biol 1990; 10:4565-4573.
48. Zerial M, Toschi L, Ryseck RP et al. The product of a novel growth factor activated gene, fos B, interacts with JUN proteins enhancing their DNA binding activity. Embo J 1989; 8:805-813.
49. Wisdom R, Yen J, Rashid D et al. Transformation by FosB requires a trans-activation domain missing in FosB2 that can be substituted by heterologous activation domains. Genes Dev 1992; 6:667-675.
50. Franza BR, Jr., Sambucetti LC, Cohen DR et al. Analysis of Fos protein complexes and Fos-related antigens by high-resolution two-dimensional gel electrophoresis. Oncogene 1987; 1:213-221.
51. Cohen DR, Curran T. fra-1: a serum-inducible, cellular immediate-early gene that encodes a fos-related antigen. Mol Cell Biol 1988; 8:2063-2069.
52. Cohen DR, Ferreira PC, Gentz R et al. The product of a fos-related gene, fra-1, binds cooperatively to the AP-1 site with Jun: transcription factor AP-1 is comprised of multiple protein complexes. Genes Dev 1989; 3:173-184.
53. Matsui M, Tokuhara M, Konuma Y et al. Isolation of human fos-related genes and their expression during monocyte-macrophage differentiation. Oncogene 1990; 5:249-255.
54. Nishina H, Sato H, Suzuki T et al. Isolation and characterization of fra-2, an additional member of the fos gene family. Proc Natl Acad Sci USA 1990; 87: 3619-3623.
55. Ryder K, Lau LF, Nathans D. A gene activated by growth factors is related to the oncogene v-jun. Proc Natl Acad Sci USA 1988; 85:1487-1491.
56. Ryder K, Lanahan A, Perez-Albuerne E et al. jun-D: a third member of the jun gene family. Proc Natl Acad Sci USA 1989; 86:1500-1503.
57. Hirai SI, Ryseck RP, Mechta F et al. Characterization of junD: a new member of the jun proto-oncogene family. Embo J 1989; 8:1433-1439.
58. Hoeffler JP, Meyer TE, Yun Y et al. Cyclic AMP-responsive DNA-binding protein: structure based on a cloned placental cDNA. Science 1988; 242:1430-1433.
59. Hai TW, Liu F, Coukos WJ et al. Transcription factor ATF cDNA clones: an extensive family of leucine zipper proteins able to selectively form DNA-binding heterodimers. Genes Dev 1989; 3: 2083-2090.
60. Foulkes NS, Borrelli E, Sassone-Corsi P. CREM gene: use of alternative DNA-binding domains generates multiple antagonists of cAMP-induced transcription. Cell 1991; 64:739-749.
61. Rehfuss RP, Walton KM, Loriaux MM et al. The cAMP-regulated enhancer-binding protein ATF-1 activates transcription in response to cAMP-dependent protein kinase A. J Biol Chem 1991; 266: 18431-18434.
62. Karpinski BA, Morle GD, Huggenvik J et al. Molecular cloning of human CREB-2: an ATF/CREB transcription factor that can negatively regulate transcription from the cAMP response element. Proc Natl Acad Sci USA 1992; 89:4820-4824.
63. Inomata Y, Kawaguchi H, Hiramoto M et al. Direct purification of multiple ATF/E4TF3 polypeptides from HeLa cell crude nuclear extracts using DNA affinity latex particles. Anal Biochem 1992; 206: 109-114.
64. Chen BP, Liang G, Whelan J et al. ATF3 and ATF3 delta Zip. Transcriptional repression versus activation by alternatively spliced isoforms. J Biol Chem 1994; 269:15819-15826.
65. Dorsey MJ, Tae HJ, Sollenberger KG et al. B-ATF: a novel human bZIP protein that associates with members of the AP-1 transcription factor family. Oncogene 1995; 11:2255-2265.
66. Benbrook DM, Jones NC. Heterodimer formation between CREB and JUN proteins. Oncogene 1990; 5:295-302.
67. Ivashkiv LB, Liou HC, Kara CJ et al. mXBP/CRE-BP2 and c-Jun form a complex which binds to the cyclic AMP, but not to the 12-O-tetradecanoylphorbol-13-acetate, response element. Mol Cell Biol 1990; 10:1609-1621.

68. Hai T, Curran T. Cross-family dimerization of transcription factors Fos/Jun and ATF/CREB alters DNA binding specificity. Proc Natl Acad Sci USA 1991; 88: 3720-3724.
69. van Dam H, Duyndam M, Rottier R et al. Heterodimer formation of cJun and ATF-2 is responsible for induction of c-jun by the 243 amino acid adenovirus E1A protein. Embo J 1993; 12:479-487.
70. Hadman M, Loo M, Bos TJ. In vivo viral and cellular Jun complexes exhibit differential interaction with a number of in vitro generated 'AP-1- and CREB-like' target sequences. Oncogene 1993; 8: 1895-1903.
71. Kim J, Struhl K. Determinants of half-site spacing preferences that distinguish AP-1 and ATF/CREB bZIP domains. Nucleic Acids Res 1995; 23:2531-2537.
72. Dony C, Gruss P. Proto-oncogene c-fos expression in growth regions of fetal bone and mesodermal web tissue. Nature 1987; 328:711-714.
73. Redemann-Fibi B, Schuermann M, Müller R. Stage and tissue-specific expression of fosB during mouse development. Differentiation 1991; 46:43-49.
74. Kouzarides T, Ziff E. The role of the leucine zipper in the fos-jun interaction. Nature 1988; 336:646-651.
75. Landschulz WH, Johnson PF, McKnight SL. The leucine zipper: a hypothetical structure common to a new class of DNA binding proteins. Science 1988; 240: 1759-1764.
76. Sassone-Corsi P, Ransone LJ, Lamph WW et al. Direct interaction between fos and jun nuclear oncoproteins: role of the 'leucine zipper' domain. Nature 1988; 336:692-695.
77. Kouzarides T, Ziff E. Leucine zippers of fos, jun and GCN4 dictate dimerization specificity and thereby control DNA binding. Nature 1989; 340:568-571.
78. Schuermann M, Neuberg M, Hunter JB et al. The leucine repeat motif in Fos protein mediates complex formation with Jun/AP-1 and is required for transformation. Cell 1989; 56:507-516.
79. Talanian RV, McKnight CJ, Kim PS. Sequence-specific DNA binding by a short peptide dimer. Science 1990; 249: 769-771.
80. Hagmeyer BM, König H, Herr I et al. Adenovirus E1A negatively and positively modulates transcription of AP-1 dependent genes by dimer-specific regulation of the DNA binding and transactivation activities of Jun. Embo J 1993; 12: 3559-3572.
81. Teurich S, Angel P. The glucocorticoid receptor synergizes with Jun homodimers to activate AP-1-regulated promoters lacking GR binding sites. Chem Senses 1995; 20:251-255.
82. Nakabeppu Y, Nathans D. The basic region of Fos mediates specific DNA binding. Embo J 1989; 8:3833-3841.
83. Cohen DR, Curran T. Analysis of dimerization and DNA binding functions in Fos and Jun by domain-swapping: involvement of residues outside the leucine zipper/basic region. Oncogene 1990; 5:929-939.
84. Gentz R, Rauscher FJ, Abate C et al. Parallel association of Fos and Jun leucine zippers juxtaposes DNA binding domains. Science 1989; 243:1695-1699.
85. O'Shea EK, Rutkowski R, Kim PS. Evidence that the leucine zipper is a coiled coil. Science 1989; 243:538-542.
86. Turner R, Tjian R. Leucine repeats and an adjacent DNA binding domain mediate the formation of functional cFos-cJun heterodimers. Science 1989; 243: 1689-1694.
87. O'Shea EK, Klemm JD, Kim PS et al. X-ray structure of the GCN4 leucine zipper, a two-stranded, parallel coiled coil. Science 1991; 254:539-544.
88. Rasmussen R, Benvegnu D, O'Shea EK et al. X-ray scattering indicates that the leucine zipper is a coiled coil. Proc Natl Acad Sci USA 1991; 88:561-564.
89. Schuermann M, Hunter JB, Hennig G et al. Non-leucine residues in the leucine repeats of Fos and Jun contribute to the stability and determine the specificity of dimerization. Nucleic Acids Res 1991; 19:739-746.
90. O'Shea EK, Rutkowski R, Kim PS. Mechanism of specificity in the Fos-Jun oncoprotein heterodimer. Cell 1992; 68:699-708.
91. O'Shea EK, Rutkowski R, Stafford WF et al. Preferential heterodimer formation by isolated leucine zippers from fos and jun. Science 1989; 245:646-648.

92. Neuberg M, Adamkiewicz J, Hunter JB et al. A Fos protein containing the Jun leucine zipper forms a homodimer which binds to the AP1 binding site. Nature 1989; 341:243-245.
93. Sellers JW, Struhl K. Changing fos oncoprotein to a jun-independent DNA binding protein with GCN4 dimerization specificity by swapping "leucine zippers". Nature 1989; 341:74-76.
94. Porte D, Oertel-Buchheit P, Granger-Schnarr M et al. Fos leucine zipper variants with increased association capacity. J Biol Chem 1995; 270:22721-22730.
95. Glover JN, Harrison SC. Crystal structure of the heterodimeric bZIP transcription factor c-Fos-c-Jun bound to DNA. Nature 1995; 373:257-261.
96. Frame MC, Simpson K, Fincham VJ et al. Separation of v-Src-induced mitogenesis and morphological transformation by inhibition of AP-1. Mol Biol Cell 1994; 5:1177-1184.
97. Patel L, Abate C, Curran T. Altered protein conformation on DNA binding by Fos and Jun. Nature 1990; 347:572-575.
98. Weiss MA, Ellenberger T, Wobbe CR et al. Folding transition in the DNA-binding domain of GCN4 on specific binding to DNA. Nature 1990; 347:575-578.
99. Abate C, Patel L, Rauscher FJ et al. Redox regulation of fos and jun DNA-binding activity in vitro. Science 1990; 249: 1157-1161.
100. Xanthoudakis S, Miao G, Wang F et al. Redox activation of Fos-Jun DNA binding activity is mediated by a DNA repair enzyme. Embo J 1992; 11:3323-3335.
101. Handel ML, Watts CK, de Fazio A et al. Inhibition of AP-1 binding and transcription by gold and selenium involving conserved cysteine residues in Jun and Fos. Proc Natl Acad Sci USA 1995; 92:4497-4501.
102. Lin A, Frost J, Deng T et al. Casein kinase II is a negative regulator of c-Jun DNA binding and AP-1 activity. Cell 1992; 70:777-789.
103. de Groot RP, Auwerx J, Bourouis M et al. Negative regulation of Jun/AP-1: conserved function of glycogen synthase kinase 3 and the *Drosophila* kinase shaggy. Oncogene 1993; 8:841-847.
104. Nikolakaki E, Coffer PJ, Hemelsoet R et al. Glycogen synthase kinase 3 phosphorylates Jun family members in vitro and negatively regulates their transactivating potential in intact cells. Oncogene 1993; 8:833-840.
105. Oehler T, Pintzas A, Stumm S et al. Mutation of a phosphorylation site in the DNA-binding domain is required for redox-independent transactivation of AP1-dependent genes by v-Jun. Oncogene 1993; 8:1141-1147.
106. Boyle WJ, Smeal T, Defize LH et al. Activation of protein kinase C decreases phosphorylation of c-Jun at sites that negatively regulate its DNA-binding activity. Cell 1991; 64:573-584.
107. Radler-Pohl A, Sachsenmaier C, Gebel S et al. UV-induced activation of AP-1 involves obligatory extranuclear steps including Raf-1 kinase. Embo J 1993; 12:1005-1012.
108. Papavassiliou AG, Treier M, Bohmann D. Intramolecular signal transduction in c-Jun. Embo J 1995; 14:2014-2019.
109. Chou SY, Baichwal V, Ferrell JE Jr. Inhibition of c-Jun DNA binding by mitogen-activated protein kinase. Mol Biol Cell 1992; 3:1117-1130.
110. Franklin CC, Sanchez V, Wagner F et al. Phorbol ester-induced amino-terminal phosphorylation of human JUN but not JUNB regulates transcriptional activation. Proc Natl Acad Sci USA 1992; 89: 7247-7251.
111. Sutherland JA, Cook A, Bannister AJ et al. Conserved motifs in Fos and Jun define a new class of activation domain. Genes Dev 1992; 6:1810-1819.
112. Pulverer BJ, Kyriakis JM, Avruch J et al. Phosphorylation of c-jun mediated by MAP kinases. Nature 1991; 353:670-674.
113. Gill G, Ptashne M. Negative effect of the transcriptional activator GAL4. Nature 1988; 334:721-724.
114. Oehler T, Angel P. A common intermediary factor (p52/54) recognizing "acidic blob"-type domains is required for transcriptional activation by the Jun proteins. Mol Cell Biol 1992; 12:5508-5515.
115. Abate C, Luk D, Curran T. Transcriptional regulation by Fos and Jun in vitro: interaction among multiple activator and regulatory domains. Mol Cell Biol 1991; 11:3624-3632.

116. Lucibello FC, Neuberg M, Jenuwein T et al. Multiple regions of v-Fos protein involved in the activation of AP1-dependent transcription: is trans-activation crucial for transformation? New Biol 1991; 3:671-677.
117. Chiu R, Angel P, Karin M. Jun-B differs in its biological properties from, and is a negative regulator of, c-Jun. Cell 1989; 59:979-986.
118. Schütte J, Viallet J, Nau M et al. jun-B inhibits and c-fos stimulates the transforming and trans-activating activities of c-jun. Cell 1989; 59:987-997.
119. Suzuki T, Okuno H, Yoshida T et al. Difference in transcriptional regulatory function between c-Fos and Fra-2. Nucleic Acids Res 1991; 19:5537-5542.
120. Deng T, Karin M. JunB differs from c-Jun in its DNA-binding and dimerization domains, and represses c-Jun by formation of inactive heterodimers. Genes Dev 1993; 7:479-490.
121. Yoshioka K, Deng T, Cavigelli M et al. Antitumor promotion by phenolic antioxidants: inhibition of AP-1 activity through induction of Fra expression. Proc Natl Acad Sci USA 1995; 92: 4972-4976.
122. Castellazzi M, Spyrou G, La Vista N et al. Overexpression of c-jun, junB, or junD affects cell growth differently. Proc Natl Acad Sci USA 1991; 88:8890-8894.
123. Metivier C, Piu F, Pfarr CM et al. In vitro transforming capacities of mouse c-jun:junD chimeric genes. Oncogene 1993; 8:2311-2315.
124. Livingstone C, Patel G, Jones N. ATF-2 contains a phosphorylation-dependent transcriptional activation domain. Embo J 1995; 14:1785-1797.
125. Lee KA, Masson N. Transcriptional regulation by CREB and its relatives. Biochim Biophys Acta 1993; 1174:221-233.
126. Parker D, Ferreri K, Nakajima T et al. Phosphorylation of CREB at Ser-133 induces complex formation with CREB-binding protein via a direct mechanism. Mol Cell Biol 1996; 16:694-703.
127. Chrivia JC, Kwok RP, Lamb N et al. Phosphorylated CREB binds specifically to the nuclear protein CBP. Nature 1993; 365:855-859.
128. Eckner R, Ewen ME, Newsome D et al. Molecular cloning and functional analysis of the adenovirus E1A-associated 300-kD protein (p300) reveals a protein with properties of a transcriptional adaptor. Genes Dev 1994; 8:869-884.
129. Arias J, Alberts AS, Brindle P et al. Activation of cAMP and mitogen responsive genes relies on a common nuclear factor. Nature 1994; 370:226-229.
130. Bannister AJ, Kouzarides T. CBP-induced stimulation of c-Fos activity is abrogated by E1A. Embo J 1995; 14:4758-4762.
131. Bannister AJ, Oehler T, Wilhelm D et al. Stimulation of c-Jun activity by CBP: c-Jun residues Ser63/73 are required for CBP induced stimulation in vivo and CBP binding in vitro. Oncogene 1995; 11:2509-2514.
132. Kamei Y, Xu L, Heinzel T et al. A CBP integrator complex mediates transcriptional activation and AP-1 inhibition by nuclear receptors. Cell 1996; 85:403-414.
133. Chakravarti D, LaMorte VJ, Nelson MC et al. Role of CBP/p300 in nuclear receptor signalling. Nature 1996; 383: 99-103.
134. Hanstein B, Eckner R, DiRenzo J et al. p300 is a component of an estrogen receptor coactivator complex. Proc Natl Acad Sci USA 1996; 93:11540-11545.
135. Smeal T, Binétruy B, Mercola D et al. Oncoprotein-mediated signalling cascade stimulates c-Jun activity by phosphorylation of serines 63 and 73. Mol Cell Biol 1992; 12:3507-3513.
136. Dévary Y, Gottlieb RA, Smeal T et al. The mammalian ultraviolet response is triggered by activation of Src tyrosine kinases. Cell 1992; 71:1081-1091.
137. Gupta S, Campbell D, Dérijard B et al. Transcription factor ATF2 regulation by the JNK signal transduction pathway. Science 1995; 267:389-393.
138. van Dam H, Wilhelm D, Herr I et al. ATF-2 is preferentially activated by stress-activated protein kinases to mediate c-jun induction in response to genotoxic agents. Embo J 1995; 14: 1798-1811.
139. Gupta S, Barrett T, Whitmarsh AJ et al. Selective interaction of JNK protein kinase isoforms with transcription factors. Embo J 1996; 15:2760-2770.

140. Cano E, Mahadevan LC. Parallel signal processing among mammalian MAPKs. Trends Biochem Sci 1995; 20:117-122.
141. Treisman R. Regulation of transcription by MAP kinase cascades. Curr Opinion Cell Biol 1996; 8:205-215.
142. Schlessinger J. How receptor tyrosine kinases activate Ras. Trends Biochem Sci 1993; 18:273-275.
143. Symons M. Rho family GTPases: the cytoskeleton and beyond. Trends Biochem Sci 1996; 21:178-181.
144. Hibi M, Lin A, Smeal T et al. Identification of an oncoprotein- and UV-responsive protein kinase that binds and potentiates the c-Jun activation domain. Genes Dev 1993; 7:2135-2148.
145. Raingeaud J, Gupta S, Rogers JS et al. Pro-inflammatory cytokines and environmental stress cause p38 mitogen-activated protein kinase activation by dual phosphorylation on tyrosine and threonine. J Biol Chem 1995; 270:7420-7426.
146. Trent JC, McConkey DJ, Loughlin SM et al. Ras signaling in tumor necrosis factor-induced apoptosis. EMBO J 1996; 15:4497-4505.
146a. van Biesen T, Luttrell LM, Hawes BE et al. Mitogenic signaling via G protein-coupled receptors. Endocrine Reviews 1996; 17:698-714.
147. Rosenberger SF, Bowden GT. Okadaic acid stimulated TRE binding activity in a papilloma producing mouse keratinocyte cell line involves increased AP-1 expression. Oncogene 1996; 12:2301-2308.
148. Schieven GL, Mittler RS, Nadler SG et al. ZAP-70 tyrosine kinase, CD45, and T cell receptor involvement in UV- and H_2O_2-induced T cell signal transduction. J Biol Chem 1994; 269:20718-20726.
149. Warmuth I, Harth Y, Matsui MS et al. Ultraviolet radiation induces phosphorylation of the epidermal growth factor receptor. Cancer Res 1994; 54:374-376.
150. Coffer PJ, Burgering BM, Peppelenbosch MP et al. UV activation of receptor tyrosine kinase activity. Oncogene 1995; 11:561-569.
151. Knebel A, Rahmsdorf HJ, Ullrich A et al. Dephosphorylation of receptor tyrosine kinases as target of regulation by radiation, oxidants or alkylating agents. Embo J 1996; 15:5314-5325.
152. Cavigelli M, Li WW, Lin A et al. The tumor promoter arsenite stimulates AP-1 activity by inhibiting a JNK phosphatase. EMBO J 1996; 15:6269-6279.
153. Dérijard B, Hibi M, Wu IH et al. JNK1: a protein kinase stimulated by UV light and Ha-Ras that binds and phosphorylates the c-Jun activation domain. Cell 1994; 76:1025-1037.
154. Kyriakis JM, Banerjee P, Nikolakaki E et al. The stress-activated protein kinase subfamily of c-Jun kinases. Nature 1994; 369:156-160.
155. Lee JC, Laydon JT, McDonnell PC et al. A protein kinase involved in the regulation of inflammatory cytokine biosynthesis. Nature 1994; 372:739-746.
156. Alessi DR, Cuenda A, Cohen P et al. PD 098059 is a specific inhibitor of the activation of mitogen-activated protein kinase kinase in vitro and in vivo. J Biol Chem 1995; 270:27489-27494.
157. Deng T, Karin M. c-Fos transcriptional activity stimulated by H-Ras-activated protein kinase distinct from JNK and ERK. Nature 1994; 371:171-175.
158. Ofir R, Dwarki VJ, Rashid D et al. Phosphorylation of the C terminus of Fos protein is required for transcriptional transrepression of the c-fos promoter. Nature 1990; 348:80-82.
159. Chen RH, Abate C, Blenis J. Phosphorylation of the c-Fos transrepression domain by mitogen-activated protein kinase and 90-kDa ribosomal S6 kinase. Proc Natl Acad Sci USA 1993; 90: 10952-10956.
160. Okazaki K, Sagata N. The Mos/MAP kinase pathway stabilizes c-Fos by phosphorylation and augments its transforming activity in NIH 3T3 cells. Embo J 1995; 14:5048-5059.
161. Gonzalez GA, Montminy MR. Cyclic AMP stimulates somatostatin gene transcription by phosphorylation of CREB at serine 133. Cell 1989; 59:675-680.
162. Taylor SS, Buechler JA, Yonemoto W. cAMP-dependent protein kinase: framework for a diverse family of regulatory enzymes. Annu Rev Biochem 1990; 59 :971-1005.
163. Sheng M, Thompson MA, Greenberg ME. CREB: a Ca^{2+}-regulated transcription factor phosphorylated by calmod-

ulin-dependent kinases. Science 1991; 252:1427-1430.
164. Dash PK, Karl KA, Colicos MA et al. cAMP response element-binding protein is activated by Ca^{2+}/calmodulin- as well as cAMP-dependent protein kinase. Proc Natl Acad Sci USA 1991; 88:5061-5065.
165. Matthews RP, Guthrie CR, Wailes LM et al. Calcium/calmodulin-dependent protein kinase types II and IV differentially regulate CREB-dependent gene expression. Mol Cell Biol 1994; 14:6107-6116.
166. Sun P, Enslen H, Myung PS et al. Differential activation of CREB by Ca^{2+}/ calmodulin-dependent protein kinases type II and type IV involves phosphorylation of a site that negatively regulates activity. Genes Dev 1994; 8:2527-2539.
167. Böhm M, Moellmann G, Cheng E et al. Identification of $p90^{RSK}$ as the probable CREB-ser^{133} kinase in human melanocytes. Cell Growth Differ 1995; 6: 291-302.
168. Xing J, Ginty DD, Greenberg ME. Coupling of the RAS-MAPK pathway to gene activation by RSK2, a growth factor-regulated CREB kinase. Science 1996; 273:959-963.
169. Tan Y, Rouse J, Zhang A et al. FGF and stress regulate CREB and ATF-1 via a pathway involving p38 MAP kinase and MAPKAP kinase-2. EMBO J 1996; 15: 4629-4642.
170. Iordanov M, Bender K, Ade T et al. CREB is activated by UVC through a p38/HOG-1 dependent protein kinase. EMBO J 1997; 16:1009-1022.
171. Ginty DD, Bonni A, Greenberg ME. Nerve growth factor activates a Ras-dependent protein kinase that stimulates c-fos transcription via phosphorylation of CREB. Cell 1994; 77:713-725.
172. Sassone-Corsi P, Lamph WW, Verma IM. Regulation of proto-oncogene fos: a paradigm for early response genes. Cold Spring Harb Symp Quant Biol 1988; 2:749-760.
173. Bading H, Ginty DD, Greenberg ME. Regulation of gene expression in hippocampal neurons by distinct calcium signaling pathways. Science 1993; 260: 181-186.
174. Shibanuma M, Kuroki T, Nose K. Inhibition of proto-oncogene c-fos transcription by inhibitors of protein kinase C and ion transport. Eur J Biochem 1987; 164:15-19.
175. Ran W, Dean M, Levine RA et al. Induction of c-fos and c-myc mRNA by epidermal growth factor or calcium ionophore is cAMP dependent. Proc Natl Acad Sci USA 1986; 83:8216-8220.
176. Bravo R, Neuberg M, Burckhardt J et al. Involvement of common and cell type-specific pathways in c-fos gene control: stable induction by cAMP in macrophages. Cell 1987; 48:251-260.
177. Büscher M, Rahmsdorf HJ, Litfin M et al. Activation of the c-fos gene by UV and phorbol ester: different signal transduction pathways converge to the same enhancer element. Oncogene 1988; 3: 301-311.
178. Gilman MZ. The c-fos serum response element responds to protein kinase C-dependent and -independent signals but not to cyclic AMP. Genes Dev 1988; 2:394-402.
179. Hollander MC, Fornace Jr AJ. Induction of fos RNA by DNA-damaging agents. Cancer Res 1989; 49:1687-1692.
180. Shah G, Ghosh R, Amstad PA et al. Mechanism of induction of c-fos by ultraviolet B (290-320 nm) in mouse JB6 epidermal cells. Cancer Res 1993; 53: 38-45.
181. Sherman ML, Datta R, Hallahan DE et al. Ionizing radiation regulates expression of the c-jun protooncogene. Proc Natl Acad Sci USA 1990; 87:5663-5666.
182. Rahmsdorf HJ, Gebel S, Krämer M et al. Ultraviolet irradiation and phorbol esters induce gene transcription by different mechanisms. In: T. G. Rossman, eds. Induced effects of genotoxic agents in eukaryotic cells. Washington, Philadelphia, London: Hemisphere Publishing Corporation, 1992:141-161
183. Treisman R. Transient accumulation of c-fos RNA following serum stimulation requires a conserved 5′ element and c-fos 3′ sequences. Cell 1985; 42:889-902.
184. Shaw PE, Schröter H, Nordheim A. The ability of a ternary complex to form over the serum response element correlates with serum inducibility of the human c-fos promoter. Cell 1989; 56:563-572.

185. Hipskind RA, Rao VN, Mueller CG et al. Ets-related protein Elk-1 is homologous to the c-fos regulatory factor p62TCF. Nature 1991; 354:531-534.
186. Dalton S, Treisman R. Characterization of SAP-1, a protein recruited by serum response factor to the c-fos serum response element. Cell 1992; 68:597-612.
187. Treisman R. Ternary complex factors: growth factor regulated transcriptional activators. Curr Opin Genet Dev 1994; 4:96-101.
188. Herrera RE, Shaw PE, Nordheim A. Occupation of the c-fos serum response element in vivo by a multi-protein complex is unaltered by growth factor induction. Nature 1989; 340:68-70.
189. Price MA, Cruzalegui FH, Treisman R. The p38 and ERK MAP kinase pathways cooperate to activate Ternary Complex Factors and c-fos transcription in response to UV. Embo J 1996; 15: 6552-6563.
190. Hill CS, Treisman R. Differential activation of c-fos promoter elements by serum, lysophosphatidic acid, G proteins and polypeptide growth factors. Embo J 1995; 14:5037-5047.
191. Prywes R, Roeder RG. Inducible binding of a factor to the c-fos enhancer. Cell 1986; 47:777-784.
192. Hayes TE, Kitchen AM, Cochran BH. Inducible binding of a factor to the c-fos regulatory region. Proc Natl Acad Sci USA 1987; 84:1272-1726.
193. Wagner BJ, Hayes TE, Hoban CJ et al. The SIF binding element confers sis/PDGF inducibility onto the c-fos promoter. Embo J 1990; 9:4477-4484.
194. Sassone-Corsi P, Visvader J, Ferland L et al. Induction of proto-oncogene fos transcription through the adenylate cyclase pathway: characterization of a cAMP-responsive element. Genes Dev 1988; 2: 1529-1538.
195. Fisch TM, Prywes R, Simon MC et al. Multiple sequence elements in the c-fos promoter mediate induction by cAMP. Genes Dev 1989; 3:198-211.
196. Berkowitz LA, Riabowol KT, Gilman MZ. Multiple sequence elements of a single functional class are required for cyclic AMP responsiveness of the mouse c-fos promoter. Mol Cell Biol 1989; 9:4272-4281.
197. Härtig E, Loncarevic IF, Büscher M et al. A new cAMP response element in the transcribed region of the human c-fos gene. Nucleic Acids Res 1991; 19: 4153-4159.
198. Sheng M, Dougan S, McFadden G et al. Calcium and growth factor pathways of c-fos transcriptional activation require distinct upstream regulatory sequences. Mol Cell Biol 1988; 8:2787-2786.
199. Nordheim A, Janknecht R, Hipskind RA. Transcriptional regulation of the human c-fos proto-oncogene. In: P. Angel and P. Herrlich, eds. The Fos and Jun families of transcription factors. Boca Raton: CRC-press, 1994:97-114
200. Robertson LM, Kerppola TK, Vendrell M et al. Regulation of c-fos expression in transgenic mice requires multiple interdependent transcription control elements. Neuron 1995; 14:241-252.
201. Schönthal A, Herrlich P, Rahmsdorf HJ et al. Requirement for fos gene expression in the transcriptional activation of collagenase by other oncogenes and phorbol esters. Cell 1988; 54:325-334.
202. Shaw PE, Frasch S, Nordheim A. Repression of c-fos transcription is mediated through p67SRF bound to the SRE. Embo J 1989; 8:2567-2574.
203. König H, Ponta H, Rahmsdorf U et al. Autoregulation of fos: the dyad symmetry element as the major target of repression. Embo J 1989; 8:2559-2566.
204. Sassone-Corsi P, Sisson JC, Verma IM. Transcriptional autoregulation of the proto-oncogene fos. Nature 1988; 334: 314-319.
205. Wilson T, Treisman R. Fos C-terminal mutations block down-regulation of c-fos transcription following serum stimulation. Embo J 1988; 7:4193-4202.
206. Gius D, Cao XM, Rauscher FJ et al. Transcriptional activation and repression by Fos are independent functions: the C terminus represses immediate-early gene expression via CArG elements. Mol Cell Biol 1990; 10:4243-4255.
207. Lucibello FC, Lowag C, Neuberg M et al. Trans-repression of the mouse c-fos promoter: a novel mechanism of Fos-mediated trans-regulation. Cell 1989; 59: 999-1007.

208. Subramaniam M, Schmidt LJ, Crutchfield CE et al. Negative regulation of serum-responsive enhancer elements. Nature 1989; 340:64-66.
209. Rivera VM, Sheng M, Greenberg ME. The inner core of the serum response element mediates both the rapid induction and subsequent repression of c-fos transcription following serum stimulation. Genes Dev 1990; 4:255-268.
210. Brach MA, Herrmann F, Yamada H et al. Identification of NF-jun, a novel inducible transcription factor that regulates c-jun gene transcription. Embo J 1992; 11:1479-1486.
211. Stein B, Angel P, van Dam H et al. Ultraviolet-radiation induced c-jun gene transcription: two AP-1 like binding sites mediate the response. Photochem Photobiol 1992; 55:409-415.
212. Herr I, van Dam H, Angel P. Binding of promoter-associated AP-1 is not altered during induction and subsequent repression of the c-jun promoter by TPA and UV irradiation. Carcinogenesis 1994; 15:1105-1113.
213. Muegge K, Vila M, Gusella GL et al. Interleukin 1 induction of the c-jun promoter. Proc Natl Acad Sci USA 1993; 90:7054-7058.
214. Rozek D, Pfeifer GP. In vivo protein-DNA interactions at the c-jun promoter: preformed complexes mediate the UV response. Mol Cell Biol 1993; 13: 5490-5499.
215. Han T-H, Lamph WW, Prywes R. Mapping of epidermal growth factor-, serum, and phorbol ester-responsive sequence elements in the c-jun promoter. Mol Cell Biol 1992; 12:4472-4477.
216. de Groot RP, Pals C, Kruijer W. Transcriptional control of c-jun by retinoic acid. Nucleic Acids Res 1991; 19: 1585-1591.
217. Yang-Yen HF, Chiu R, Karin M. Elevation of AP1 activity during F9 cell differentiation is due to increased c-jun transcription. New Biol 1990; 2:351-361.
218. Kitabayashi I, Kawakami Z, Chiu R et al. Transcriptional regulation of the c-jun gene by retinoic acid and E1A during differentiation of F9 cells. Embo J 1992; 11:167-175.
219. Goldberg Y, Treier M, Ghysdael J et al. Repression of AP-1-stimulated transcription by c-Ets-1. J Biol Chem 1994; 269: 16566-16573.
220. Pankov R, Neznanov N, Umezawa A et al. AP-1, ETS, and transcriptional silencers regulate retinoic acid-dependent induction of keratin 18 in embryonic cells. Mol Cell Biol 1994; 14:7744-7757.
221. Bassuk AG, Leiden JM. A direct physical association between ETS and AP-1 transcription factors in normal human T cells. Immunity 1995; 3:223-237.
222. Treier M, Bohmann D, Mlodzik M. JUN cooperates with the ETS domain protein pointed to induce photoreceptor R7 fate in the *Drosophila* eye. Cell 1995; 83: 753-760.
223. Schüle R, Evans RM. Cross-coupling of signal transduction pathways: zinc finger meets leucine zipper. Trends Genet 1991; 7:377-381.
224. Beato M, Herrlich P, Schütz G. Steroid hormone receptors: many actors in search of a plot. Cell 1995; 83:851-857.
225. Mordacq JC, Linzer DI. Co-localization of elements required for phorbol ester stimulation and glucocorticoid repression of proliferin gene expression. Genes Dev 1989; 3:760-769.
226. Diamond MI, Miner JN, Yoshinaga SK et al. Transcription factor interactions: selectors of positive or negative regulation from a single DNA element. Science 1990; 249:1266-1272.
227. Zhang XK, Dong JM, Chiu JF. Regulation of α-fetoprotein gene expression by antagonism between AP-1 and the glucocorticoid receptor at their overlapping binding site. J Biol Chem 1991; 266: 8248-8254.
228. Miner JN, Yamamoto KR. The basic region of AP-1 specifies glucocorticoid receptor activity at a composite response element. Genes Dev 1992; 6:2491-2501.
229. Lu B, Rothnagel JA, Longley MA et al. Differentiation-specific expression of human keratin 1 is mediated by a composite AP-1/steroid hormone element. J Biol Chem 1994; 269:7443-7449.
230. Jonat C, Rahmsdorf HJ, Park KK et al. Antitumor promotion and antiinflammation: down-modulation of AP-1 (Fos/

Jun) activity by glucocorticoid hormone. Cell 1990; 62:1189-2104.
231. Nicholson RC, Mader S, Nagpal S et al. Negative regulation of the rat stromelysin gene promoter by retinoic acid is mediated by an AP1 binding site. Embo J 1990; 9:4443-4454.
232. Schüle R, Rangarajan P, Kliewer S et al. Functional antagonism between oncoprotein c-Jun and the glucocorticoid receptor. Cell 1990; 62:1217-1226.
233. Yang-Yen HF, Chambard JC, Sun YL et al. Transcriptional interference between c-Jun and the glucocorticoid receptor: mutual inhibition of DNA binding due to direct protein-protein interaction. Cell 1990; 62:1205-1215.
234. Shemshedini L, Knauthe R, Sassone-Corsi P et al. Cell-specific inhibitory and stimulatory effects of Fos and Jun on transcription activation by nuclear receptors. Embo J 1991; 10:3839-3849.
235. König H, Ponta H, Rahmsdorf HJ et al. Interference between pathway-specific transcription factors: glucocorticoids antagonize phorbol ester-induced AP-1 activity without altering AP-1 site occupation in vivo. Embo J 1992; 11:2241-2246.
236. Pearce D, Yamamoto KR. Mineralocorticoid and glucocorticoid receptor activities distinguished by nonreceptor factors at a composite response element. Science 1993; 259:1161-1165.
237. Heck S, Kullmann M, Gast A et al. A distinct modulating domain in glucocorticoid receptor monomers in the repression of activity of the transcription factor AP-1. Embo J 1994; 13:4087-4095.
238. Starr DB, Matsui W, Thomas JR et al. Intracellular receptors use a common mechanism to interpret signaling information at response elements. Genes Dev 1996; 10:1271-1283.
239. Göttlicher M, Heck S, Doucas V et al. Interaction of the Ubc9 human homologue with c-Jun and with the glucocorticoid receptor. Steroids 1996; 61: 257-262.
240. Fanjul A, Dawson MI, Hobbs PD et al. A new class of retinoids with selective inhibition of AP-1 inhibits proliferation. Nature 1994; 372:107-111.
241. Chen JY, Penco S, Ostrowski J et al. RAR-specific agonist/antagonists which dissociate transactivation and AP1 transrepression inhibit anchorage-independent cell proliferation. Embo J 1995; 14:1187-1197.
241a. Heck S, Bender U, Kullmann M et al. IκBα-independent downregulation of NF-κB activity by glucorticoid receptor. EMBO J 1997; 16:4698-4707.
242. Stein B, Baldwin AS, Jr., Ballard DW et al. Cross-coupling of the NF-kappa B p65 and Fos/Jun transcription factors produces potentiated biological function. Embo J 1993; 12:3879-3891.
243. Bengal E, Ransone L, Scharfmann R et al. Functional antagonism between c-Jun and MyoD proteins: a direct physical association. Cell 1992; 68:507-519.
244. Li L, Chambard J-C, Karin M et al. Fos and Jun repress transcriptional activation by myogenin and MyoD: The amino terminus of Jun can mediate repression. Genes & Dev 1992; 6:676-689.
245. Su H, Bos TJ, Monteclaro FS et al. *Jun* inhibits myogenic differentiation. Oncogene 1991; 6:1759-1766.
246. Caldenhoven E, Liden J, Wissink S et al. Negative cross-talk between RelA and the glucocorticoid receptor: a possible mechanism for the antiinflammatory action of glucocorticoids. Mol Endocrinol 1995; 9:401-412.
247. Ciechanover A. The ubiquitin-proteasome proteolytic pathway. Cell 1994; 79:13-21.
248. Murray A. Cyclin ubiquitination: the destructive end of mitosis. Cell 1995; 81:149-152.
249. Scheffner M, Nuber U, Huibregtse JM. Protein ubiquitination involving an E1-E2-E3 enzyme ubiquitin thioester cascade. Nature 1995; 373:81-83.
250. Stancovski I, Gonen H, Orian A et al. Degradation of the proto-oncogene product c-Fos by the ubiquitin proteolytic system in vivo and in vitro: identification and characterization of the conjugating enzymes. Mol Cell Biol 1995; 15:7106-7116.
251. Wang ZQ, Ovitt C, Grigoriadis AE et al. Bone and haematopoietic defects in mice lacking c-fos. Nature 1992; 360:741-745.

252. Johnson RS, Spiegelman BM, Papaioannou V. Pleiotropic effects of a null mutation in the c-fos proto-oncogene. Cell 1992; 71:577-586.
253. Grigoriadis AE, Wang ZQ, Cecchini MG et al. c-Fos: a key regulator of osteoclast-macrophage lineage determination and bone remodeling. Science 1994; 266: 443-448.
254. Grigoriadis AE, Wang ZQ, Wagner EF. Fos and bone cell development: lessons from a nuclear oncogene. Trends Genet 1995; 11:436-441.
255. Okada S, Wang ZQ, Grigoriadis AE et al. Mice lacking c-fos have normal hematopoietic stem cells but exhibit altered B-cell differentiation due to an impaired bone marrow environment. Mol Cell Biol 1994; 14:382-390.
256. Haas S, Kaina B. c-Fos is involved in the cellular defence against the genotoxic effect of UV radiation. Carcinogenesis 1995; 16:985-991.
257. Hilberg F, Aguzzi A, Howells N et al. c-jun is essential for normal mouse development and hepatogenesis. Nature 1993; 365:179-181.
258. Brown JR, Ye H. A defect in nurturing in mice lacking the immediate early gene fosB. Cell 1996; 86:297-309.
259. Saez E, Rutberg SE, Mueller E et al. c-fos is required for malignant progression of skin tumors. Cell 1995; 82:721-732.
260. Wang ZQ, Grigoriadis AE, Mohle-Steinlein U et al. A novel target cell for c-fos-induced oncogenesis: development of chondrogenic tumours in embryonic stem cell chimeras. Embo J 1991; 10: 2437-2450.
261. Grigoriadis AE, Schellander K, Wang ZQ et al. Osteoblasts are target cells for transformation in c-fos transgenic mice. J Cell Biol 1993; 122:685-701.
262. Wang ZQ, Liang J, Schellander K et al. c-fos-induced osteosarcoma formation in transgenic mice: cooperativity with c-jun and the role of endogenous c-fos. Cancer Res 1995; 55:6244-6251.
263. Blendy JA, Kaestner KH, Schmid W et al. Targeting of the CREB gene leads to up-regulation of a novel CREB in RNA isoform. EMBO J 1996; 15:1098-1106.

CHAPTER 6

Steroid, Thyroid and Retinoid Receptors as Transcription Factors

Aria Baniahmad, Martin Eggert and Rainer Renkawitz

INTRODUCTION AND OVERVIEW

Regulation of gene transcription by nuclear hormone receptors is mediated by binding of these receptors to specific DNA sequences located in the vicinity of the regulated genes. Nuclear receptors exemplify a shortcut in signal transduction: extracellular signals (lipophilic hormones) are directly transferred to the regulated genes. These ligands play a major role in the intercellular communication within complex organisms. Their mode of action involves the binding to their receptors within the target cells and thereby direct activation or inhibition of gene expression at the chromatin level is achieved. Each of these steps has been extensively studied. The cloning of the cDNA coding for many of these receptors has revealed that they belong to a still increasing multigene superfamily. This family is comprised of different classes of genes coding for known receptors, which bind true steroids or other lipophilic hormones. In addition so called orphan receptors are known, which either function without a ligand or for which a ligand has not yet been found. Here, we summarize the current knowledge on receptor function in the first part of this chapter. The biological and medical importance of nuclear receptors is reviewed in the second part by discussing their malfunctions in human diseases. Finally, in a third section we list the possible mechanisms controlling the transcription initiation complex.

Historical Aspects

Historically, the function of steroid receptors as transcription factors was indicated by the analysis of insect metamorphosis. The observation that ecdysone induces puffing in polytene chromosomes of insect salivary glands[1] suggested that steroids might act directly at the level of gene activation. Ecdysone induced puffs have been grouped into those which appear early after induction and others, which belong to a second (late) group.[2] Finally, with the cloning of the ecdysone receptor cDNA,[3] the close relationship to the vertebrate steroid receptors was obvious. The dramatic impact of steroids on multiple functions in vertebrate

Transcription Factors in Eukaryotes, edited by Athanasios G. Papavassiliou.

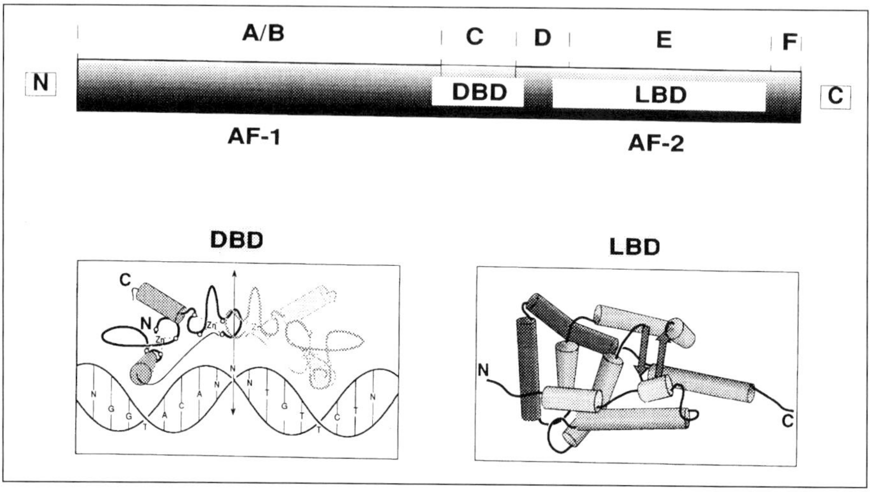

Fig. 6.1. Nuclear receptor tripartite structure. The tripartite structure of nuclear receptors is shown schematically with the homology regions A to F (see text) divided into the activation functions AF-1 and AF-2, the DNA-binding domain (DBD) and the ligand binding domain (LBD). The two inserts depict structural information on the DBD of the glucocorticoid receptor GR with the bound DBD of a second GR shown in light gray, and on the LBD of RXR in the absence of ligand.

organisms has been known a long time. The effect on sex differentiation by estradiol or testosterone, on uterus differentiation during pregnancy by progesterone, the effect on the metabolism of carbohydrates by cortisol, or the maintenance of water and iron balance by aldosterone are well established examples of steroid activities. These already show an important impact on development, cell differentiation or physiology. Similarly, nonsteroid hormones, like retinoids, thyroid hormone and vitamin D regulate a number of important physiological processes in vertebrates. In fact, based on the importance of these modulators, analysis of the mechanism of hormone action was always a focus of research. Therefore it is not surprising that the first transcription factor to be isolated and studied in detail was the glucocorticoid receptor (GR). Among the many nuclear receptors the GR was the first one to be cloned.[4-6]

The DNA-Binding Domain

All of the identified nuclear receptors show a tri-partite structure, with the DNA-binding domain (DBD) in the center and the ligand binding domain in the C-terminus. Sequence comparison of the human and chicken estrogen receptor with the glucocorticoid receptor and with the retroviral homologue of the thyroid receptor (v-erbA) revealed a more refined substructure of homology domains denoted A to F (Fig. 6.1) with the most conserved domain, C, harboring the central part of the functional DNA-binding domain.[7] This DNA-binding domain contains two motifs of the "zinc finger" type. The stability of the zinc finger structure is achieved by the coordination of one zinc atom to four cysteins. The structure of the DBD bound to DNA has been determined for several nuclear receptors and shows remarkable similarity.[8-11] Several functional features of the DBD could be determined from the structure analysis as well as from the analysis of mutants affecting binding. The results showed that one of the two α-helices within a DBD monomer serves as a recognition helix located in the major groove of the DNA (Fig. 6.1 and P-box in Fig. 6.2). In addition, the so-called D-box of the C-terminal finger provides an interaction domain with the corresponding

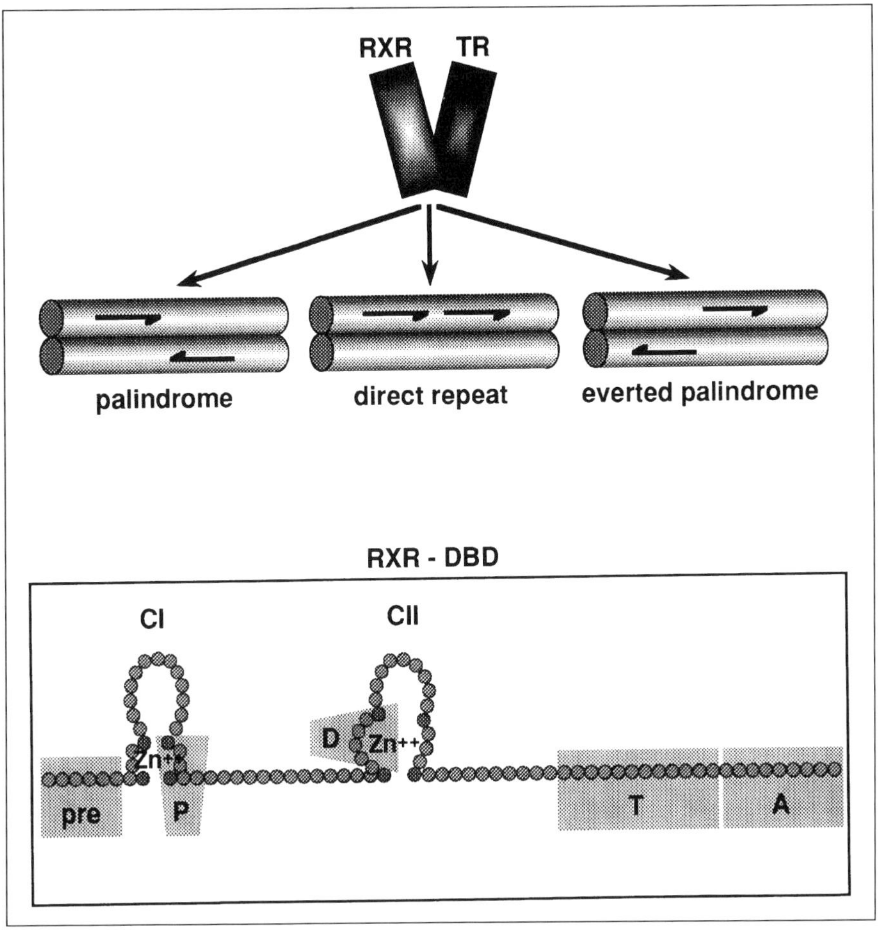

Fig. 6.2. Flexibility in DNA binding of the RXR/TR heterodimer. RXR/TR heterodimer binding to dimeric DNA half sites arranged as a palindrome, as direct repeats and as an everted palindrome is indicated. The insert shows schematically the amino acids of the RXR-DBD organized in two zinc fingers (CI and CII) with the functional boxes (shaded areas) of the prefinger region (pre), the P-, D-, T-, and A-box (P, D, T, A) as discussed in the text.

region of a second DBD within a receptor homodimer (Fig. 6.1). Initially, the finding that RXR heterodimers bind to response elements organized in a direct repeat structure (for a review see ref. 12) was quite puzzling, especially since heterodimeric receptors can bind several response elements, such as direct repeated structures, inverted (palindromic) repeats and everted repeated elements (Fig. 6.2). This flexibility argued for additional dimerization surfaces in the DNA-binding domain. Indeed, for the RXR/TR heterodimer bound on a direct repeated element, it was found[11] that TR contacts the D-box of RXR by residues from the first zinc finger (CI), from the prefinger region and from the T-box (Fig. 6.2). The spacing between the direct repeat half sites is determined by the A-box. Whether and how the DNA-binding domain of heterodimerized receptors contact each other on everted palindromic response elements has still to be shown.

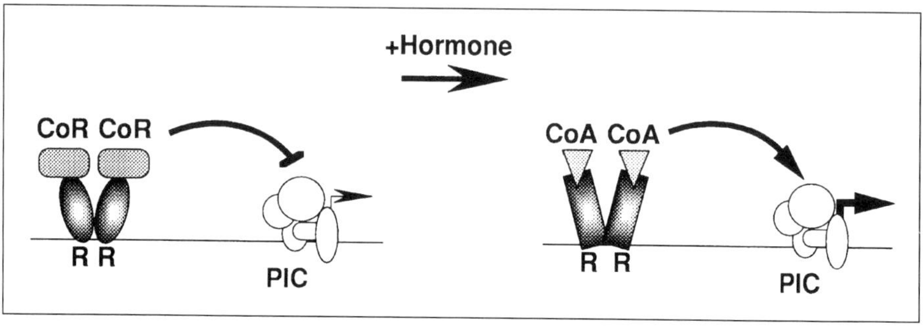

Fig. 6.3. Functional switch from repression to induction. Several nonsteroid nuclear receptors in the absence of ligand bind to corepressors (CoR) and mediate transcriptional repression on the preinitiation complex (PIC). After hormone addition the corepressors are exchanged for coactivators (CoA) activating the preinitiation complex.

Transcriptional Activation and the Role of Hormone

Transcriptional activation functions (AF) have been localized in the N-terminal part of some of the receptors as well as in the C-terminal part. The function of the N-terminal AF (AF-1) seems to be influenced by tissue specific factors,[13,14] whereas the C-terminal part of several receptors mediates multiple functions. The activation function (AF-2), the nuclear localization function, the ligand binding activity as well as an additional dimerization domain are all found interspersed within the C-terminal region. Based on the crystal structure of this domain of three nuclear receptors,[115-17] it can be seen that functional regions, apparently extending over quite a large region on the linear molecule, are in close contact in the three dimensional structure (Fig. 6.1) (for a review see refs. 18-20). Although the structure of the ligand binding domain in the absence and presence of ligand has not been shown for a single receptor, it is conceivable to assume that the ligand induced conformational changes are conserved within the nuclear receptors.[20] From this it follows that the ligand is trapped in a hydrophobic pocket, with the landmark features of two anti-parallel β-sheets and the C-terminal α-helix (helix 12) flipped back closing the pocket (Fig. 6.1).

What might be the role of hormones? For the homodimeric steroid receptors (for a review see ref. 21) the release from heat shock proteins (chaperones), the induction of DNA binding and the association with coactivators have been determined. In contrast, the heterodimeric receptors (for a review see ref. 12) are not associated with heat shock proteins and bind to their response elements in the absence of hormone. Rather the binding to corepressors is released by hormone (Fig. 6.3) and exchanged for the binding of coactivators. For both types of receptors protein phosphorylation has been shown to modulate several of the hormone mediated effects (see last section of this chapter).

The Superfamily of Nuclear Receptors

Sequence comparison and alignment has not only identified conserved domains, but has also helped in understanding the evolution of the nuclear receptor gene superfamily.[22] As far as being analyzed, nuclear receptors are found throughout the animal kingdom. Homology comparisons and evolutionary tree constructions point to a complex evolutionary history involving gene duplication events and individual domain shuffling.[22] Functional conservation has been observed as well. For example the *Drosophila* seven-up gene is highly related to the vertebrate COUP orphan receptor and shows a similar development specific expression pattern to that of the zebrafish homologue (for a review see ref. 23). Similarly, *Drosophila* and vertebrate HNF4 orphan receptors are expressed in liver, intestine and kidney of vertebrates and in functionally related organs of flies, the Malpighian tubules and midgut.[23] Another striking example of functional homology is seen with the *Drosophila* ecdysone receptor and its heterodimerization partner ultraspiracle. Apparently, ultraspiracle serves identical heterodimerization functions as the vertebrate RXR.

This large superfamily of nuclear receptors, now roughly numbering up to 150 members and still increasing, can be ordered by different means. Maybe the most practical division into subclasses is based on the DNA-binding and dimerization function.[24] Such a division distinguishes four classes of receptors: homodimeric receptors binding to inverted repeats as in the classical steroid hormone receptors (class I), receptors heterodimerizing with RXR and preferentially binding to direct repeats (class II), the RXR receptor itself and others binding as homodimers to direct repeat sequences (class III), and finally receptors which bind as monomers to monomeric response elements (class IV). So far analyzed, the so called orphan receptors, which bind unknown ligands or which function without any ligand, group into the class III and IV receptors.

Subcellular Localization

The subcellular distribution of many of the receptors has been analyzed and found to be nuclear, although a continuous shuttling between the nucleus and the cytoplasm has been demonstrated.[25] For the glucocorticoid receptor the reports on the localization in the absence of ligand are contradictory. The classical concept of a cytoplasmic receptor being translocated to the nucleus after binding of hormone is both supported and questioned by recent publications. Possible artifacts in the immune localization of the glucocorticoid receptor may arise from the permeabilization and/or from the fixation of the cell and the nucleus. To circumvent problems with the permeabilization, Brink et al[26] used semithin cryosections for immune fluorescent labeling of GR and found nuclear staining in the presence as well as in the absence of ligand. In contrast, Ogawa et al[27] used a receptor fusion with the green fluorescent protein from the jellyfish *Aequorea victoria*, which allows fluorescence microscopic analysis with living cells. These authors describe the cytoplasmic localization in the absence of ligand and a nuclear localization in the presence of a glucocorticoid. In addition it has been shown that GR colocalizes and copurifies with both microtubules and tubulin, while GR is distributed in cytoplasm and nucleus irrespective of the presence of hormone.[28] Within the nuclear compartment, colocalization of GR with the nuclear matrix has been demonstrated.[29]

Receptor and Chromatin

Obviously, nuclear localization and DNA binding are not sufficient for receptor function. The nuclear receptors, like any other transcription factor, have to cope with the DNA being packaged into chromatin and an optimal interaction with the transcription initiation complex has to be achieved. As recently discussed,[21] DNA wrapping around a nucleosome enables simultaneous binding of receptors and neighboring transcription factors, which could not be bound on naked linear DNA. Thus, the receptor has to be able to bend DNA or at least bind to bent DNA (on a nucleosome). Receptor induced DNA bending has been shown for the estrogen receptor (ER), the thyroid hormone receptor (TR), the progesterone receptor (PR) and the orphan receptor RORα.[30-33] Similarly, the silencer protein NeP1, binding next to the TR on the chicken lysozyme silencer,[34] which synergizes in repression in the absence of ligand, induces a strong bend in the DNA.[35] Although several functions for bent DNA have been proposed, the nucleosomal arrangement is clearly affected by bent DNA.[36] In the case of the TR/RXR heterodimer, chromatin assembly augments transcriptional repression in the absence of ligand, whereas the addition of thyroid hormone leads to the disruption of the chromatin structure by the chromatin-bound receptor.[37]

Finally, after nuclear translocation, possible homo- or heterodimerization, binding to chromatin, ligand induced conformational changes and possible modifications (e.g., phosphorylation), the receptor has to somehow interact with the transcriptional initiation complex. As discussed in the last section of this chapter, such an interaction may be conferred by transcriptional coactivators, by transcriptional corepressors, or by direct interaction with the initiation complex. The switch from repression (Figure 6.3) with bound corepressors to activation with bound coactivators especially is not yet fully understood.

DISEASES DUE TO MUTATIONS IN NUCLEAR HORMONE RECEPTOR GENES

The biological importance of nuclear hormone receptors is envisaged in various diseases due to receptor mutations. A brief overview of each of the syndromes caused by mutant nuclear hormone receptors will be given. Thereby, the effect of the mutation on the receptors' overall activity will be pointed out. Furthermore, possible pathways missing or changed by mutant receptors will be discussed in order to understand the molecular basis of the syndromes. Thus, this part of the review will emphasize the clinical role of these receptors as transcription factors in the human endocrine system.

A set of diseases were found to be associated with mutations in genes coding for nuclear hormone receptors. Individuals carrying those mutant receptors show distinct malfunctions of the endocrine system (Table 6.1). So far, receptors for both types of ligands—the steroids and the nonsteroids—were shown to be involved in human diseases. This includes receptors for androgens, glucocorticoids and mineralocorticoids, estrogens and progestins as well as thyroid hormone and vitamin D. It is quite possible that even more human diseases will be characterized on the molecular and genetic level with the link to mutations in nuclear hormone receptors. This may also include the growing number of newly isolated orphan receptors.

Most of the diseases are characterized by "resistance" to a particular hormone (Table 6.1). That is, the patients metabolism is not responding to the treatment of a specific hormone. Similarly, cells derived from the patients' tissue seem to resist the otherwise expected hormonal response. It is noteworthy that patients with hormone resistance syndrome show elevated blood levels of that specific hormone. Feedback control mechanisms for hormonal production are probably missing, leading to higher production of the hormone.

In addition to the involvement of nuclear hormone receptors in hormone resistance syndrome, mutant receptors are also involved in cancer. The correlation between cancer and nuclear hormone receptor mutants is especially obvious in tissues responding to the endocrine system. Those tissues, such as breast, endometrium or prostate are growth regulated by steroid hormones. Thereby, estrogens, progestins and androgens play an essential role in both growth and differentiation of these tissues, which are essential for reproduction. Cancers derived from these tissues have been shown in many cases to contain receptor mutants.

Hormone Resistance Syndromes

It is possible to classify receptor mutants by their functional activities. In general, mutations of nuclear hormone receptors can be divided into three major types:

Type one is the group of malfunctional receptors. These mutant receptors cause a weaker cellular response to hormones as compared to normal. Usually point mutations in the ligand binding domain can be detected for these cases. These patients are expected to show weak symptoms.

Type two are the nonfunctional receptors. Mutations lead to a complete loss of response of the receptor to hormonal stimuli. The second wild type allele may compensate fully or partly for the receptor deficiencies. Usually these mutations abolish DNA binding or have lost coding sequence by large deletions or by generation of a stop codon. Patients may have symptoms with

Table 6.1. An overview of diseases caused by mutations in genes coding for members of the nuclear hormone receptor superfamily

Mutant Receptors	Disease	Remarks[1]
AR	Androgen insensitivity syndrome (AIS)	Complete AIS leads to a complete female phenotype
GR	Glucocorticoid resistance syndrome	Symptoms of fatigue, hirsutism and hypertension
MR	Mineralocorticoid resistance syndrome	Rare autosomal disease, with symptoms of hyponatraemia and dehydration
ER	Estrogen resistance syndrome	Only one case with complete resistance has been reported. Symptoms of incomplete epiphysal closure, prolonged linear growth.
VDR	Vitamin D3 resistance syndrome	Severe rickets, secondary hypocalcemia, hyperparathyroidism
TR	Resistance to thyroid hormone syndrome	Caused by mutations in the β-receptor gene. Goiter, learning disabilities, hearing defects, delayed bone maturation, mental retardation.
	Erythroleukemia	Mutant form of the α-receptor. Block of erythrocyte differentiation by the oncogene v-erbA.
RAR	Acute promyelocytic leukemia	Various chromosomal translocations involving the gene for RAR-alpha and generation of RAR-fusion proteins.
DAX	Adrenal hypoplasia congenita and hypogonadism	X-chromosomally inherited and leads to profound hormonal deficiences.

[1]For references, see text.

various degrees of expression. In this case the disease is expected to be inherited as a recessive mode (except for AR in males).

Type three represents the dominant negative receptor mutants. Those have lost part of their function and in addition inhibit the wild type receptor pathway. Therefore, mutations in one allele can affect the function of the gene product of the second allele. For that reason the phenotype of such an individual is more severe as compared to other mutations. It is expected that the diseases based on type three mutations are inherited in a dominant manner. Dominant negative receptor mutants are found predominantly for the thyroid hormone receptor.

Androgen resistance

The androgen receptor (AR) is responsible for male sexual development. AR causes masculinization of the developing male fetus. Abnormalities of receptor function result in the syndrome called "androgen resistance," and alternatively the "androgen insensitivity syndrome". The syndrome is characterized by an incomplete or total absence of normal development of the male sexual phenotype. The degree of the symptoms shows a great variation ranging from male infertility, prostate and breast cancer to motor neuron disease (Kennedy's disease)[38] and complete testicular feminization (for a review see refs. 39-41). Individuals with complete testicular feminization have the male XY genotype; however, they develop a complete female phenotype. It should be noted that the gene coding for AR is localized on the X chromosome and therefore in males any malfunction of AR cannot be compensated by a normal allele. Thus, in males a defect in one gene can result in a profound presentation of the above described clinical symptoms.

Patients with androgen insensitivity syndrome (AIS) have a mutation in the AR gene. Genetic analysis of tissues derived from patients revealed that the generation of AIS is based on a variety of different mutations of the AR gene, which reflects the heterogeneity of the clinical symptoms.[40] Mutations were found to be localized in exons of the gene coding for the ligand binding domain and the DNA-binding domain. Also, alterations in the splicing of receptor mRNA have been reported.[42] Mutations in the receptors' ligand binding domain affect hormone binding. Those receptor mutants have a weaker affinity for their hormone, which leads to a weaker response in receptor action to regulate gene activity. The patients usually have a milder form of AIS, which is referred to as partial AIS.[41]

Various mutations have been identified in the DNA-binding region which reduc the DNA-binding affinity.[40] This leads to a more or less complete loss of androgen receptor function. Similarly, the complete deletion of the AR gene has also been reported.[43-44] Two cases have been reported in which just one amino acid exchange, either in the DNA-binding region[45,46] or in the hormone binding region,[47] is sufficient to cause complete androgen insensitivity. Thus, patients carrying those receptor gene mutations present the complete AIS resulting in a complete female phenotype.

Glucocorticoid resistance

There are few reported cases of glucocorticoid resistance. It is a rare hereditary disorder characterized by elevated cortisol levels in blood and the absence of stigmata of Cushings´ syndrome. The symptoms vary from fatigue and hirsutism to severe hypertension. The patient's body compensates for the inability of glucocorticoids to exert their effect by increasing the blood levels of adrenocorticothrophic hormone (ACTH, corticotropin) and cortisol.[48-50] Higher levels of ACTH induce hyperandrogenism in women and can lead to acne, hirsutism and infertility.[51] In addition, reproductive dysfunctions in males and females have been reported.[52] The syndrome is caused by mutation of the glucocorticoid receptor gene. The receptor mutants are malfunctional; each mutation can affect different levels of receptor activity. Decreased number of receptors,[53] thermolabile receptors[54] or decreased binding affinity for glucocorticoids[52,55] have been described. Thus, mutations in the receptor gene lead to decreased receptor number in the cell or to a weaker functional activity, which results in a lower capacity of gene regulation by hormone.

Mineralocorticoid resistance

The syndrome of mineralocorticoid resistance is similar to that of glucocorticoid resistance in that it is a rare autosomal hereditary disease. It causes wasting of body salt in young children. It is characterized by episodes of dehydration and hyponatraemia, although elevated levels of aldosterone can be mesure.[50,56] Patients are treated by salt supplementation. The cause of mineralocorticoid resistance is a functionally altered mineralocorticoid receptor. The receptor has reduced binding affinity to its ligand, aldosterone, but no mutation of the receptor cDNA has yet been found.[50,57]

Estrogen resistance

Recently one patient has been described expressing complete estrogen resistance syndrome.[58] This man was reported to have the following symptoms: incomplete epiphysal closure, continued linear growth in adulthood despite normal pubertal development, about 10-fold increased serum levels of estrogens, hyperinsulinoma and no detectable response to estrogens. In addition his bone mineral density was lower than normal. The patient was homozygous for a mutation in the estrogen receptor gene creating a premature stop codon in the amino-terminal receptor part. This leads to a nonfunctional receptor, lacking DNA and hormone binding function.

Vitamin D3 resistance

Patients with the disease of vitamin D resistance exhibit conditions including severe rickets, hypocalceamia, secondary hyperparathyroidism and elevated levels of free 1,25-dihydroxy-cholecalciferol.[59-61] The syndrome is also referred to as hypocalcaemic vitamin D resistant rickets.[62-63] The disease is hereditarily transmitted in an autosomal recessive manner. Mutations have been found in the gene coding for the vitamin D receptor. Resistance to vitamin D can be due either to lack of the vitamin D receptor or to structural abnormalities, both in the hormone binding and DNA recognition domains.[61-62,64] Thus, receptor mutations lead to a failure to regulate gene expression in the appropriate manner.

Thyroid hormone resistance and the oncogene v-ErbA

There are two forms of the thyroid hormone receptor, the α- and the β-form, encoded by two different genes. For both receptors naturally occurring mutants have been described (Fig. 6.4). The chicken retroviral oncogene product v-erbA is a mutant of the α-receptor, whereas mutations in the human β-receptor can lead to the syndrome of "Generalized Resistance to Thyroid Hormone" (GRTH).

GRTH is a genetic disorder characterized by the lack of response of the peripheral and pituitary tissue to the action of thyroid hormones.[65-67] The syndrome is manifested in a large pattern of variable presentations. The main characteristics are elevated levels of serum thyroid hormones and inappropriate levels of thyrotropin. Other frequently associated symptoms are goiter, attention deficit, learning disabilities and hearing defects, delayed bone maturation and mental retardation. Furthermore, other characteristics have been described such as speech impediment, frequent ear, nose and throat infection, and "low-weight-for-height" children.[68,69]

Genetic analysis has clearly linked this disorder to mutations in the c-erbA gene coding for the human thyroid hormone receptor β.[68] With one exception, the syndrome of GRTH is inherited as a dominant trait. A large number of hTRβ-genes from patients with GRTH have been analyzed.[67] All mutations identified so far alter the protein sequence in the carboxy-terminal part (C-terminus) of the receptor. The receptor C-terminus has various functions such as hormone binding, dimerization, transactivation and silencing (Figs. 6.1, 6.3, 6.4). The mutations primarily affect the hormone binding and transactivation function of the

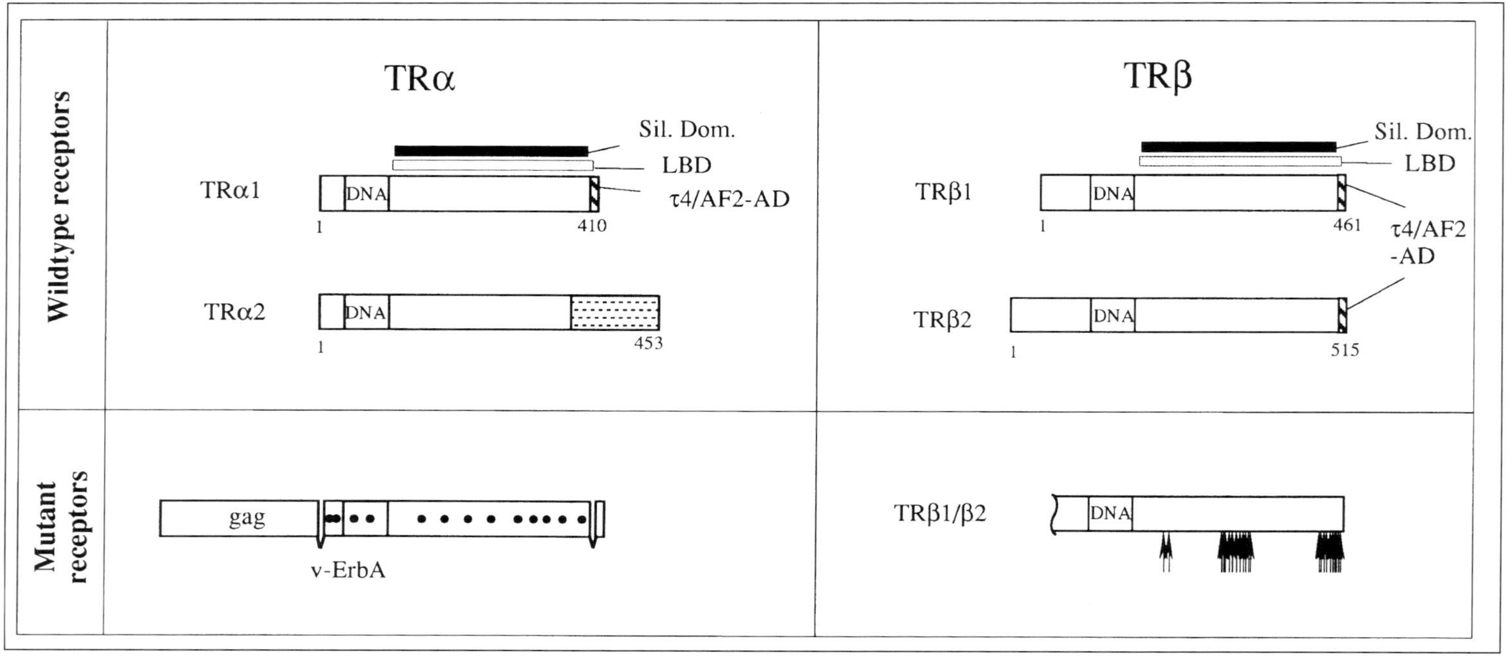

Fig. 6.4. Schematic presentation of the two thyroid hormone receptors including their isoforms and naturally occurring mutants. There are two genes (α and β) coding for thyroid hormone receptors, located on chromosome 3 and 17. The isoforms are generated by alternative splicing. TRα2 has lost its ability to bind thyroid hormone. The extent of both the ligand binding and the silencing domain are indicated. Also, the position of one of the transactivation domains (τ4/AF2-AD) is shown. Filled circles show the position of the amino acid exchanges between the α1-receptor and the oncogene v-ErbA. v-ErbA is a fusion with the retroviral gag-protein, which has lost hormone binding, but not silencing function. Each of the arrows for the β-receptor represents one mutant receptor isolated from a patient with GRTH-syndrome. Mutations in the β-receptor gene affect both of the isoforms β1 and β2.

receptor. The mutant receptors show reduced or complete abolished thyroid hormone binding affinity. Although silencing function of the receptor is greatly overlapping the hormone binding region and all mutants are also localized within the silencing domain, silencing function is in general not impaired by these mutations.[70,77]

Also mutations have been found which affect directly the recently identified transactivation domain τ4[71] also called AF2-AD.[72-74] Those mutations create a premature stop codon or a frame shift, and thus the mutant receptors lack the far C-terminal transactivation domain τ4/AF2-AD.[75-77] The presence of this transactivation domain, however, is essential for relief of the silencing function by hormone.[71] Interestingly, the mutations in the C-terminus of the receptor show a clustered arrangement. There are basically three clusters: one in the so called hinge-(D)-region, a second (aa 310 to 350) is located close to the transcativation domain τ3 and the third cluster localized at the far C-terminus, adjacent to the C-terminal transactivation domain τ4/AF2-AD.[67,70]

Thus, mutations mostly affect the activation function, while the silencing function remains active (Fig. 6.4). This results in a mutant receptor which passively blocks the action of wild type receptor and therefore represents a dominant negative acting receptor. Since the receptor mutants are able to actively repress target genes due to their remaining silencing function even in the presence of hormone, they act as a very potent inhibitor of hormonal response.

The disease of GRTH is inherited mostly in an autosomal dominant mode. It is noteworthy that two rare events of homozygous patients were described in the literature: The Refetoff patient has a complete deletion in both alleles of the hTRβ-gene coding region. The homozygous patient shows symptoms such as stippled epiphysis, dysmorphic features and deaf mutism.[78] In this case the inheritance of GRTH is transmitted in recessive mode.

The Bercu patient harbors two alleles of the mutant receptor with just one amino acid deletion, also referred to as the kindred S receptor.[68,79] The receptor mutant is unable to bind hormone. The individual has severe symptoms such as very high levels of thyroid hormones, profound mental retardation and a severe delay in bone development. At the age of 2 3/4 years the patient's bone age was 4 months.[80] In this case the inheritance of the disease followed a dominant mode. The mechanism of dominant inheritance is likely to be due to the dominant negative regulation by mutant receptors found in cell culture systems. Thereby, mutant receptors have the ability to repress wild type receptor function.

The mechanisms of negative dominance are mediated by receptor heterodimer formation between mutant and wild type receptor and/or occupancy of the thyroid hormone binding site by receptor mutants.[67,70,81-83] Furthermore, there may be an additional level of interference of mutant receptors in the pathway of thyroid hormone receptors. Since thyroid hormone can autoregulate the expression of its own receptor, dominant negatively acting receptors with residual silencing function will decrease the level of thyroid hormone receptor molecules in a cell.[84-85]

There are striking differences in the phenotypic expression of the two homozygous patients. The Bercu patient is affected in a more severe way. This is likely to be due to the dominant negatively acting kindred S receptor. It could be demonstrated that the receptor mutant not only affects the action of both wild type TRs,[80] but also the closely related retinoic acid receptor (RAR),[70] indicating a more broad spectrum of interference in gene regulation pathways. In the case of the Refetoff patient it is thought that due to the absence of any β-receptor at least part of the hTRβ pathways can be compensated by the α-receptor.[68] This could easily explain the milder symptoms and a recessive mode of inheritance.

Thus, analysis of receptor mutants and their transcriptional behavior in cell culture provides a good model system for understanding the mechanisms of recessive or dominant inheritance.

The v-erbA oncogene product was isolated from the avian erythroleukemia virus. It is a mutated form of the α-thyroid hormone receptor with point mutations distributed along the molecule.[86-87] In addition it contains small deletions, one being in the far C-terminal end and deleting the transactivation domain τ4/AF2-AD (Fig. 6.4). The oncogene interferes with erythroid differentiation.[88] It is believed that important genes necessary for development and differentiation are repressed.[89] V-ErbA has lost hormone binding[90] and is a constitutive transcriptional silencer in similar extent to the kindred S receptor.[70] However, v-ErbA is a much stronger dominant negative acting protein of thyroid hormone and the retinoic acid pathway than the kindred S receptor.[70] This fact may at least in part explain the oncogenic potential of v-ErbA. One cause of the more potent dominant negative effect may be due to mutations in the DNA-binding domain. It has been reported that a back mutation of the zinc finger sequence of v-ErbA to that of cellular TR impaired erythroid transformation.[91]

Thus, there is good evidence that neoplastic transformation by v-ErbA is due to inhibition of TR and RAR mediated gene regulation.[90,92]

DAX1 and adrenal hypoplasia congenita and hypogonadism

A novel member of the nuclear hormone receptor gene superfamily is DAX1.[93,94] However, DAX1 represents an unusual member of this family, since its DNA-binding domain is completely different and does not have the characteristic Zn finger motif. The homologies between other members and DAX1 is restricted to the C-terminus of the DNA-binding domain, with highest homologies to the orphan receptor Ear-2 COUP-TF and the RXRs. The disease is X-chromosomally inherited. It results in profound hormonal deficiencies and is lethal if untreated. Patients have low serum levels of glucocorticoids, mineralocorticoids and androgens and do not response to ACTH.[95] DAX1 was shown to interfere with the retinoic acid pathway.[94] However, the molecular mechanisms by which mutations of DAX1 cause the syndrome are still unclear.

Nuclear Hormone Receptor in Association with Cancer

Mutations in nuclear hormone receptors may also be involved in the development of cancer and in resistance to treatment of cancer by antihormones.

Acute promyelocytic leukemia

Acute promyelocytic leukemia (APL) results from a malignant process that leads to the accumulation of myeloid precursor cells in blood and bone marrow. The precursor cells are characterized by abnormal behavior and differentiation arrest. APL is caused by various chromosomal translocations: t[15:17];[96,97] t[11:17];[98,99] and t[5:17].[100] In all of these translocation events the RARα gene on chromosome 17 is involved and a fusion gene is generated.

The chromosomal translocation t[15:17] involves the PML gene, which is fused in frame to the coding sequences of RARα, encoding a fusion protein. The generation of this fusion gene is the key event in leukogenesis of APL.[101] PML is a growth suppressor gene which is induced by interferons,[102] and is localized in the nucleus in so called POD (PML oncogenic domain)-structures. However, the PML-RAR hybrid is found in a different pattern of multiple small spots in the nucleus. Treatment with retinoic acid changes the pattern of nuclear localization of the oncoprotein, which is then reverted and reorganized to generate the normal organization.[96,97,103] Similarly, overexpression of PML-RAR in white blood cells has been demonstrated to block their differentiation, while high doses of all-trans retinoic acid force their maturation. This is in concordance with the clinical treatment of APL patients. Therapy of individuals with APL is focused on all-trans retinoic acid.[101,104-105] Thus, the delocalization of PML-RAR or their associated factors may be a molecular cause for the generation of APL. The relocation and nuclear reorganization followed by retinoic acid treatment corresponds to the complete clinical remission rate of 90% of newly diagnosed patients.[105]

Another cause for APL is associated with the translocation t[11:17]. Thereby, a fusion protein is generated between the RARα and a Krüppel-like Zn finger protein encoded by the PLZF gene. The PLZF protein is normally localized in a punctate speckled pattern distinct from that of the PML nuclear bodies.[98] However, this syndrome seems to fail to respond to all-trans therapy.[99,106,107] In patient's leukemic cells both the RAR-PLZF fusion and the wild type RARα from the unaffected allele are expressed. One effect of the RAR-PLZF fusion could be demonstrated by which it can interfere with the cellular retinoid pathway. The fusion protein was shown to antagonize the normal RARα activity in the cell.[98]

A further variant of APL is associated with the chromosomal translocation t[5:17]. In these cases the genes coding for nucleolar nucleophosmin and the RARα are fused together.[100] The sequence of the protein part of RAR is identical to those found in PML-RAR and in PLZF-RAR fusion proteins.

Thus, future experiments are required to demonstrate the detailed molecular genetic cascade involved in the differentiation of myeloid cell by retinoic acid and to identify genes, which are misregulated.

Hormone resistant tumors

Another aspect of involvement of nuclear hormone receptor mutations in cancer is focused on tumors that become resistant to endocrine therapy. Some tissues require steroid hormones for growth and differentiation, such as breast, endometrium, gonads, prostate. Cancer derived from those tissues are being treated with anti-hormones and adjuvans. However, there are also hormone resistant tumors which do not respond to an anti-hormone therapy. Furthermore, during endocrine drug therapy tumor cells evolve which become hormone resistant. In those cases of hormone resistant cancers mutations in genes encoding nuclear hormone receptors have been identified.

Breast tumors and meningiomas which became resistant to hormonal therapy have been found to contain ER variants (for a review see refs. 108, 109). They are generated by aberrant splicing mechanisms of the precursor RNA of ER or by introduction of premature stop codons in mutation events. Missing exons in the mature mRNA can lead to dominant negative receptors. Similarly to the hormone resistance syndromes, these receptor mutants lack a hormonal response and in addition they can interfere with the remaining normal ER action.

Androgen is required for the development of prostate and also prostate cancer.[110] Androgen receptor variants have been detected in some prostate cancers that exhibit resistance to androgen therapies.[108,111] Similarly, mutations in the glucocorticoid receptor cause resistance to lysis by glucocorticoids in leukemia cells and lymphoma.[112-114]

Endocrine tissues require hormone for growth and development. However, the molecular basis by which endocrine derived cancers are relieved from this hormonal requirement is largely unknown.

MECHANISMS MODULATING TRANSCRIPTION

Phosphorylation of Nuclear Hormone Receptors

Nearly all of the nuclear hormone receptors with known ligands are reported to be phosphorylated (reviewed in ref. 115). Receptor phosphorylation and possible effects on transactivation are best characterized for GR and PR; in addition we summarize some studies about the influence of phosphorylation on transactivation by ER, VDR, TR and RAR.

GR-phosphorylation

The GR has been found in two phosphorylated forms:[116,117]

1) The basal phosphorylated form, which appears when the inactive GR is complexed with heat shock proteins.
2) The hyperphosphorylated form, which is found in the activated nuclear GR.

The GR phosphorylation is hormone-dependent, since the phosphorylation state of the receptor increased up to five fold by adding hormone to NIH 3T3 cells.[118,182] Seven main phosphorylation sites of the hyperphosphorylated mGR have been identified by microsequencing, five of them are conserved in rat and man. A proline directed kinase like the MAP-kinase or the cdc2-kinase might be responsible for the phosphorylation of the serine residues 212, 220 and 232. The phosphorylated serine residue 122 is positioned in a putative casein kinase II motif. Site directed mutagenesis of all seven phosphorylation sites did not lead to dramatic changes in receptor transactivation.[119] Taken together, the functional importance of GR phosphorylation remains to be shown.

PR-phosphorylation

Phosphorylation of PR-A and PR-B is enhanced by hormone.[120,121] In further studies a hormone independent phosphorylation of cPR-B at Ser-211 and Ser-260 and the ligand dependent phosphorylation at Ser-367 (aminoterminal to the DBD) and Ser-530 (hinge region) was shown.[122] Phospho-amino acid analysis revealed that PR is phosphorylated exclusively at serine residues. To date 5 phosphorylation sites of human PR-B (serine residues 81, 102, 162, 294 and 345) are identified, 3 phosphorylation sites (serine residues 81, 102, 162) are located within the unique N-terminal segment of PR-B[123] and 3 of them are phosphorylated in response to hormone (serine residues 102, 294, 345). The responsible kinase for these 3 phosphorylation sites, might be the MAP-kinase or the p34cdc2 kinase, since the sites are contained in a S/T-P motif. Serine residue 81 is positioned in a CK II motif and could be phosphorylated by CK II in vitro.[124]

PR phosphorylation seems to function in 3 distinct steps:[123] 1.The basal phosphorylation in the absence of hormone 2. A rapid increase in phosphorylation a few minutes after adding of progestine 3. A slower (40-60 minutes), less substantial increase in phosphorylation that promoters a decrease in electrophoretic mobility in SDS gels.

The serine residues 81and 162 are basally phosphorylated sites, which exhibit a general increase in phosphorylation in response to hormone (Step 1 and 2) whereas the serine residues 102, 294 and 345 are only phosphorylated in the later phase, in the presence of hormone (Step 3).[123] Further experiments revealed that the use of the antagonist ZK 98299 results in a minimal increase in overall phosphorylation of PR, whereas the antagonist RU 486 leads to the same rapid increase of phosphorylation as the agonist RU 5020.[125]

Thus, two antagonists completely differing in their potency to inhibit PR-phosphorylation, support the concept of two different mechanistic classes of antiprogestins.

ER-phosphorylation

Analysis of ER phosphorylation yielded, that serine residues 104, 106 and 118 are estrogen inducible phosphorylation sites and that phosphorylation of these sites is involved in ER transactivation in COS-1 cells.[126,127] Recently more detailed studies confirmed that phosphorylation of Ser-118 is required for full activity of the activation function 1 (AF-1) and that the MAPK is the responsible kinase, since it phosphorylated Ser-118 in vitro and in vivo.[128] In addition, phosphorylation of Ser-167 was identified as the major estradiol induced phosphorylation site on hER.[129] In this study it could be shown that estradiol treatment of MCF-7 cells leads to a 4 fold increase of phosphorylation of Ser-167 of ER. Furthermore, CKII was found to phosphorylate recombinant hER on Ser-167 in vitro in a hormone dependent manner. Lahooti et al[130] identified 4 phosphorylation sites in mER, including Ser-122 which lies in the AF-1 region. Site directed mutagenesis revealed that mutation of serine 122 to alanine results in a 30-40% decrease of ER dependent transactivation in transient transfection assays. Stronger effects were observed when serine 122 was mutated to alanine, in the context of mutations in a putative amphipatic helix involved in AF-2 activity. These results support the idea that phosphorylation of serine 122 of mER is required for the recently reported cooperation of AF-1 and AF-2.[131]

Phosphorylation of other members of the nuclear hormone receptor family

TR-phosphorylation could be shown at two major sites when labeling intact transformed erythroblasts with radioactive orthophosphate.[132,133] One site, in the N-terminal region of the receptor near to the DBD, was shown be phosphorylated by PKA in vitro; the other at serine 12 could be phosphorylated by CKII in vitro. Further studies revealed that phosphorylation of hTRβ, enhances the target gene sequence dependent dimerization of TR with RXRβ.[134] The cotransfection of TR, RXR and TRE-CAT reporter into CV-1 cells in the presence of ocadaic acid lead to a 10-fold increase of TR phosphorylation and to 2- to 3-fold increase in TR mediated CAT activity, compared to the same experiments without the phosphatase inhibitor. Investigations with TRα2, a TR variant which differs in the C-terminus, cannot bind T3 and is exclusively expressed in the brain, resulted in the identification of two putative phosphorylation sites.[135] The authors demonstrated that in vitro translated TRα2 binds only poorly to DNA, whereas dephosphorylation restored its DNA binding. Site directed mutagenesis of 9 serine residues between amino acids 461-475 yielded the same effect. In vitro phosphorylation of TRα2 by CKII revealed two phosphoacceptor sites at serines 474 and 475. Since the only known function for TRa2 is the inhibition of TR dependent transactivation, phosphorylation may provide a mechanism for cell specific modulation of the expression of T3 responsive genes.

Phosphorylation of VDR has been studied in transfection experiments, suggesting a major hormone dependent phosphorylation site within the HBD of VDR.[136] Jurutka et al[137] could show that receptor phosphorylation and transcriptional activity are dependent on the presence of vitamin D3 in intact cells and that CKII can phosphorylate the VDR in vitro. Transfection studies in Cos cells indicated that hVDR is phosphorylated by CKII and that this phosphorylation at serine residue 208 potentiated transcriptional activation.[138] Replacement of Ser-208 revealed that it was not obligatory for receptor action, but coexpression of wtVDR and CKII elicited a dose dependent enhancement of vitamin D3 dependent transcription. This supports the hypothesis that phosphorylation at Ser-208 of the VDR modulates its transactivation potency.

RAR was reported to be phosphorylated by PKA in vitro.[139] In cotransfection studies in HeLa cells, using a RAR expression plasmid, an expression plasmid encoding the catalytic subunit β of PKA and a RARE-CAT reporter plasmid, revealed a strong reporter gene activation in the absence or presence of ligand. Further investigations[140] confirmed these results and identified serine residue 369 as a phosphoacceptor site in vitro and in vivo. In addition it could be demonstrated that a point mutation of Ser-369 to alanine decreases, whereas a mutation of Ser-369 to glutamic acid increases, the binding of RAR/RXR heterodimers to a DR 5 response element. Thus, a PKA phosphorylation dependent increase of retenoic acid driven gene transcription, provides evidence for a coupling of the RA and cAMP signaling pathways.

Taken together, there is good evidence that phosphorylation affects nuclear receptor action, most likely by causing a conformational change which modulates receptors transactivation potency.

Interacting Partners

Nuclear hormone receptors are regulating gene transcription in cooperation with other nuclear factors (interacting partners). These factors can be divided into 3 groups: 1. other nonreceptor transcription factors, that are able to interact with the receptors and can bind to DNA; 2. basal factors comprising the transcription pre-initiation complex (PIC), which are directly interacting with the receptors; 3. interacting factors, which cannot bind to DNA by themselves. They associate with the receptors, most likely by acting as coactivators or corepressors of nuclear hormone receptors.

Interacting transcription factors

One of the best characterized interactions between nuclear hormone receptors and other transcription factors is between GR and AP-1.This interaction is essential for the modulation of transcription of several genes. In some cases GR induced transcription has been shown to be inhibited by AP-1.[141-144] In principle this regulation is based on a direct physical interaction, as shown by coimmunoprecipitation of activated GR and AP-1.[141] Other studies revealed that the DBD of GR is needed for inhibition of AP-1 binding to DNA, whereas the DNA binding of GR is not influenced. Moreover, it could be shown that the direct physical interaction takes place between GR and jun.[143] The interaction seemed to be cell specific, since treatment with the phorbol ester TPA inhibited GR driven transcription in NIH 3T3 cells, but enhanced glucocorticoid response in several T cell lines .[142]

Previous experiments lead to the hypothesis that the GR/AP1 crosstalk is based on competition for adapter molecules,[144] which is confirmed by the recent identification of the adapter CBP[145] (see below). In this study[144] the regulation of the tyrosine aminotransferase gene has been investigated. In vivo footprint analysis revealed that the GR occupied its binding site under phorbol ester treatment, without inducing gene transcription, indicating that inhibition occurs at a later step necessary for transactivation.

The GR/AP1 crosstalk is essential for glucocorticoid induced apoptosis in lymphocytes,[146,147] which could be blocked by the hormone antagonist RU 486.[146] The region of GR important for AP-1 repression could be assigned to amino acids 336-796 of mGR, suggesting that the transactivation domain AF-1 of GR is dispensable for GR/AP-1 crosstalk.[147] AP-1 interference with other nuclear receptors has been shown as well (for discussion see ref. 145).

Another prominent transcription factor, known to interact with nuclear hormone receptors is the nuclear factor κB (NF-κB). Direct physical interactions between GR and NF-κB have been shown in coimmunoprecipitation experiments.[148-150] There is further evidence for direct interaction between ER and NF-κB as shown by GST-pull down experiments. The interacting region of the ER consists of the hinge region and the LBD. In transfection studies a NF-κB driven promoter, the IL-6 promoter, was inhibited by estrogens, although this promoter lacks an ERE, suggesting an in vivo protein-protein interaction.[151] Furthermore NF-κB is able to inhibit GR driven transcription on the MMTV-promoter, as shown by transfection studies with overexpressed NF-κB in HeLa cells.[148] The interacting regions were mapped to the DBD and LBD of the GR.[149,150] GST NF-κB pull down experiments revealed that GR/ NF-κB interaction is hormone-independent.

Thus, the crosstalk between nuclear hormone receptors and other types of transcription factors via protein-protein interaction seems to play an important role in cell response to external signals. In this way, the coupling of different signal transduction pathways results in activation or repression of gene transcription.

Basal factors

One hypothesis for how nuclear receptors activate or repress gene expression is based on the idea that they are able to interact directly with components of PIC.

Interactions between GR and the basal transcription machinery were postulated, because the expression of the GR transactivation domain AF-1 inhibits transcription from a basal promoter in a cell free transcription system. Since this inhibition could not be reversed by recombinant TBP, the target protein of GR-AF-1 was postulated to be a part of the general transcription apparatus distinct from TBP.[152]

The first basal factor reported to be directly interacting with progesterone and estrogen receptors was the transcription factor IIB ($TF_{II}B$).[153] Further studies revealed an in vitro and in vivo interaction between the TRβ and $TF_{II}B$, shown in transfection and GST pull down assays.[154] The responsible domains of the TR could be mapped to the N-terminal region (amino acids

1-181) and to part of the C-terminal ligand binding domain (LBD). The strength of interaction was decreased by ligand, suggesting that the hormone is involved in preventing transcriptional repression.[154] The binding site seemed to be located in the C-terminal part of $TF_{II}B$, since mutants lacking this area were unable to bind to hTRα.[155]

Previous studies revealed an interaction between nuclear hormone receptors and the TATA binding protein,[155,156] suggesting a functional role of TBP in receptor dependent transactivation. In vivo binding experiments, using the two hybrid system, demonstrated that the ligand binding domain of human RXR is able to interact with the conserved C-terminal part of human TBP in the presence ligand. The ligand binding domains of TR and RAR failed to interact in vivo.[156] In contrast, in vitro studies exhibited TBP binding to hTRα in the absence of ligand, which was inhibited by hormone. In GST-pull down experiments the conserved C-terminal part of TBP was able to interact with nearly the whole hTRα except the hinge region.[155] A hormone independent association of TBP and $TF_{II}B$ with ER in vitro has been reported by Cavailles et al,[157,158] using the GST-system.

Interactions between nuclear hormone receptors and TBP associated factors (TAFs) have been reported as well. *Drosophila* $TAF_{II}110$ is able to interact with the LBDs of RXR and TR in the presence of ligand.[156]

An in vitro interaction between $TAF_{II}30$ and the AF-2 domain of ER was shown by pull down experiments.[159] Further experiments revealed that an antibody against $TAF_{II}30$ inhibits transcriptional activation by a fusion protein consisting of the GAL4-DBD and the E and F domains of ER, on a minimal adenovirus major late promoter, containing 5 upstream GAL4 binding sites. The same antibody had no influence on transactivation by a GAL-VP16 fusion protein, using the same promoter. This indicates that there are different TAFs responsible for interaction with different nuclear transcription factors.

Taken together, these results indicate that in some cases nuclear receptors are able to interact directly with the basal transcription machinery and thereby modulate gene transcription. Probably an indirect interaction between receptors and PIC via bridging proteins is another possibility for signal transmission to the PIC. Eventually both pathways may be used simultaneously.

Interacting factors (putative intermediary factors)

In the last 3 years many interacting factors, which may be intermediary factors transmitting the signal from the nuclear hormone receptors to the PIC, have been identified (Table 6.2). Some of them are able to modulate receptor dependent transactivation and are therefore designated coactivators or corepressors. These factors exhibit at least two characteristics:

1) They are unable to bind to specific DNA sequences.
2) They can bind to nuclear hormone receptors.

One of the first proteins reported to interact with GR was the SWI3 protein of yeast.[161] It belongs to the SWI/SNF protein complex, known to be involved in regulating transcription. Evidence for interaction was shown by coimmunoprecipitation of GR and SWI3 from the protein extract of GR transfected yeast. In contrast, studies investigating other SWI-homologues in *Drosophila*, mouse and man did not detect a direct binding of a SWI-protein to any receptor.[163,165,166]

The isolation of the receptor associated factor RAF 130, which directly interacts with AR and GR, demonstrated that some of the interacting factors might function by enhancing receptor-DNA binding and thereby modulating receptors' transactivation potency.[168] The protein calreticulin was found to interact with the DBD of AR, GR and RAR in vivo; more precisely, the interacting area of the receptors could be assigned to a conserved amino acid sequence adjacent to the first zinc finger, shared by all members of the nuclear hormone receptor family.[170,172] Whether this association is of biological relevance remains to be shown.

Table 6.2. Nuclear hormone receptor interacting factors.

Interacting factors	Molecular weight	Ligand effect on binding	Interacting receptors	Interacting receptor domain	Effect on receptor transcriptional activity	Reference
Calreticulin	60 kDa	no	AR, GR, RXR	C	yes[2]	170, 172
CBP	300 kDa	yes	ER, GR, RAR, RXR, TR	E	yes	145
GRIP1	90 kDa	yes	GR, ER, AR	D + E + F	yes	177
GRIP170	170 kDa	no	GR	?	yes	179
N-CoR/RIP13	270 kDa	yes	RAR, TR	D, part of E	?	162
RAF130	130 kDa	?	AR, GR	?	?[1]	168
RAP46	46 kDa	no	ER, GR, TR	?	?	175
RIP110	? kDa	yes	TR, (RAR)	?	?	173
RIP140	140 kDa	yes	ER	E (AF-2)	yes	157, 158, 160
RIP160	160 kDa	yes	ER, RXR, TR	E (AF-2)	?	158, 160
SMRT	168 kDa	yes	TR, RAR, (RXR)	D + E	yes	180
SRC-1	125 kDa	yes	ER, GR, PR, RXR, TR	D + E	yes	176
SUG1/Trip1	46 kDa	yes	ER, RAR, RXR, TR, VDR	E (AF-2)	?	167, 169
SWI3	95 kDa	?	GR	?	?	161
TIF1	112 kDa	yes	ER, RAR, RXR	E (AF-2)	?	171
TRIPs 2-11	? kDa	yes	RXR, TR	D + E	?	164
TRIPs 12-15	? kDa	yes	TR	D + E	?	164
TRUP	30 kDa	no	RAR, TR	D + part of E	yes[2]	174

[1]enhances DNA binding
[2]abolishes DNA binding

Two groups identified the receptor interacting protein RIP160[158,160] which was shown to bind to ER, TR and RXR in the presence of ligand.[158,160,162] RIP160 bound to the activation domain AF-2 of nuclear receptors.[157] Interestingly, RIP160 is able to bind to the interacting protein CBP (see below).

Several TR interacting proteins (Trips) were identified by Lee at al.[164] In this study the two hybrid system was used for screening a HeLa cDNA library. 15 proteins that interact with a LexA/TR fusion protein in the presence or absence of hormone were isolated. Most of the 15 Trips were also able to interact with a LexA/RXR fusion protein in a hormone-dependent manner, except for Trip12, Trip13 and Trip14. None of the isolated Trips was able to interact with GR. Three Trips revealed sequences predicting specific functional roles: Trip7 turned out to be a member of a class of nonhistone chromosomal proteins and Trip12 contained a highly conserved domain associated with ubiquitination of specific target proteins. Whether these proteins are of functional relevance for hormone dependent activation or repression of transcription awaits further investigation. Trip1 is a human homologue of the yeast transcription factor SUG1 and could be shown to functionally substitute for SUG1.[167] The murine SUG1 protein is able to interact with the RARα in vitro and in vivo in a ligand dependent manner. The interaction domain of RARα could be assigned to the AF-2 activation domain and further experiments revealed in vivo interaction of SUG1 with ER, VDR, RXR and TR.[169] An interaction with the AF-2 activation domain of nuclear hormone receptors was also described for the transcriptional intermediary factor (TIF1). In several experiments, in vivo and in vitro interactions have been demonstrated in the presence of hormone. TIF1 is also able to interact with ER in mammalian cells, as shown by coimmunocytofluorescence in COS cells transfected with TIF1 and an ER deficient for nuclear translocation.[171] Comparison of TIF1 and SUG1 revealed differences in their binding properties. Whereas TIF1 interacted efficiently with RXR, but to a lesser extent with TR, SUG1 bound TR more efficiently than RXR. There is evidence that SUG1 is a true intermediary factor, since it is able to bind to the TBP associated factor $TAF_{II}30$.[169]

Seol et al[173] identified 2 intermediary proteins which interact with the RXR independent of ligand, named RIP13 and RIP110. These proteins have been identified, using the interaction trap to isolate cDNAs encoding proteins that specifically interact with the LBD of hRXRα. RIP13 and RIP110 were also able to interact with TR in a hormone-independent manner (RIP 13) or in the presence of hormone (RIP110) and both with RAR independent of ligand. Recent studies revealed that RIP13 is a homolog of the coactivator N-CoR (see below).[162]

The TR uncoupling protein (TRUP) binds to the hinge region and to a N-terminal portion of the LBD of TR in a hormone-independent manner. DNA transfections indicated that TRUP inhibits transactivation by TR and RAR, but not RXR and ER in vivo. In bandshift experiments it was shown that TRUP interferes with the ability of TR and RAR to bind DNA, whereas RXR and ER were unaffected.[174] Taken together this reveals that TRUP belongs to a new class of proteins, which prevent a subclass of nuclear receptors from binding to their response elements.

The human receptor associated protein RAP46 was found by screening a human liver cDNA expression library with GR.[175] In pull down assays RAP46 was demonstrated to bind to GR, TR and ER in a hormone-independent manner.

In summary, all the proteins listed above are able to bind to nuclear hormone receptors. For many of these it remains to be shown whether they are responsible for modulation of gene transcription. To date there are only a few interacting proteins, for which such a coactivator or corepressor function have been shown.

Coactivators

The CREB binding protein (CBP), is a factor reported to interact with nuclear hormone receptors, most likely by serving as an integrator of multiple signal transduction pathways. This hypothesis is based on a study showing that CBP can interact with the LBDs of RAR, RXR, GR, ER and TR in vivo and in vitro, as shown by transfection studies and pull-down assays. Additional experiments revealed that AP-1 inhibition by nuclear receptors is based at least in part on competition for limiting amounts of CBP.[145] Interestingly this investigation also pointed out that CBP is able to interact with RIP 160, a cofactor known to interact with ER, TR and RXR (see above), suggesting an integrator function for CBP in multiple cellular signaling pathways.

The steroid receptor coactivator-1 (SRC-1) is a protein recently reported to stimulate hormone-induced in vivo transactivation of ER, GR, PR, TR and RXR.[176] Furthermore, SRC-1 was able to reverse the ability of ER to squelch transcriptional activation by PR. The N-terminal truncated form of SRC-1, when coexpressed with PR or TR, acted as a dominant negative repressor of receptor driven transcription, most likely by competition with the endogenous SRC-1 molecules. GST pull-down experiments indicated that the PR-LBD and the hinge region are the interacting domains. The binding was increased by progesterone and decreased by the antagonist RU 486.

GRIP1 was identified in a yeast two hybrid screen by binding to the LBD of GR, AR and ER in a hormone-regulated manner. GRIP1 contains an activation domain, since coexpression of GAL4/GR-LBD fusion protein and GRIP1 lead to strong activation of a UAS driven reporter, whereas the fusion protein alone had no effect.[177]

Sequence comparison revealed that GRIP1, SRC-1 and RIP160 (see above) are partially homologous, indicating that these proteins belong to a new protein family.

The receptor interacting protein RIP140 was detected by far Western blotting, using radioactive ER-LBD as probe.[157,160] In cotransfection studies RIP140 enhanced the ER dependent transcription in the presence of ligand. This enhancement was agonist dependent, since the antagonist 4-hydroxytamoxifen abolished the effect.[157]

The yeast SPT6 protein is another coactivator protein described by Baniahmad et al.[178] In transfection assays SPT6 potentiated hER activity in yeast. The LBD of ER was identified as an interacting domain in vivo. GST pull-down experiments confirmed these data, where specific interaction between ER-LBD and SPT6 was shown. However, the physiological function of SPT6 remains unknown and no mammalian homolog has been reported until now.

The glucocorticoid receptor interacting protein GRIP170 is a yet unknown protein of 170 kDa; it was detected in HeLa nuclear extracts as a protein interacting with a GR crosslinked to a double GRE. A cell-free in vitro transcription assay showed that a fraction highly enriched for GRIP170 was able to enhance glucocorticoid dependent transcription.[179]

Thus, a new class of proteins, which enhance nuclear receptor driven transcription, is another important step in mediating or modulating gene expression.

Corepressors

Intermediary factors that are functioning as corepressors have been described.[162,180] The nuclear receptor corepressor N-CoR[162] is a 270 kDa protein which interacts, only in the absence of hormone, with TR and RAR. It contains a nuclear receptor interaction domain in the distal C-terminus and a transferable repressor domain in the N-terminus. These domains were identified by GST pull-down assays. In a yeast two hybrid screen, N-CoR was shown to interact with TR and RAR but not with VDR, ER or GR. The responsible receptor domain seemed to be the hinge region, since specific mutations in this region of TR abolished in vitro binding to N-CoR. These mutations also eliminate the ligand independent repression function of TR, suggesting that N-CoR mediates ligand independent transcriptional repression. Further studies[181] pointed out that RAR/N-CoR interaction also requires the hinge region of the nuclear receptor and that the important part is an α-helical structure encompassing amino acids 181-198 of RAR.

Independently of N-CoR another silencing mediator for retinoic and thyroid hormone receptors SMRT was identified. It is a 168 kDa protein associating with these nuclear receptors in solution or when bound to DNA.[180] SMRT exhibited strong interaction with TR and RAR in the absence of ligand in vitro and in vivo, whereas addition of ligand decreased binding.

Like the coactivators, the corepressors seem to represent a new family of proteins, which is able to influence nuclear receptor activity. In fact, N-CoR and SMRT have sequence homologies throughout the whole protein.

References

1. Clever U, Karlson P. Induktion von Puff-Veränderungen in den Speicheldrüsenchromosomen in Chironomus tentans durch Ecdyson. Expl Cell Res 1960; 20:623-626.
2. Ashburner M, Chihara C, Meltzer P, Richards G. Temporal control of puffing activity in polytene chromosomes. Cold Spring Harbor Symp Quant Biol 1974; 38:655-662.
3. Koelle MR, Talbot WS, Segraves WA et al. The *Drosophila* EcR gene encodes anecdysone receptor, a new member of the steroid receptor superfamily. Cell 1991; 67:59-77.
4. Hollenberg SM, Weinberger C, Ong ES et al. Primary structure and expression of a functional human glucocorticoid receptor cDNA. Nature 1985; 318:635-41.
5. Miesfeld R, Rusconi S, Godowski PJ et al. Genetic complementation of a glucocorticoid receptor deficiency by expression of cloned receptor cDNA. Cell 1986; 46:389-99.
6. Green S, Walter P, Kumar V et al. Human oestrogen receptor cDNA: sequence, expression and homology to v-erb-A. Nature 1986; 320:134-9.
7. Krust A, Green S, Argos P et al. The chicken oestrogen receptor sequence: homology with v-erbA and the human oestrogen and glucocorticoid receptors. EMBO J 1986; 5:891-7.
8. Härd T, Kellenbach E, Boelens R, Maler BA, Dahlman K, Freedman LP, Carlstedt-Duke J, Yamamoto KR, Gustafsson JA, Kaptein R. Solution structure of the glucocorticoid receptor DNA-binding domain. Science 1990; 249:157-60.
9. Luisi BF, Xu WX, Otwinowski Z et al. Crystallographic analysis of the interaction of the glucocorticoid receptor with DNA. Nature 1991; 352:497-505.
10. Schwabe JW, Chapman L, Finch JT, Rhodes D. The crystal structure of the estrogen receptor DNA-binding domain bound to DNA: how receptors discriminate between their response elements. Cell 1993; 75:567-78.
11. Rastinejad F, Perlmann T, Evans RM, Sigler PB. Structural determinants of nuclear receptor assembly on DNA direct repeats. Nature 1995; 375:203-11.
12. Mangelsdorf DJ, Evans RM. The RXR heterodimers and orphan receptors. Cell 1995; 83:841-850.
13. Tora L, White J, Brou C et al. The human estrogen receptor has two independent non-acidic transcriptional activation functions. Cell 1989; 59:477-87.
14. Bocquel MT, Kumar V, Stricker C, Chambon P, Gronemeyer H. The contribution of the N- and C-terminal regions of steroid receptors to activation of transcription is both receptor and cell-specific. Nucleic Acids Res 1989; 17:2581-95.
15. Bourguet W, Ruff M, Chambon P, Gronemeyer H, Moras D. Crystal structure of the ligand-binding domain of the human nuclear receptor RXR-α. Nature 1995; 375:377-82.
16. Renaud JP, Rochel N, Ruff M et al. Crystal structure of the RAR-gamma ligand-binding domain bound to all-transretinoic acid. Nature 1995; 378:681-9.
17. Wagner RL, Apriletti JW, McGrath ME, West BL, Baxter JD, Fletterick RJ. A structural role for hormone in the thyroid hormone receptor. Nature 1995; 378:690-7.
18. Parker MG, White R. Nuclear receptors spring into action. Nat Struct Biol 1996; 3:113-115.
19. Schwabe WR. Transcriptional control: How nuclear receptors get turned on. Current Biology 1996; 6:372-374.

20. Wurtz JM, Bourguet W, Renaud JP et al. A canonical structure for the ligand-binding domain of nuclear receptors. Nat Struct Biol 1996; 3:87-94.
21. Beato M, Herrlich P, Schutz G. Steroid hormone receptors: many actors in search of a plot. Cell 1995; 83:851-7.
22. Laudet V, Hanni C, Coll J, Catzeflis F, Stehelin D. Evolution of the nuclear receptor gene superfamily. EMBO J 1992; 11:1003-13.
23. Thummel CS. From embryogenesis to metamorphosis: The regulation and function of *Drosophila* nuclear receptor superfamily members. Cell 1995; 83:871-877.
24. Mangelsdorf DJ, Thummel C, Beato M et al. The nuclear receptor superfamily: the second decade. Cell 1995; 83:835-9.
25. Guichon-Mantel A, Delabre K, Lescop P, Milgrom E. Intracellular traffic of sterpoid hormone receptors. J Steroid Biochem Molec Biol 1996; 56:3-9.
26. Brink M, Humbel BM, de Kloet ER, van Driel R. The unliganded glucocorticoid receptor is localized in the nucleus, not in the cytoplasm. Endocrinology 1992; 130:3575-3581.
27. Ogawa H, Inouye S, Tsuji FI, Yasuda K, Umesono K. Localization, trafficking, and temperature-dependence of the Aequoreagreen fluorescent protein in cultured vertebrate cells. Proc Natl Acad Sci USA 1995; 92:11899-903.
28. Akner G, Wikstrom AC, Gustafsson JA. Subcellular distribution of the glucocorticoid receptor and evidence for its association with microtubules. J Steroid Biochem Mol Biol 1995; 52:1-16.
29. van Steensel B, Brink M, van der Meulen K et al. Localization of the glucocorticoid receptor in discrete clusters in the cell nucleus. J Cell Sci 1995; 108:3003-11.
30. Nardulli AM, Grobner C, Cotter D. Estrogen receptor-induced DNA bending: orientation of the bend and replacement of an estrogen response element with an intrinsic DNA bending sequence. Mol Endocrinol 1995; 9:1064-76.
31. Shulemovich K, Dimaculangn, Katz D et al. DNA bending by thyroid hormone receptor: influence of half-site spacing and RXR. Nucleic Acids Res 1995; 23:811-8.
32. Prendergast P, Pan Z, Edwards DP. Progesterone receptor-induced bending of its target DNA: distinct effects of the A and B receptor isoforms. Molecular Endocrinology 1996; 10:393-407.
33. McBroom LD, Flock G, Giguere V. The non-conserved hinge region and distinct amino-terminal domains of the ROR α orphan nuclear receptor isoforms are required for proper DNA bending and RORα-DNA interactions. Mol Cell Biol 1995; 15:796-808.
34. Baniahmad A, Steiner C, Kohne AC, Renkawitz R. Modular structure of a chicken lysozyme silencer: Involvement of an unusual thyroidhormone receptor binding site. Cell 1990; 61:505-14.
35. Arnold R, Burcin M, Kaiser B et al. DNA bending by a silencer protein NeP1 is modulated by TR and RXR. NAR 1996; in press.
36. Li Q, Wrange O. Accessibility of a glucocorticoid response element in a nucleosome depends on its rotational positioning. Mol Cell Biol 1995; 15:4375-84.
37. Wong J, Shi YB, Wolffe AP. A role for nucleosome assembly in both silencing and activation of the Xenopus TRβ A gene by the thyroid hormone receptor. Genes Dev 1995; 9:2696-711.
38. La Spada AR, Wilson EM, Lubahn DB et al. Androgen receptor gene mutation in X-linked spinal and bulbar mascular atrophy. Nature 1991; 352, 77-79.
39. McPhaul MJ, Marcelli M, Tilley WD et al. Androgen resistance caused by mutations in the androgen receptor gene. FASEB J 1991; 5:2910-2915.
40. McPhaul MJ, Marcelli M, Zoppi MS, Griffin JE, Wilson JD. Genetic basis of endocrine disease 4: The spectrum of mutations in the androgen receptor gene that cause androgen resistance. J Clin Endocrinol Metab 1993; 76:17-23.
41. MacLean HE, Warne GL, Zajac JD. Defect of androgen receptor function: from sexual reversal to motor neuron disease. Mol Cell Endocrinol 1995; 112:133-141.
42. Lubahn DB, Brown TR, Simental et al. Sequence of the intron/exon junctions of the coding region of the human androgen receptor gene and identification of a point mutant in a family with complete

androgen insensitivity. Proc Natl Acad. Sci USA 1989; 86:9534-9538.

43. Quigley CA, Frieman KJ, Johnson A et al. Complete deletion of the androgen receptor gene: Definition of the null phenotype of the androgen insensitivity syndrome and determination of the carrier status. J Clin Endocrinol Metab 1992; 74: 927-933.
44. Trifiro M, Gottlieb BL, Pinsky L et al. The 56/58 kDa androgen-binding protein in male genital skin fibroblasts with a deleted androgen receptor gene. Mol Cell Endo 1991; 75:37-47.
45. Marcelli M, Tilley WD, Wilson et al. Definition of the human androgen gene structure permits the identification of mutations that cause androgen resistance: Premature termination of the receptor protein at amino acid residue 558 causes complete androgen resistance. Mol Endocrinol 1990; 4:1105-1116.
46. Marcelli M, Zoppi S, Grino PB et al. A mutation in the DNA-binding domain of the androgen receptor gene causes complete testicular feminization in a patient with receptor positive androgen resistance. J Clin Invest 1991; 87:1123-1126.
47. Brown TR, Lubahn DB, Wilson EM, French FS, Mieron CJ, Corden JL. Functional characterization of naturally occurring mutant androgen receptors from patients with complete androgen insensitivity. Mol Endocrinol 1990; 4:1759-1772.
48. Brandom DD, Markwick AJ, Chrousos GP, Loriaux D. Glucocorticoid resistance in human and non-human primates. Cancer Res 1989; 49:2203-2213.
49. Lamberts SWJ, Koper JW, Blemond P, denHolder FH, deJong FH. Cortisol receptor resistance. The variability of its clinical presentation and response to treatment. J Clin Endocrinol Metab 1992; 74:313-321.
50. Arai K, Chrousos GP. Syndromes of glucocorticoid and mineralocorticoid resistance. Steroid 1995; 60:173-179.
51. Werner S, Thoren M, Gustafsson JA, Bronnegard M. Glucocorticoid receptor abnormalities in fibroblasts from patients with idiopathic resistance to dexamethasone diagnosed when evaluated for adrenal disorders. J Clin Endocrinol Metab 1992; 75:1005-1009.
52. Malchoff CD, Javier EC, Malchoff DM et al. Primary cortisol resistance presenting as isosexual precocity. J Clin Endocrinol Metab 1990; 70:503-507.
53. Iida S, Gomi M, Moriwaki K et al. Primary cortisol resistance accompanied by a reduction in glucocorticoid receptors in two members of the same family. J Clin Endocrinol Metab 1985; 60:967-971.
54. Bronnegard M, Werner S, Gustafsson. Primary cortisol resistance associated with thermolabile glucocorticoid receptor in a patient with fatigue as the only symptom. J Clin Invest 1989; 78:1270-1278.
55. Malchoff DM, Brufsky A, Reardon et al. A mutation of the glucocorticoid receptor in primary cortisol resistance. J Clin Invest 1993; 91:1918-1925.
56. Komeseroff PA, Funder JW, Tuller PJ. Hormone-nuclear receptor interaction in health and disease. Baillieres Clin Endocrinol Metab 1994; 8:333-355.
57. Komesaroff PA. Pseudohypoaldosteronism: options for consideration. Steroids 1995; 60:168-172.
58. Smith EP, Boyd J, Frank GR et al. Estrogen resistance caused by a mutation in the estrogen receptor gene in man. N Engl J Med 1994; 331:1056-1061.
59. Hughes MR, Malloy PJ, Kieback DG et al. Point mutations in the human vitamin D receptor gene associated with hypocalcemic rickets. Science 1988; 242:1702-1725.
60. McKusick VA. Mendelian inheritance in man: catalogs of autosomal dominant, autosomal recessive and X-linked phenotypes. Johns Hopkins University Press: NewYork, Baltimore,1988.
61. Hewison M, O'Riordian JL. Hormone-nuclear receptor interaction in health and disease. Vitamin D resistance. Clin Endocrinol Metab 1994; 8:305-315.
62. Malloy PJ, Hochberg Z, Tiosano D, Pike JW, Hughes MR, Feldman D. The molecular basis of heriditary 1,25-dihydroxyvitamin D3 resistant rickets in seven related families. J Clin Invest 1990; 86:2071-2079.
63. Hughes MR, O'Malley BW. Genetic defects of receptors involved in disease. In:

Nuclear Hormone Receptors. M Parker ed. Academic Press 1991; 321-353
64. Ritchie HH, Hughes MR, Thompson ET et al. An ochre mutation in the vitamin D receptor gene causes heriditary 1, 25-dihydroxyvitamin D resistant rickets in three families. Proc Natl Acad Sci USA 1989; 86:9783-9787.
65. Usala SJ, Bale AE, Gesundheit N et al. Tight linkage between the syndrome of generalized thyroid hormone resistance and the human c-erbAβ gene. Mol Endocrinol 1988; 2:1217-1220.
66. Refetoff S, Weiss RE, Usala SJ. The syndrome of resistance to thyroid hormone. Endocr Rev 1993; 14:348-399.
67. Refetoff S. Resistance to thyroid hormone: an historical overview. Thyroid 1994; 4:345-349.
68. Usala. Molecular diagnosis and characterization of thyroid hormone resistance syndromes. Thyroid 1991; 1:361-367.
69. Brucker-Davis F, Skarulis MC et al. Genetic and clinical features of 42 kindreds with resistance to thyroid hormone. Ann Intern Med 1995; 123:572-583.
70. Baniahmad A, Tsai SY, O'Malley BW, Tsai M-J. The kindred S thyroid hormone receptor is an active and constitutive silencer and a repressor for thyroid hormone and retinoic acid. Proc Natl Acad Sci USA 1992; 89:10633-10637.
71. Baniahmad A, Leng X, Burris TP, Tsai SY, Tsai M-J, O'Malley BW. The t4 activation domain of the thyroid hormone receptor is required to release a corepressor(s) necessary for transcriptional silencing. Mol Cell Biol 1995; 15:76-86.
72. Danielian P, White SR, Lees JA, Parker GM. Identification of a conserved region required for hormone dependent transcriptional activation by steroid hormone receptors. EMBO J 1992; 11:1025-1033.
73. Barretino D, Ruiz MdMV, Stunnenberg HG. Characterization of the ligand-dependent transactivation domain of thyroid hormone receptor. EMBO J 13: 3039-3049.
74. Durand B, Saunders M, Gaudon C, Roy B, Chambon P. Activation domain2 (AF-2) of RAR and RXR: presence of a conserved autonomous constitutive activating domain and influence of the nature of the response element on AF-2 activity. EMBO J 1994; 13:5370-5382.
75. Groenhout E, Dorin RI. Generalized thyroid hormone resistance due to a deletion of the carboxy terminus of the c-erbAβ receptor. Mol Cell Endocr 1994; 99:81-88.
76. Collingwood TN, Adams M, Tone Y, Chatterjee VKK. Spectrum of transcriptional, dimerization, and dominant negative properties of twenty different mutant thyroid hormone β-receptors in thyroid hormone resistance syndrome. Mol Endocrinol 1994; 8:1262-1277.
77. Piedrafita FJ, Ortiz MA, Pfahl M. Thyroid hormone receptor-β mutants associated with generalized resistance to thyroid hormone show defects in their ligand-sensitive repression. Mol Endocrinol 1995; 9:1533-1548.
78. Takeda K, Sakurai A, DeGroot LJ, Refetoff S. Recessive inheritance of thyroid hormone resistance caused by a complete deletion of the protein-coding region of the thyroid hormone receptor-β gene. J Clin Endocrinol Metab 1992; 74:49-55.
79. Ono S, Schwartz ID, Mueller OT, Root AW, Usala SJ, Bercu BB. Homozygosity for a dominant negative thyroid hormone receptor gene responsible for generalized resistance to thyroid hormone. J Clin Endocrinol Metab 1991; 73:990-994.
80. Usala SJ, Menke JB, Watson TL et al. A homozygous deletion in the c-erbAβ thyroid hormone receptor gene in a patient with generalized thyroid hormone resistance: Isolation and characterization of the mutant receptor. Mol Endocrinol 1991; 5:327-335.
81. Sakurai A, Miyamoto T, Refetoff S, DeGroot LJ. Dominant negative transcriptional regulation by a mutant thyroid hormone receptor β in a family with generalized resistance to thyroid hormone. Mol Endocrinol 1990; 4:1988-1994.
82. Chatterjee VKK, Nagaya T, Madison LD, Datta S, Rentoumis A, Jameson JL. Thyroid hormone resistance syndrome. Inhibition of normal receptor function by mutant thyroid hormone receptors. J Clin Invest 1991; 87:1977-1984.

83. Jameson JL. Mechanisms by which thyroid hormone receptor mutations cause clinical syndromes of resistance to thyroid hormone. Thyroid 1994; 4:485-492.
84. Hodin RA, Lazar MA, Wintman BJ et al. Identification of a thyroid hormone receptor that is pituitary specific. Science 1989; 244:76-79.
85. Lazar MA, Chin WW. Induction of two messenger ribonucleic acids in rat GH3 cells by thyroid hormone. Mol Endocrinol 1988; 2:479-484.
86. Damm K, Thompson CC, Evans RM. Protein encoded by v-erbA functions as a thyroid hormone receptor antagonist. Nature 1989; 339:593-597.
87. Sap J, Munoz A, Schmitt J et al. Repression of transcription mediated at a thyroid hormone response element by the v-erbA oncogene product. Nature 1989; 340:242-244.
88. Grandillon O, Jurdic P, Pain B et al. Expression of the v-erbA product, an altered nuclear hormone receptor, is sufficient to transform erythrocytic cells in vitro. Cell 1989; 58:115-121.
89. Schroeder C, Raynoschek C, Fuhrman U et al. The v-erbA oncogene causes repression of erythrocyte-specific genes and an immature, aberrant differentiation phenotype in normal erythroid progenitors. Oncogene 1990; 5:1445-1453.
90. Zenke M, Munoz A, Sap J, Vennstrom B, Beug H. V-erbA oncogene activation entails the loss of hormone-dependent regulator activity of c-erbA. Cell 1990; 61:1035-1049
91. Bonde B, Sharif M, Privalsky ML. Ontogeny of the v-erbA oncoprotein from the thyroid hormone receptor: an alteration in the DNA binding domain plays a role crucial for v-erbA function. J Virol 1991; 65:2037-2046
92. Scharif M, Privalsky ML. V-erbA oncogene function in neoplasia correlates with its ability to repress retinoic acid receptor action. Cell 1991; 66:885-893.
93. Muscatelli F, Strom TM, Walker et al. Mutations in the DAX1 gene give rise to both X-linked adrenal hypoplasia congenita and hypogonadotropic hypogonadism. Nature 1994; 372:672-676.
94. Zanaria E, Muscatelli F, Bardoni B et al. An unusual member of the nuclear hormone receptor superfamily responsible foe X-linked adrenal hypoplasia congenita. Nature 1994; 372:635-641.
95. Kletter GB, Kelch RP. Disorders of puberty in boys. Endocrin Metab Clin North Am 1993; 22:425-477.
96. Lavau C, Dejean A. The t[15:17] translocation in acute promyelocytic leukemia. Leukemia 1994; 8:9-15.
97. Weis K, Ramband S, Lavau et al. Retinoic acid regulates aberrant nuclear localization of PML-RAR α in acute promyeloticytic leukemia cells. Cell 1994; 76: 345-356.
98. Licht JD, Shaknoviel R, English MA et al. Reduced and altered DNA binding and transcriptional prperties of the PLZF-retinoic acid receptor α chimera generated in t[11:17] associated acute promyelocytic leukemia. Oncogene 1996; 12:323-336.
99. Licht JD, Chomienne C, Goy A et al. Clinical and molecular characterization of a rare syndrome of acute promyelocytic leukemia associated with translocation (11;17). Blood 1995; 85:1083-1094.
100. Redner RL, Rush EA, Faas S, Rudert WA, Corey SJ. The t[5:17] variant of acute promyelocytic leukemia expresses a nucleoplasmin-retinoic acid receptor fusion. Blood 1996; 87:882-886.
101. Takatsuki H, Umemura T, Nishimura J, Nawata H. Which progenitor is the target cell in the development of acute promyelotcytic leukemia? Lek Lymphoma 1995; 17:251-254.
102. Stadler M, Chelbi A, Lix MK et al. Transcriptional induction of the PML growth suppressor gene is induced by interferons is mediated through an ISRE and GAS element. Oncogene 1995; 11:2565-2573.
103. Dyck JA, Maul GG, Miller WH Jr. A novel macronuclear structure is a target of the promyelocyte-retinoic acid receptor oncoprotein. Cell 1994; 76:333-343.
104. Cornic M, Delva C, Castaigne S et al. In vitro all-trans retinoic acid (ATRA) sensitivity and cellular retinoic acid binding protein (CRABP) levels in relapse leukemic cells after remission induction by ATRA in acute promyelocytic leukemia. Leukemia 1994; 8:914-917.

105. Chen Z, Chen SJ, Wang ZY. Retinoic acid and acute promyelocytic leukemia: a model for targeting treatment for human cancer. CRAcad Sci III 1994; 317:1135-1141.
106. Guidez F, Huang W, Tong JH et al. Poor response to all-trans reitnoic acid therapy in a t[11:17] PLZF-RARα patient. Leukemia 1994; 8:312-317.
107. Chen SJ, Wang ZJ, Chen Z. Acute promyelocytic leukemia: from clinical to molecular biology. Stem Cells Dayt 1995; 13:22-31.
108. Sluyser. Hormone resistance in cancer: the role of abnormal steroid receptors. Crit Rev Oncog 1994; 5:539-554.
109. Traish AM, Newton AW, Styperek K et al. Estrogen receptor functional status in human breast cancer. Diagn Mol Pathol 1995; 4:220-228.
110. Henderson BE, Ross RK, Pike MC. Toward the primary prevention of cancer. Science 1991; 254:1131-1138.
111. Newmark JR, Hardy DO, Tonb DC et al. Androgen receptor gene mutations in human prostate cancer. Proc Natl Acad Sci USA 1992; 89:6319-6323.
112. Powers JH, Hillman AG, Tong DC, Haman JM. Cloning and expression of mutant glucocorticoid sensitive and resistant human leukemic cells. Cancer Res 1995; 53:4059-4065.
113. Nazareth LV, Thompson EB. Leukemic cells apoptosis caused by constitutively active mutant glucocorticoid receptor fragment. Recent Prog Horm Res 1995; 50:417-421.
114. Stresser-Wozak EM, Hattmannstorfer EM, Hala M et al. Splice site mutation in the glucocorticoid receptor gene causes resistance to glucocorticoid induced apoptosis in human acute leukemic cell line. Cancer Res 1995; 55: 348-353.
115. Orti E, Bodwell J, Munck A et al. Phosphorylation of steroid hormone receptors. Endocrin Rev 1992; 13:105-128.
116. Orti E, Mendel DB, Smith LI et al. Agonist-dependent phosphorylation and nuclear dephosphorylation of glucocorticoid receptors in intact cells. J Biol Chem 1989; 264:9728-9731.
117. Eggert M, Muller M, Renkawitz R et al. The glucocorticoid hormone receptor. Ind Gen Exp 1994; 2:131-156.
118. Hoeck W, Rusconi S, Groner B et al. Down-regulation and phosphorylation of glucocorticoid receptors in cultured cells. J Biol Chem 1989; 264:14396-14402.
119. Mason SA, Hously PR et al. Site-directed mutagenesis of the phospohrylation sites in the mouse glucocoritcoid receptor. J Biol Chem 1993; 268:21501-21504.
120. Sullivan W, Smith DF, Beito TG et al. Hormone-dependent procressing of the avian progesterone receptor. J Cell Biochem 1988; 36:103-119.
121. Denner L, Schrader W, O'Malley B et al. Hormonal regulation and identification of chicken progesterone receptor phosphorylation sites. J Bio Chem 1990; 265:16548-16555.
122. Poletti A, Weigel N et al. Idendification of a hormone-dependent phosphorylation site adjacent to the DNA-binding domain of the chicken progesterone receptor. Mol Endocrin 1992; 7:241-246.
123. Zhang Y, Beck CA, Poletti A et al. Identification of a group of ser-pro motif hormone-inducible phosphorylation sites in the human progesterone receptor. Mol Endocrin 1995; 9:1029-1040.
124. Zhang Y, Beck CA, Poletti A et al. Identification of phosphorylation sites unique to the B form of human progesterone receptor: in vitro phosphorylation by casein kinase II. J Biol Chem 1994; 269:31034-31040.
125. Beck CA, Zhang Y, Weigel NL et al. Two types of anti-progestins have distinct effects on site-specific phosphorylation of human progesterone receptor. J Biol Chem 1996; 271:1209-1217.
126. Le Goff P, Montana MM, Schodin DJ et al. Phosphorylation of the human estrogen receptor: identification of hormone-regulated sites and examination of their influence on transcriptional activity. J Biol Chem 1994; 269:4458-4466.
127. Ali S, Metzger D, Bonert JM et al. Modulation of transcriptional activation by ligand-dependent phosphorlylation of the human estrogen receptor A/B region. EMBO J 1993; 12:1153-1160.
128. Kato S, Endoh H, Masuhiro Y et al. Activation of the estrogen receptor through phosphorylation by mitogen-activated protein kinase. Science 1995; 270: 1491-1494.

129. Arnold SF, Obourn JD, Jaffe H et al. Serine 167 is the major estradiol-induced phosphorylation site on the human estrogen receptor. Mol Endocrin 1994; 8:1208-1214.
130. Lahooti H, White R, Hoare SA et al. Identification of phosphorylation sites in the mouse oestrogen receptor. J Steroid Biochem Mol Biol 1995; 55:305-313.
131. Kraus WL, Mcinerney E, Katzenellenbogen BS et al. Ligand-dependent, transcriptionally productive association of the amino- and carboxyl-terminal regions of a steroid hormone nuclear receptor. Proc Natl Acad Sci 1995; 92:12314-12318.
132. Goldberg Y, Glineur C, Gesquiere JC et al. Activation of protein kinase C or cAMP-dependent protein kinase increases phosphorylation of the c-erbA-encoded thyroid hormone receptor and of the v-erbA-encodend protein. EMBO J 1989; 7:2425-2433.
133. Glineur C, Bailly M, Ghysdael J et al. The c-erbAa-encoded thyroid hormone receptor is phosphorylated in its amino terminal domain by casein kinase II. Oncogene 1989; 4:1247-1254.
134. Bhat MK, Ashizawa K, Cheng SY et al. Phosphorylation enhances the target gene sequence-dependent dimerization of thyroid hormone receptor with retinoid X receptor. Proc Natl Acad Sci 1994; 91: 7927-7931.
135. Katz D, Reginato M, Lazar M et al. Functional regulation of thyroid hormone receptor variant TRα2 by phosphorylation. Mol Cell Biol 1995; 15:2341-2348.
136. Jones BB, Jurutka PW, Haussler CA et al. Vitamin D receptor phosphorylation in transfected ROS17/2.8 cells is localized to the N-terminal region of the hormonebinding domain. Mol Endocrin 1991; 5:1137-1146.
137. Jurutka PW, Terpening CM, Haussler MR et al. The 1,25-dihydroxy-vitamin D3 receptor is phosphorylated in response to 1,25-dihydroxy-vitamin D3 and 22-oxacalcitriol in rat osteoblasts, and by casein kinase II, in vitro. Biochem 1993; 32: 8184-8192.
138. Jurutka PW, Hsieh JC, Nakajima S et al. Human vitamin D receptor phosphorylation by casein kinase II at ser-208 potentiates transcriptional activation. Proc Natl Acad Sci 1996; 93:3519-3524.
139. Huggenvik JI, Collard MW, Kim YW et al. Modification of the retinoic acid signaling pathway by the catalytic subunit of protein kinase-A. Mol Endocrin 1993; 7:543-550.
140. Rochette-Egly C, Oulad-Abdelghani M, Staub A et al. Phosphorylation of the retinoic acid receptor-α by protein kinase A. Mol Endocrin 1995; 9:860-871.
141. Jonat C, Stein B, Herrlich P et al. Positive and negative regulation of collagenase gene expression. Matrix Suppl 1992, 1:145-155.
142. Maroder M, Farina AR, Vacca A et al. Cell-specific bifunctional role of Jun oncogene family members on glucocorticoid receptor-dependent transcription. Mol Endocrinol 1993; 7:570-584.
143. Kerppola TK, Luk D, Curran T. Fos is a preferential target of glucocorticoid receptor inhibition of AP-1 activity in vitro. Mol Cell Biol 1993; 13:3782-3791.
144. Reik A, Stewart F, Schütz G. Cross-talk of signal transdution pathways: Two mechanisms are involved in the control of tyrosine aminotransferase gene expression by phorbol esters. Mol Endocrinol 1994; 8:490-497.
145. Kamei Y, Xu L, Heinzel T et al. A CBP integrator complex mediates transcriptional activation and AP-1 inhibition by nuclear receptors. Cell 1996; 85:403-414.
146. Jondal M, Okret S, McConkey D et al. Killing of immature CD4+ CD8+ thymocytes in vivo by anti CD3 or 5′-(N-ethyl)-carboxamide adenosine is blocked by glucocorticoid receptor antagonist RU-486. Eur J Immunol 1993; 23: 1246-1250.
147. Helmberg A, Auphan N, Caelles C et al. Glucocorticoid-induced apotosis of human leukemic cells is caused by the repressive function of the glucocorticoid receptor. EMBO J 1995; 3:452-460.
148. Ray A, Prefontaine K et al. Physical association and functional antagonism between the p65 subunit of transcription factor NF-κB and the glucocorticoid receptor. Proc Natl Acad Sci 1994; 91: 752-756.

149. Scheinman R, Gualberto A, Jewell C et al. Characterization of mechanisms involed in transrepression of NF-κB by activated glucocorticoid receotors. Mol Cell Bio 1995; 15:943-953.
150. Caldenhoven E, Liden J, Wissink S et al. Negative cross-talk between RelA and the glucocorticoid receptor: a possible mechanism for the antiinflammatory action of glucocorticoids. Mol Endocrinol 1995; 9:401-412.
151. Stein B, Yang M et al. Repression of the interleukin-6 promotor by estrogen receptor is mediated by NF-κB and C/EBPb. Mol Cell Biol 1995; 9:4971-4979.
152. McEwan IJ, Wright APH, Dahlman-Wright et al. Direct interaction of the t1 transactivation domain of the human glucocorticoid receptor with the basal transcriptional machinery. Mol Cell Biol 1993; 13:399-407.
153. Ing NH, Beekman JM, Tsai SY et al. Members of the steroid hormone superfamily interact with TFIIB. J Biol Chem 1992; 267:17617-17623.
154. Baniahmad A, Ha I, Reinberg D et al. Interaction of human thyroid receptor β with transcription factor TFIIB may mediate target gene derepression and activation by thyroid hormone. Proc Natl Acad Sci 1993; 90:8832-8836.
155. Fondell JD, Brunel F, Hisatake K et al. Unliganded thyroid hormone receptor a can target TATA-binding protein for transcriptional repression. Mol Cell Biol 1996; 16:281-287.
156. Schulman IG, Chakravarti D, Juguilon H et al. Interactions between the retinoic X receptor and a conserved region of the TATA-binding protein mediate hormone-dependent transactivation. Proc Natl Acad Sci 1995; 92:8288-8292.
157. Cavaillès V, Dauvois S, L'Horset F et al. Nuclear factor RIP140 modulates transcriptional activation by the estrogen receptor. EMBO J 1995; 14:3741-3751.
158. Cavaillès V, Dauvois S, Danielian PS et al. Interaction of proteins with transcriptionally active estrogen receptors. Proc Natl Acad Sci 1994; 91:10009-10013.
159. Jacq X, Brou C, Lutz Y et al. Human TAF II30 is present in a distinct TF IID complex and is required for transcriptional activation by the estogen receptor. Cell 1994; 79:107-117.
160. Halachmi S, Marden E, Martin G et al. Estrogen receptor-associated proteins: possible mediators of hormone-induced transcription. Science 1994; 264: 1455-1458.
161. Yoshinaga SK, Peterson CL, Herskowitz I et al. Roles of SWI1, SWI12 and SWI13 proteins for transcriptional enhancement by steroid receptors. Science 1992; 258: 1598-1604.
162. Hörlein A, Näär AM, Heinzel T et al. Ligand-independent repression by the thyroid hormone receptor mediated by a nuclear receptor co-pressor. Nature 1995; 377:397-404.
163. Tamkun JW, Deuring R, Scott MP et al. A regular of *Drosophila* homeotic genes structurally related to the yeast transcriptional activator SNF2/SW12. Cell 1992; 68:561-572.
164. Lee J, Choi HS, Gyurist J et al. Two classes of protein dependent on either the presence or absence of thyroid hormone for interaction with thyroid hormone receptor. Mol Endocrin 1995; 9:234-254.
165. Muchard C, Yaniv M et al. A human homologue of Saccharomyces cerevisiae SNF2/SW12 and drosophila brm genes potentiates transcriptional activation by the glucocorticoid receptor. EMBO J 1993; 12:4279-4290.
166. Khavari PA, Peterson CL, Tamkun JW et al. BRG1 contains a conserved domain of the SWI2/SNF2 family necessary for normal mitotic growth and transcription. Nature 1993; 366:170-174.
167. Lee J, Ryan F, Swaffield J et al. Interaction of the thyroid-hormone receptor with a conserved transcriptional mediator. Nature 1995; 347:91-94.
168. Kupfer SR, Marschke KB, Wilson EM et al. Receptor accessory factor enhances specific DNA binding of androgen and glucocorticoid receptors. J Biol Chem 1993; 268:17519-17527.
169. Vom Baur E, Zechel C, Heery D et al. Differential ligand-dependent interactions between the AF-2 activating domain of nuclear receptors and the putative transcriptional intermediary factors

mSUG1 and TIF1. EMBO J 1996; 15: 110-124.
170. Burns K, Duggan B, Atkinson EA et al. Modulation of gene expression by calreticulin binding to the glucocorticoid receptor. Nature 1994; 367:476-480.
171. Le Douarin B, Zechel C, Garnier JM et al. The n-terminal part of TIF1, a putative mediator of the ligand-dependent activation function (AF-2) of nuclear receptors, is fused to B-raf in the oncogenic protein T18. EMBO J 1995; 14: 2020-2033.
172. Dedhar S, Rennie PS, Shago M et al. Inhibition of nuclear hormone receptor activtiy by calreticulin. Nature 1994; 367:480-483.
173. Seol W, Choi HS, Moore D et al. Isolation of proteins that interact specifically with the retinoid X receptor: two novel orphan receptors. Mol Endocrin 1995; 9:72-85.
174. Burris T, Nawaz Z, Tsai MJ et al. A nuclear hormone receptor-associated protein that inhibits transactivation by the thyroid hormone and retinoic acid receptors. Proc Natl Acad Sci 1995; 92:9525-9529.
175. Zeiner M, Gehring U et al. A protein that interacts with members of the nuclear hormone receptor family: identification and cDNA cloning. Proc Natl Acad Sci 1995; 92:11465-11469.
176. Onate SA, Sophia Y, Tsai MJ et al. Sequence and characterization of a coactivator for the steroid hormone receptor superfamily. Science 1995; 270: 1354-1357.
177. Hong H, Kohli K, Trivedi A et al. GRIP1, a novel mouse protein that serves as a transcriptional coactivator in yeast for the hormone binding domains of steroid receptors. Proc Natl Acad Sci 1996; 93:4948-4952.
178. Baniahmad C, Nawarz Z, Baniahmad A et al. Enhancement of human estrogen receptor activity by SPT6: a potential coactivator. Mol Endocrin 1995; 9:34-43.
179. Eggert M, Möws CC, Tripier D et al. A fraction enriched in a novel glucocorticoid receptor-interacting protein stimulates receptor-dependent transcription in vivo. J Biol Chem 1995; 270(51): 30755-30759.
180. Don Chen J, Evans RM et al. A transcriptional co-repressor that interacts with nuclear hormone receptors. Nature 1995; 377:454-457.
181. Kurokawa R, Söderström M, Hörlein A et al. Polarity-specific activities of retinoic acid receptors determined by a co-repressor. Nature 1995; 377:451-454.
182. Bodwell JE, Orti E, Coull JM et al. Identification of phosphorylated sites in the mouse glucocorticoid receptor. J Bio Chem 1991; 266:7549-7555.

CHAPTER 7

The STAT Family of Transcription Factors

Michael Gilman and Marie-Luce Vignais

The meteoric rise of STAT (*s*ignal *t*ransducers and *a*ctivators of *t*ranscription) proteins from relative obscurity to their present prominence is breathtaking, even by modern standards of rough-and-tumble molecular and cellular biology. While we've resisted constructing a graph of STAT-related publications over the last several years, it is instructive to recognize that prior to 1992, these proteins were but finicky and unusual DNA-binding activities on mobility-shift gels, largely the preserve of a single research group which had worked for many years on interferon signaling. The stunning succession of revelations concerning STAT structure, mechanism of activation, and ubiquity in signal transduction has brought the STATs much notoriety. But they have not given up their secrets readily.

This chapter will focus on the general themes of STAT structure and function and key unanswered questions. For more detailed and scholarly reviews, the reader is referred to refs. 1, 2. The real significance of the STAT proteins is the very direct connection they provide between cell surface and the genome. And while the physical means by which STATs transfer information to the nucleus is elegantly simple and clear, how the specific message conveyed by the original signal is embodied in active STAT proteins and relayed to appropriate target genes is still not well understood. In T cells, for example, interferon-γ (IFN-γ) activates Stat1, IL-12 activates Stat4, IL-2 activates Stat5 (and sometimes Stat3), and IL-4 activates Stat6, and each cytokine elicits a different biological response. How is a particular STAT selected by a particular receptor? How do different STATs target different genes, especially given very similar DNA-binding specificities?

Also of interest: What are the biological functions of the different STAT proteins? For some STATs, whose expression and activation are highly restricted, predicting their function is straightforward, and knock-out mice have brought confirmation. But for other STATs, which are ubiquitously expressed and activated by a multitude of external signals, decoding their critical functions is much more difficult.

And, finally, structural analysis of STAT proteins will certainly yield new surprises. These proteins house many different—and important—biochemical activities. They can select both a specific phosphopeptide sequence in a receptor and a specific DNA sequence in a target

Transcription Factors in Eukaryotes, edited by Athanasios G. Papavassiliou.

gene. They can be alternately in the cytoplasm or in the nucleus. They can exist as monomers, homodimers, heterodimers, or in larger multicomponent complexes. They can activate transcription or repress. They are substrates for several protein kinases. Yet, with the exception of an SH2 homology, the structure and organization of STAT proteins remain mysterious.

HISTORY

The modern history of the STAT proteins began in the mid-1980s, with the discovery of the first genes regulated by interferon-α (IFN-α). Analysis of the promoters of these genes, principally by James Darnell and collaborators, pointed to a common regulatory element, the ISRE, and eventually to protein factors that bound this element.[3-5] One of these factors, ISGF-3, had the striking property of being rapidly induced, apparently as a primary consequence of engagement of the interferon receptor.[6-8] ISGF-3 was a complex of multiple polypeptide chains.[9] When cDNAs encoding these proteins were finally cloned, two of them shared sequence homology throughout their entire length.[10,11] These became the founding members of the STAT family—Stat1 and Stat2. It was soon discovered that activation of ISGF-3 was associated with the rapid phosphorylation of Stat1 and Stat2 on tyrosine.[12-17]

At around the same time, Brent Cochran and coworkers described a factor termed SIF that bound to a sequence in the enhancer of the *c-fos* proto-oncogene.[18] Like ISGF-3, SIF activity was rapidly induced by an extracellular ligand, in this case platelet-derived growth factor (PDGF), and induction was a primary response to PDGF that occurred without requirement for new protein synthesis. With the availability of molecular reagents derived from the first STAT clones, SIF was shown to be composed of Stat1 and a distinct STAT-like protein, later found to be Stat3.[19-22]

What was remarkable about these observations was not only the novelty of the mechanism—most signal transduction pathways known at the time took a more circuitous route to the nucleus—but also the idea that two very distinct types of receptors could share a common pathway. The PDGF receptor is a single polypeptide chain possessing intrinsic protein tyrosine kinase activity, whereas the interferon receptors, though incompletely understood at the time, were believed to be multiprotein complexes devoid of intrinsic catalytic activity. The observation that both classes of receptors activated the same transcription factor, in a strikingly direct way, was difficult to explain and even a bit troubling, since these receptors generally exerted quite different biological effects on cells. Indeed, the issue of specificity of STAT signaling, while brought into sharper focus in recent years, still remains largely unresolved.

Interestingly, both of these early STAT sightings turned out to be nonparadigmatic. ISGF-3, with its multisubunit structure, is much more complicated structurally than STAT factors characterized subsequently, which are generally simple homo- or heterodimers. Further, the biological significance and generality of STAT activation by receptor tyrosine kinases (RTKs), such as the PDGF receptor, is not nearly as clear today as it is for noncatalytic cytokine receptors.

A parallel line of research, also with its roots in the interferon system, unearthed another protein family now closely linked to STAT activation, the JAKs. These are large nonreceptor protein tyrosine kinases, first cloned by sequence homology to other kinases.[23-26] They were without assigned function until an ambitious project to use somatic cell genetics to dissect the interferon signaling pathway[27] revealed that the JAK-family protein Tyk2 was essential for signaling by IFN-α and IFN-β.[28] Soon thereafter, the roles of Jak1 and Jak2 in interferon signaling were also established in the same system.[29,30] Given the structural relationship of the IFN-γ receptor, in particular, to the broader family of cytokine receptors, the JAK connection was rapidly generalized to virtually all cytokine receptors; hand in hand came the recognition that STAT activation was also a common feature of cytokine action.

Table 7.1. STAT proteins and the signals that activate them.

Stat1	Stat2	Stat3	Stat4	Stat5	Stat6
IFNα/β	IFNα/β	IFNα/β	IL-12	IL-2	IL-4
IFNγ		IFNγ		IL-3	IL-13
IL-2		IL-2		IL-7	
IL-3		IL-4		IL-9	
IL-5		IL-6		IL-15	
IL-7		IL-7		Prolactin	
IL-11		IL-9		EPO	
LIF		IL-11		TPO	
Prolactin		IL-12		GH	
GH		IL-13		GM-CSF	
GM-CSF		IL-15		EGF	
PDGF		G-CSF			
EGF		TPO			
		GH			
		GM-CSF			
		PDGF			
		EGF			

The list is not exhaustive. What is most significant, however, is the pattern. Stat1, Stat3 and Stat5 are activated by a broad spectrum of signals, whereas Stat1, Stat2 and Stat4 are each coupled to only one or a few receptors. Stat5 is encoded by two distinct but closely linked and closely related genes. Most studies have not determined which of the two Stat5 isoforms is activated by a particular stimulus.

A STAT PRIMER

We now know of seven STAT proteins in mammalian cells (Table 7.1). Stat1 and Stat2 were the founding members of the family, originally identified as components of the IFN-α-induced ISGF-3 complex.[10,11] ISGF-3 is composed of at least one molecule each of Stat1 and Stat2 plus an unrelated 48 kDa subunit.[31] Stat1 was also identified as the IFN-γ-induced protein that bound to a so-called GAS site distinct in sequence from ISREs.[12] In contrast to the multicomponent nature of ISGF-3, Stat1 binds to GAS sites apparently as a simple homodimer. Stat1 activation is essential for the biological function of all three interferons.[32] It is activated by multiple cytokine receptors and RTKs, although the functional relevance of Stat1 activation for any ligand other than the interferons is not yet obvious.

Stat2 remains unique among the STAT proteins in that it does not appear to bind DNA on its own. Stat2 activation is restricted to the IFN-α pathway, and Stat2 only binds DNA in the context of ISGF-3.[33] But its role in ISGF-3 function—and in the biological activity of IFN-α and -β—is critical in that it provides the transcriptional activation function to ISGF-3 and also plays a key role in the activation of Stat1.[34,35]

Stat3, like Stat1, is both ubiquitously expressed and activated. Its role is perhaps best characterized in conjunction with the IL-6 class of cytokines, whose receptors share a common signaling subunit, gp130. Biochemically, Stat3 was first recognized as a component of the PDGF-induced SIF protein and as an activity termed APRF, which mediates induction of acute-phase genes.[19,36,37] Biochemical purification of APRF led to the cloning of the Stat3 cDNA[36] at approximately the same time as it yielded to homology cloning.[38] Stat3 is also activated by multiple cytokine and growth factor

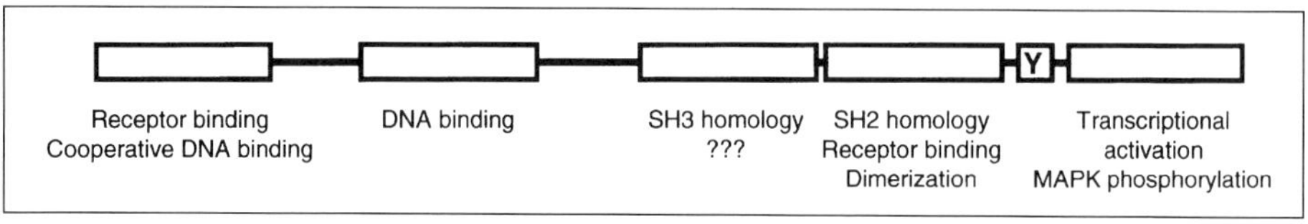

Fig. 7.1. STAT topography and functional units. The open boxes along the linear representation of STAT primary amino acid sequence indicate areas of high sequence similarity among all known STATs. The rightmost box represents the C-terminal tail, which is not significantly homologous among STATs. Indicated below each box are any known domain homologies and reported functions for each region. The box carrying a "Y" represents the unique tyrosine residue that is phosphorylated upon activation.

receptors, as well as by IFN-α. Stat3 forms both homodimers and heterodimers with Stat1. The critical biological function of Stat3, if any, is not yet known.

Stat4 was identified initially by homology cloning and remained an orphan for some time.[38,39] Unlike previously characterized Stats, Stat4 exhibited a highly restricted pattern of expression, limited to hematopoietic cells and spermatagonia. Stat4 was eventually recognized as a target for activation by IL-12 in lymphoid cells.[40,41] Stat4 remains one of the most selectively activated of the STAT proteins and it plays a crucial role in controlling the Th1 response in T cells which drives cellular immunity.

Stat5 is encoded by two distinct but closely linked and highly related genes. It was first identified as a biochemical activity termed MGF, for mammary gland factor.[42,43] MGF is induced by prolactin in mammary cells where it is responsible for the upregulation of milk protein genes. Like Stat1 and Stat3, however, Stat5 activation is observed in many different cells and in response to an astonishing array of stimuli, including growth hormone, erythropoeitin, thrombopoietin, leptin, GM-CSF, multiple interleukins, PDGF and CSF-1.

Stat6 resembles Stat4 in that its expression pattern is restricted and its activity appears to be dedicated to one (or a few) cytokine(s). Stat6 was first identified as an activity induced by IL-4[44] and was purified and cloned on this basis.[45] It was also identified on the basis of an expressed sequence tag.[46] Stat6 is activated by a few other stimuli, including IL-3 and PDGF, but its crucial function appears to be on the IL-4 pathway controlling the Th2 response in T cells.

At least one STAT protein is known in *Drosophila*, where it functions in conjunction with a previously identified *Drosophila* JAK family member, hopscotch.[47,48] The fly STAT appears to mediate the activation of certain pair-rule genes during embryonic development and is involved in the proliferation of blood cells. The receptor/ligand systems that activate the JAK/STAT pathway in *Drosophila* are not known.

STAT TOPOGRAPHY AND THE ACTIVATION CYCLE

All STAT proteins share significant sequence homology through their entire length (Fig. 7.1). Homology is concentrated in four regions. The N-terminal regions of the STAT proteins exhibit high homology and appear to harbor multiple functions. In Stat2, this region is required for binding to the IFN-α receptor, and even small N-terminal deletions inactivate Stat2.[49] In Stat1 and Stat4, this region is required for cooperative interactions between STAT dimers.[50,51] A highly conserved internal domain of roughly 100 amino acids houses a DNA-binding domain of unknown structure.[52] Near the C-terminus, all STAT proteins contain an SH2 domain,[14] which, as described below, is the key to the STAT activation cycle. Immediately C-terminal to the SH2 domain lies the single tyrosine residue that is phosphorylated in response to incoming signals.[16] The proteins diverge in sequence at the very C-terminus. In several proteins, this region carries a transcriptional

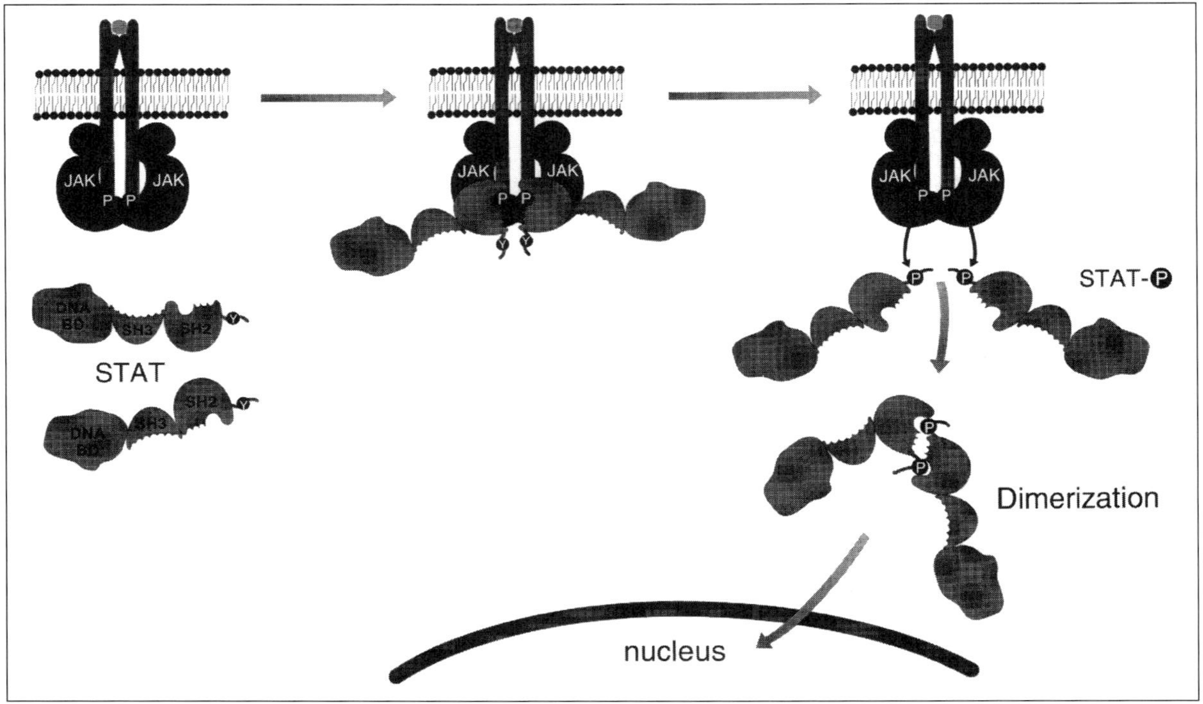

Fig. 7.2. STAT activation. The scenario shown is the simplest case, involving a single-component receptor. Examples include erythropoietin, growth hormone, and prolactin. In these cases, each receptor monomer is associated with a single JAK. Ligand binding induces receptor dimerization, leading to phosphorylation of the cytoplasmic tail of the receptor proteins on tyrosine residues. These phosphotyrosine-containing peptide motifs are recognized by the SH2 domains of latent cytoplasmic STAT monomers. The receptor-bound STATs are phosphorylated on unique C-terminal tyrosine residues. Phosphorylated STATs dissociate from the receptor, form dimers through mutual SH2-phosphotyrosine interactions, and translocate to the nucleus. In the case of interferons and interleukins whose receptors are composed of multiple subunits, each subunit can associate with a distinct JAK or STAT, but the basic elements of activation are the same. In some cases, STATs may be recruited to a tyrosine-phosphorylated JAK or STAT instead of the receptor. Although shown in the figure, the functional significance of the STAT SH3 homology is not clear. Not represented in the figure is the STAT N-terminal domain, which in some cases may participate in receptor binding and in cooperative DNA binding (see text).

activation domain.[32,49,53] This is also the site of signal-mediated serine phosphorylation in certain STAT proteins.[54] An additional region of homology just N-terminal to the SH2 domain has been suggested to be an SH3 domain,[14] but no evidence for any hallmark SH3-like activities has yet emerged.

This topography is consistent with a generally accepted model for STAT activation (Fig. 7.2). Upon binding of ligand to receptor, the latter becomes phosphorylated on one or more tyrosine residues. In the case of cytokine receptors, phosphorylation is presumably performed by receptor-associated JAKs. Phosphorylation of a receptor tyrosine residue within a particular sequence context creates a docking site for the SH2 domain of a STAT protein, resulting in the recruitment of an inactive STAT monomer to the receptor. Once bound to the receptor, the STAT protein is subject to phosphorylation on its critical C-terminal tyrosine residue, again presumably by a receptor-associated JAK. Phosphorylated STAT proteins dissociate from the receptor, form dimers mediated by mutual SH2-phosphotyrosine interactions,[55] translocate to the nucleus and bind target genes. Subsequent phosphorylation on serine, presumably by MAP kinases, may further modulate STAT activity. Dephosphorylation reverses STAT activation and returns STAT proteins to an inactive state in the cytoplasm. In the next few sections, we address details of each of these steps, with particular attention to understanding how signaling specificity may be preserved during the activation cycle.

HOW DOES A RECEPTOR SELECT A STAT PROTEIN FOR ACTIVATION?

Since most cells contain multiple STAT proteins, how does a given receptor select a particular STAT for activation? The first hint came from studies of the IFN-γ receptor. Schreiber and coworkers identified a critical tyrosine residue in the α chain of the IFN-γ receptor that is phosphorylated upon ligand binding and is absolutely required for Stat1 activation.[56] They showed that a synthetic phosphopeptide encompassing this sequence could select Stat1 out of a cell lysate and that such peptides could block Stat1 activation by IFN-γ in a cell-free system. Importantly, this peptide did not select Stat2 out of the same lysates. In contrast, a distinct phosphopeptide sequence derived from the IFN-α R1 subunit is able to select Stat2 and not Stat1.[35] These studies, and similar ones performed with Stat6 and phosphopeptides derived from the IL-4 receptor,[57] suggested that ligand-induced receptor phosphorylation on tyrosine creates recruitment sites for STATs and that the sequence context surrounding the phosphotyrosine residue determines which STAT is selected.

This conclusion was further supported by analysis of gp130, the signaling component of several cytokine receptors, including the receptor for IL-6. IL-6 activates Stat3 and, to a lesser extent, Stat1. Activation of Stat3 requires four functionally redundant tyrosine residues in gp130, which share the common motif YXXQ.[58] Appending such a motif to the erythropoietin receptor, which does not normally activate Stat3, endows that receptor with this activity. Thus, specific phosphotyrosine-containing peptides are not only necessary but also sufficient to specify which STAT is recruited to and activated by a given receptor. Interestingly, two of the four YXXQ motifs in gp130 select only Stat3 for activation whereas the other two select both Stat3 and Stat1.[59] By comparing the Stat3-only recruitment sites in gp130 with the Stat1-only recruitment site previously identified in the IFN-γ receptor, Gerhartz et al[59] were able to reprogram the selectivity of the two receptors with just two amino acid substitutions in each motif, a dramatic demonstration that short peptide motifs in the cytoplasmic domains of cytokine receptors select specific Stats for activation.

The role for phosphotyrosine-containing peptide motifs in selecting STATs for activation clearly implicates the STAT SH2 domain as the partner in this interaction. Indeed, replacing the SH2 domain of Stat2 with the homologous region of Stat1 redirects the chimeric Stat2 protein to the IFN-γ receptor.[60] Similarly, transferring the Stat1 SH2 domain to Stat3 restricts activation of the chimeric protein to the subset of gp130 sites able to activate both Stat1 and Stat3 and imparts to the chimera the ability to respond to a hybrid receptor containing the tyrosine phosphorylation motif of the IFN-γ receptor.[61] Thus, the SH2 domains of STAT proteins direct them to particular phosphotyrosine-containing sequence motifs in activated receptors. It is the inherent sequence specificity of SH2-phosphopeptide interactions that in large part determines STAT selection by cytokine receptors.

This missing piece of data in support of this elegant model is the lack of demonstrable sequence selectivity of isolated STAT SH2 domains. This behavior contrasts with that of most other SH2 domains which exhibit readily detectable selectivity for certain phosphotyrosine-containing peptides.[62] This observation raises the possibility that the SH2 domain alone is not sufficient for high-affinity interaction with activated receptors and that cooperation with another part of the STAT protein is required.

Other observations also leave loose ends not neatly tied up by this model. For example, Stat1 is activated by IFN-α despite the absence of a binding site on any of the receptor subunits. Here, the evidence suggests that Stat2 acts as an intermediary in Stat1 recruitment.[34,35,63] There are also cases in which all tyrosine residues can be deleted from the intracellular domains of cytokine receptors without affecting STAT activation. The hypothesis in this case is that a receptor-associated JAK protein provides the recruitment site. In the case of the growth hormone receptor, a YXXQ motif in Jak2 is suspected to recruit Stat3 to the receptor complex.[64,65]

A final wrinkle is that cell-type-specific factors and differences in the growth state of a given cell can modulate the selection of STATs by a particular receptor. For example, IL-15 induces both Stat3 and Stat5 in T cells but only Stat5 in mast cells. This difference is due to distinct subunit compositions of the IL-15 receptor in the two cell types.[66] IL-2 normally induces only Stat5 in resting peripheral T cells but induces Stat3 as well in cells pretreated with phytohemagglutinin.[67] In another interesting scenario, certain mammary tumor cells are resistant to growth inhibition by IFN-β; these cells express very little detectable Stat1. Treatment of the cells with retinoic acid upregulates Stat1 expression and restores IFN-β sensitivity.[68] Thus, a variety of other mechanisms can affect the receptor-STAT selection process.

HOW DOES AN ACTIVATED STAT SELECT A TARGET GENE?

STAT proteins generally bind DNA as simple dimers, so most known STAT binding sites are palindromic. Initial work with naturally occurring sites and variants thereof suggested a surprising degree of overlap in the recognition specificities of different STAT proteins. Indeed, this overlap has been borne out by systematic analysis of STAT specificity based on selection and amplification of binding sites from random oligonucleotide pools.[52] With the exception of Stat2 (which does not appear to bind DNA on its own), all known STATs, including the *Drosophila* protein, bind sequences related to TTCCNGGAA, where the central base pair N is the only obvious basis for selectivity. The spacing of the palindromic half-sites can impart a degree of selectivity to DNA binding: While the conventional TTN_5AA motif binds most STATs very well, TTN_4AA and TTN_6AA exhibit selectivity for Stat3 and Stat6, respectively.[69] Thus, the selectivities of different STAT complexes are demonstrable but small and are likely amplified by other interactions at target genes.

Perhaps the most interesting set of interactions is among the STATs themselves. When STATs 1, 4 and 5, whose recognition specificities are nearly identical, were presented with a regulatory region derived from the IFN-γ gene, each of the proteins generated a distinct footprint.[50] In addition, each footprint encompassed multiple degenerate STAT recognition motifs, suggesting that the extended binding sites arose from cooperative interactions between STAT dimers at adjacent sites. Indeed, such an interaction can be demonstrated in vitro and, in Stat1 and Stat4, requires the N-terminal homology region.[50,51] Thus, although all of the proteins display a similar binding consensus for single high-affinity sites, deviations from consensus and perhaps the spacing of adjacent binding sites imparts selectivity to DNA recognition.

STATs are also capable of interacting with unrelated transcription factors, implying that selectivity is further created by the sequence context in which a recognition site resides. In this regard, of course, STATs are no different from other sequence-specific transcription factors. Therefore, a few representative examples should suffice. One interesting case is the interaction of Stat3β with Jun, a member of the AP-1 complex (chapter 5). Stat3β is a C-terminally truncated derivative of Stat3, lacking the activation domain. Stat3β interacts with Jun in a yeast two-hybrid assay, and Stat3β, but interestingly not full-length Stat3, is capable of cooperatively activating a reporter gene in conjunction with Jun.[70] One natural gene target in which STATs likely function in conjunction with other neighboring proteins is the protooncogene *c-fos*. The STAT binding site in the *c-fos* enhancer is a very low-affinity site in vitro.[19,71] Nevertheless, in vivo footprinting indicates that this site is inducibly occupied in cells treated with growth factor.[72] Thus, it is likely that STAT binding requires a cooperative interaction with other proteins bound nearby. Indeed, the activity of the STAT site in the *c-fos* promoter shows a strong dependence on the binding site for serum response factor (SRF), located about 30 bp downstream (see chapter 14).[73,74] Functional interactions have also been reported between various STAT proteins and nuclear factor I, C/EBP, Sp1, PU.1 and NFκB. Clearly, such interactions are likely to contribute to the selectivity of STATs toward their target genes.

TRANSCRIPTIONAL REGULATION BY STAT PROTEINS

The transcriptional activation domains that have been characterized to date in STAT proteins reside in the C-terminal tails beyond the SH2 domain and tyrosine phosphorylation site. Several STATs are found in alternatively spliced forms that lack the activation domain. In the case of Stat1, when the deleted form (Stat1β) is introduced into cells lacking any Stat1 protein, it is unable to restore activation of IFN-γ target genes, presumably due to the absence of an activation domain from the Stat1 homodimers that bind to these genes.[32] In contrast, Stat1β does reconstitute ISGF-3 activity and induction of target genes by IFN-α, indicating that another component of ISGF-3 provides the activation function. Indeed, this function resides in the C-terminus of Stat2.[49,75] Stat3 also comes in a truncated form.[53,70] As noted earlier, deletion of the C-terminus of Stat3 endows it with the ability to interact with Jun,[70] but also turns it into a repressor of conventional Stat3 sites, presumably just by competing with full-length Stat3 dimers.[53]

The C-terminal activation domains of several STAT proteins contain potential phosphorylation sites for MAP kinases. In the case of Stat1 and Stat3, there is good evidence that these sites are indeed inducibly phosphorylated in vivo and that this phosphorylation is required for the fully induced transcriptional activity of the proteins.[54] That MAP kinases are in fact responsible for this phosphorylation is suggested by the observation that activated MAPK can be coimmunoprecipitated from IFN-β-stimulated cells.[76] Phosphorylation of STATs on serine residues is also reported to enhance DNA-binding activity.[77-79] Although this effect is not reproducibly seen, it may well be relevant in certain cell types and especially on lower-affinity binding sites. In any case, the observation that STAT proteins are inducibly phosphorylated on both tyrosine and serine residues indicates that they may serve an additional important function of signal integration.

BIOLOGICAL FUNCTION OF STAT PROTEINS

Several different experimental approaches provide insight into the biological function of STATs. Although gene knockouts in mice have recently grabbed the limelight, much of the fundamental information on the specific roles of JAK and STAT proteins in the interferon response came from an elegant exercise in somatic cell genetics. Stark and Kerr and colleagues created cell lines in which selectable markers were expressed under the control of IFN-inducible promoters.[27,80-82] They mutagenized the cell lines and selected variant lines that were unable to activate marker genes following IFN treatment. The cell lines were sorted into complementation groups. Among these were cell lines deficient in individual JAKs and STATs. Analysis of these cell lines established that Stat1 was required for response to both IFN-α/β and IFN-γ, whereas Stat2 was required only for IFN-α/β response.[16,32,34]

In the case of Stat1, therefore, it was no surprise when the first knock-out animals turned out to be defective in mounting an IFN-mediated antiviral response.[83,84] What was surprising, however, was that nothing else was wrong with these animals, despite widespread expression of Stat1 and induction of Stat1 by a broad range of cytokines and growth factors. This would imply that Stat1 induction by these stimuli is either incidental or redundant to other signaling events.

The phenotypes of Stat4 and Stat6 knockouts were pleasingly consistent with their biology in cell culture. Stat4-deficient mice fail to mount any detectable response to IL-12.[85,86] In particular, they fail to launch a so-called Th1 response, which is the cornerstone of cellular immunity. Indeed, lymphocytes from these animals exhibit a tendency to move down the alternative Th2 pathway, which drives humoral immunity. In contrast, Stat6-deficient mice have the reciprocal phenotype.[87,88] Lymphocytes from these animals fail to mount a Th2 response, normally triggered by IL-4. Thus, Stat4 and Stat6 stand at a critical crossroads of the inflammatory system.

As of this writing, results are not yet in for other STAT knockouts. It seems reasonable to expect that Stat2 knockouts will be selectively defective in the IFNα/β response but will respond normally to IFN-γ. It will be of great interest to contrast the biological phenotype of these animals with Stat1-deficient animals. The phenotypes of Stat3 and Stat5 knockouts are anybody's guess. Given the pleiotropic pattern of expression and induction, the phenotype could range from embyronic-lethal to wild-type, with virtually any possibility in between. In the case of Stat5, however, critical roles in lactation and/or cytokine function might be anticipated.

The ability to selectively mutate STAT docking sites in particular cytokine receptors has, in some cases, allowed the role of STAT activation to be evaluated in cell culture systems. It is probably unwise to generalize from these isolated observations, but a few examples are worth noting. In the IL-2 receptor, deletion of the portion of the β chain required for activation of Stat5 did not affect IL-2-induced proliferation in transfected BAF/3 cells.[89] Similarly, mutation of a single tyrosine residue in the erythropoietin receptor completely eliminated Stat5 activation without affecting the mitogenic response.[90] Interestingly, in Stat6-deficient mice, the proliferative response of lymphocytes to IL-4 was only partially reduced, even though other IL-4 responses were completely abolished.[87] If one were tempted to generalize, these observations might suggest that STATs contribute primarily to effector functions induced by cytokines, rather than to proliferative responses. But this correlation is not universal.

STATS AS TARGETS FOR DRUG DEVELOPMENT

Recent progress in understanding STAT function in vivo, particularly the roles of Stat4 and Stat6 in controlling the flow of activated T cells into the competing Th1 and Th2 pathways, suggests that STATs may be attractive targets for new classes of drugs that modulate the immune and inflammatory system. This idea is reinforced by the repeated observation that STATs primarily control effector functions rather than cell proliferation, suggesting that the effects of STAT inhibitors would be rapid and reversible. Inhibitors of Stat4 and Stat6 might also prove to be more precise and selective than the more broadly-acting immunosuppressants in use today. A Stat4 inhibitor, for example, would blunt the cell-based immune response that underlies many autoimmune diseases without affecting the humoral response. In contrast, a Stat6 inhibitor might be effective against the humoral foundation of asthma and allergic diseases without compromising cellular immunity.

While the biology of a specific protein might suggest that it is an excellent drug target, it is not always obvious how to make such a drug. In the case of the STATs, however, one possibility comes immediately to mind. Many studies have now shown that phosphopeptides derived from STAT docking sites on cytokine receptors can block STAT activation in cell-free systems, presumably by blocking recruitment to activated receptors, and can dissociate active STAT dimers, in both cases by binding competitively to the STAT SH2 domains.[55-57] Indeed, STAT activation in a cell-free system can be selectively blocked by phosphotyrosine,[19] establishing the principle that a small-molecule ligand directed at a STAT SH2 domain can block STAT activation. Given the strong evidence for the recognition of distinct phosphopeptide docking sites by different STATs, one can be cautiously optimistic about the prospect for making SH2-directed drugs that are selective for a single STAT.

Acknowledgments

We thank Ian MacNeil and Dorre Grueneberg for comments on the manuscript; Ian MacNeil and Renate Hellmiss for help with figures; Chris Stearns and Mandy Cunningham for preparation of the manuscript. M.-L.V. was supported by a Human Frontier Science Program fellowship.

References

1. Schindler C, Darnell JE. Transcriptional responses to polypeptide ligands: the JAK-STAT pathway. Annu Rev Biochem 1995; 64:621-51.
2. Ihle JN, Witthuhn BA, Quelle FW et al. Signaling through the hematopoietic cytokine receptors. Annu Rev Immunol 1995; 13:369-98.
3. Reich N, Evans B, Levy DE et al. Interferon-induced transcription of a gene encoding a 15 kDa protein depends on an upstream enhancer element. PNAS 1987; 84:6394-6398.
4. Levy DE, Kessler DS, Pine R et al. Interferon-induced nuclear factors that bind a shared promoter element correlate with positive and negative control. Genes and Dev 1988; 2:383-393.
5. Kessler DS, Levy DE, Darnell JE Jr. Two interferon-induced nuclear factors bind a single promoter element in interferon-stimulated genes. Proc Natl Acad Sci USA 1988; 85:8521-8525.
6. Levy DE, Kessler DS, Pine RI et al. Cytoplasmic activation of ISGF3, the positive regulator of interferon-α stimulated transcription, reconstituted in vitro. Genes Dev 1989; 3:1362-1372.
7. Dale TC, Rosen JM, Guille MJ et al. Overlapping sites for constitutive and induced DNA binding factors involved in interferon-stimulated transcription. EMBO J 1989; 8:831-839.
8. Kessler DS, Veals SA, Fu X-Y et al. Interferon-α regulates nuclear translocation and DNA-binding affinity of ISGF-3, a multimeric transcriptional ativator. Genes Dev 1990; 4:1753-1765.
9. Fu X-Y, Kessler DS, Veals SA et al. ISGF3, the transcriptional activator induced by interferon α, consists of multiple interacting polypeptide chains. Proc Natl Acad Sci USA 1990; 87:8555-8559.
10. Schindler C, Darnell JE Jr. Proteins of transcription factor ISGF-3: One gene encodes the 91- and 84 kDa ISGF-3 proteins that are activated by interferon α. Proc Natl Acad Sci USA 1992; 89: 7836-7839.
11. Fu X-Y, Schindler C, Improta T et al. The proteins of ISGF-3: the interferon α-induced transcriptional activator, define a gene family involved in signal transduction. Proc Natl Acad Sci USA 1992; 89:7840-7843.
12. Shuai K, Schindler C, Presiozo VR et al. Activation of transcription by IFNγ: tyrosine phosphorylation of a 91 kDa DNA binding protein. Science 1992; 258: 1808-1812.

13. Schindler C, Shuai K, Prezioso VR et al. Interferon-dependent tyrosine phosphorylation of a latent cytoplasmic transcription factor. Science 1992; 257:809-813.
14. Fu X-Y. A transcription factor with SH2 and SH3 domains is directly activated by interferon-α-induced cytoplasmic protein tyrosine kinase(s). Cell 1992; 70:323-335.
15. Gutch MJ, Daly C, Reich NC. Tyrosine phosphorylation is required for activation of an α-interferon-stimulated transcription factor. Proc Natl Acad Sci USA 1992; 89:11411-11415.
16. Shuai K, Stark GR, Kerr IM et al. A single phosphotyrosine residue of Stat91 required for gene activation by interferon-γ. Science 1993; 261:1744-1746.
17. Shuai K, Ziemiecki A, Wilks AF et al. Polypeptide signalling to the nucleus through tyrosine phosphorylation of Jak and Stat proteins. Nature 1993; 366: 580-583.
18. Hayes TE, Kitchen AM, Cochran BH. Inducible binding of a factor to the c-fos regulatory region. Proc Natl Acad Sci USA 1987; 84:1272-1276.
19. Sadowski HB, Shuai K, Darnell JE Jr et al. A common nuclear signal transduction pathway activated by growth factor and cytokine receptors. Science 1993; 261:1739-1744.
20. Ruff-Jamison S, Chen K, Cohen S. Induction by EGF and interferon-γ of tyrosine phosphorylated DNA binding proteins in mouse liver nuclei. Science 1993; 261:1733-1736.
21. Silvennoinen O, Schindler C, Schlessinger J et al. Ras-independent growth factor signaling by transcription factor tyrosine phosphorylation. Science 1993; 261: 1736-1739.
22. Fu X-Y, Zhang J-J. Transcription factor p91 interacts with the epidermal growth factor receptor and mediates activation of the c-fos gene promoter. Cell 1993; 74:1135-1145.
23. Wilks AF. Two putative protein-tyrosine kinases identified by application of the polymerase chain reaction. Proc Natl Acad Sci USA 1989; 86:1603-1607.
24. Firmbach-Kraft I, Byers M, Shows T et al. Tyk2, prototype of a novel class of non-receptor tyrosine kinase genes. Oncogene 1990; 5:1329-1336.
25. Wilks AF, Anders AD, Ziemiecki A et al. Two novel protein-tyrosine kinases, each with a second phosphotransferase-related catalytic domain define a new class of protein kinase. Mol Cell Biol 1991; 11:2057-2065.
26. Harpur AG, Andres AC, Ziemiecki A et al. JAK2, a third member of the JAK family of protein tyrosine kinases. Oncogene 1992; 7:1347-1353.
27. Pellegrini S, John J, Shearer M et al. Use of a Selectable Marker Regulated by Alpha Interferon To Obtain Mutations in the Signaling Pathway. Mol Cell Biol 1989; 9:4605-4612.
28. Velazquez L, Fellous M, Stark GR et al. A protein tyrosine kinase in the interferon α/β signaling pathway. Cell 1992; 70:313-322.
29. Müller M, Briscoe J, Laxton C et al. The protein tyrosine kinase JAK1 complements defects in interferon-α/β and -gamma signal transduction. Nature 1993; 366:129-135.
30. Watling D, Guschin D, Muller M et al. Complementation by the protein tyrosine kinase JAK2 of a mutant cell line defective in the interferon-gamma signal transduction pathway. Nature 1993; 366:166-170.
31. Veals SA, Schindler C, Leonard D et al. Subunit of an Alpha-interferon-responsive transcription factor is related to interferon regulatory factor and Myb families of DNA-binding protein. Mol Cell Biol 1992; 12:3315-3324.
32. Müller M, Laxton C, Briscoe J et al. Complementation of a mutant cell line: central role of the 91 kDa polypeptide of ISGF3 in the interferon-α and -γ signal transduction pathways. EMBO J 1993; 12:4221-4228.
33. Qureshi SA, Salditt GM, Darnell JJ. Tyrosine-phosphorylated Stat1 and Stat2 plus a 48-kDa protein all contact DNA in forming interferon-stimulated-gene factor 3. Proc Natl Acad Sci USA 1995; 92:3829-33.
34. Leung S, Qureshi SA, Kerr IM et al. Role of STAT2 in the α interferon signaling pathway. Mol Cell Biol 1995; 15:1312-7.
35. Yan H, Krishnan K, Greenlund A et al. Phosphorylated interferon-α receptor 1

subunit (IFNαR1) acts as a docking site for the latent form of the 113 kDa STAT2 protein. EMBO J 1996; 15: 1064-1074.

36. Akira S, Nishio Y, Inoue M et al. Molecular cloning of APRF, a novel IFN-stimulated gene factor 3 p91-related transcription factor involved in the gp130-mediated signaling pathway. Cell 1994; 77:63-71.
37. Zhong Z, Wen Z, Darnell JE Jr. Stat3: a STAT family member activated by tyrosine phosphorylation in response to epidermal growth factor and interleukin-6. Science 1994; 264:95-8.
38. Zhong Z, Wen Z, Darnell JE Jr. Stat3 and Stat4: members of the family of signal transducers and activators of transcription. Proc Natl Acad Sci USA 1994; 91:4806-10.
39. Yamamoto K, Quelle FW, Thierfelder WE et al. Stat4, a novel gamma interferon activation site-binding protein expressed in early myeloid differentiation. Mol Cell Biol 1994; 14:4342-9.
40. Bacon CM, Petricoin EF, Ortaldo JR et al. Interleukin 12 induces tyrosine phosphorylation and activation of STAT4 in human lymphocytes. Proc Natl Acad Sci USA 1995; 92:7307-7311.
41. Jacobson NG, Szabo SJ, Weber NR et al. Interleukin 12 signaling in T helper type 1 (Th1) cells involves tyrosine phosphorylation of signal transducer and activator of transcription (Stat)3 and Stat4. J Exp Med 1995; 181:1755-62.
42. Wakao H, Gouilleux F, Groner B. Mammary gland factor (MGF) is a novel member of the cytokine regulated transcription factor gene family and confers the prolactin response. EMBO J 1995; 14:854-855.
43. Liu X, Robinson GW, Gouilleux F et al. Cloning and expression of Stat5 and an additional homologue (Stat5b) involved in prolactin signal transduction in mouse mammary tissue. Proc Natl Acad Sci USA 1995; 92:8831-5.
44. Kotanides H, Reich NC. Requirement of tyrosine phosphorylation for rapid activation of a DNA binding factor by IL-4. Science 1993; 262:1265-1267.
45. Hou J, Schindler U, Henzel WJ et al. An interleukin-4-induced transcription factor: IL-4 Stat. Science 1994; 265: 1701-1706.
46. Quelle FW, Shimoda K, Thierfelder W et al. Cloning of murine Stat6 and human Stat6, Stat proteins that are tyrosine phosphorylated in responses to IL-4 and IL-3 but are not required for mitogenesis. Mol Cell Biol 1995; 15:3336-43.
47. Yan R, Small S, Desplan C et al. Identification of a Stat gene that functions in Drosophila development. Cell 1996; 84:421-430.
48. Hou SX, Melnick MB, Perrimon N. Marelle acts downstream of the Drosophila HOP/JAK kinase and encodes a protein similar to the mammalian STATs. Cell 1996; 84:411-419.
49. Qureshi SA, Leung S, Kerr IM et al. Function of Stat2 protein in transcriptional activation by α interferon. Mol Cell Biol 1996; 16:288-293.
50. Xu X, Sun YL, Hoey T. Cooperative DNA binding and sequence-selective recognition conferred by the STAT amino-terminal domain. Science 1996; 273: 794-797.
51. Vinkemeier U, Cohen SL, Moarefi I et al. DNA binding of in vivo activated Stat1α, STAT1β and truncated Stat1: interaction between NH_2-terminal domains stabilizes binding of two dimers to tandem DNA sites. EMBO J 1996; 15: 5616-5626.
52. Horvath CM, Wen Z, Darnell JJ. A STAT protein domain that determines DNA sequence recognition suggests a novel DNA-binding domain. Genes Dev 1995; 9:984-94.
53. Caldenhoven E, van Dijk TB, Solari R et al. STAT3b, a splice variant of transcription factor STAT3, is a dominant negative regulator of transcription. J Biol Chem 1996; 271:13221-13227.
54. Wen Z, Zhong Z, Darnell JJ. Maximal activation of transcription by Stat1 and Stat3 requires both tyrosine and serine phosphorylation. Cell 1995; 82:241-50.
55. Shuai K, Horvath CM, Huang LH et al. Interferon activation of the transcription factor Stat91 involves dimerization

through SH2-phosphotyrosyl peptide interactions. Cell 1994; 76:821-8.
56. Greenlund AC, Farrar MA, Viviano BL et al. Ligand-induced IFN gamma receptor tyrosine phosphorylation couples the receptor to its signal transduction system (p91). Embo J 1994; 13:1591-1600.
57. Hou J, Schindler U, Henzel WJ et al. An interleukin-4-induced transcription factor: IL-4 Stat. Science 1994; 265:1701-6.
58. Stahl N, Farruggella TJ, Boulton TG et al. Choice of STATs and other substrates specified by modular tyrosine-based motifs in cytokine receptors. Science 1995; 267:1349-53.
59. Gerhartz C, Hemmann U, Heesel B et al. Differential activation of acute phase response factor/Stat3 and Stat1 via the cytoplasmic domain of the interleukin 6 signal transducer gp130 I. Definition of a novel phosphotyrosine motif mediating Stat1 activation. J Biol Chem 1996; 271:12991-12998.
60. Heim MH, Kerr IM, Stark GR et al. Contribution of STAT SH2 groups to specific interferon signaling by the Jak-STAT pathway. Science 1995; 267:1347-9.
61. Hemmann U, Gerhartz C, Heesel B et al. Differential activation of acute phase response factor/Stat3 and Stat1 via the cytoplasmic domain of the interleukin 6 signal transducer gp130 II. Src homology Sh2 domains define the specificity of Stat factor activation. J Biol Chem 1996; 271:12999-13007.
62. Songyang Z, Shoelson SE, Chaudhuri M et al. SH2 domains recognize specific phosphopeptide sequences. Cell 1993; 72:767-778.
63. Gupta S, Yan H, Wong LH et al. The SH2 domains of Stat1 and Stat2 mediate multiple interactions in the transduction of IFN-α signals. Oxford University Press 1996; 1075-1081.
64. Sotiropoulos A, Moutoussamy S, Binart N et al. The membrane proximal region of the cytoplasmic domain of the growth hormone receptor is involved in the activation of Stat 3. Febs Lett 1995; 369: 169-72.
65. Han Y, Leaman DW, Watling D et al. Participation of JAK and STAT proteins in growth hormone-induced signaling. J Biol Chem 1996; 271:5947-5952.
66. Tagaya Y, Bamford RN, DeFilippis AP et al. IL-15: A pleiotropic cytokine with diverse receptor/signaling pathways whose expression is controlled at multiple levels. Immunity 1996; 4:329-336.
67. Lin JX, Migone TS, Tsang M et al. The role of shared receptor motifs and common Stat proteins in the generation of cytokine pleiotropy and redundancy by IL-2, IL-4, IL-7, IL-13, and IL-15. Immunity 1995; 2:331-9.
68. Kolla V, Lindner DJ, Weihua X et al. Modulation of interferon (IFN)-inducible gene expression by retinoic acid. J Biol Chem 1996; 271:10508-10514.
69. Seidel HM, Milocco LH, Lamb P et al. Spacing of palindromic half sites as a determinant of selective STAT (signal transducers and activators of transcription) DNA binding and transcriptional activity. Proc Natl Acad Sci USA 1995; 92:3041-5.
70. Schaefer TS, Sanders LK, Nathans D. Cooperative transcriptional activity of Jun and Stat3 β, a short form of Stat3. Proc Natl Acad Sci USA 1995; 92: 9097-101.
71. Wagner BJ, Hayes TE, Hoban CJ et al. The SIF binding element confers sis/PDGF inducibility onto the c-fos promoter. EMBO J 1990; 9:4477-4484.
72. Herrera RE, Shaw PE, Nordheim A. Occupation of the c-fos serum response element in vivo by a multi-protein complex is unaltered by growth factor induction. Nature 1989; 340:68-70.
73. Gilman MZ. The c-fos serum response element responds to protein kinase C-dependent and -independent signals but not to cyclic AMP. Genes Dev 1988; 2:394-402.
74. Hill CS, Treisman R. Differential activation of c-fos promoter elements by serum, lysophosphatidic acid, G proteins and polypeptide growth factors. Embo J 1995; 14:5037-47.
75. Bhattacharya S, Eckner R, Grossman S et al. Cooperation of Stat2 and p300/CBP in signalling induced by interferon-α. Nature 1996; 383:344-347.
76. David M, Petricoin E III, Benjamin C et al. Requirement for MAP kinase (ERK2) activity in interferon α- and interferon

β-stimulated gene expression through STAT proteins. Science 1995; 269:1721-3.
77. Zhang X, Blenis J, Li HC et al. Requirement of serine phosphorylation for formation of STAT-promoter complexes. Science 1995; 267:1990-4.
78. Eilers A, Georgellis D, Klose B et al. Differentiation-regulated serine phosphorylation of STAT1 promotes GAF activation in macrophages. Mol Cell Biol 1995; 15:3579-86.
79. Boulton TG, Zhong Z, Wen Z et al. STAT3 activation by cytokines utilizing gp130 and related transducers involves a secondary modification requiring an H7-sensitive kinase. Proc Natl Acad Sci USA 1995; 92:6915-9.
80. McKendry R, John J, Flavell D et al. High-frequency mutagenesis of human cells and characterization of a mutant unresponsive to both α and gamma interferons. PNAS 1991; 88: 111455-11459.
81. Mao C, Davies D, Kerr IM et al. Mutant human cells defective in induction of major histocompatibility complex class II genes by interferon gamma. PNAS 1993; 90:2880-2884.
82. John J, McKendry R, Pellegrini S et al. Isolation and characterization of a new mutant human cell line unresponsive to α and β interferons. Mol Cell Biol 1991; 11:4189-4195.
83. Durbin JE, Hackenmiller R, Simon MC et al. Targeted disruption of the mouse *Stat1* gene results in compromised innate immunity to viral disease. Cell 1996; 84:443-450.
84. Meraz MA, White JM, Sheehan KCF et al. Targeted disruption of the *Stat1* gene in mice reveals unexpected physiologic specificity in the JAK-STAT signaling pathway. Cell 1996; 84:431-442.
85. Kaplan MH, Sun Y-L, Hoey T et al. Impaired IL-12 responses and enhanced development of Th2 cells in Stat4-deficient mice. Nature 1996; 382:174-177.
86. Thierfelder WE, van Deursen J, Yamamoto K et al. Requirement for Stat4 in interleukin-12-mediated responses of natural killer and T cells. Nature 1996; 382:171-174.
87. Shimoda K, van Deursen J, Sangster MY et al. Lack of IL-4-induced Th2 response and IgE class switching in mice with disrupted *Stat6* gene. Nature 1996; 380: 630-633.
88. Takeda K, Tanaka T, Shi W et al. Essential role of Stat6 in IL-4 signalling. Nature 1996; 380:627-630.
89. Fujii H, Nakagawa Y, Schindler U et al. Activation of Stat5 by interleukin 2 requires a carboxyl-terminal region of the interleukin 2 receptor β chain but is not essential for the proliferative signal transmission. Proc Natl Acad Sci USA 1995; 92:5482-6.
90. Quelle FW, Wang D, Nosaka T et al. Erythropoietin induces activation of Stat5 through association with specific tyrosines on the receptor that are not required for a mitogenic response. Mol Cell Biol 1996; 16:1622-1631.

CHAPTER 8

Transcription Factors and Cell Differentiation: The C/EBP Paradigm

Erika C. Shugart and Robert M. Umek

INTRODUCTION

Transcription factors regulate the expression of genes in a spatial and temporal manner in order to differentiate tissues. The CCAAT enhancer binding protein (C/EBP) family of transcription factors plays a role in the differentiation of such diverse tissues as adipose, liver and granulocytes. C/EBPα, C/EBPβ (also known as LAP,[1] NF-IL6,[2] IL-6DBP,[3] AGP/EBP,[4] CRP2,[5] and SF-B[6]), C/EBPδ (also called CRP3,[5] CELF,[7] and NF-IL6β[8]) and C/EBP homologous protein (CHOP,[9] also known as gadd153)[10] have different patterns of expression in each of these tissues.[11-13] This differential expression may be important in specifying cell type. C/EBPγ (also known as IL/EBP[14] and CRP1[5]) is ubiquitously expressed in all tissues, and probably does not have a role in differentiation.[14] The C/EBPs regulate the differentiation of tissues at the level of gene transcription. The modulation of gene expression by the C/EBP family of transcription factors is important for the terminal differentiation of tissues and appropriate metabolic regulation in specialized cell types.

C/EBPα, the first member of the C/EBP family described, was isolated from rat liver nuclei by McKnight and colleagues.[15] This transcription factor was first identified by virtue of its binding affinity for the CCAAT box as well as the enhancer core homology of several mammalian viruses.[16] All members of the C/EBP family contain a DNA-binding domain comprised of a basic region adjacent to a leucine zipper, often called a bZIP domain.[17] The leucine zipper is an α helix with a heptad repeat of leucine residues at regularly spaced intervals. The leucines on two protein monomers interact and thus define a dimerization interface.[18] The α-helical basic region is contiguous with the leucine zipper and directly contacts DNA.[16] Since C/EBPs must bind DNA as dimers, mutations in either region disrupt DNA binding.[18] The C/EBP consensus sequence RTTGCGYAAY[19](where Y=C or T and R=A or G) is bound with varying affinity; C/EBPβ binds most avidly, followed by C/EBPα, then C/EBPδ.[11] CHOP has a naturally occurring proline in the basic region and as a consequence fails to bind the consensus sequence. Furthermore, CHOP can heterodimerize with other

Transcription Factors in Eukaryotes, edited by Athanasios G. Papavassiliou.

C/EBPs and, by sequestering them, act as a transcriptional repressor.[9] Recently, Ron and coworkers described a novel DNA-binding site for CHOP-C/EBP heterodimers.[20] The biological significance of this sequence remains to be established. The C/EBPs share homology in the C-terminal, bZIP domain. However, the amino acid sequences are divergent in the N-terminal regions that contain the transcriptional trans-activation domains.[11] Since the DNA-binding properties are similar, with the exception of CHOP, the trans-activation domains are likely to account for differences in the physiological roles of the C/EBP isoforms.

Additional diversity is generated in the C/EBP family as a result of alternative translation products. C/EBPα and C/EBPβ both have alternate translation products termed p30[21] and LIP,[22] respectively. These shorter translation products are due to internal initiation and lack the N terminal regions of the full-length protein. p30 is initiated at the third in frame AUG, giving rise to a 30 kd C/EBPα protein as opposed to a 42 kd protein. p30 lacks a potent, N-terminal trans-activation domain of C/EBPα.[21] Similarly, liver inhibitory protein (LIP) is initiated at the third in frame AUG of C/EBPβ giving rise to a 20 kd protein instead of a 39 kd protein.[22] These alternate translation products have distinct biological properties relative to the full-length isoforms. p30 and LIP dimerize with C/EBPα and C/EBPβ and reduce trans-activation.[21,22] Furthermore, while p30 is a weak transcriptional trans-activator, it lacks the anti-mitotic activity observed for full-length C/EBPα.[23,24]

In general, the C/EBP isoforms are expressed in the same tissue types. However, tissue culture models and animal studies suggest that the C/EBP isoforms are expressed in an ordered fashion during differentiation. The order of expression is tissue specific and may be important for specifying cell type. We postulate that an individual isoform, or subset, acts in a discrete period and, in concert with other transcription factors, activates genes necessary for the stepwise differentiation of cells. Once cells are terminally differentiated, C/EBP isoforms also modulate gene expression in response to changes in physiological states.

ADIPOCYTE DEVELOPMENT

There are several cultured cell lines available for studying conversion of adipoblasts to adipocytes, including 3T3-L1,[25] 3T3-422A[26] and Ob1771.[27] These cell lines will differentiate in response to external stimuli. The 3T3-L1 cell line has been extensively studied. 3T3-L1 cells progress through an ordered series of growth states as a function of adipocyte differentiation.[28] In the standard differentiation protocol,[25] 3T3-L1 adipoblasts are cultured to growth arrest at contact inhibition. The cells reenter the cell cycle and progress through limited mitotic divisions in response to the addition of differentiation inducing hormones[25] during the period termed clonal expansion.[29] The cells arrest again after the removal of the differentiation inducing hormones and remain growth arrested throughout terminal differentiation. The growth states of 3T3-L1 cells throughout adipocyte development are distinguished by the differential expression of C/EBP isoforms (Fig. 8.1A).[11] C/EBPα is undetectable in logarithmically growing and contact inhibited adipoblasts, while C/EBPβ and C/EBPδ levels are detectable but low. C/EBPβ and C/EBPδ levels increase to maximum during clonal expansion. Thereafter, C/EBPδ levels decline to preinduction levels while C/EBPβ levels decline to an abundance intermediate between adipoblasts and mature adipocytes (Fig. 8.1). C/EBPα levels begin to increase coincident with decreases in C/EBPβ and C/EBPδ and continue to increase throughout phenotypic differentiation.[11] Several lines of evidence suggest that sequential expression of C/EBP isoforms drives adipocyte differentiation.

The addition of adipogenic hormones initiates clonal expansion of 3T3-L1 cells and commitment to adipocyte differentiation. Dexamethasone, methyl-isobutyl-xanthine (MIX), and insulin[25] are required for the maximal mitogenic response and maximal differentiation.[30] Cao et al demonstrated that dexamethasone induced C/EBPδ

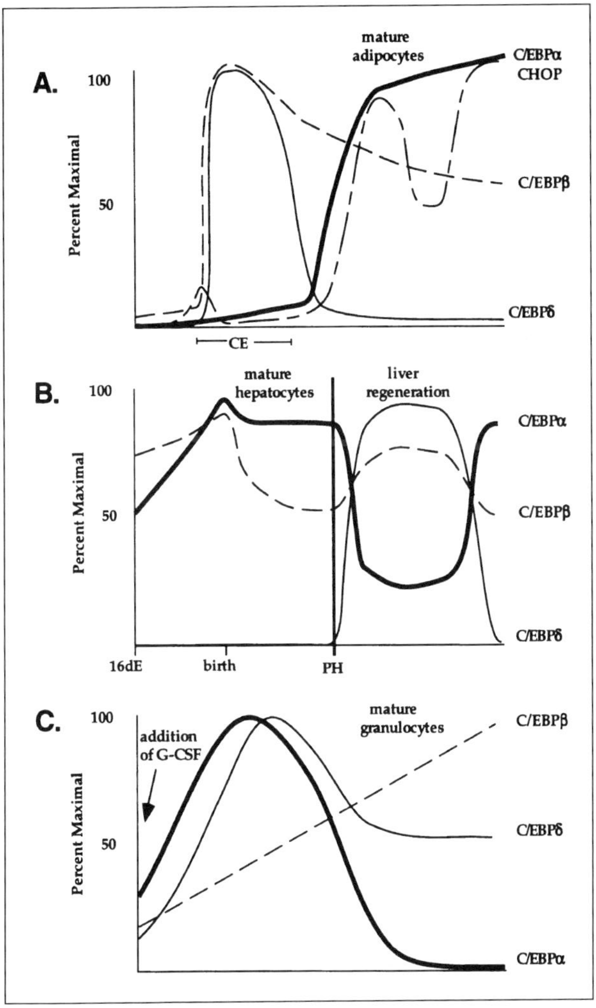

Fig. 8.1. Expression of C/EBP isoforms during adipogenesis, liver development and myelomonocytic differentiation. A. Relative levels of the C/EBP isoforms during 3T3-L1 differentiation. C/EBPβ and C/EBPδ protein levels increase during the period of clonal expansion (CE), then decrease after the removal of the differentiation inducers. C/EBPα protein and CHOP mRNA have the highest level of expression after cells have begun adipocyte conversion. The levels of C/EBPβ and C/EBPδ decline during this period. B. Relative mRNA levels of C/EBPα and C/EBPβ during mouse liver development. Both C/EBPα and C/EBPβ are expressed in the liver at day 16 of embryonic development. The expression levels of both isoforms peak at birth. The protein levels of the isoforms change after a partial hepatetomy (PH). C/EBPα protein levels decline during liver regeneration. C/EBPβ and C/EBPδ protein levels increase during this period with C/EBPδ increasing most dramatically. C. Relative protein levels of the C/EBP isoforms during differentiation of 32D C13 myelomonocytic cells. Cells were induced to differentiate by the addition of granulocyte colony-stimulating factor (G-CSF). The levels of all of the isoforms increase after the addition of G-CSF. C/EBPα and C/EBPδ decline during the conversion to granulocytes while the expression of C/EBPβ increases. In each case the absolute levels of protein and mRNA are unknown and is reported as a percentage of the maximal expression.

mRNA and MIX induced C/EBPδ mRNA respectively. This transcriptional activation of the C/EBP genes is direct and is reflected in protein accumulation.[11] Thus, the post-confluent, proliferative response of 3T3-L1 cells is characterized by coordinate accumulation of C/EBPβ and C/EBPδ. Elimination of individual hormones from the standard cocktail of hormones reveals the significance of each to adipose conversion. In the absence of dexamethasone, C/EBPδ is not induced and C/EBPβ expression is minimal and transient. Under this regimen, C/EBPα does not accumulate and adipogenesis is nearly undetectable.[31] Perhaps C/EBPδ stabilizes the expression of C/EBPβ or activates a pathway required to sustain C/EBPβ accumulation. In the absence of MIX, all of the isoforms are expressed, albeit with delayed expression profiles.[31] C/EBPδ accumulation may drive C/EBPβ expression independent of MIX, which ultimately manifests C/EBPα expression. McKnight and colleagues interpreted these observation to suggest a cascade relationship among the isoforms in adipocyte development. Specifically, they postulate that C/EBPδ acts to stabilize C/EBPβ and, in turn, C/EBPβ induces transcription factors required to activate adipocyte-specific genes, including C/EBPα. This model obviates the need for any of the isoforms to act as a developmental "master switch" and is consistent with the expression of the C/EBP isoforms in numerous cell types. Instead, combinatorial use of the C/EBP isoforms at specific ratios may drive differentiation. In support of this hypothesis, one observes that C/EBP isoform ratios vary from cell type to cell type as a function of differentiation (Fig. 8.1).

The ability of C/EBP isoforms to direct adipocyte development can be evaluated by ectopic expression in pluripotent cells. Forced expression of C/EBPα in numerous fibroblastic cells lines concomitant with exposure to differentiation inducers results in overt adipocyte-specific morphology and gene expression.[32,33] Similarly, expression of C/EBPβ in pluripotent NIH-3T3 cells and exposure to dexamethasone, MIX, insulin and fetal bovine serum induces adipogenesis.[31,34] Notably, C/EBPβ-expressing NIH-3T3 adipocytes do not express C/EBPα.[31] The latter observation suggests that C/EBPβ can substitute C/EBPα in the stimulation of adipocyte-specific genes. Why then does anti-sense C/EBPα RNA inhibit the differentiation of 3T3-L1 cells? Perhaps the level of C/EBPβ in 3T3-L1 cells is insufficient to drive adipogenesis. Furthermore, C/EBPβ levels may be maintained in 3T3-L1 cells through a C/EBPα-dependent mechanism that is circumvented with the vector used for ectopic expression.[31,34] C/EBPβ expression in NIH-3T3 cells stimulates expression of another adipogenic transcription factor, peroxisome proliferator-activated receptor γ (PPARγ), which may mediate some of the adipogenic effects of C/EBPβ.[34] Differentiation is greater when ectopic C/EBPβ expression is accompanied by culturing the cells in the presence of a PPARγ ligand.[34] Spiegelman and colleagues have demonstrated that ectopic expression of PPARγ and the adipocyte determination and differentiation-dependent factors 1 (ADD1/SREBP1) also convert NIH-3T3 cells to adipocytes.[35,36] Thus, the known adipogenic transcription factors include C/EBP isoforms and proteins outside the C/EBP family.

What is the relationship among the C/EBP isoforms, PPARγ and ADD1? PPARγ is exclusively expressed in adipose and ADD1 is restricted to adipose and liver.[35,37] These observations are consistent with PPARγ and ADD1 playing an important role in specifying the adipocyte cell lineage. However, in 3T3-L1 cells, C/EBPβ and C/EBPδ are direct targets of the differentiation inducing hormones and are maximally expressed before maximal expression of PPARγ and ADD1. We speculate that elevated accumulation of C/EBPβ and C/EBPδ in 3T3-L1 adipoblasts results in the unique mitotic divisions that determine 3T3-L1 to be adipocytes. In fact, we have recently discovered an autocrine signal transduction cascade that is initiated by C/EBPδ (E.S. and R.U., unpublished results). Activation of *ppar*γ and *add*1 gene transcription may be a consequence of this unique growth

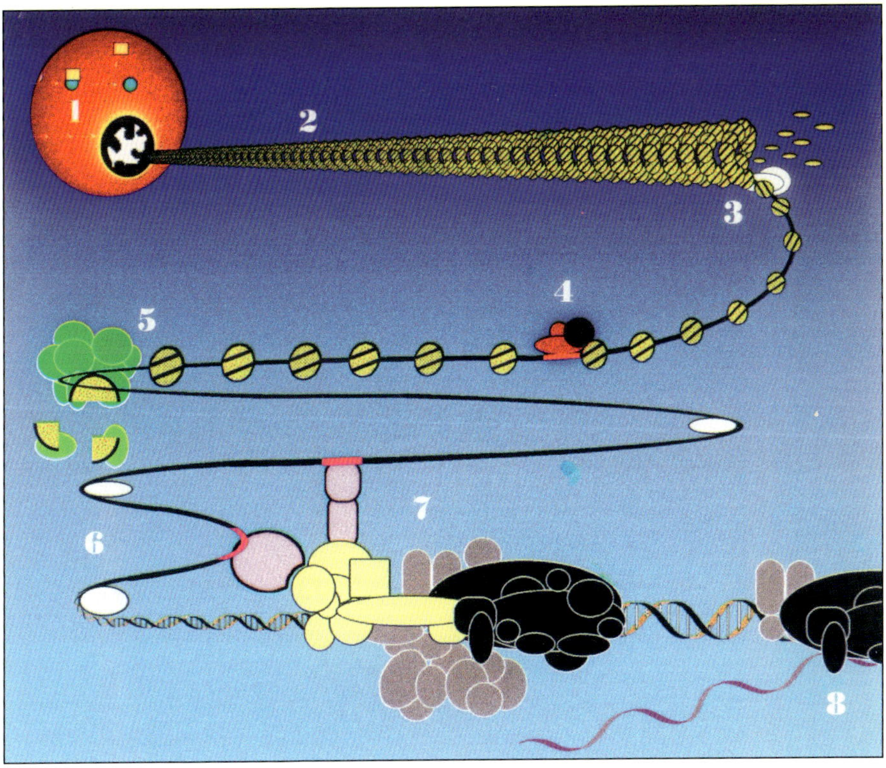

Fig. 3.1. Molecular events in the expression of a gene. 1. Transduction of signals from the cell surface to the nucleus. 2. 30 nm chromatin fiber. 3. Break-up of 30 nm fiber to 10 nm fiber by transcription factors. 4. Translational repositioning of nucleosomes by a sequence-specific regulator. 5. Nucleosome disassembly by SWI/SNF complexes and histone chaperones. 6. Spatial positioning of DNA by architectural transcription regulators. 7. A preinitiation complex (PIC) containing sequence-specific transcriptional activators (pink), a TBP-containing complex (yellow), basal initiation factors (gray), and pol II (black). 8. Gene transcription by RNA polymerase II (pol II).

Transcription Factors in Eukaryotes, edited by Athanasios G. Papavassiliou.

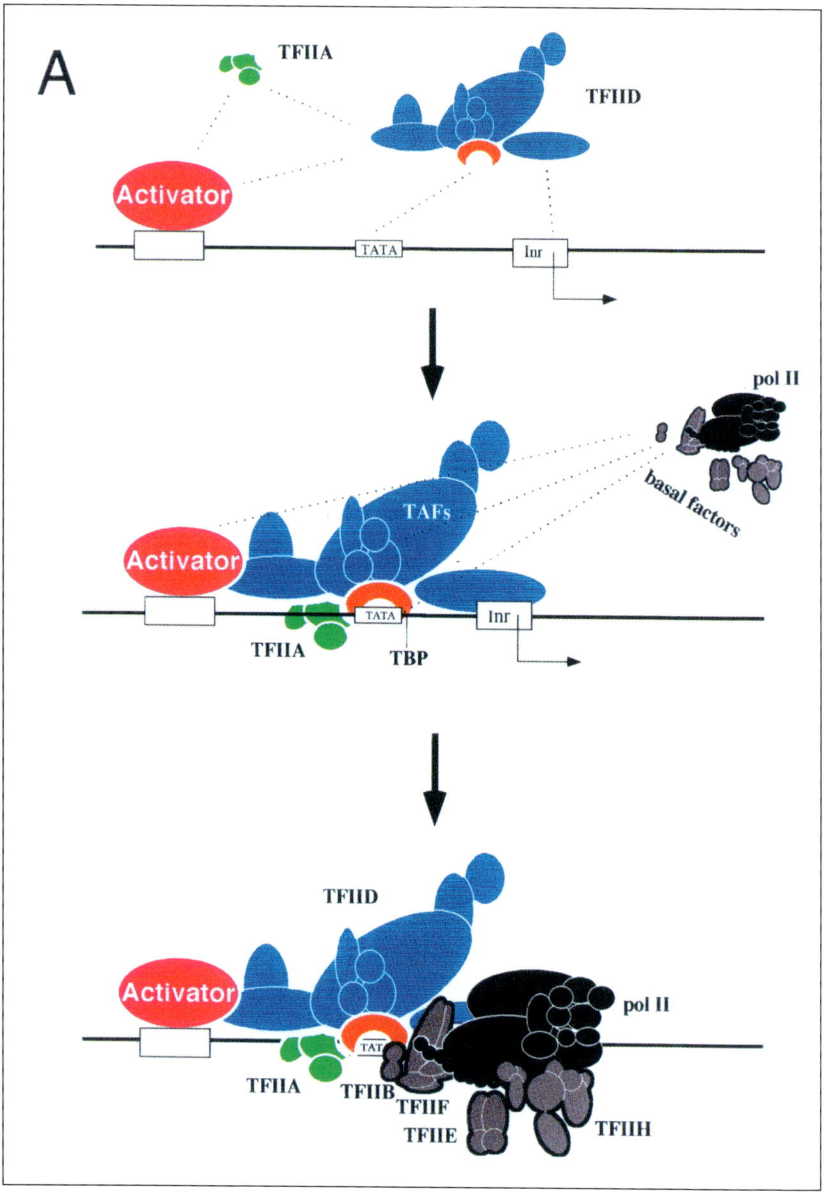

Fig. 3.2. Two related, but not exclusive views on transcription complex assembly. Each view has two major kinetic segments which differ only in detail. (A) Promoter bound activators (fuchsia), in conjunction with TFIIA (green) recruit the TBP-TAF TFIID complex (orange and blue, respectively) to TATA and/or initiator (Inr) elements. The activator-TFIIA-TFIID complex then incorporates additional basal initiation factors (TFIIB, TFIIE, TFIIF, and TFIIH shown in gray) and pol II (black) stepwise into the developing PIC. (B) (Opposite) The same as in A except that TAFs are replaced with the SRB/mediator complex (yellow), which interacts with TBP. The mediator, basal factors (gray), and the SWI/SNF complex (gray) are components of the pol II holoenzyme. The pol II holoenzyme is shown here to be preassembled and incorporated into the developing PIC in a single step.

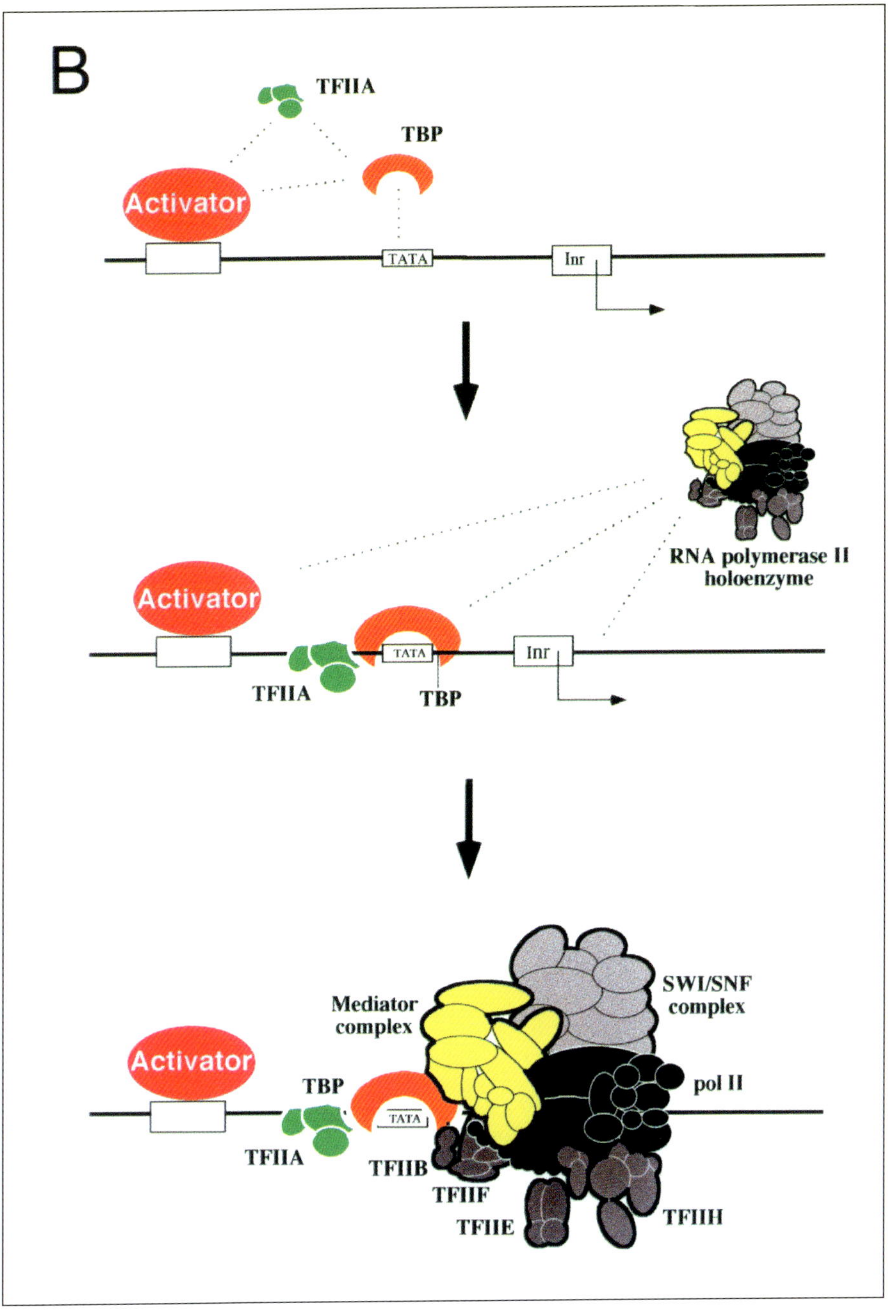
B
TFIIA
TBP
Activator
TATA
Inr
RNA polymerase II
holoenzyme
Activator
TATA
Inr
TFIIA
TBP
Mediator
complex
SWI/SNF
complex
Activator
TBP
TATA
pol II
TFIIA
TFIIB
TFIIF
TFIIE
TFIIH

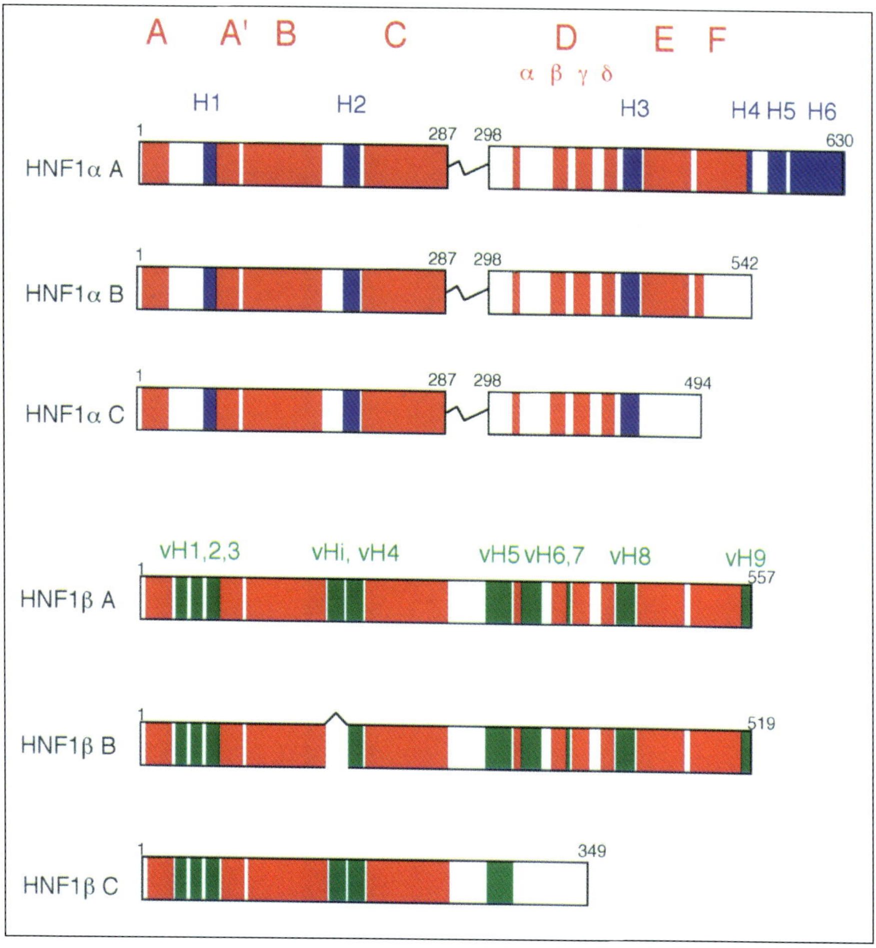

Fig. 10.3. Alternatively spliced forms of HNF1-α and HNF1-β in tetrapodians. Regions conserved in both HNF1-α and HNF1-β members are in red. Regions only conserved among HNF1-α members are in blue. Regions only conserved among HNF1-β members are in green. The nomenclature of these regions is that of refs. 18, 38.

state. Alternatively, these adipogenic transcription factors may provide parallel pathways for adipocyte differentiation. In either case, it is clear that the combined activity of multiple adipogenic transcription factors results in more efficient differentiation than is afforded by individual factors.[38] All, in fact, are expressed in mature 3T3-L1 adipocytes. Notably, ectopic expression of C/EBPδ does not convert NIH-3T3 cells to adipocytes.[31] Furthermore, C/EBPδ is expressed at very low levels in mature 3T3-L1 adipocytes (comparable to levels in adipoblasts). Thus, C/EBPδ may be unique among the isoforms in being strictly involved in determination and not differentiation of adipocytes.

The variation in temporal expression of the adipogenic transcription factors may provide clues as to their specific functions in adipogenesis. C/EBPα is expressed late in adipogenesis. Furthermore, C/EBPα trans-activates numerous adipogenic genes including; stearoyl-CoA desaturase (SCD1),[39] insulin stimulated glucose transporter 4 (GLUT4),[40] 422/aP2m,[39] phosphoenol carboxykinase (PEPCK),[41] the insulin receptor[42] and the obese gene (ob).[43] In each case, a proximal promoter element has been identified that mediates C/EBPα trans-activation. The 422 gene also contains a PPARγ-responsive sequence in a distal enhancer.[35] We speculate that C/EBPα is required to drive maximal expression of genes previously assembled into active complexes by enhancer-binding proteins (e.g., PPARγ). Two observations are consistent with this notion. First, Lin and Lane demonstrated that expression of anti-sense C/EBPα mRNA in 3T3-L1 cells during adipogenesis reduced the accumulation of triglycerides in these cells to approximately 15% of normal.[44] Second, Darlington and co-workers demonstrated that C/EBPα knockout mice have adipose tissue but are incapable of accumulating lipid droplets.[45] If adipocytes have been determined in the C/EBPα knockout mice, in situ staining of adipose should reveal PPARγ and ADD1 expression. We presume that lipid storage requires maximal expression of adipocyte-specific genes driven by C/EBPα trans-activation.

In addition to promoting adipogenesis, C/EBPα also suppresses mitotic growth. McKnight and colleagues constructed a 3T3-L1 subline expressing an estrogen-dependent chimeric form of C/EBPα and the estrogen receptor hormone binding domain, termed C/EBP-ER.[23] When logarithmically growing C/EBP-ER cells were exposed to estrogen, mitotic growth was arrested. Mitotic growth resumed when estrogen was removed from the medium. Growth inhibition by C/EBPα requires the same functional domains as those required for gene activation.[23] The latter observation led the authors to hypothesize that C/EBPα stimulates the transcription of growth arrest-associated and phenotype-associated genes in adipocytes. Two recent discoveries fulfill their hypothesis. First, we have determined that C/EBPα regulates the growth arrest and DNA damage inducible gene *gadd*45 in postmitotic adipocytes. *gadd*45 is preferentially expressed in mature adipocytes and its promoter is trans-activated by C/EBPα in a transient transfection assay.[46] Second, Darlington and co-workers found that levels of the cell cycle inhibitor p21 (encoded by *waf*1/*cip*1/*sdi*1) are elevated in cells growth arrested by ectopic expression of C/EBPα.[47] C/EBPα increases the expression of p21 through transcriptional activation and posttranslational stabilization.[47] Notably, both *gadd*45 and *waf*1/*cip*1 are trans-activated by the tumor suppressor p53 (chapter 16) in response to DNA damage.[48,49] However, introduction of a dominant negative p53 mutant into 3T3-L1 cells does not alter the developmental expression of GADD45.[46] Thus, C/EBPα regulation of *gadd*45 seems to be p53 independent.

The expression of GADD45 discriminates postmitotic adipocytes from other growth arrest states adopted by 3T3-L1 cells.[46] We conclude that C/EBPα promotes a specific growth arrest state that is compatible with differentiation. This growth arrest state permissive for differentiation had been hypothesized by Scott.[50] Subsequently, Freytag showed that constitutive expression of Myc (chapter 13) precludes 3T3-L1 cells from entering this unique growth arrest state.[51] Furthermore, he, along with his colleagues, demonstrated that forced expression of C/EBPα overcomes this Myc-mediated block of differentiation.[52] Freytag concluded that C/EBPα and Myc have a reciprocal relationship in 3T3-L1 cells wherein C/EBPα promotes phenotypic differentiation and Myc affords proliferative potential.[52] Our findings suggest a more direct antagonism between the two transcription factors: C/EBPα stimulates postmitotic growth arrest while Myc sustains proliferative potential. The antagonistic relationship between these transcription factors is manifest in their regulation of *gadd*45. Specifically, C/EBPα stimulates *gadd*45, while Myc inhibits C/EBP-mediated trans-activation.[46] The regulation of growth arrest-associated genes and phenotype-associated genes by C/EBPα provides a mechanism for coordinating differentiation with exit from the mitotic division cycle.

A third fundamental role for C/EBP isoforms may be the modulation of transcriptional activity after cells are terminally differentiated. Holbrook and colleagues have shown that CHOP mRNA levels fluctuate in response to glucose deprivation in 3T3-L1 cells.[53] As the level of glucose in the media decreases, CHOP mRNA increases. Increases in CHOP expression may act to inhibit adipogenesis. In fact, if CHOP is expressed constitutively in 3T3-L1 cells, adipogenesis is inhibited.[54] Inhibition by CHOP may be a result of a failure to produce C/EBPα homodimers at the appropriate time. CHOP dimerizes with both C/EBPα and C/EBPβ and, as a consequence, inhibits DNA binding.[9] Holbrook postulated that formation of inactive heterodimers modulates the expression of adipocyte specific genes as a consequence of glucose deprivation.[53] This interpretation is supported by the observation that the over-expression of C/EBPα can induce differentiation in CHOP-containing cells.[54] Insulin also regulates the expression of C/EBP isoforms in mature adipocytes.[55] C/EBPβ and C/EBPδ increase in response to insulin while C/EBPα rapidly declines. The down regulation of C/EBPα mRNA correlates with a decrease in the expression of GLUT4 mRNA, a target of C/EBPα.[55] The fluctuations in the activity of C/EBP isoforms in response to insulin and glucose provide a mechanism whereby the C/EBPs can modulate transcriptional activation in differentiated cells.

HEPATOCYTE DIFFERENTIATION

The role of the C/EBP isoforms in hepatocyte differentiation is less well understood due to the lack of established cell lines that can be differentiated in culture. Instead, combined observation from developing liver in vivo, cultured hepatoma cell lines, and regenerating liver after partial hepatectomy provide a model for the role of C/EBP isoforms in hepatocytes. The analysis of gene expression in vivo throughout neonatal and postnatal development affords an opportunity to examine temporal expression of C/EBP isoforms in the organ. However, these studies do not discriminate cell types or domains within the organ. Hepatoma cell lines retain varying subsets of hepatocyte-specific gene expression and permit analysis of C/EBP isoform-dependent regulation. However, these cells lines do not traverse determination and differentiation stages in culture. After the removal of approximately 2/3 of the liver, the remaining cells move out of G0 into the G1 stage and proceed through the cell cycle in order to regenerate the original mass in 9 to 12 days.[56] This is a unique state in which the liver is growing but still capable of performing normal physiological functions.[57] Thus, the regenerating liver may or may not be a reliable model for differentiation during organogenesis.

Studies done to date suggest that the general pattern of temporal expression of the C/EBP isoforms is similar in hepatocytes and adipocytes (Fig. 8.1). Specifically, C/EBPα expression increases throughout neonatal development and is localized to terminally differentiated hepatocytes in adult liver.[58] Furthermore, C/EBPα levels decline after partial hepatectomy,[59] consistent with the observed restriction of C/EBPα expression to nonproliferating adipocytes. The

C/EBPβ and C/EBPδ isoforms also show expression patterns reminiscent of their expression in adipocytes. C/EBPδ expression is restricted to the proliferative period following partial hepatectomy.[59] Although its expression hasn't been examined in liver throughout neonatal development, we speculate that C/EBPδ expression will be associated with proliferating precursor cells. Thus, C/EBPδ seems to be a counterpart to C/EBPα in adipocytes and hepatocytes; maximal C/EBPα expression is restricted to the differentiated, postmitotic state, while maximal C/EBPδ expression coincides with the proliferation of immature cells. Although both isoforms are expressed in regenerating liver, analysis on an individual cell basis might reveal a stoichiometric distribution of immature dividing cells and mature hepatocytes. C/EBPβ is expressed at half-maximal levels in adult liver[58] similar to its half-maximal expression in mature 3T3-L1 adipocytes. Additionally, increases in C/EBPβ accompany the proliferation following partial hepatectomy.[59] Unlike C/EBPα and C/EBPδ, C/EBPβ does not exhibit a strong association with growing or nongrowing states. There may be numerous explanations for this observation. One possibility is that cellular responses differ depending on the level of C/EBPβ present. Alternatively, the activity of C/EBPβ may differ depending on its partner in heterodimeric complexes. If the isoforms heterodimerize as readily in vivo as observed in solution,[11] C/EBPβ-C/EBPδ dimers should be more abundant in proliferating cells and C/EBPβ-C/EBPα heterodimers would be more abundant in differentiating cells. Further characterization of the formation of heterodimers in vivo and their biological properties will be crucial to a better understanding of their roles in development.

Similar to its activity in adipocytes, C/EBPα stimulates the transcription of phenotype-associated genes to drive hepatocyte differentiation. C/EBPα trans-activates the promoters of several hepatocyte genes, including albumin,[60] α1-antitrypsin,[61] transthyretin,[62] vitellogenin,[63] rat α-fetoprotein,[64] apoB-100,[65] aldolase B,[66] and human class I alcohol dehydrogenase.[67] C/EBPα acts on these promoters in conjunction with other transcription factors. The albumin promoter illustrates these relationships. The albumin promoter is bound by C/EBPα, Hepatic nuclear factor 1 (HNF-1) (chapter 10), D-site binding protein (DBP), C/EBPβ and the ubiquitous factors NF-1 and NF-Y.[68] Zaret and co-workers documented that the presence of NF-Y stabilizes C/EBPα binding to its recognition site in the albumin promoter.[69] Notably, this interaction impairs direct promoter binding by NF-Y while generating a stable transcription complex. Thus, protein-protein interactions with a more ubiquitous transcription factor contribute to C/EBPα-dependent, liver-specific gene expression. The leucine zipper of C/EBPα may be crucial to these protein-protein interactions. Nerlov and Ziff substituted the leucine zipper of the yeast trans-activator GCN4 for the naturally-occurring leucine zipper of C/EBPα and evaluated the consequences for liver-specific gene activation.[70] While wild-type C/EBPα preferentially trans-activates the albumin promoter in hepatoma-derived HepG2 cells as compared to HeLa cells, the hybrid protein trans-activates the albumin promoter equally in both cell types.[70] The interactions described above support the notion that C/EBPα assembles into liver-specific, transcriptional regulatory complexes through interactions with other transcription factors bound nearby in the proximal promoter region. We anticipate that these complexes also involve enhancer binding proteins similar to the interactions described in adipocytes.

The three members of the alcohol dehydrogenase (ADH) family of proteins offer an example of how the temporal expression of genes can be regulated by a series of transcription factors. ADH is encoded by three genes whose expression is activated at different times in development.[71] ADH1 is expressed throughout the development of the fetal liver. ADH2 is expressed late in fetal liver development. ADH3 expression is added during postnatal fetal development. All of the *adh* genes have a similar core promoter sequence that can bind HNF-1,

C/EBP, Sp1 and USF.[72] The *adh*3 promoter includes an additional retinoic acid response element (RARE).[73] HNF-1 potently trans-activates the *adh*1 promoter in CAT assays, but is ineffective in the trans-activation of *adh*2 and *adh*3. Notably, HNF-1 is expressed during early fetal development. Increases in C/EBPα accumulation follow HNF-1 accumulation late in fetal development. C/EBPα homodimers potently trans-activate the promoter of *adh*2 with greater efficiency than the promoters of *adh*1 or *adh*3. Late in liver development, coincident with ADH3 expression, C/EBPβ levels increase allowing for the formation of C/EBPβ homodimers and C/EBPα-C/EBPβ heterodimers which can trans-activate all of the *adh* genes. DBP is also expressed late in differentiation and can induce transcription from all of the *adh* promoters.[72] The *adh* family of genes illustrates the temporal regulation of closely related, liver-specific genes by C/EBP and non-C/EBP transcription factors expressed as a function of liver development.

The phenotype of C/EBPα knockout mice suggests a similar role for this transcription factor in hepatocyte and adipocyte differentiation. C/EBPα is expressed late in liver development and is important for full liver function.[45] C/EBPα knockout mice are indistinguishable from normal mice at the level of gross morphology. However eight hours after birth the mice die due to a lack of glycogen stores in the liver. The knockout mice lack glycogen synthase, a target of C/EBPα, and are unable to produce glycogen. Thus, C/EBPα knockout mice exhibit normal liver organogenesis but are incapable of executing normal metabolic functions.[45] We conclude that C/EBPα stimulates the transcription of genes in the liver to levels required for viability.

C/EBPα may have an anti-proliferative effect in hepatocytes as well as adipocytes. Ectopic expression of C/EBPα in the hepatoma-derived cell line Hep3B2 induces mitotic arrest.[74] This observation contrasts with an earlier observation that C/EBPα does not affect proliferation in HepG2 cells, another hepatoma cell line.[75] Farmer and colleagues observed that the expression of C/EBPα decreases in rat liver after a partial hepatectomy.[74] This corresponds to the period of time when the cells are moving out of G0 and into G1. The downregulation of C/EBPα is reciprocal with the expression of the immediate early and early genes: c-*jun*, c-*fos* and c-*myc*.[74] Perhaps C/EBPα and *myc* promote antagonistic growth states in hepatocytes as they do in adipocytes. C/EBPα gene expression also declines in primary cultures of hepatocytes grown on rat tail collagen, which induces dedifferentiation. This downregulation of C/EBPα is abolished if cells are growth arrested by adherence to reconstituted basement membrane gel.[76] Combined, these observations suggest that C/EBPα may coordinate differentiation and mitotic arrest in hepatocytes as it does in 3T3-L1 adipocytes.

The formation of heterodimeric complexes of C/EBP isoforms may also be important to gene regulation in hepatocytes. The ratio of C/EBPβ/LIP increases five-fold during rat liver differentiation.[22] C/EBPβ activity in hepatocytes is modulated by its alternative translation product LIP. Descombes and Schibler demonstrated that increasing amounts of LIP reduced the ability of C/EBPβ to transcriptionally activate reporter constructs in HepG2 cells.[22] Presumably, changes in LIP abundance impact C/EBPβ gene regulation in vivo as well. LIP inactivates C/EBPβ through the formation of heterodimers. Therefore, decrease in LIP throughout liver development should lead to an increase in C/EBPβ homodimers. C/EBPβ has been shown to be an inhibitor of proliferation in hepatoma cells. Specifically, ectopic expression of C/EBPβ arrests HepG2 cells at the G1/S boundary.[75] LIP appears to be antagonistic to the C/EBPβ induced mitotic arrest.[75] LIP may be inhibiting the transcriptional activation of growth arrest-associated genes downstream of C/EBPβ. Heterodimeric complexes may also be important during liver regeneration. Farmer and coworkers have examined DNA-binding activity in quiescent and regenerating liver using a C/EBP-responsive sequence from the rat *c/ebpα* gene promoter.[77] The authors documented that the predominant DNA-binding activity changes from C/EBPα/C/EBPβ heterodimers to homodimers of C/EBPβ during the initial G1 phase of regeneration. These findings support the notion that isoform-specific incorporation into transcriptional complexes is important to gene regulation and may involve the discrimination of heterodimers from homodimers.

C/EBP isoforms also modulate gene expression in differentiated hepatocytes. Specifically, C/EBPβ and C/EBPδ regulate changes in gene expression during the acute phase response. The liver responds to inflammation by producing acute phase responsive proteins such as serum amyloid A[78] and a1 acid glycoprotein (AGP)[79] at high levels. The levels of other proteins such as albumin drop during this period.[80] The expression of the C/EBP isoforms also changes in response to inflammation. C/EBPα mRNA expression decreases within four hours to half of its normal expression. Both C/EBPβ and C/EBPδ mRNA expression increase, with C/EBPδ having the greatest increase to peak at a level 70 fold higher than the basal level of expression in the liver.[81] The murine *agp* promoter contains an acute phase responsive element (APRE) that is bound by C/EBPα during normal liver functions. At the onset of the acute phase in response to lipopolysaccarides, the APRE is bound by C/EBPβ and AGP expression increases.[82] In rabbits treated with turpentine to induce the acute phase response, both C/EBPδ and C/EBPβ bind the APRE of the AGP promoter.[79] Additionally, C/EBPβ and C/EBPδ are the primary inducers of the serum amyloid A gene.[78] CHOP also shows an increase in expression during the acute phase response, although delayed in relation to C/EBPβ and C/EBPδ.[83] CHOP may be stimulated by C/EBPb and C/EBPd through its C/EBP binding site. Consistent with this notion, C/EBPβ has been show to trans-activate the CHOP promoter in HepG2 cells.[83] The delayed nature of the CHOP expression and its inhibition of C/EBP activation through dimerization suggests that CHOP may dampen the acute phase response through inactivation of C/EBPβ and C/EBPδ. These observations suggest that the role of CHOP is similar in both hepatocytes and adipocytes; inhibition of C/EBP trans-activators through heterodimerization.

MYELOMONOCYTIC DIFFERENTIATION

The temporal expression of the C/EBP isoforms throughout granulocyte differentiation differs significantly from that observed in adipocytes and hepatocytes. Friedman and colleagues examined the expression of the C/EBP isoforms during the differentiation of 32D C13 cells, a murine myelomonoblastic line that differentiates into granulocytes in response to granulocyte colony stimulating factor (G-CSF).[13] C/EBPα is detectable in proliferating, immature blast cells, rises transiently after addition of G-CSF and declines throughout terminal differentiation. Thus, the profile of C/EBPα expression during granulocyte differentiation is similar to the profile of C/EBPδ in adipocytes. C/EBPδ increases significantly after G-CSF addition and then declines to a half-maximal level in mature granulocytes, a profile reminiscent of C/EBPβ in adipocytes. C/EBPβ is expressed at low levels in immature blast cells and increases linearly throughout the conversion from myeloblasts to granulocytes, similar to the C/EBPα expression profile in adipocytes. Thus, the C/EBP isoforms seem to have switched roles in myelomonocytic differentiation compared to adipocyte and hepatocyte development. C/EBPα has been shown to activate transcription of the macrophage colony-stimulating factor receptor (M-CSF receptor).[84] This activity of C/EBPα is consistent with a role in determination as opposed to differentiation. In contrast, C/EBPβ (NF-M) stimulates differentiation-specific expression in eosinophils.[85] The expression profile of C/EBPβ in 32D C13 cells suggest that it trans-activates differentiation-specific genes in that lineage as well. It will be interesting to see whether C/EBPβ regulates growth arrest associated genes in this developmental transition and thus reflects the spectrum of activities of C/EBPα in adipocytes.

The distinctly different expression profiles and activities of C/EBP isoforms in the development of blood cell lineages forces speculation on the mechanisms that underlie these observations. Presumably, the transcription factors that C/EBP isoforms collaborate with dictate which isoform or heterodimer can be incorporated into transcriptional activation complexes on a given gene. Perhaps the milieu of transcription factors in myeloid cells is sufficiently different from adipocytes and hepatocytes to have inverted roles of the isoforms. The latter is not difficult to imagine given the significant differences in the metabolic properties of these cells. Notably, liver and fat play an important role in energy metabolism. Based upon early observations, McKnight and colleagues speculated that C/EBPα might coordinate energy homeostasis through gene regulation in these tissue.[86] Further work has demonstrated the expression of the C/EBP isoforms in a wide variety of cell types. We speculate more broadly that the specific temporal expression of the C/EBP isoforms will be conserved among cell types with integrated metabolic functions. Furthermore, we predict that lineage-specific enhancer binding proteins will package promoters into an active configuration and as a consequence dictate which C/EBP isoform or heterodimer will be required to drive maximal expression.

FUTURE DIRECTIONS

A more detailed understanding of the roles of C/EBP isoforms in development will require a better grasp of the in vivo significance of heterodimers. The leucine zipper dimerization interface theoretically affords the opportunity to use combinatorial interactions among a limited number of transcription factors to differentially regulate a larger set of genes. However, our understanding of the in vivo formation of heterodimers is very limited. Furthermore, other than the sequestration of certain trans-activators into inactive heterodimers, biological differences among various heterodimer combinations remain uncharacterized.

An appreciation for the role of C/EBP isoforms in cellular growth control has only recently emerged. Future work should illuminate the significance of C/EBP stimulation of growth-arrest associated genes in postmitotic cells. Presumably, this regulation must be circumvented in certain neoplastic diseases. The temporal expression profiles of the C/EBP isoforms also suggest an isoform-specific, cell type-specific association with proliferative states. The identification of targets genes of C/EBP isoforms in proliferating cells will help clarify the relationship between these transcription factors and those among the proto-oncogene family of trans-activators (chapter 12). Such studies will undoubtedly improve our understanding of the mechanisms that couple cellular growth control and differentiation.

Acknowledgments

Research in the authors' laboratory is supported by the American Cancer Society through an award to RMU.

References

1. Descombes P, Chojkier M, Lichtsteiner S et al. LAP, a novel member of the C/EBP gene family, encodes a liver-enriched transcriptional activator protein. Genes & Dev 1990; 4:1541-1551.
2. Akira S, Isshiki H, Sugita T et al. A nuclear factor for IL-6 expression (NF-IL6) is a member of the C/EBP family. EMBO J 1990; 9:1897-1906.
3. Poli V, Mancini FP, Cortese R. IL-6DBP, a nuclear protein involved in interleukin-6 signal transduction, defines a new family of leucine zipper proteins related to C/EBP. Cell 1990; 63:643-653.
4. Chang CJ, Chen TT, Lei HY et al. Molecular cloning of a transcription factor, AGP/EBP, that belongs to members of the C/EBP gene family. Mol Cell Biol 1990; 10:6642-6653.
5. Williams SC, Cantwell CA, Johnson PF. A family of C/EBP-related proteins capable of forming covalently linked leucine zipper dimers in vitro. Genes Dev 1991; 5:1553-1567.
6. Imagawa M, Osada S, Koyama Y et al. SF-B that binds to a negative element in

glutathione transferase P gene is similar or identical to trans-activator LAP/IL6-DBP. BBRC 1991; 179:293-300.
7. Kageyama R, Sasai Y, Nakannishi S. Molecular characterization of transcription factors that bind to the cAMP responsive region of the substance P precursor gene. JBC 1991; 266:15525-15531.
8. Kinoshita S, Akira S, Kishimoto T. A member of the C/EBP family, NF-IL6 β, forms a heterodimer and transcriptionally synergizes with NF-IL6. Proc Natl Acad Sci 1992; 89:1473-1476.
9. Ron D, Habener JF. CHOP, a novel developmentally regulated nuclear protein that dimerizes with transcription factors C/EBP and LAP and functions as a dominant-negative inhibitor of gene transcription. Genes Dev 1993; 6: 439-453.
10. Fornace AJ Jr, Alamo I Jr, Hollander MC. DNA damage-inducible transcripts in mammalian cells. Proc Natl Acad Sci USA 1988; 85: 8800-8804.
11. Cao Z, Umek RM McKnight SL. Regulated expression of three C/EBP isoforms during adipose conversion of 3T3-L1 cells. Genes Dev 1991; 9:1538-1552.
12. Magy P, Bisgaard C, Thorgeirsson SS. Expression of hepatic transcription factors during liver development and oval cell differentiation. JCB 1994; 126(1): 223-233.
13. Scott LM, Civin CI, Rorth P et al. A novel temporal expression pattern of three C/EBP family members in differentiating myelomonocytic cells. Blood 1992; 80(7):1725-1735.
14. Roman C, Platero JS, Shuman J et al. Ig/EBP-1: a ubiquitously expressed immunoglobulin enhancer binding protein that is similar to C/EBP and heterodimerizes with C/EBP. Genes Dev 1990; 4: 1404-1415.
15. Johnson PF, Landschultz WH, Graves BJ et al. Identification of a rat liver nuclear protein that binds to the enhancer core element of three animal viruses. Genes Dev 1987; 1:133-146.
16. Landschultz WH, Johnson PF, Adashi EY et al. Isolation of a recombinant copy of the gene encoding C/EBP. Genes Dev 1988; 2:786-800.
17. Landschulz WH, Johnson PF, McKnight SL. The leucine zipper: A hypothetical structure common to a new class of DNA binding proteins. Science 1988; 240: 1759-1764.
18. Landschulz WH, Johnson PF, McKnight SL. The DNA binding domain of the rat liver nuclear protein C/EBP is bipartite. Science 1989; 243:1681-1688.
19. Osada S, Yamamoto H, Nishihara T et al. DNA binding specificity of the CCAAT/enhancer-binding protein transcription factor family. JBC 1996; 271: 3891-3896.
20. Ubeda M, Wang X, Zinszner H et al. Stress-induced binding of the transcription factor CHOP to a novel DNA control element. Mol Cell Biol 1996; 16(4): 1479-1489.
21. Ossipow V, Descombes P, Schibler U. CCAAT/enhancer-binding protein mRNA is translated into multiple proteins with different transcription activation potentials. Proc Natl Acad Sci 1993; 90:8219-8223.
22. Descombes P, Schibler U. A liver-enriched transcriptional activator protein, LAP, and a transcriptional inhibitory protein, LIP, are translated from the same mRNA. Cell 1991; 67:569-579.
23. Umek RM, Friedman AD, McKnight SL. CCAAT/enhancer binding protein: A component of a differentiation switch. Science 1991; 251:288-292.
24. Lin FT, MacDougald OA, Diehl AM et al. A 30-kD alternative translation product of the CCAAT/enhancer binding protein α message: transcriptional activator lacking antimitotic activity. Proc Natl Acad Sci USA 1993; 90: 9606-1960.
25. Green H, Kehinde O. Sublines of mouse 3T3 cells that accumulate lipid. Cell 1974; 1:113-116.
26. Green H, Kehinde O. Spontaneous heritable changes leading to increased adipose conversion in 3T3 cells. Cell 1976; 7:105-113.
27. Negrel R, Grimaldi P, Ailhaud G. Establishment of preadipocyte clonal line from epididymal fat pad of ob/ob mouse that responds to insulin and to lipolytic hormones. PNAS 1978; 75:6054-6058.

28. Shugart EC, Levenson AS, Constance CM et al. Differential expression of gas and gadd genes at distinct growth arrest points during adipocyte development. Cell Growth & Diff 1995; 6:1541-1547.
29. Vassuer-Cognet M, Lane MD. Trans-acting factors involved in adipogenic differentiation. Cur Opin Gen Dev 1993; 3:238-245.
30. Russell TR, Files N, Ingram M. Mitosis of contact-inhibited 3T3 preadipocytes precedes chemically induced differentiation into adipocytes. PSEBM 1983; 173: 471-474.
31. Yeh WC, Cao Z, Classon M et al. Cascade regulation of terminal adipocyte differentiation by three members of the C/EBP family of leucine zipper proteins. Genes Dev 1995; 9:168-181.
32. Lin F, Lane MD. CCAAT/enhancer binding protein α is sufficient to initiate the 3T3-L1 adipocyte differentiation program. PNAS 1994; 91:8757-8761.
33. Freytag SO, Paielli DL, Gilbert JD. Ectopic expression of the CCAAT/enhancer-binding protein α promotes the adipogenic program in a variety of mouse fibroblastic cells. Genes & Dev 1994; 8:1654-1663.
34. Wu Z, Xie Y, Bucher NLR et al. Conditional ectopic expression of C/EBPβ in NIH-3T3 cells induces PPARγ and stimulates adipogenesis. Genes Dev 1995; 9:2350-2363.
35. Totonoz P, Hu E, Graves RA et al. mPPARγ2: a tissue-specific regulator of an adipocyte enhancer. Genes Dev 1994; 8:1224-1234.
36. Kim JB, Spiegelman BM. ADD1/SREBP1 promotes adipocyte differentiation and gene expression linked to fatty acid metabolism. Genes Dev 1996; 10:1096-1107.
37. Totonoz P, Kim JB, Graves RA. ADD1: a novel helix-loop-helix transcription factor associated with adipocyte determination and differentiation Mol Cell Biol 1993; 13:4753-4759.
38. Hu E, Tontonoz P, Spiegelman BM. Transdifferentiation of myoblasts by the adipogenic transcription factors PPAR gamma and C/EBP α. PNAS 1995; 92:9856-9860.
39. Christy RJ, Yang VW, Ntambi JM et al. Differentiation-induced gene expression in 3T3-L1 preadipocytes: CCAAT/enhancer binding protein interacts with and activates the promoters of two adipocyte-specific genes. Genes Dev 1994; 3: 1323-1331.
40. Kaestner KH, Christy RJ, Lane MD. Mouse insulin-responsive glucose transporter gene: characterization of the gene and trans-activation by the CCAAT/enhancer binding protein. PNAS 1990; 87:251-255.
41. Park EA, Roesler WJ, Liu J et al. The role of the CCAAT/enhancer-binding protein in the transcriptional regulation of the gene for phosphoenolpyruvate carboxykinase (GTP). Mol Cell Biol 1990; 10: 6264-6272.
42. McKeon C, Pham T. Transactivation of the human insulin receptor gene by the CCAAT/enhancer binding protein. BBRC 1991; 174:721-728.
43. Hwang C, Mandrup S, MacDougald OA et al. Transcriptional activation of the mouse obese (ob) gene by CCAAT/enhancer binding protein α. PNAS 1996; 93:873-877.
44. Lin FT, Lane MD. Antisense CCAAT/enhancer-binding protein RNA suppresses coordinate gene expression and triglyceride accumulation during differentiation of 3T3-L1 preadipocytes. Genes Dev 1992; 6, 533-544.
45. Wang N, Finegold MJ, Bradley A et al. Impaired energy homeostasis in C/EBPα knockout mice. Science 1995; 269: 1108-1112.
46. Constance CM, Morgan JI, Umek RM. C/EBPα regulation of the growth arrest associated gene *gadd*45. Mol Cell Biol 1996; 6:3878-3883.
47. Timchenko NA, Wilde M, Nakanishi M et al. CCAAT/enhancer-binding protein a (C/EBPa) inhibits cell proliferation through the p21 (WAF-1/CIP-1/SCI-1) protein. Genes Dev 1996; 10:804-815.
48. el-Deiry WS, Tokino T, Velculescu VE et al. WAF1, a potential mediator of p53 tumor suppression. Cell 1993; 75: 817-25.
49. Kastan MB, Zhan Q, el-DeiryWS et al. A mammalian cell cycle checkpoint pathway utilizing p53 and GADD45 is defective in ataxia-telangiectasia. Cell 1992; 71:587-597.

50. Scott RE, Florine DL, Willie JJ et al. Coupling of growth arrest and differentiation at a distinct state in the G_1 phase of the cell cycle: GD. Proc Natl Acad Sci 1982; 79:845-849.
51. Freytag SO. Enforced expression of the c-myc oncogene inhibits cell differentiation by precluding entry into a distinct predifferentiation state in G_0/G_1. Mol Cell Biol 1990; 8:1614-1624.
52. Freytag SO, Geddes TJ. Reciprocal regulation of adipogenesis by Myc and C/EBP α. Science 1992; 256:379-382.
53. Carlson SG, Fawcett TW, Bartlett JD et al. Regulation of the C/EBP-related gene *gadd*153 by glucose deprivation. Mol Cell Biol 1993; 13:4736-4744.
54. Batcharova N, Wang XZ, Ron D. Inhibition of adipogenesis by the stress-induced protein CHOP (Gadd153). EMBO J 1995; 14(19):4654-4661.
55. MacDougland OA, Cornelius P, Liu R et al. Insulin regulates transcription of the CCAAT/enhancer binding protein (C/EBP)a, b and d genes in fully-differentiated 3T3-L1 adipocytes. JBC 1995; 270(2):647-654.
56. Higgins GM, Anderson RM Experimental pathology of the liver. I. Restoration of the liver of the white rat following partial surgical removal. Arch Pathol 1931; 12:186-202.
57. Greenbaum LE, Cressman DE, Haber BA et al. Coexistence of C/EBP α, β, growth-induced proteins and DNA synthesis in hepatocytes during liver regeneration. Implications for maintenance of the differentiated state during liver growth. J Clin Invest 1995; 96:1351-1365.
58. Nagy P, Bisgaard C, Thorgeirsson SS. Expression of hepatic transcription factors during liver development and oval cell differentiation. JCB 1994; 126(1): 223-233.
59. Flodby P, Antonson P, Barlow C et al. Differential patterns of expression of three C/EBP isoforms, HNF-1, and HNF-4 after partial hepatectomy in rats. Exp Cell Res 1993; 208:248-256.
60. Friedman AD, Landshulz WH, McKnight SL. CCAAT/enhancer binding protein activates the promoter of the serum albumin genes in cultured hepatoma cells. Genes Dev 1989; 3:1314-1322.
61. Grayson DR, Costa RH, Xanthopoulous KG et al. A cell-specific enhancer of the mouse a_1-anti trypsin gene has multiple functional regions and corresponding protein-binding sites. Mol Cell Biol 1988; 8:1055-1066.
62. Costa RH, Lai E, Grayson DR et al. The cell-specific enhancer of the mouse transthyretin (prealbumin) gene binds a common factor at one site and a liver specific factor(s) at two other sites. Mol Cell Biol 1988; 8:81-90.
63. Cardinaux JR, Chapel S, Wahli W. Complex organization of CTF/NF-1, C/EBP and HNF3 binding sites within the promoter of the liver-specific vitellogenin gene. JBC 1994; 269:32947-32956.
64. Bois-Joyeux B, Danan JL. Members of the CAAT/enhancer-binding protein, hepatocyte nuclear factor-1 and nuclear factor-1 families can differentially modulate the activities of the rat α-fetoprotien promoter and enhancer. Biochem J 1994; 301:49-55.
65. Metzger S, Leff T, Breslow JL. Nuclear factors AF-1 and C/EBP bind to the human ApoB gene promoter and modulate its transcriptional activity in hepatic cells. JBC 1990; 265:9978-9983.
66. Gregori C, Kahn A, Pichard AL. Competition between transcription factors HNF1 and HNF3, and alternative cell-specific activation by DBP and C/EBP contribute to the regulation of the liver-specific aldolase B promoter. Nucleic Acids Res 1993; 21:897-903.
67. Stewart MJ, Shean ML, Paeper BW et al. The role of the CCAAT/enhancer-binding protein in the differential transcriptional regulation of a family of human liver alcohol dehydrogenase genes. JBC 1991; 266:11594-11603.
68 Herbst RS, Friedman N, Darnell JE et al. Positive and negative regulatory elements in the mouse albumin enhancer. PNAS 1989; 86:1553-1557.
69. Milos PM, Zaret KS. A ubiquitous factor is required for C/EBP-related proteins to form stable transcription complexes on an albumin promoter segment in vitro. Genes Dev 1992; 6:991-1004.
70. Nerlov C, Ziff EB. Three levels of functional interaction determine the activity of CCAAT/enhancer binding protein-α

on the serum albumin promoter. Genes Dev 1994; 8:350-362.
71. Smith M. Genetics of human alcohol and aldehyde dehydrogenases. Adv Hum Genet, 1986; 15:249-290.
72. van Ooij C, Snyder RC, Paeper BW et al. Temporal expression of the human alcohol dehydrogenase gene family during liver development correlates with differential promoter activation by hepatocyte nuclear factor 1, CCAAT/enhancer-binding protein α, liver activator protein, and D-element-binding protein. Mol Cell Biol 1992; 12(7):3023-3031.
73. Duester G, Shean ML, McBride MS et al. Retinoic acid response element in the human alcohol dehydrogenase gene ADH3: implication for regulation of retinoic acid synthesis. Mol Cell Biol 1991; 11:1638-1646.
74. Mischoulon D, Rana B, Bucher NLR et al. Growth-dependent inhibition of CCAAT enhancer-binding protein (C/EBPa) gene expression during hepatocyte proliferation in the regenerating liver and in culture. Mol Cell Biol 1992; 12(6): 2553-2560.
75. Buck M, Tuler H, Chojkier M. LAP (NF-IL-6), a tissue-specific transcriptional activator, is an inhibitor of hepatoma cell proliferation. EMBO J 1994; 13(4): 851-860.
76. Rana B, Mischoulon D, Xie Y et al. Cell-extracellular matrix interactions can regulate the switch between growth and differentiation in rat hepatocytes: reciprocal expression of C/EBPα and immediate-early growth response transcription factors. Mol Cell Biol 1994; 14(9): 5858-5869.
77. Rana B, Xie Y, Mischoulon D et al. The DNA binding activity of C/EBP transcription factors is regulated in the G1 phase of the hepatocyte cell cycle. JBC 1995; 270(30):18123-18132.
78. Ray A, Ray BL. Serum amyloid A gene expression under acute-phase condition involves participation of inducible C/EBPβ and C/EBPδ and their activation by phosphorylation. Mol Cell Biol 1994; 14(6):4324-4332.
79. Ray BK, Ray A. Expression of the gene encoding a1-acid glycoprotein in rabbit liver under acute-phase condition involves induction and activation of β and δ CCAAT-enhancer-binding proteins. FEBS 1994; :891-900.
80. Riley JH, Bathurst IC, Edbrooke MR et al. Alpha-1-antitrypsin and serum albumin mRNA accumulation in normal, acute phase and ZZ human liver. Febs Lett 1985; 185:361-366.
81. Alam T, An MR, Papaconstantinou J. Differentiation expression of three C/EBP isoforms in mulitple tissues during the acute phase response. JBC 1992; 267(8): 5021-5024.
82. Alam T, An MR, Mifflin RC et al. trans-activation of the a1-acid glycoprotein gene acute phase responsive element by multiple isoforms of C/EBP and glucocorticoid receptor. JBC 1993; 268(21): 15681-15688.
83. Sylvester SL, ap Rhys CMJ, Luethy-Martindale JD et al. Induction of GADD153, a CCAAT/enhancer-binding protein (C/EBP)-related gene, during the acute phase response in rats. JBC 1994; 269(31):20119-20125.
84. Hohaus S, Petrovick MS, Voso MT et al. PU.1 (Spi-1) and C/EBP α regulate expression of the granulocyte-macrophage colony-stimulating factor receptor α gene. Mol Cell Biol 1995; 15:5830-5845.
85. Muller C, Kowenz-Leutz E, Grieser-Ade S et al. NF-M (chicken C/EBP β) induces eosinophilic differentiation and apoptosis in a hematopoietic progenitor cell line. EMBO J 1995; 14:6127-6135.
86. McKnight SL, Lane MD, Gluecksohn-Waelsch S. Is CCAAT/enhancer-binding protein a central regulator of energy metabolism? Genes Dev 1989; 3:2021-2024.

CHAPTER 9

The NF-κB/Rel System of Transcriptional Activators

Ralf Schreck, Barbara Kistler and Thomas Wirth

INTRODUCTION

NF-κB was originally described as a B cell-specific nuclear factor binding to the intronic κ light chain enhancer. The factor was constitutively active in mature B and plasma cells. In pre-B cell lines as well as in many other cell types, it was present in an inactive but inducible form. Multiple different extracellular stimuli and conditions of cellular stress all resulted in the activation of NF-κB. An important clue towards the understanding of NF-κB regulation was derived from the observation that the inactive form of NF-κB was anchored in the cytoplasm due to its interaction with an inhibitory protein, termed IκB. Binding sites for NF-κB have now been identified in a wide variety of genes, and it was demonstrated that NF-κB is a key-regulator of inducible gene expression in many different cell types. The molecular and biochemical basis of this rapid response system has been worked out to a large degree over the past years. The functions of NF-κB under both physiological and pathophysiological conditions have begun to emerge. In this article, we will describe the main features of this unique regulatory system and discuss the involvement of the NF-κB/IκB system in normal cellular development and function as well as its contribution to diseases.

STRUCTURE AND FUNCTION OF THE NF-κB/REL AND IκB PROTEINS

NF-κB/Rel Proteins

Earlier biochemical analysis of NF-κB had already suggested that this factor is a multimeric DNA-binding protein. Two subunits of 50 kDa and 65 kDa were identified.[1] Cloning of the genes for these subunits revealed that they belong to a family of related proteins that all share significant homology to *v-rel*, a well-known viral oncogene.[2-5] The *v-rel*-oncogene had been originally isolated from an acutely transforming avian retrovirus that causes lymphomas in chicken and other birds. Today, five cellular family members are known, which are collectively called the NF-κB/Rel family of transcription factors. These are NF-κB1 (p105/

Transcription Factors in Eukaryotes, edited by Athanasios G. Papavassiliou.

p50), NF-κB2 (p100/p52), RelA, RelB and c-Rel.[6,7] They all share an extensive region of homology, a 300 aa domain that has been termed the *R*el-*H*omology-*D*omain (RHD) (Fig. 9.1). The RHDs share about 50% homology between individual family members.

The RHD represents an important functional domain mediating both dimerization of NF-κB/Rel-proteins and specific DNA recognition. In addition, the RHD is also essential for the interaction with the IκB inhibitory proteins (see below). The structure of a specific NF-κB/Rel homodimer (p50/p50) binding to its cognate DNA motif has recently been solved by crystallography.[8,9] The structure revealed several unique features. First of all, the whole Rel-Homology-Domain is involved in contacting the binding site. A total of 10 loops, five from each subunit of the dimer, decorate every nucleotide of the 10-base pair recognition site by making exclusively major groove contacts. The overall structure resembles a butterfly, with each of the two subunits representing a wing. The dimer interface is formed by about 100 aa of the C-terminal part of the RHD, which form a β-stranded sandwich. Comparison of the two independently solved structures of p50 homodimers bound either to a symmetrical consensus site (5′ GGGAATTCCC 3′) or to a natural binding site (5′ GGGAATCCCC 3′) revealed that the homodimer recognizes half-sites spaced by three and four base pairs, respectively. This is achieved through conformational adaptations in the DNA and domain shifts in the protein, as well as small changes in amino acid side chain conformations.[10] These different binding modes may be of relevance in vivo with respect to recognition of different binding sites, specific heterodimer combinations as well as interactions with additional factors (see below).

Two of the family members, NFKB1 and NFKB2, are produced as precursor proteins that need limited proteolytical processing in order to generate mature proteins. The precursor proteins, p105 and p100, give rise to p50 and p52 mature proteins, respectively.[2-5,11-13] Although the exact mechanism of this processing has not been worked out yet, recent experiments suggest that the glycine-rich hinge region, which is present in both of these proteins N-terminal to the processing site, directs a novel type of protease that measures the distance from this motif to the actual processing site.[14] The mature p50 and p52 proteins are mainly made up of the RHD and they lack potent transcriptional activation domains. Therefore, homodimers of p50 and p52 bind to their target sites but they fail to activate transcription efficiently.[15-18] However, in vitro transcription experiments as well as studies performed in yeast suggest that at least some gene activation function can be observed with p50 homodimers.[19-21] Furthermore, these experiments indicated that upon binding to specific target sites a conformational change of the p50 homodimer is induced, which may be responsible for exposing a weak transactivation function.[20]

The most potent transcriptional activators of the family are RelA, RelB and c-Rel. Synthesis of these proteins does not require proteolytic processing and, in addition to the RHD, they contain potent transcriptional activation domains. In case of RelA and c-Rel, these domains are localized in the C-terminal region of the protein, whereas transactivation function was mapped to both the N- and C-terminal region of RelB.[16,22-24] The RelA transactivation domain has been characterized in more detail.[25,26] It represents a composite of two separable transactivation domains that synergize to yield full level transcriptional activity. Secondary structure predictions of one of these subdomains suggested that it adopts an α-helical conformation. However, an α-helical structure was detected only by NMR in the presence of hydrophobic solvents.[25] The authors proposed that the primarily unstructured domain is converted into an α-helix upon interaction with respective target proteins. In vitro transcription experiments and protein-protein interaction studies revealed that the transactivation domain is in direct contact with the basal transcriptional machinery. Both TBP and TFIIB interact with the RelA activation do-

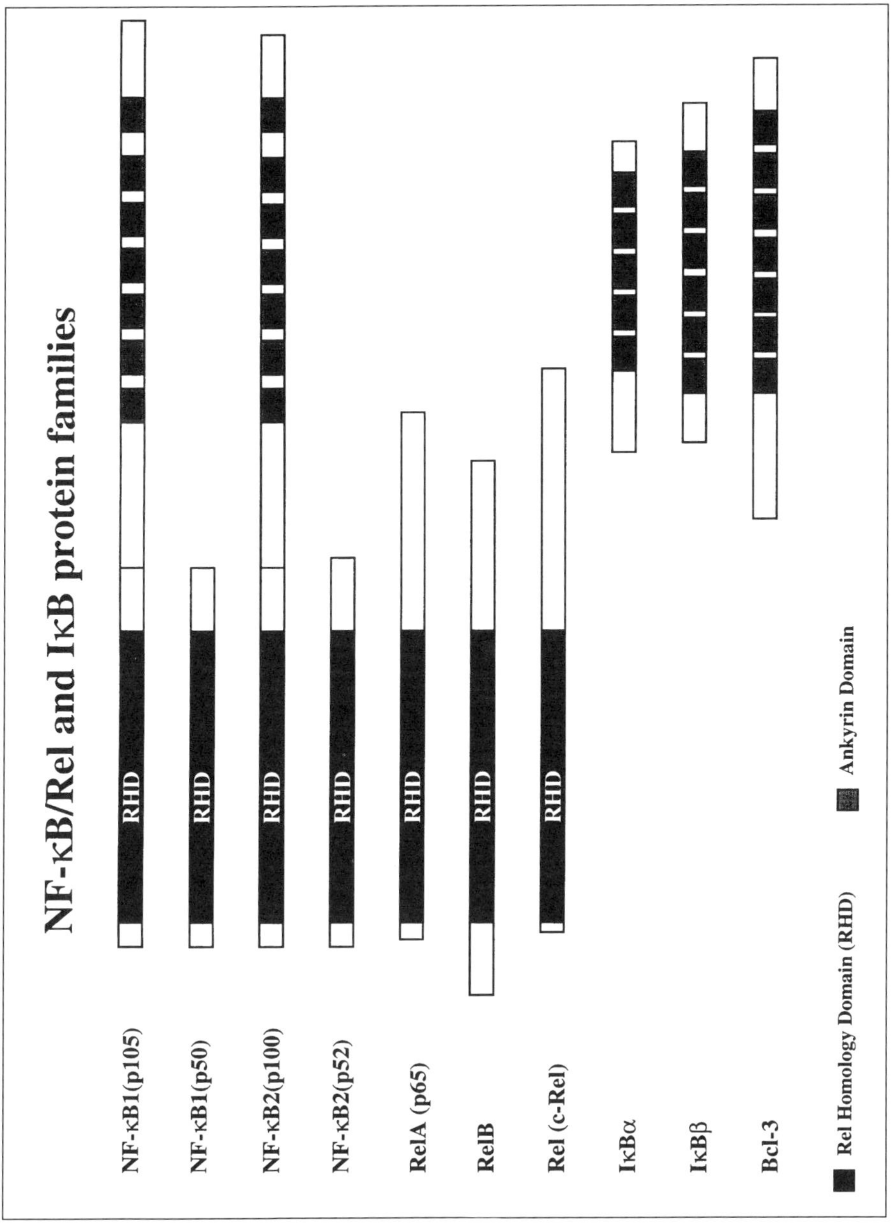

Fig. 9.1. Schematic outline of the structures of mammalian NF-κB/Rel and IκB proteins. In case of NF-κB1 and NF-κB2, both precursor and mature proteins are shown. Only the Rel Homology Domains and the Ank repeat domains are highlighted. See text for further details.

main and the recently identified positive coactivator PC1 is required for efficient in vitro transcriptional stimulation by RelA.[27] Similarly, c-Rel has been shown to physically interact with TBP, although some discrepancy exists with respect to the part of the c-Rel protein involved in this interaction.[28,29]

As noted earlier, NF-κB/Rel-proteins bind the cognate DNA-binding side as dimers. A large degree of promiscuity allows virtually any combination of homo- and heterodimers to be formed within this family, at least in vitro. There is one notable exception, however. RelB is apparently exquisitely restrictive with respect to its interaction partners. It can not homodimerize and forms functional heterodimers only with the p50 and p52 proteins.[30] The different homo- and heterodimers have been demonstrated to preferentially recognize distinct target sites. The most common NF-κB complex, the p50/RelA heterodimer, interacts with the consensus motif 5′GGGRNNYYCC 3′.[31] The left half-site of the motif is in contact with the p50 subunit, whereas RelA interacts with the right half-site.[32] A variant of this consensus was found in several NF-κB target genes specifically expressed in monocytes and endothelial cells (see below). The promoters of these genes contain the consensus 5′ G/TGGARNYYCC 3′, which is recognized by the RelA/c-Rel heterodimer.[33,34] This suggests an interesting mode of combinatorial gene regulation, by which different heterodimeric complexes are involved in mediating distinct gene activation profiles. Such combinatorial mechanisms of gene regulation due to differential heterodimerization have previously also been noted for other dimerizing transcription factors, e.g., members of the leucine zipper family of transcription factors.[35]

A domain with some homology to the RHD is present in the family of NFAT transcription factors that are induced upon activation of T cells.[36-39] Cloning of different members of the NFAT family revealed that they share a domain with approximately 20% homology to the RHD. This domain in NFAT proteins is required for specific DNA-binding. So far, no evidence for direct heterodimerization between NFAT and NF-κB-family members has been obtained.

IκB Proteins

An important early finding was that the activity of NF-κB is regulated to a large extent by specific inhibitory proteins. Treatment of the inactive cytoplasmic complex in vitro with dissociating agents resulted in the separation of NF-κB and IκB proteins.[40,41] Initial purification studies revealed the presence of at least two inhibitory proteins, which were called IκBα and IκBβ.[42] The genes for both proteins have been cloned and the structure revealed that they belong to a family of proteins[43,44] (Fig. 9.1). The hallmark of IκB proteins is a central repeat of ankyrin motifs (ANK repeat), a 33 amino acid long domain that is conserved in proteins of quite distinct functions from yeast to mammals.[6,45] This ANK repeat domain is largely responsible for the specific interaction of IκB with the NF-κB/Rel proteins. The N- and C-terminal domains play an important role in regulating the stability of IκBα and IκBβ (see below). Interestingly, the C-terminal region of the of NFKB1 and NFKB2 precursor proteins (p105 and p100) also contain such ANK repeats. Indeed, it could be demonstrated that these precursor proteins function as inhibitors of NF-κB by heterodimerization with Rel subunits.[46-48] Furthermore, in murine B cell lines an alternate transcript was detected encoding almost exclusively the C-terminal ANK repeats of NFKB1. This protein was termed IκBγ.[49] A comparable transcript for NFKB2 encoding a hypothetical IκBδ has not been identified yet. Very recently, a new member of the IκB-family, IκBε, sharing structural homology to IκBα and IκBβ, was cloned in a yeast two-hybrid screening.[49] Interestingly, the *bcl-3* gene, which is translocated in some malignant lymphomas, also encodes an ankyrin repeat protein.[50] In vitro experiments revealed that bcl-3 can inhibit DNA-binding of NF-κB proteins.[51,52] Other ankyrin repeat proteins, such as IκBR and the translocation-associated Notch homologue (TAN-1) were also found to inhibit DNA-binding of NF-κB/Rel-proteins.[53,54]

There at least two mechanisms by which IκB-proteins interfere with NF-κB-function. IκB can retain NF-κB/Rel-protein complexes in the cytoplasm.[40] This is most likely achieved by masking the nuclear localization sequence (NLS), which is localized in the C-terminal part of the RHD.[55-57] NLS masking excludes recognition by the nuclear import machinery, therefore interfering with the active transport of these proteins to the nucleus. Indeed, mutations altering the NLS region of RelA affected the interaction with IκB.[55,56] The second mode of interference is a direct inhibition of sequence-specific DNA binding by IκB proteins.[41,42] The exact molecular mechanism for this interference is not clearly understood; however, it may play a role when the IκB–associated NF-κB complexes become temporarily nuclear after mitosis, as a consequence of breakdown of the nuclear envelope.

Is there a need for different IκB proteins? The most likely reason for the existence of multiple IκB proteins is that they allow for an increased versatility in using this highly developed rapid response system. Clearly, having multiple inhibitors allows the formation and specific regulation of distinct multimeric complexes. Furthermore, the release of different NF-κB DNA-binding subunits can proceed via different pathways for individual IκBs. In addition, cell-type-specific expression of IκB and NF-κB proteins would generate additional combinations and regulatory circuits. The combination of all these mechanisms results in a plethora of different responses all mediated by this system.

Indeed, available evidence suggests that all of these mechanisms are in fact utilized. Most information is available on the dimer-specificity of IκB-proteins. IκBα is a potent inhibitor for complexes containing the RelA, the RelB or c-Rel protein, but barely inhibits p50 or p52 homodimers.[43,58-60] Interestingly however, some experiments propose that not all these complexes are targeted by IκBα with the same efficiency. Heterodimers of p52 and RelB have a reduced sensitivity to IκBα compared to other heterodimers.[61] IκBβ is a potent inhibitor of RelA-containing dimers. It has also been shown to be physically associated with c-Rel, but it is clearly much less effective in the inhibition of c-Rel containing complexes.[44] Furthermore, RelB heterodimers are barely inhibited at all by IκBβ (Kistler et al, unpublished). IκBε has not been characterized in all details, but seems to be specific for RelA and c-Rel and heterodimers thereof.[49a] IκBγ, Bcl-3 and IκBR preferentially target p50 or p52 subunits or homodimers.[49,51-53] Although Bcl-3 was able to inhibit p50/p52 DNA-binding in vitro, bcl-3 has been suggested to function also as a transcriptional coactivator together with p50 and p52 homodimers in vivo.[62,63] This model suggests that Bcl-3 is forming a metastable complex with p50 and p52 homodimers, which results in transcriptional activation before pulling off the complexes from the DNA. Along these lines, it was demonstrated that Bcl-3 functions as a transcriptional activator when fused artificially to the GAL4 DNA-binding domain. In addition, transgenic mice overexpressing Bcl-3 in thymocytes showed an increased binding of p50 to DNA.[64] This induction of DNA-binding is apparently achieved by a posttranslational mechanism involving Bcl-3.

The detailed mechanism of IκB release will be described in a later section. It is important to note however, that not all IκB proteins follow the exact same route of inactivation. Thompson and colleagues demonstrated that, whereas IκBα was inactivated by treatment of cells with a variety of stimuli, only a subset of these stimuli resulted in degradation of IκBβ.[44] Furthermore, compared with the kinetics of inactivation of IκBα, IκBβ was significantly more slowly degraded suggesting that different biological consequences, e.g., short-term and long-term effects by NF-κB, could be regulated by these two types of inhibitors. In addition, recent data suggest that during primary B cell differentiation, slow and long-lasting activation of NF-κB is correlated with the kinetics of IκBβ inactivation (B. Kistler and T. Wirth, unpublished observations). The pathways of inactivation for other IκB proteins have not been elucidated in detail yet.

The various IκB molecules may also be responsible for generating cell-type-specific responses. Several of the inhibitors show tissue- or cell-type-specific expression patterns. IκBγ is exclusively found in B cells, Bcl-3 is enriched in lymphoid cells, whereas IκBε was recently found to be most abundant in a monocytic cell line[49,65] (A. Israel, personal communication). Furthermore, the regulation of IκB inhibition itself maybe under tissue-specific control. For instance, RelB containing heterodimers escape IκBα inhibition in B cells, most likely due to a specific modification of the RelB protein or an associated factor.[66] Furthermore, as the different NF-κB/Rel family members also show cell-type-specific differences in their expression,[62,67,68] a wide variety of combinations is generated with the potential for differential regulation.

Evolutionary Conservation

NF-κB/Rel and IκB-proteins have been detected in a variety of mammalian and vertebrate species. Interestingly, both the genes and the main regulatory features of the system, have been conserved between vertebrates and invertebrates. The best example for this conservation is found in the establishment of the dorso-ventral pattern during early *Drosophila* embryogenesis. More than 10 maternal genes are required to finally establish a morphogenetic gradient along the dorso-ventral axis.[69,70] At the end of this genetic hierarchy, the transcription factor Dorsal is specifically localized to the nucleus on the ventral side of the embryo. Dorsal shares significant homology to Rel proteins.[71] It is synthesized constitutively in all cells, but kept cytoplasmic due to its interaction with the inhibitor protein Cactus, a member of the IκB family.[72] Interestingly, other members of the signaling cascade that have been identified in *Drosophila* such as the transmembrane receptor Toll and the serine/threonine kinase Pelle have mammalian homologues, such as the IL1-receptor and the recently identified kinase, IRAK, respectively.[73,74]

NF-κB related proteins also have additional functions in insects. Regulatory motifs resembling NF-κB binding sites have been identified in a variety of promoters from inducible insect immunity genes, such as cecropins and diptericin. An important regulator of these genes that interacts with these motifs is Dif, a further member of the NF-κB/Rel-family with significant homology to dorsal and mammalian Rel proteins.[75] Very recently, another member of the family called Relish, has been identified with significant structural homology to the precursor proteins of NF-κB1 and NF-κB2.[75a] All these findings suggest, that the signaling cascades leading to NF-κB activation developed early in evolution.

NF-κB-Target Genes and Combinatorial Regulation

The classical target gene of NF-κB is the κ light chain of the immunoglobulin genes.[76] The NF-κB binding site within the intronic enhancer has been shown to be essential for B cell specific enhancer function.[77] The interaction of NF-κB with this motif is not only important for the transcriptional activation of this gene, recent experiments further suggest that NF-κB also plays an important role in establishing transcriptional competence of the locus. B cell-specific demethylation of the κ-locus depends on the presence of a functional enhancer element.[78] It was shown that the intronic κ-enhancer depended on an intact κB-binding site for inducing demethylation. Furthermore, the presence of nuclear NF–κB proteins was a prerequisite for demethylation to occur in B cells.[79]

The Igκ-locus is an exceptional NF-κB target, as it is highly cell-type-specific for the B cell lineage, and after transition from the pre-B to the immature B cell stage it is constitutively expressed. In contrast, most of the other NF-κB targets are expressed in a exclusively inducible fashion. The list of target genes is very long and summarized in Table 9.1. A large number of target genes play a role in cellular immune responses and inflammatory processes. In addition, several viruses such as the human

Table 9.1. NF-κB target genes.

Immunoreceptors	Immunoglobulin κ light chain (Igκ)
	Immunoglobulin heavy chain (IgH 3'enhancer)
	T cell receptor β chain (TCR β)
	Major histocompatibility complex (MHC) class I ($H\text{-}2K^b$)
	MHC class II ($E\alpha^d$)
	β_2-microglobulin
	MHC class II invariant chain
	B7.1
	Tissue Factor-1
	Interleukin-2 receptor α chain
	CD7
Cytokines and growth factors	Interleukin-1α
	Interleukin-1β
	Tumor necrosis factor α (TNF-α)
	Lymphotoxin (LT) (TNF-β)
	Interleukin-8
	IP-10
	Gro-α, -β and -γ
	MIP-1α
	MCP-1/JE
	RANTES
	Interleukin-2
	Interleukin-6
	Interleukin-12
	Macrophage colony-stimulating factor (M-CSF)
	Granulocyte/macrophage colony-stimulating factor (GM-CSF)
	Granulocyte colony-stimulating factor (G-CSF)
	Erythropoietin
	Interferon-β
	Interferon-γ
	Proenkephalin
	Transforming growth factor-β2 (TGF-β2)
Adhesion molecules	Endothelial-leukocyte adhesion molecule-1 (ELAM-1)
	Vascular cell adhesion molecule-1 (VCAM-1)
	Intracellular cell adhesion molecule-1 (ICAM-1)
Acute phase proteins	Angiotensinogen
	Serum amyloid A precursor
	Complement factor B
	Complement factor C4
	Urokinase-type plasminogen activator
	Matrix Metalloprotease-9

(Cont. on next page)

Table 9.1. NF-κB target genes. (con't)

Viruses	Human immunodeficiency virus 1 (HIV-1) Human immunodeficiency virus 2 (HIV-2) Simian immunodeficiency virus (macaques) (SIV mac) Cytomegalovirus (CMV) Adenovirus Herpes simplex virus 1 (HSV-1) Human neurotropic virus (JCV) Simian virus 40 (SV-40)
Transcription factors and regulators	c-Rel NFKB1 NFKB2 IκB-α c-Myc c-Myb Interferon regulatory factor 1 (IRF-1) Interferon regulatory factor 2 (IRF-2) A-20
Others	inducible NO-synthetase (iNOS) TAP1 LMP2 Apolipoprotein CIII type IV Gelatinase Perforin Vimentin Decorin α-V Integrin

Modified after refs. 31, 201 and 272. Additional references: Interleukin-12,[273] c-Myb,[274] TAP1 + LMP2,[275] Gelatinase,[276] IL-1,[277] 3'IgH enhancer,[278] B7.1,[279] Matrix Metalloprotease-9,[280] α-V Integrin.[281]

immunodeficiency virus (HIV) and the human cytomegalovirus utilize NF-κB transcription factors to regulate their gene expression. Typically, NF-κB is a critical component of the inducible transcriptional regulation of those genes. Additional transcription factors interacting with NF-κB proteins at a given promoter or enhancer element determine the tissue- and stage-specific expression pattern of the respective target genes.

One of the best studied examples of such combinatorial regulation involving NF-κB was outlined for the interferon-β (IFN-β) gene. The promoter/enhancer of this gene contains binding sites NF-κB, IRF (interferon response factor) and ATF2/c-jun in close proximity to each other. In addition, three binding sites for the HMG I(Y) protein are present and one of the HMG I(Y) sites overlaps the binding site for NF-κB.[80-82] HMG I(Y) is an abundant nuclear protein, which has been generally associated with transcriptionally active chromatin.[83] It interacts with A/T-rich DNA sequences with a moderate specificity and its DNA contacts lie exclusively within the minor groove of the DNA. In case of the IFN-β promoter it was shown that NF-κB and HMG I(Y) interact with the DNA simultaneously in a synergistic manner.[80] In fact, in the absence of HMG I(Y), NF-κB fails to bind to this specific target site. Binding of HMG I(Y) to DNA results in a strong bending of the DNA or, alternatively, if HMG I(Y) binds to prebend DNA it can reverse the bend.[84] The role of HMG I(Y) within transcription complexes is therefore predominantly regarded as that of an architectural protein, which by altering the topology of the DNA allows the assembly of specific protein-DNA complexes. The contribution to the assembly of higher order protein-DNA complexes is further enhanced by the ability of HMG I(Y) to make specific protein-protein interactions with NF-κB as well as other transcription factors. Ultimately, all these protein-protein and protein-DNA interactions result in the formation of a defined stereospecific three-dimensional "enhancosome" complex, whose integrity is a prerequisite for efficient transcription of the IFN-β gene.[85,86]

Many other examples for combinatorial interactions between NF-κB and other transcription factors have been described, including members of the C/EBP and AP1 families, IRF-1, IRF-2 and others.[87-90] The result of these interactions can be either cooperative gene induction or transrepression. One of the most interesting interactions is exemplified by the crosstalk between NF-κB and the glucocorticoid receptor. As mentioned before, NF-κB is a positive regulator of many genes involved in inflammatory responses. Glucocorticoids, working via the glucocorticoid-receptor (GR), are prominent anti-inflammatory drugs. At least one of the molecular mechanisms for this anti-inflammatory effect of glucocorticoids is their interference with the NF-κB activation pathway.[91-93] This is achieved at two levels. The first one being a direct physical interaction between GR and p50/RelA, the main component of the inducible NF-κB system.[92,93] This protein-protein interaction results in a downmodulation of the transactivation potential of NF-κB. Furthermore, GR directly upregulates the expression of IκBα thereby interfering with NF-κB function via a second route.[94,95]

Feedback Regulation Systems

Several of the NF-κB/IκB gene members are themselves regulated by NF-κB. This establishes an important autoregulatory feedback loop. Direct transcriptional regulation by NF-κB was demonstrated for NFKB1, NFKB2, *c-rel* as well as the IκBα gene.[96-103] Furthermore, the relB gene is most likely also directly regulated by NF-κB.[79,104] This autoregulatory feedback loop can be used to establish two distinct consequences in this response system. First of all, it can be used to ensure that NF-κB activation results in a temporal response which is rapidly reversible. Secondly, in specific situations this can be exploited to establish long-lasting responses. The rapid transcriptional induction of IκBα as a consequence of NF-κB activation results in a rapid replenishment of the IκBα pool. This newly-synthesized IκBα can then first of all capture activated, but still cytoplasmic, NF-κB complexes. Furthermore, it has been suggested that IκBα can also migrate to the nucleus and remove DNA-bound NF-κB.[105] Indeed early experiments have demonstrated that IκBα is able to reduce the half-life of DNA-bound NF-κB complexes significantly.[41,42] Therefore, this feedback loop will typically ensure a temporal induction of NF-κB.

Transcriptional induction of the NF-κB/Rel-family members such as RelB or c-Rel could be responsible for establishing a long-lasting NF-κB response. As discussed earlier, some of the RelB and c-Rel containing complexes are differentially regulated by IκB proteins and, depending on the cell type, are not targeted by IκBα.[66] Therefore, transcriptional induction of these proteins may be responsible for establishing an NF-κB response of extended time.

The autoregulatory feedback loop is not only confined to the transcriptional induction of different family members. The regulation of a stable amount of IκBα is also controlled to a large extent by specific protein stabilization. In the absence of NF-κB, IκBα is rapidly turned over, whereas the interaction with NF-κB results in an significant stabilization of the protein.[106-108] This was initially shown in cotransfection experiments, but more recently nicely demonstrated in a transgenic mouse system. In mice that overexpress the RelA protein in T cells, all of the RelA protein was still found trapped in the cytoplasm by IκBα.[109] Interestingly, this was not due to an increased synthesis of IκBα, but rather to a mere increase of IκBα at the protein level, suggesting a mechanism of posttranslational stabilization. This result was specific for RelA, as a similar transgenic mouse expressing the RelB protein did not show this IκBα stabilization.[110]

ACTIVATION OF NF-κB

Initial experiments suggested that the DNA-binding activity of NF-κB is activated by stimulation of pre-B cells with bacterial lipopolysaccharide or phorbol ester involving a posttranslational mechanism.[111] Subsequently, many highly diverse stimuli activating NF-κB in different cell types were described. These included many mitogens and cytokines, infection by certain viruses, bacteria and parasites, physical and oxidative stress, as well as various chemical agents (see Table 9.2). For a while it remained obscure how these completely distinct agents might all result in ultimately the same consequence—the activation of NF-κB. Protein kinases were immediately suspected to play a critical role in the activation process. Indeed, partially purified protein kinase C, an enzyme that is directly activated by phorbol esters, was shown to activate NF-κB in vitro.[48,112,113] Because protein kinase A and heme-regulated eIF-2 kinase were also activating NF-κB in vitro,[113] the specificity of the phosphorylation and activation by PKC remained unclear. Moreover, the depletion of protein kinase C by chronic phorbol ester treatment or the incubation of cells with specific protein kinase inhibitors, such as H7 or staurosporine, had no effect on the activation of NF-κB by other activating agents, such as tumor necrosis factor (TNF) or interleukin 1 (IL-1), arguing against a direct phosphorylation of IκB by PKC as a common activation mechanism in vivo.[53,56,56a,124] In the meantime, several additional kinases, such as doublestranded RNA-activated kinase,[114,115] a newly discovered PKC isoform, PKCζ,[116,117] and Raf-1,[118,119,299] have been implicated in the activation of NF-κB in vitro and in vivo, but no direct link was established between an extracellular stimulus, the given kinase and NF-κB.

The Raf serine/threonine kinase is a cytoplasmic component of the classical mitogen-activated protein kinase (MAPK) cascade, which is primarily activated by mitogenic signals and is linking extracellular stimuli to nuclear shuttle kinases.[125] Although this pathway is clearly involved in the regulation of κB-dependent transcription, recent evidence suggested that the activation of Raf alone by this pathway is not sufficient to yield NF-κB activation.[122,123] In addition to the MAPK cascade at least two more pathways exist that are activated by distinct stimuli—the JNK MAPK cascade and the p38/RK MAPK cascade.[125] Activated Jun-N-terminal kinase (JNK) contributes to stimulation of κB-dependent transcription. Overexpression of MEK kinase, an upstream activator of JNK, induced κB-dependent reporters, whereas a kinase-negative MEK kinase partially suppressed TNFα-mediated κB activation.[126] Additionally, JNK1 was found to physically interact with c-Rel in a yeast two-hybrid assay and by coimmunoprecipitation.[127] Yet, JNK1 does not directly phosphorylate c-Rel, NF-κB or IκB, indicating that c-Rel may serve as a docking molecule to allow JNK1 phosphorylation of certain Rel-associated molecules. The stress-activated p38/RK MAP kinase was also shown to regulate inducible expression of NF-κB target genes, such as the IL-6 and GM-CSF.[128] Induction of these genes by TNFα was efficiently downregulated in fibroblasts and a T cell hybridoma cell line by the specific p38 MAP kinase inhibitor SB203580. However, TNF-induced DNA binding of NF-κB or phosphorylation of NF-κB/IκB subunits were not impaired by the inhibitor. This implies that the p38/MAPK pathway is not directly targeting NF-κB, but may regulate factors that contribute to the transactivation potential of NF-κB.

The signaling properties of TNFα, a very potent and rapid activator of NF-κB, were extensively studied over the last few years. TNFα is a pro-inflammatory cytokine regulating various biological functions such as immunoregulatory activities, antiviral response, cytotoxicity, and reprogramming of gene expression. TNFα has been shown to activate a whole range of signaling pathways including mitogen-activated and stress-activated cascades discussed above. In addition, significant progress has recently been made in the understanding of membrane-proximal events triggered upon binding of TNF to TNF receptors (for a review see refs. 120 and 129) The two known TNF receptors, p55 (TNFR1) and p75 (TNFR2), belong to the TNF receptor superfamily and binding of TNF to either of these receptors triggers receptor oligomerization and thereby intracellular downstream signaling events. Using the intracellular part of these receptors as a bait in yeast two-hybrid screens, receptor-associated proteins were isolated and the effect of these proteins on NF-κB was analyzed. The p55 TNF receptor (TNFR1) was found to be associated with a 34 kDa protein called TRADD (for TNFR1-associated death domain protein), and overexpression of TRADD triggered the activation of NF-κB.[130] Upon stimulation with TNF, a 74 kDa serine-threonine kinase called RIP (for receptor interacting protein) is recruited by TRADD into the TNFR1 signaling complex. Although overexpression of RIP activated NF-κB, no direct phosphorylation of IκB was observed.[131] TRAF (for TNF receptor-associated factor) proteins have been identified as candidate signal transducers in p75 TNF receptor (TNFR2) signaling. Overexpression of TRAF2, but not TRAF1 or TRAF3, was sufficient to activate NF-κB and it was shown that the RING finger domain of TRAF2 mediates the activation.[132] TRAF2 functionally interacts with TANK (for TRAF family member associated NF-κB activator) and the two proteins synergize in NF-κB activation.[133] Interestingly, TRAF2 is also involved in NF-κB induction via TNFR1. It could be demonstrated that TRADD, which interacts specifically with TNFR1, is able to recruit TRAF2 into the signaling complex, thereby mediating activation of NF-κB.[134] Further studies have to address how these newly identified signaling components are linked to the signaling cascades described above and to the activation of NF-κB.

Redox Regulation as a Unifying Concept of NF-κB Activation

An important observation in pursuit of a general pathway for NF-κB activation was made about five years ago, when it was demonstrated that the antioxidant *N*-acetylcysteine is able to block TNF-induced activation of NF-κB.[135,136] Furthermore, in Jurkat T cells NF-κB was efficiently induced by micromolar amounts of hydrogen peroxide.[136] This led to a novel concept implicating the generation of reactive oxygen intermediates (ROIs) as an essential step in the activation of NF-κB. This concept has received ample support over the last few years. Most importantly, the activation of NF-κB by almost any stimulus analyzed so far was potently blocked by structurally unrelated substances with antioxidative or metal-chelating properties such as *N*-acetylcysteine,[135] dithiocarbamates,[137] and vitamin E derivatives.[138] Furthermore, many conditions that result in NF-κB activation also increase the intracellular concentration of ROIs. Treatment of cells with either phorbol ester, lipopolysaccharide or okadaic acid as well as crosslinking of the CD28 surface receptor stimulated the formation of ROIs.[137,139-141] In addition, NF-κB was shown to be activated in response to hydrogen peroxide or butylperoxide in many cell lines including HeLa cells, L929 fibroblasts or 70Z preB cells.[137,142,143] Importantly, overexpression of cytosolic Cu/Zn-dependent superoxide dismutase, an enzyme that generates H_2O_2 within the cell, efficiently costimulates NF-κB induction by TNF or okadaic acid.[144] In contrast, overexpression of the two major H_2O_2-detoxifying enzymes, catalase and seleno-glutathione peroxidase, resulted in decreased NF-κB activation in response to TNF, hydrogen peroxide or okadaic acid.[144,145]

Table 9.2. NF-κB activating stimuli

Bacteria and bacterial products	*Shigella flexneri*	
	Mycobacterium tuberculosis	
	Toxoplasma gondii	
	Cryptococcus neoformans	
	Candida albicans	
	Plasmodium falciparum	
	Cell wall products:	
	Lipopolysaccharide (LPS)	
	Muramyl peptides	
	Toxins:	
	Staphylococcus enterotoxin A and B (SEA and SEB)	
	Toxic shock syndrome toxin-1 (TSST-1)	
	Cholera toxin	
	Glycosylphosphatidylinositol	
Viruses and viral products	Human T cell leukemia virus-1 (HTLV-1): Tax	
	Hepatitis B virus (HBV): HBx, MHBs^{t}	
	Epstein-Barr virus (EBV): EBNA-2, LMP	
	Adenovirus 5: E1A	
	Cytomegalovirus (CMV)	
	Human immunodeficiency virus-1 (HIV-1)	
	Herpes simplex virus-1 (HSV-1)	
	Human herpes virus-6 (HHV-6)	
	Newcastle disease virus	
	Moloney murine leukemia virus	
	Rhinovirus	
	Sendai virus	
	double-stranded RNA	
Eukaryotic parasite	*Theileria parva*	
Cytokines	Tumor necrosis factor-α (TNF-α)	
	Lymphotoxin (LT) (TNF-β)	
	Interleukin-1 α and β (IL-1 α and β)	
	Interleukin-2 (IL-2)	
	Leukemia inhibitory factor (LIF)	
	Nerve growth factor (NGF)	
Mitogens	Antigen	anti-αβ T cell receptor
	Allogeneic stimulation	anti-CD2
	Lectins	anti-CD3
	Phorbol esters	anti-CD28
	Calcium ionophores	anti-surface IgM
	Serum	anti-CD30

(Con't on next page)

Table 9.2. NF-κB activating stimuli (con't)

Other biological mediators	Leukotriene B4 Prostaglandin E2 Insulin	p39 (CD40 ligand) gp160 (CD4 ligand) Fas ligand (CD95 ligand, APO-1 ligand)
Oxidative and physical stress	Hydrogen peroxide Butyl peroxide hypoxia/reoxygenation UV light asbestos	Ozone oxidized lipids cigarette smoke ionizing radiation
Chemical agents and drugs	Calyculin A Okadaic acid Pervanadate Ceramide Glutamate Cycloheximide Emetine	Anthralin Taxol Ouabain Hyaluronan fragments Kainate Anisomycin
others	partial hepatectomy Amyloid β protein hemorrhage	

Modified after refs. 31, 201 and 272. Additional references: *Toxoplasma gondii*,[282] *Cryptococcus neoformans*,[283] *Candida albicans*,[283] *Plasmodium falciparum*,[284] Glycosylphosphatidylinositol,[284]Moloney murine leukemia virus,[285] Rhinovirus,[286] NGF,[251] anti-CD30,[287] gp160,[301] Fas ligand,[288] Ozone,[289] reoxygenation,[290] cigarette smoke,[291] asbestos,[292] Anthralin,[270] Taxol, [142]Ouabain,[293] Hyaluronan fragments,[294] Glutamate,[252,253] Kainate,[253] Amyloid β protein,[257]hemorrhage.[295]

A novel redox-dependent signal transduction pathway from the endoplasmic reticulum (ER) to the nucleus involving NF-κB was described recently. ER-overload was shown to result in a specific activation of NF-κB.[146,147] This pathway was suggested to play a critical role in initiating a cellular response as a consequence of overexpression of certain viral proteins in the ER, such as hemagglutinin of Influenza virus or the E3 protein of adenovirus. It could be demonstrated that internal oxidative stress caused by the accumulation of these proteins in the endoplasmic reticulum activates DNA binding of NF-κB and induces κB-dependent gene expression.[147a]

In summary, all available data are consistent with the notion that activation of the NF-κB transcription factor depends on a redox-regulated step. The principal intracellular messenger of the activation process seems to be hydrogen peroxide, whereas superoxide or other reactive oxygen intermediates may only be indirectly involved, i.e., as a source for H_2O_2.[144] Although some experimental evidence suggests that the phosphorylation status, the degradation of IκB or the processing of the NFKB1 precursor (see below) are directly or indirectly controlled by a redox-dependent mechanism,[122,148] the exact redox-regulated step in the activation of NF-κB is still unknown.

Nevertheless, the above mentioned findings have had major impact on studying redox-regulated gene expression under physiological and pathophysiological conditions and initiated a search for therapeutical drugs with antioxidative properties selectively blocking the activation of NF-κB.

Release of NF-κB Requires IκB Degradation

As already mentioned above, the ultimate target for the activation of NF-κB is the inhibitory subunit IκB (Fig. 9.2). Originally, phosphorylation of IκB was considered to be sufficient for NF-κB activation. This model suggested that signal transduction pathways activate one or several kinases that phosphorylate IκB and that the phosphorylation triggers the dissociation of IκB from NF-κB.[149,150] Indeed, IκBα was found to be phosphorylated in vivo after stimulation of cells with agents that activate NF-κB, and several sites of phosphorylation have been mapped within the IκBα protein.[151-154] Initial work focused on the constitutive phosphorylation of the C-terminal PEST-domain of IκBα. PEST domains are polypeptide sequences that are enriched in proline, glutamate, serine and threonine, and are implicated to target proteins for rapid destruction.[155] Within the C-terminal region of IκB, five potential phosphorylation sites for casein kinase II (CKII) were identified.[156-159] Mutation of all CKII sites doubled the half-life of IκBα, suggesting that the constitutive phosphorylation by CKII is part of a degradation pathway removing free IκB that is not associated with NF-κB/Rel proteins.[159] However, mutant IκB proteins that are no longer phosphorylated by CKII can still be inducibly degraded, indicating that these sites are not essential for stimulation-dependent degradation. From systematic mutagenesis studies it is now known that degradation of IκBα in vivo requires site-specific phosphorylation of two N-terminal serines, Ser-32 and Ser-36, and mutation of these two sites to either alanine or even threonine blocks inducible degradation.[160-164]

Convincing evidence showing that phosphorylation per se is not sufficient for NF-κB activation was derived from studies using cell-permeable protease inhibitors, such as peptide aldehydes and serine protease inhibitors.[153,165-168] The incubation of cells with some inhibitors lead to an accumulation of phosphorylated IκB upon stimulation,[165,167-171] and phosphorylated IκBα was found to be still associated with NF-κB in coimmunoprecipitation experiments.[165,166,168,170,171] This suggested that phosphorylation of IκB is a necessary prerequisite, but by itself is not sufficient for activation. Rather, the subsequent proteolytical degradation of IκB is the essential event in the activation of NF-κB. The protease inhibitors that interfere with NF-κB activation are known inhibitors of the proteasome complex and suggest that the proteasome might be responsible for IκB degradation. The proteasome is a 26S multisubunit complex that functions as an ATP-dependent protease in the ubiquitin pathway, which has been implicated in the degradation of abnormal or short-lived regulatory proteins as well as in antigen presentation.[172,173] In the ubiquitin pathway, proteins are targeted for degradation by the covalent addition of multiple ubiquitin molecules. Subsequently, poly-ubiquitinated proteins are specifically recognized and degraded by the 26S proteasome. Indeed, IκBα is ubiquitinated in vivo after stimulation of cells with tumor necrosis factor.[174] Moreover, phosphorylation of IκB is necessary for ubiquitination and ubiquitinated IκBα is degraded by the proteasome while still associated with NF-κB.[174] The current model of activation is as follows:[150] IκBα is degraded by the proteasome pathway in response to stimulation. The degradation of IκBα in vivo requires site-specific phosphorylation of two N-terminal serines, Ser-32 and Ser-36. The phosphorylation triggers ubiquitination at the N-terminal lysine residues Lys-21 and Lys-22 of IκBα.[175,176] The ubiquitinated intermediates of IκB are then recognized by the proteasome. This model was further supported by experiments using methylated

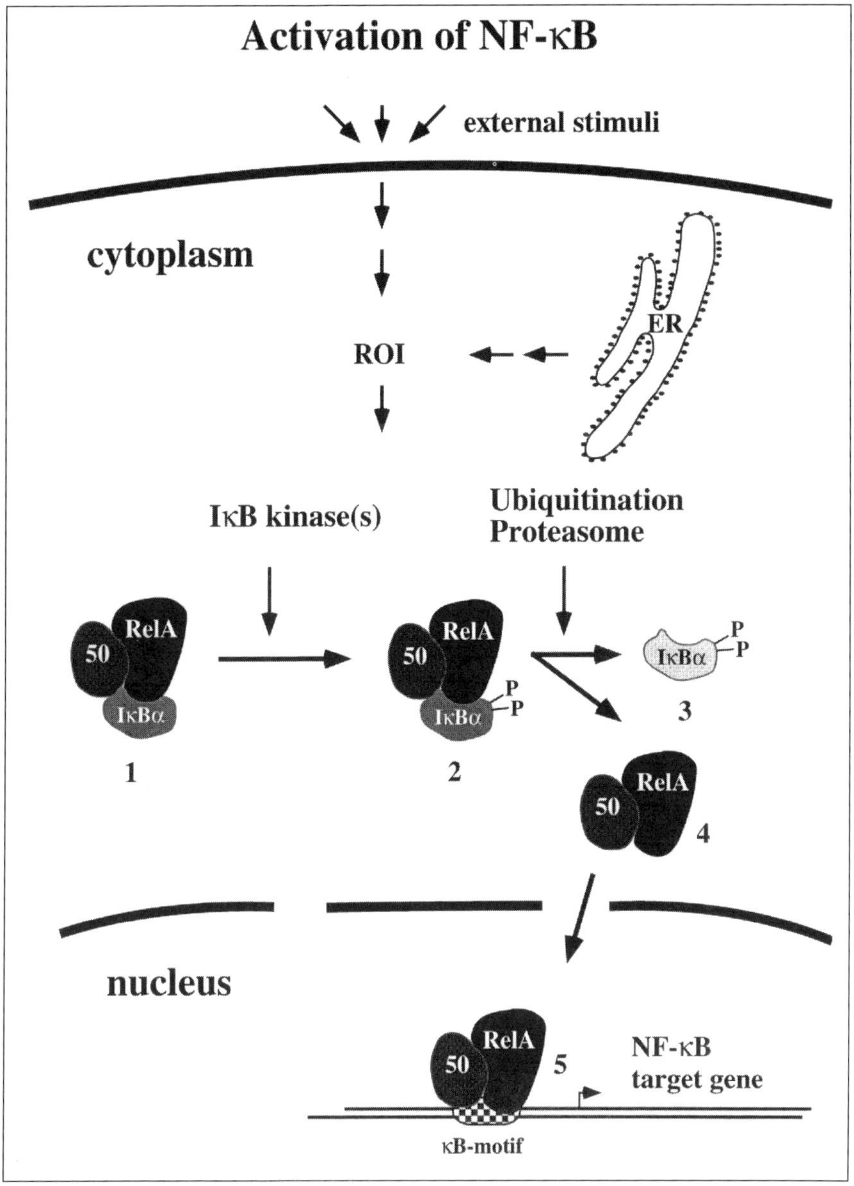

Fig. 9.2. Overview of the NF-κB activation pathway. Many external stimuli or signals originating from the ER (ER overload) are propagated by different intracellular signaling cascades involving reactive oxygen intermediates (ROI) and finally activate the IκB kinase(s). The IκB protein within the inhibited cytoplasmic NF-κB complexes (1) (the p50/RelA heterodimer is shown as an example), is specifically phosphorylated (2), Phosphorylated IκB is subsequently poly-ubiquitinated and degraded by the proteasome (3), Released NF-κB (4) is translocated to the nucleus, binds to κB-motifs in regulatory regions of respective target genes (5), and activates their transcription. See text for more details.

ubiquitin that prevented polyubiquitination and also efficient degradation of IκBα.[168] Dominant-negative variants of IκBα were generated either by N-terminal truncation, mutation of Ser-32 and Ser-36 or mutation of Lys-21 and Lys-22. These IκBα proteins are resistant to degradation and block inducible activation of NF-κB.[164,175-178] The dominant-negative IκB proteins are useful tools to further dissect the complex activation process.

Recently, a large (> 700 kDa) multisubunit IκB kinase has been described that phosphorylates Ser-32 and Ser-36 of IκB. Interestingly, this kinase requires ubiquitination to become activated.[179] This indicates that ubiquitination is not only regulating the final degradation step but, moreover, is also involved in regulation of the initial phosphorylation event that triggers IκBα degradation. Since most of the studies examining the activation of NF-κB have focused so far on IκBα, future experiments have to address whether NF-κB complexes containing other IκB family members are activated in the same way.

Aside from the ubiquitination-proteasome pathway, at least one other mechanism for regulating the activity of the NF-κB transcription factor exists. Very recently, a tyrosine kinase pathway has been implicated in the activation of NF-κB.[180-182] Upon blocking of T cell activation by the protein tyrosine phosphatase inhibitor pervanadate, IκBα was found to be phosphorylated on tyrosine residue 42. This triggered the release of IκB, but not the subsequent degradation of the inhibitor. Moreover, Jurkat T cells that are deficient for p56 lck tyrosine kinase showed a defect in both pervanadate-induced IκBα phosphorylation and NF-κB activation.[182a] Taken together, these data suggest that multiple pathways exist that regulate the release or degradation of IκB molecules and thereby the activation of NF-κB.

Regulation of NF-κB Precursor Processing

Various reports have suggested that the p105 and p100 NF-κB precursor proteins are associated with other NF-κB/Rel proteins in the cytoplasm, thus serving as IκB-like inhibitory subunits.[46-48] Therefore, functional NF-κB/Rel proteins might be released either by complete or limited proteolysis of the precursor molecules. The limited processing of the p105 precursor was originally proposed by Riviere et al, showing that increased nuclear levels of p50 are correlated with reduced levels of cytoplasmic p105 during acute HIV-1 infection.[183] The direct precursor-product relationship was demonstrated first in vivo by Fang and Maniatis.[184] They were able to demonstrate that the p50 subunit is generated by limited ATP-dependent proteolysis of the precursor protein. Subsequently, various groups provided experimental evidences for a stimulation-dependent enhancement of the processing rate of the precursor molecules. Treatment of cells by either phorbolester, doublestranded RNA, hydrogen peroxide, lipopolysaccharide or tumor necrosis factor resulted in an increased level of the p50 subunit.[47,148,153,185,186] Whether this increase is only due to the processing of the accumulated precursor molecule, or whether this might reflect in addition increased precursor synthesis in combination with preferential processing of newly synthesized precursor molecules needs further investigation.[187] Phosphorylation has also been implicated in upregulation of p105 processing following stimulation with tumor necrosis factor or a combined phorbolester/ionomycin treatment,[148,153,300] but the underlying mechanism is poorly understood. Indeed, serine-phosphorylation in the PEST-region of p105 is absolutely required for processing induced by treatment of cells with phorbolester/ionomycin or okadaic acid and specific mutations of Ser-894 and Ser-908 in p105 retarded the processing in vivo.[186,296] These data suggest that processing of the precursor molecules is controlled at least in part by phosphorylation, thus protecting unphosphorylated PEST sequences within the precursor molecule from extensive processing.

Several lines of evidence support the conclusion that processing of the p105 precursor also involves the ubiquitin-proteasome pathway.[169] Ubiquitinated full length p105 can be found prior to processing, degradation of p105 is an ATP-dependent process, various protease inhibitors reduce the processing of p105 in vivo and in vitro, and generation of p50 from p105 is prevented in yeast strains with mutations in the proteasome. Direct demonstration for p105 processing by the ubiquitin-proteasome pathway was achieved with the help of a cell-free system, and the requirement for distinct components of the ubiquitination machinery such as the ubiquitin carrier protein E2-F1 and a 320 kDa ubiquitin-protein ligase was described.[188] Interestingly, the HTLV-I Tax protein, a potent activator of NF-κB, acts as a bridging molecule between the precursor molecule p105 and the proteasome subunit HC9, thus accelerating the proteolytical maturation of p105.[189]

LESSONS FROM TRANSGENIC AND KNOCKOUT MICE

Although there is extensive knowledge of the NF-κB/Rel/IκB system from biochemical studies, we still have only a limited understanding of the common and distinct roles of the various subunits with respect to development, differentiation and additional physiological functions. Ongoing transgenic and knock-out studies help to address these questions in more detail.

Given the fact that many manipulations of cells in vitro result in the activation of NF-κB it was important to analyze NF-κB activity in mice containing the full set of NF-κB/Rel family members. Therefore, transgenic mice were generated containing reporter constructs under the control of NF-κB/Rel binding sites.[190,191] These mice allowed investigation of the developmental regulation and tissue specificity of NF-κB/Rel activity in vivo. A reporter under the control of multimerized κB motifs from the immunoglobulin kappa chain enhancer was constitutively expressed in lymphoid organs and the expression of the reporter gene could be correlated with the presence of constitutively nuclear p50/RelB heterodimers in those tissues.[190] The use of a NLS-lacZ reporter gene under the control of a promoter fragment of the NFKB1 gene containing three κB-binding sites allowed the analysis of NF-κB activity during embryonic development.[191] Interestingly, no NF-κB/Rel dependent activity could be detected in early development prior to implantation or during gastrulation. The earliest expression was localized on day E12.5 in the rhombencephalon, the spinal medulla and some blood vessels. On day E13.5 expression in the thymus started. After birth, the NF-κB/Rel activity in thymus remained and was additionally found in bone marrow, spleen and lymph nodes. The authors concluded that NF-κB/Rel proteins are probably not involved in the processes of early development and differentiation, but rather in maintenance functions once tissues are differentiated.

Transgenic approaches were also used to investigate how primary cells and organs would be affected by ectopic overexpression of members of the NF-κB or IκB families. In these studies, transgene expression was directed to the thymus cortex by use of the mouse *lck* proximal promoter. Surprisingly, the overexpression of RelA in transgenic mice did not increase the overall amount of NF-κB binding activity in thymocytes.[109] This was explained by a concomitant increase in the level of IκBα protein, indicating that IκBα may be stabilized by association with overexpressed RelA (see above). In contrast to the results obtained with RelA, overexpression of RelB resulted in a dramatic increase in overall κB-binding activity with no increase in the level of the IκBα.[110] Taken together, these results indicate that within the same cell type RelA and RelB complexes are differentially regulated by IκBα, supporting earlier results suggesting that a lymphoid-specific post-translational modification is responsible for the reduced susceptibility of RelB to IκBα.[66] Constitutive overexpression of Bcl-3 had no phenotype with respect to thymus morphology, T cell development, susceptibility to infections or incidence of tumors.[64] At the molecular level, an increased binding of p50 but not p52 was detected in extracts from transgenic mice, but the expression level of presumptive target genes for NF-κB remained unchanged. This suggests that, in contrast to the overexpression of Bcl-3 found in B cell leukemias, additional genetic alterations have to occur in thymocytes to develop T cell leukemia.

The first member of the family to be mutated by homologous recombination was the NFKB1 gene.[192] No gross developmental abnormalities were apparent in homozygous mutant mice, but multifocal defects in immune responses involving B lymphocytes and nonspecific responses to infection were detected. Mutant B cells did not proliferate in response to lipopolysaccharide and showed a reduced synthesis of antibodies. Purified resting B cells from these mice exhibited a normal proliferative response to soluble but not to membrane-bound CD40 ligand, as well as normal steady state levels of germ-line $C_H\gamma3$ and $C_H\varepsilon$ RNA upon appropriate stimulation. Substantial switching to IgG_1, but reduced switching to IgG_3, IgE and IgA, was observed.[193] Mice lacking NFKB1 were not able to clear infections by certain pathogens such as *Listeria monocytogenes* or *Streptococcus pneumoniae* and were more resistant to infection by encephalomyocarditis virus. These data support the role of NF-κB in specific and nonspecific immune responses, but do not indicate an early developmental role for the factor.

Disruption of the gene encoding RelA resulted in a much more dramatic phenotype.[194] Homozygous mutant embryos die around day E15 of embryonic development. These RelA (-/-) mice showed massive liver degeneration due to widespread apoptosis shortly before death. NF-κB regulated target genes such as GM-CSF or IκBα were no longer induced by tumor necrosis factor in fibroblasts isolated from mutant embryos, supporting the idea that RelA primarily regulates inducible, but not basal, transcription.

RelB-deficient mice developed normally, but displayed characteristic defects in hematopoiesis and specific immune responses.[195,196] The function of antigen presentation was impaired and thymic medullary UEA-1^+ cells did not develop in mutant mice. Moreover, the number of granulocytes and macrophages was increased and immigration of mixed inflammatory cells in several organs was also detected.

Mice lacking c-Rel developed normally with no obvious phenotypic abnormalities.[197] They displayed an impaired humoral response, reduced antibody synthesis in response to antigenic challenge and reduced interleukin-2 expression. B cells did not proliferate in response to stimulation by lipopolysaccharide, CD40 or anti-IgM. T cells showed a proliferative response to phorbolester in combination with ionomycin but not to costimulation by CD28 and the T cell receptor, indicating that T cell proliferation occurs by Rel-dependent and -independent mechanisms. T cells deficient for *c-rel* exhibit normal surface marker expression after mitogenic stimulation, but the levels of IL-3, IL-5, GM-CSF, TNFα and γ-IFN were lower compared to normal T cells.[198] Exogenous IL-2 was able to restore the synthesis of IL-5, TNFα and γ-IFN, but not IL-3 or GM-CSF expression, indicating that c-Rel is absolutely required for the inducible synthesis of IL-3 and GM-CSF. In contrast to T cells, macrophages lacking c-Rel display higher than normal levels of GM-CSF after lipopolysaccharide treatment. These results suggest that c-Rel can activate and inhibit NF-κB/Rel dependent gene expression in a tissue or promoter specific fashion.

Mice deficient for IκBα were apparently normal at birth, but exhibited severe runting, skin defects and extensive granulopoiesis postnatally.[199,200] Typically, IκBα (-/-) mice died around day 8 after birth. The skin lesions of these mice were related to those of psoriasis, showing lymphocytic infiltration and hyperkeratosis. In hematopoietic tissues of IκBα-deficient animals NF-κB was constitutively activated, and in thymocytes some target genes such as G-CSF, VCAM-1 and MIP-2 were upregulated whereas others like IL-2, IL2Rα, IL-6, GM-CSF or *c-myc* remained unaffected. IκBα-mutant fibroblasts showed minimal constitutive NF-κB activity and a delayed poststimulation repression. This suggests that at least in some cell types IκBα is primarily involved in postinduction repression, whereas IκBβ is required for signal-dependent activation of NF-κB. In fibroblasts the absolute levels of NFKB1(p105/p50), NFKB2(p100/p52), RelA, c-Rel and IκBβ were similar to those of wildtype fibroblasts. No increase of NF-κB binding activity in the nuclear fraction was seen, indicating that other IκB family members may compensate for the loss of IκBα. Nevertheless, an increase in constitutive NF-κB activity is most likely the major cause of lethality in these mice. When double mutant mice lacking both NFKB1 and IκBα were analyzed, greatly reduced levels of NF-κB DNA-binding activity were detected. These mice displayed increased survival and the onset of phenotypic abnormalities was delayed.

NF-κB AND DISEASE

NF-κB/Rel family members have received considerable attention in medicine because they are induced by a wide range of pathogenic conditions and are important transcriptional regulators of genes encoding cytokines, hematopoetic growth factors, immunoreceptors, acute phase proteins and cell adhesion molecules (for a review see ref. 201). Therefore, NF-κB/Rel family members have been discussed in the context of various diseases, such as malignant transformation, AIDS, neurological disorders, arthritis, atherosclerosis, septic shock and various autoimmune diseases.

Oncogenesis

NF-κB/Rel and IκB family members have been implicated in the development of cancer (for a review see ref. 202). For example, the viral counterpart of c-Rel, v-Rel, was originally isolated from an acutely transforming reticuloendotheliosis virus and is highly oncogenic in birds. Furthermore, expression of the *v-rel* oncogene in transgenic mice under the control of the *lck* promoter induces malignant T cell leukemia/lymphoma.[203] The mechanism of transformation by v-Rel is not fully understood.[202,204] In cotransfection experiments v-Rel functions as a competitive inhibitor of NF-κB induced transcription.[205] At least three additional modes of action are discussed for the cellular members of the NF-κB/Rel family including chromosomal alterations such as translocations or amplification, deregulated expression of single subunits and activation of NF-κB/Rel by infection of transforming viruses.

Translocation and amplification of *c-rel* and relA genes is often detected in diffuse lymphoma with a large cell compound (DLLC) and in primary mediastinal B cell lymphoma, indicating that chromosomal aberrations of Rel proteins constitute at least a tumor progression-associated marker of certain lymphomas.[206] The translocation t(14;19)(q32;q13.1) of the protooncogene *bcl-3* to the highly expressed immunoglobulin heavy chain locus, was found in some cases of chronic lymphocytic leukemia.[50,207] In case of the NFKB2/Lyt10 gene, highly diverse chromosomal aberrations were detected, such as translocations in B cell lymphoma, reciprocal translocation in multiple myeloma and small chromosomal deletions in T cell cutaneous lymphoma and B cell chronic lymphocytic leukemia.[208-210] Most of these aberrations target the carboxy-terminal ankyrin-encoding domain of NFKB2, resulting in unprocessed, constitutively nuclear transactivators.[211,212] A novel molecule that was found to interfere with the NF-κB dependent gene expression was TAN-1, the human Notch homologue that is often translocated in T cell leukemia.[213] The cytoplasmic portion of TAN-1 that contains six ankyrin repeats acts like IκB, preventing the specific binding of p50-containing complexes and also κB-dependent transactivation in transient reporter gene assays.[54] This suggested that TAN-1 is able to modulate NF-κB dependent gene expression during T cell leukemia.

Up or downregulation of the expression of single subunits within the NF-κB/Rel or IκB family has been described in various types of cancer. Cell lines established from patients with Hodgkin's disease as well as primary H/Reed-Sternberg cells display abundant constitutive nuclear NF-κB/Rel binding activity, containing primarily p50 and RelA subunits.[214] This is in contrast to the usual constitutive B cell activity consisting of p50 and c-Rel or RelB subunits, suggesting that deregulated expression of Rel family members may contribute to the molecular defects of Hodgkin's lymphoma. In addition, overexpression of p100/p52 and p105/p50 was detected in human breast cancer cells[215] and non-small lung carcinomas,[216] respectively, suggesting that NF-κB may be involved in carcinogenesis by sequestering other NF-κB/Rel proteins in the cytoplasmic compartment. Finally, malignant transformation as measured by saturation density, growth in soft agar and tumorigenicity in nude mice was efficiently induced by overexpression of IκBα antisense RNA in NIH3T3 cells.[217] This is in line with the enhanced granulopoiesis found in IκBα deficient mice described above and indicates that disruption of the NF-κB/IκB autoregulatory loop may lead to cellular transformation.

NF-κB/Rel proteins are activated by a wide variety of viral proteins, among them proteins with transforming properties, such as the Tax protein of human T cell leukemia virus type I (HTLV-1) and proteins from Hepatitis B virus (HBV) and Epstein-Barr virus (EBV). HTLV-1 is the etiological agent of adult T cell leukemia (ATL), and although the molecular basis of transformation by HTLV-1 is not fully understood, the viral Tax protein seems to be directly involved in the transformation process (for a review see ref. 218). Surprisingly, recombinant Tax protein activates NF-κB if added to the culture medium,[219] and the activation of NF-κB by overexpressing Tax is efficiently blocked by antioxidants.[220] A wide variety of pathways and mechanisms have been described that activate NF-κB in response to Tax, such as direct interactions of Tax with NF-κB/Rel family members, the induction of phosphorylation and subsequent release or degradation of IκB and enhanced processing of the precursor molecule p105 (for a review see ref. 221). The subsequent increase in expression of NF-κB dependent target genes, such as interleukin 2, GM-CSF, TNF-β and *c-myc*,[222] might be part of a program that causes T cell malignancy. Epidemiological evidences are linking the hepatitis B virus to the development of hepatocellular carcinoma, and expression of the viral regulatory gene HBx under the control of its own promoter in transgenic mice induces liver cancer.[297] Both HBx and a second transactivating viral protein called MHBs^t are potent activators of the NF-κB transcription factor.[223-225] Since HBx also induces tumor promoter signaling pathways[226] and deregulates cell cycle checkpoints,[227] future studies have to address the exact contribution of NF-κB to the initiation and progression of hepatocarcinogenesis. Epstein-Barr virus (EBV) is a human herpes virus that is closely associated with a number of lymphoid and epithelial malignancies including endemic forms of Burkitt's lymphoma, several T cell lymphomas, subtypes of Hodgkin's disease and certain carcinomas. Two viral gene products of EBV have been shown to activate NF-κB, LMP1, the latent membrane protein 1,[228] and EBNA2, the nuclear antigen 2.[229] The activation of NF-κB seems to be mediated by targeting the inhibitory subunit for degradation[230] and by an increase of p105 processing.[231] How the activation of NF-κB is contributing to carcinogenesis and immortalization by EBV remains unclear at the moment.

AIDS

The human immunodeficiency virus (HIV) is the etiologic agent for the progressive degeneration of the immune system that leads to the development of AIDS, the acquired immunodeficiency syndrome (for a review see ref. 232). About ten years ago, Nabel and Baltimore established a role for NF-κB in controlling the transcriptional activation of the HIV-1 provirus that harbors two adjacent binding sites for NF-κB in its core enhancer within the long terminal repeat.[233] This original finding received considerable experimental support over the years and it is now established that inducible transcriptional activation of the HIV-1 LTR depends on NF-κB/Rel proteins although NF-κB independent pathways of HIV activation clearly exist (for a review see ref. 234). Several open questions remain to be addressed. First, the exact contribution of NF-κB to the retroviral life cycle in different cell types, such as T lymphocytes, macrophages and peripheral blood monocytes, is not fully understood. For example, NF-κB activity seems to be required in peripheral T lymphocytes and primary macrophages for the initiation of LTR transcriptional activity and the establishment of HIV-1 infection.[235,236] Infection of monocytes or macrophages by HIV is itself activating NF-κB,[237-239] and permanent NF-κB activation within these cell types is required for long-term maintenance of proviral transcription.[240] These findings suggest that NF-κB is necessary at early and late time points of infection and indicates that NF-κB is a key player in the self-perpetuation of viral transcription and replication in the monocytic lineage. Second, it is not known whether NF-κB is engaged at advanced disease stages that are characterized by secondary infections and AIDS-associated malignancies. Finally, it is still a matter of debate whether the inhibition of NF-κB is a useful tool for anti-HIV therapy. One way to interfere with NF-κB dependent HIV-1 activation might be the use of highly specific protease inhibitors that prevent processing of the precursor molecules or degradation of the inhibitory subunits (see above). Another way of intervention results from the observation that the intracellular redox state is modulating NF-κB activation as well as HIV-1 replication.[135,136,241] In addition, decreased glutathione levels are observed in the sera of HIV-1 infected individuals,[242] and the regulatory Tat protein of HIV-1 is influencing the cellular redox state.[243] Along these lines, oxidant stress has also been implicated in AIDS-Kaposi's sarcoma pathogenesis.[244] Taken together these data suggest that drugs that normalize the intracellular redox state, such as antioxidants, might be candidate therapeutical drugs.[245,246] On the other hand, since NF-κB is a key factor in the immune response as well as a general regulator of cellular homeostasis, downregulation of its activity might provoke or enhance side effects and thus be disadvantageous for the individual.

Neurological Diseases

Almost four years ago it was speculated that NF-κB may be involved in a variety of neurological disorders.[247] In the meantime, data in support of this idea have been accumulated. A variety of reports described the presence of NF-κB/Rel family members in the nervous system by immunostaining and cell fractionation techniques. Constitutive nuclear NF-κB/Rel is localized in adult neurons,[248,249] and inducible forms of NF-κB/Rel are present in neurons and in synapses of 3 to 7 day-old mice.[121,250] In addition, NF-κB/Rel dependent transcriptional activity in the developing CNS was investigated in transgenic mice carrying NF-κB/Rel reporter constructs.[191] Drastic changes in the abundance and activity of NF-κB/Rel proteins were observed in the first three weeks after birth during maturation of cortex and cerebellum.

The functional relevance of NF-κB/Rel proteins in brain is underscored by novel specific inducers and by various brain disease models. For example, NF-κB is activated by nerve growth factor,[251] and the neurotransmitter glutamate.[252,253] In various rat disease models, such as traumatic brain injury,[254] the ischemic brain[255] and convulsant-induced seizures,[256] an increase in NF-κB/Rel DNA-binding activity was measured. It was also demonstrated that the neurotoxic amyloid β protein that is associated with plaques in the brains of Alzheimer's patients induces NF-κB.[257] Using an antisense approach, it was shown that the activation of NF-κB is necessary and sufficient to protect primary hippocampal neurons form the toxicity of amyloid β protein.[258] Moreover, a binding site for NF-κB was identified in the promoter region of the amyloid precursor protein which can be alternatively processed to neuroprotective or neurotoxic peptides.[259] In summary, it is highly likely that NF-κB/Rel proteins are involved in a neuroprotective response by increasing the expression of target genes that counteract neurotoxic effects.

Rheumatoid Arthritis

Rheumatoid Arthritis (RA) is a chronic destructive inflammatory polyarthropathy. Using immunohistochemical staining, nuclear p50 and RelA proteins were detected in synovial tissue from patients with RA, but only weak, diffuse staining was seen in control subjects.[260,261] The staining was found in specific cell types such as type A synoviocytes and vascular endothelium. Moreover, the majority of CD14-positive cells within the synovium also displayed p50/RelA heterodimers in the nuclear fraction, indicating that NF-κB is involved in the expression of synovial, monocyte-derived cytokines. In support of a possible contribution of NF-κB to arthritis, the drugs Aurothioglucose and salicylate, which are often used in the treatment of RA patients, were shown to inhibit NF-κB activation.[262,263]

Atherosclerosis

Atherosclerosis is a vascular disease of the large arteries causing heart disease and stroke; it is characterized by features of endothelial dysfunction, infiltration of inflammatory cells into the vessel wall and vascular remodeling. Recently, it was demonstrated in a mouse model that susceptibility to aortic atherosclerotic lesion formation was associated with the accumulation of lipid peroxidation products, induction of inflammatory genes and the activation of NF-κB/Rel-like transcription factors.[264] Furthermore, activated NF-κB was indeed detected in human atherosclerotic tissue.[265] Using immunofluorescence and immunohistochemical techniques, activated RelA was detected in fibrotic-thickened intima/media and atheromatous areas of the atherosclerotic lesion, whereas little or no activated p65 was found in vessels lacking atherosclerosis. Additionally, in an arterial injury model it was demonstrated that NF-κB is activated in the aorta of mice and rats by limited endothelial denudation.[266] Within 45 minutes after wounding, NF-κB was activated in endothelial cells at the edge of the wound and increased synthesis of NF-κB subunits persisted in replicating endothelium until complete regeneration. Taken together, these data suggest that NF-κB is contributing to events in vascular pathophysiology.

Drugs and NF-κB

Instead of looking for NF-κB/Rel activation under pathogenic conditions, another way to address whether NF-κB/Rel is contributing to disease is to analyze whether the activity of NF-κB/Rel family members is influenced by well-known therapeutic drugs. We will give a few examples for this direction of research.

Glucocorticoids have been used as immunosuppressive and anti-inflammatory drugs for decades. As already described above, glucocorticoids have been shown to interfere with the activation of NF-κB via two different mechanisms, direct interaction of the activated glucocorticoid receptor and NF-κB and induction of IκB synthesis. Also, nonsteroidal anti-inflammatory drugs such as sodium salicylate and aspirin have been demonstrated to inhibit NF-κB activation at the millimolar concentration range.[263] Phosphorylation and subsequent degradation of IκB is efficiently blocked by salicylate.[298] FK506, another immunosuppressive drug, blocked c-Rel expression and activation in T and B cells,[267,268] whereas NF-κB/Rel-dependent DNA-binding activity and target gene expression was enhanced in nonlymphoid cells, such as fibroblasts and renal mesangial cells, indicating cell type-specific effects of the drug.[269]

Anthralin is a drug for the treatment of psoriasis, a skin disease characterized by hyperproliferation and hyperkeratosis. The drug has been shown to induce an anti-inflammatory response involving cytokine expression and synthesis of cell adhesion molecules. NF-κB is potently activated by micromolar concentrations of Anthralin, suggesting that NF-κB might contribute to the anti-psoriatic activity of the drug.[270]

The potent anticancer drug taxol activates the NF-κB transcription factor in a murine pre-B cell line, indicating that there may be a link between the state of tubulin polymerization and NF-κB activity.[142] Furthermore, it was demonstrated that taxol is able to mimic the effects of bacterial lipopolysaccharides on murine macrophages and that NF-κB is only activated by taxol in macrophages derived from LPS-responsive mice.[271] This helps to explain the failure to completely reverse the activation by taxol by depolymerizing agents such as nocodazole[142] and suggests that more than one mechanism is involved in the activation of NF-κB/Rel dependent gene expression that may contribute to the antitumor effects of this drug.

OUTLOOK

The NF-κB/IκB-system of regulated gene expression represents a highly developed rapid cellular response mechanism utilized in many normal and pathogenic situations. The observation that multiple NF-κB heterodimer combinations differentially interact with distinct IκB molecules allows for the integration of quite diverse response pathways. Still, the identity of the IκB kinase targeting IκB for degradation represents a mystery. In this respect it will be interesting to determine whether there is more than a single kinase, i.e., distinct kinases regulating different IκBs. Additionally, there might be differences in the subunit composition of the multisubunit kinase, i.e., with cell type-specific subunit combinations. The analyses of combined mutations of individual components of this system will shed more light on the biological functions of NF-κB dependent gene regulation for normal cellular development and cell function. In this respect, the generation of conditional mutations that allow a regulated inactivation/activation of individual family members should provide additional insights. Ultimately, a better understanding of the molecular details of NF-κB function will be a prerequisite for the development/identification of highly specific and potent inhibitory drugs.

REFERENCES

1. Baeuerle PA, Baltimore D. A 65-kD subunit of active NF-κB is required for inhibition of NF-κB by IκB. Genes Dev 1989; 3:1689-1698.
2. Kieran M, Blank V, Logeat F et al. The DNA binding subunit of NF-κB is identical to factor KBF1 and homologous to the *rel* oncogene product. Cell 1990; 62:1007-1018.
3. Ghosh S, Gifford AM, Riviere LR et al. Cloning of the p50 DNA binding subunit of NF-κB: homology to *rel* and *dorsal*. Cell 1990; 62:1019-1029.
4. Bours V, Villalobos J, Burd PR et al. Cloning of a mitogen-inducible gene encoding a κB DNA-binding protein with homology to the *rel* oncogene and to cell-cycle motifs. Nature 1990; 348:76-80.
5. Meyer R, Hatada EN, Hohmann H-P et al. Cloning of the DNA-binding subunit of human nuclear factor κB: The level of its mRNA is strongly regulated by phorbol ester or tumor necrosis factor α. Proc Natl Acad Sci USA 1991; 88: 966-970.
6. Blank V, Kourilsky P, Israel A. NF-κB and related proteins: Rel/dorsal homologies meet ankyrin-like repeats. Trends Biochem 1992; 17:135-140.
7. Nabel GJ, Verma IM. Proposed NF-κB/IκB family nomenclature. Genes Dev 1993; 7:2063.
8. Ghosh G, van Duyne G, Ghosh S et al. Structure of the NF-κB p50 homodimer bound to a κB site. Nature 1995; 373: 303-310.
9. Müller CW, Rey FA, Sodeoka M et al. Structure of the NF-κB p50 homodimer bound to DNA. Nature 1995; 373: 311-317.
10. Müller CW, Rey FA, Harrison SC. Comparison of two different DNA-binding modes of the NF-κB p50 homodimer. Nature Struct Biol 1996; 3:224-227.
11. Neri A, Chang C-C, Lombardi L et al. B cell lymphoma-associated chromosomal translocation involves candidate oncogene *lyt*-10, homologous to NF-κB p50. Cell 1991; 67:1075-1087.
12. Schmid RM, Perkins ND, Duckett CS et al. Cloning of an NF-κB subunit which stimulates HIV transcription in synergy with p65. Nature 1991; 352:733-736.
13. Bours V, Burd PR, Brown K et al. A novel mitogen-inducible gene product related to p50/p105-NF-κB participates in transactivation through a κB site. Mol Cell Biol 1992; 12:685-695.

14. Lin L, Ghosh S. A glycine-rich region in NF-κB p105 functions as a processing signal for the generation of the p50 subunit. Mol Cell Biol 1996; 16:2248-2254.
15. Kang S-M, Tran A-C, Grilli M et al. NF-κB subunit regulation in nontransformed CD4+ T lymphocytes. Science 1992; 256:1452-1456.
16. Schmitz ML, Baeuerle PA. The p65 subunit is responsible for the strong transcription activating potential of NF-κB. EMBO J 1991; 10:3805-3817.
17. Israel A, Le Bail O, Hatat D et al. TNF stimulates expression of mouse MHC class I genes by inducing an NFκB-like enhancer binding activity which displaces constitutive factors. EMBO J 1989; 8: 3793-3800.
18. Franzoso G, Bours V, Park S et al. The candidate oncoprotein Bcl-3 is an antagonist of p50/NF-κB mediated inhibition. Nature 1992; 359:339-342.
19. Kretzschmar M, Meisterernst M, Scheidereidt C et al. Transcriptional regulation of the HIV-1 promoter by NF-κB in vitro. Genes Dev 1992; 6:761-774.
20. Fujita T, Nolan GP, Ghosh S et al. Independent modes of transcriptional activation by the p50 and p65 subunits of NF-κB. Genes Dev 1992; 6:775-787.
21. Moore PA, Ruben SM, CA R. Conservation of transcriptional activation functions of the NF-κB p50 and p65 subunits in mammalian cells and *Saccharomyces cerevisiae*. Mol Cell Biol 1993; 13: 1666-1674.
22. Bull P, Morley KL, Hoekstra MF et al. The mouse c-*rel* protein has an N-terminal regulatory domain and a C-terminal transcriptional transactivation domain. Mol Cell Biol 1990; 10:5473-5485.
23. Dobrzanski P, Ryseck R-P, Bravo R. Both N- and C-terminal domains of RelB are required for full transactivation: role of the N-terminal leucine zipper-like motif. Mol Cell Biol 1993; 13:1572-1582.
24. Ballard DW, Dixon EP, Pfeffer NJ et al. The 65-kDa subunit of human NF-κB functions as potent transcriptional activator and a target for v-Rel-mediated repression. Proc Natl Acad Sci USA 1992; 89:1875-1879.
25. Schmitz ML, dos Santos Silva MA, Altmann H et al. Structural and functional analysis of the NF-κB p65 C-terminus. J Biol Chem 1994; 269: 25613-25620.
26. Schmitz ML, dos Santos Silva MA, Baeuerle PA. Transactivation domain 2 (TA2) of p65 NF-κB. J Biol Chem 1995; 270:15576-15584.
27. Schmitz ML, Stelzer G, Altmann H et al. Interaction of the COOH-terminal transactivation domain of p65 NF-κB with TATA-binding protein, transcription factor IIB and coactivators. J Biol Chem 1995; 270:7219-7226.
28. Kerr LD, Ransone LJ, Wamsley P et al. Association between proto-oncoprotein Rel and TATA-binding protein mediates transcriptional activation by NF-κB. Nature 1993; 365:412-419.
29. Xu X, Prorock C, Ishikawa H et al. Functional interaction of the v-Rel and c-Rel oncoproteins with the TATA-binding protein and association with transcription factor IIB. Mol Cell Biol 1993; 13:6733-6741.
30. Ryseck R-P, Novotny J, Bravo R. Characterization of elements determining the dimerization properties of RelB and p50. Mol Cell Biol 1995; 15:3100-3109.
31. Grilli M, Chiu JJ-S, Lenardo MJ. NF-κB and Rel: Participants in a multiform transcriptional regulatory system. Int Rev Cytol 1993; 143:1-62.
32. Urban MB, Schreck R, Baeuerle PA. NF-κB contacts DNA by a heterodimer of the p50 and p65 subunit. EMBO J 1991; 10:1817-1825.
33. Hansen SK, Baeuerle PA, Blasi F. Purification, Reconstitution and IκB association of the c-Rel/p65 (RelA) complex, a strong activator of aranscription. Mol Cell Biol 1994; 14:2593-2603.
34. Parry GCN, Mackman N. A set of inducible genes expressed by activated human monocytic and endothelial cells contain κB-like sites that specifically bind c-Rel/ p65 heterodimers. J Biol Chem 1994; 269:20823-20825.
35. Lamb P, McKnight SL. Diversity and specificity in transcriptional regulation: the benefits of heterotypic dimerization. Trends Biochem Sci 1991; 16:417-422.
36. Crabtree GR, Clipstone NA. Signal transmission between the plasma membrane and nucleus of T lymphocytes. Annu Rev Biochem 1994; 63:1045-1083.

37. Rao A. NF-ATp: a transcription factor required for the coordinate induction of several cytokine genes. Immunol Today 1994; 15:274-281.
38. Northrop JP, Ho SN, Chen L et al. NF-AT components define a family of transcription factors targeted in T-cell activation. Nature 1994; 369:497-502.
39. Jain J, Burgeon E, Badalian TM et al. A similar DNA-binding motif in NF-AT family proteins and the Rel homology region. J Biol Chem 1995; 270:4138-4145.
40. Baeuerle PA, Baltimore D. Activation of DNA-binding activity in an apparently cytoplasmic precursor of the NF-κB transcription factor. Cell 1988; 53:211-217.
41. Baeuerle PA, Baltimore D. IκB: a specific inhibitor of the NF-κB transcription factor. Science 1988; 242:540-546.
42. Zabel U, Baeuerle PA. Purified human IκB can rapidly dissociate the complex of the NF-κB transcription factor with its cognate DNA. Cell 1990; 61:255-265.
43. Haskill S, Beg AA, Tompkins SM et al. Characterization of an immediate-early gene induced in adherent monocytes that encodes IκB-like activity. Cell 1991; 65:1281-1289.
44. Thompson JE, Phillips RJ, Erdjument-Bromage H et al. IκB-β regulates the persistent response in a biphasic activation of NF-κB. Cell 1995; 80:573-582.
45. Michaely P, Bennett V. The ANK repeat: a ubiquitous motif involved in macromolecular recognition. Trends Cell Biol 1992; 2:127-129.
46. Rice NR, MacKichan ML, Israel A. The precursor of NF-κB p50 has IκB-like functions. Cell 1992; 71:243-253.
47. Mercurio F, DiDonato JA, Rosette C et al. p105 and p98 precursor proteins play an active role in NF-κB-mediated signal transduction. Genes & Dev 1993; 705-718.
48. Naumann M, Wulczyn FG, Scheidereit C. The NF-κB precursor p105 and the proto-oncogene product Bcl-3 are IκB molecules and control nuclear translocation of NF-κB. EMBO J 1993; 12: 213-222.
49. Inoue J-I, Kerr LD, Kakizuka A et al. IκBγ, a 70 kd protein identical to the C-terminal half of p110 NF-κB: a new member of the IκB family. Cell 1992; 68:1109-1120.
49a. Whiteside ST, Epinat JC, Rice NR, Israel A. IκBε, a novel member of the IκB family, controls RelA and cRel NF-κB activity. EMBO J 1997; 16:1413-1426.
50. Ohno H, Takimoto G, McKeithan T. The candidate proto-oncogene bcl-3 is related to genes implicated in cell lineage determination and cell cycle control. Cell 1990; 60:991-997.
51. Wulczyn FG, Naumann M, Scheidereit C. Candidate proto-oncogene *bcl-3* encodes a subunit-specific inhibitor of transcription factor NF-κB. Nature 1992; 358:597-599.
52. Hatada EN, Nieters A, Wulczyn FG et al. The ankyrin repeat domains of the NF-κB precursor p105 and the proto-oncogene *bcl-3* act as specific inhibitors of NF-κB DNA binding. Proc Natl Acad Sci USA 1992; 89:2489-2493.
53. Ray P, Zhang DH, Elias J et al. Cloning a differentially expressed IκB-related protein. J Biol Chem 1995; 270:10680-10685.
54. Guan E, Wang J, Laborda J et al. T cell leukemia-associated human Notch translocation-associated Notch homologue has IκB-like activity and physically interacts with Nuclear Factor-κB proteins in T cells. J Exp Med 1996; 183:2025-2032.
55. Beg AA, Ruben SM, Scheinman RI et al. IκB interacts with the nuclear localization sequences of the subunits of NF-κB: a mechanism for cytoplasmic retention. Genes & Develop 1992; 6:1899-1913.
56. Ganchi PA, Sun S-C, Greene WC et al. IκB/MAD-3 masks the nuclear localization signal of NF-κB p65 and requires the transactivation domain to inhibit NF-κB p65 DNA-binding. Mol Cell Biol 1992; 12:1339-1352.
56a. Meichle A, Schutze S., Hensel G, Brunsing D, Kronke M. Protein kinase C-independent activation of nuclear factor kappa B by tumor necrosis factor. J Biol Chem 1990; 265:8339-8343.
57. Zabel U, Henkel T, dos Santos Silva M et al. Nuclear uptake control of NF-κB by MAD-3, an IκB protein present in the nucleus. EMBO J 1993; 12:201-211.
58. Davis N, Ghosh S, Simmons DL et al. Rel-associated pp40: An inhibitor of the rel family of transcription factors. Science 1991; 253:1268-1271.

59. Kerr LD, Inoue J-I, Davis N et al. The Rel-associated pp40 protein prevents DNA binding of Rel and NF-κB: relationship with IκBβ and regulation by phosphorylation. Genes Dev 1991; 5: 1464-1476.
60. Tewari M, Dobrzanski P, Mohn KL et al. Rapid induction in regenerating liver of RL/IF-1 (an IκB that inhibits NF-κB, RelB-p50, and c-Rel-p50) and PHF, a novel κB site-binding complex. Mol Cell Biol 1992; 12:2898-2908.
61. Dobrzanski P, Ryseck R-P, Bravo R. Differential interactions of Rel-NF-κB complexes with IκBα determine pools of constitutive and inducible NF-κB activity. EMBO J 1994; 13:4608-4616.
62. Bours V, Franzoso G, Azarenko V et al. The oncoprotein Bcl-3 directly transactivates through κB motifs via association with DNA-binding p50B homodimers. Cell 1993; 72:729-739.
63. Fujita T, Nolan GP, Liou HC et al. The candidate proto-oncogene *bcl-3* encodes a transcriptional coactivator that activates through NF-κB p50 homodimers. Genes & Dev 1993; 7:1354-1363.
64. Gaamano JH, Perez P, Lira SA et al. Constitutive expression of Bcl-3 in thymocytes increases the DNA binding of NF-κB1 (p50) homodimers in vivo. Mol Cell Biol 1996; 16:1342-1348.
65. Lenardo M, Siebenlist U. Bcl-3-mediated nuclear regulation of the NF-κB *trans*-activating factor. Immunol Today 1994; 15:145-147.
66. Lernbecher T, Kistler B, Wirth T. Two distinct mechanisms contribute to the constitutive activation of RelB in lymphoid cells. EMBO J 1994; 13:4060-4069.
67. Carrasco D, Ryseck RP, Bravo R. Expression of *rel*B transcripts during lymphoid organ development: specific expression in dendritic antigen-presenting cells. Development 1993; 118:1221-1231.
68. Carrasco D, Weih F, Bravo R. Developmental expression of the mouse c-*rel* proto-oncogene in hematopoietic organs. Development 1994; 120:2991-3004.
69. Govind S, Steward R. Dorsoventral pattern formation in *Drosophila*: signal transduction and nuclear targeting. Trends Genet 1991; 7:119-125.
70. St Johnston D, Nüsslein-Volhard C. The origin of pattern and polarity in the Drosophila embryo. Cell 1992; 68: 201-219.
71. Steward R. *Dorsal*, an embryonic polarity gene in *Drosophila*, is homologous to the vertebrate proto-oncogene c-*rel*. Science 1987; 238:692-694.
72. Geisler R, Bergmann A, Hiromi Y et al. *cactus*, a gene involved in dorsoventral pattern formation of drosophila, is related to the IκB gene family of vertebrates. Cell 1992; 71:613-621.
73. Wasserman SA. A conserved signal transduction pathway regulating the activity of the rel-like proteins dorsal and NF-κB. Mol Biol Cell 1993; 4:761-771.
74. Cao Z, Henzel J, Gao X. IRAK: a kinase associated with the interleukin-1 receptor. Science 1996; 271:1128-1131.
75. Ip YT, Reach M, Engstrom Y et al. *Dif*, a dorsal-related gene that mediates an immune response in *Drosophila*. Cell 1993; 75:753-763.
75a. Dushay MS, Asling B, Hultmark D. Origins of immunity: Relish, a compound Rel-like gene in the antibacterial defense of Drosophila. Proc Natl Acad Sci USA 1996; 93:10343-10347.
76. Sen R, Baltimore D. Multiple nuclear factors interact with the immunoglobulin enhancer sequences. Cell 1986; 46: 705-716.
77. Lenardo M, Pierce JW, Baltimore D. Protein-binding sites in Ig gene enhancers determine transcriptional activity and inducibility. Science 1987; 236: 1573-1577.
78. Lichtenstein M, Keini G, Cedar H et al. B cell-specific demethylation: a novel role for the intronic κ chain enhancer sequence. Cell 1994; 76:913-923.
79. Kirillov A, Kistler B, Mostoslavsky R et al. A role for nuclear NF-κB in B-cell-specific demethylation of the Igκ locus. Nature Genetics 1996; 13:435-441.
80. Thanos D, Maniatis T. The High Mobility Group Protein HMG I(Y) is required for NF-κB-dependent virus induction of the human IFN-β gene. Cell 1992; 71: 777-789.
81. Du W, Thanos D, Maniatis T. Mechanisms of transcriptional synergism between distinct virus-inducible enhancer elements. Cell 1993; 74:887-898.

82. Du W, Maniatis T. The high mobility group protein HMG I(Y) can stimulate or inhibit DNA binding of distinct transscription factor ATF-2 isoforms. Proc Natl Acad Sci USA 1994; 91: 11318-11322.
83. Bustin M, Lehn DA, Landsman D. Structural features of the HMG chromosomal proteins and their genes. Biochem Biophys Acta 1990; 1049:231-243.
84. Falvo JV, Thanos D, Maniatis T. Reversal of intrinsic DNA bends in the IFN β gene enhancer by transcription factors and the architectural protein HMG I(Y). Cell 1995; 83:1101-1110.
85. Tjian R, Maniatis T. Transcriptional activation: a complex puzzle with few easy pieces. Cell 1994; 77:5-8.
86. Thanos D, Maniatis T. Virus induction of human IFN-β gene expression requires the assembly of an enhancosome. Cell 1995; 83:1091-1100.
87. Stein B, Cogswell PC, AS BJ. Functional and physical associations between NF-κB and C/EBP family members: a Rel domain-bZIP interaction. Mol Cell Biol 1993; 13:3964-3974.
88. Stein B, Baldwin AS Jr, Ballard DW et al. Cross-coupling of the NF-κB p65 and Fos/Jun transcription factors produces potentiated biological function. EMBO J 1993; 12:3879-3891.
89. Neish AS, Read MA, Thanos D et al. Endothelial interferon regulatory factor 1 cooperates with NF-κB as a transcriptional activator of vascular cell adhesion molecule 1. Mol Cell Biol 1995; 15: 2558-2569.
90. Drew PD, Franzoso G, Carlson LM et al. Interferon regulatory factor-2 physically interacts with NF-κB in vitro and inhibits NF-κB induction of major histocompatibility class I and β2-microglobulin gene expression in transfected human neuroblastoma cells. J Neuroimmunol 1995; 63:157-162.
91. Mukaida N, Morita M, Ishikawa Y et al. Novel mechanism of glucocorticoid-mediated gene repression. Nuclear factor-κB is target for glucocorticoid-mediated interleukin 8 gene repression. J Biol Chem 1994; 269:13289-13295.
92. Ray A, Prefontaine A. Physical association and functional antagonism between the p65 subunit of transcription factor NF-κB and the glucocorticoid receptor. Proc Natl Acad Sci USA 1994; 91: 752-756.
93. Scheinman RI, Gualberto A, Jewell CM et al. Characterization of mechanisms involved in transrepression of NF-κB by activated glucocorticoid receptors. Mol Cell Biol 1995; 15:943-953.
94. Scheinman RI, Cogswell PC, Lofquist AK et al. Role of transcriptional activation of IκBα in mediation of immunosuppression by glucocorticoids. Science 1995; 270:283-286.
95. Auphan N, DiDonato JA, Rosette C et al. Immunosuppression by glucocorticoids: inhibition of NF-κB activity through induction of IκB synthesis. Science 1995; 270:286-290.
96. Ten RM, Paya CV, Israel N et al. The characterization of the promoter of the gene encoding the p50 subunit of NF-κB indicates that it participates in its own regulation. EMBO J 1992; 11: 195-203.
97. Cogswell PC, Scheinman RI, Baldwin AS Jr. Promoter of the human NF-κB p50/p105 gene. Regulation by NF-κB subunits and by c-Rel. J Immunol 1993; 150:2794-2804.
98. Liptay S, Schmid RM, Nabel EG et al. Transcriptional regulation of NF-κB2: evidence for IκB-mediated positive and negative autoregulation. Mol Cell Biol 1994; 14:7695-7703.
99. Lombardi L, Ciana P, Capellini C et al. Structural and functional characterization of the promoter regions of the NFKB2 gene. Nucl Acids Res 1995; 23: 2328-2336.
100. Grumont RJ, Richardson IB, Gaff C et al. Rel/NF-κB nuclear complexes that bind κB sites in the murine c-rel promoter are required for constitutive c-rel transcription in B-cells. Cell Growth Differ 1993; 4:731-743.
101. Le Bail O, Schmidt-Ullrich R, Israel A. Promoter analysis of the gene encoding the IκB-α/MAD3 inhibitor of NF-κB: positive regulation by members of the rel/NF-κB family. EMBO J 1993; 12: 5043-5049.
102. Cheng Q, Cant CA, Moll T et al. NF-κB subunit-specific regulation of the IκBα

promoter. J Biol Chem 1994; 269: 13551-13557.
103. Ito CY, Kazantsev AG, Baldwin AS Jr. Three NF-κB sites in the IκBα promoter are required for induction of gene expression by TNFα. Nucl Acids Res 1994; 22:3787-3792.
104. Ryseck R-P, Bull P, Takamiya M et al. RelB, a new Rel family transcription activator that can interact with p50-NF-κB. Mol Cell Biol 1992; 12:674-684.
105. Arenzana-Seisdedos F, Thompson J, Rodriguez MS et al. Inducible nuclear expression of newly synthesized IκBα negatively regulates DNA-binding and transcriptional activities of NF-κB. Mol Cell Biol 1995; 15:2689-2696.
106. Scott ML, Fujita T, Liou H-C et al. The p65 subunit of NF-κB regulates IκB by two distinct mechanisms. Genes & Dev 1993; 7:1266-1276.
107. Sun S-C, Ganchi PA, Ballard DW et al. NF-κB controls expression of inhibitor IκB: Evidence for an inducible autoregulatory pathway. Science 1993; 259: 1912-1915.
108. Rice NR, Ernst MK. In vivo control of NF-κB activation by IκBα. EMBO J 1993; 12:4685-4695.
109. Perez P, Lira SA, Bravo R. Overexpression of RelA in transgenic mouse thymocytes: Specific increase in levels of the inhibitor protein IκBα. Mol Cell Biol 1995; 15:3523-3530.
110. Weih F, Lira SA, Bravo R. Overexpression of RelB in transgenic mice does not affect IκB-α levels: Differential regualtion of RelA and RelB by the inhibitor protein. Oncogene 1996; 12:445-449.
111. Sen R, Baltimore D. Inducibility of κ immunoglobulin enhancer-binding protein NF-κB by a posttranslational mechanism. Cell 1986; 47:921-928.
112. Shirakawa F, Mizel SB. In vitro activation and nuclear translocation of NF-κB catalyzed by cyclic AMP-dependent protein kinase and protein kinase. Mol Cell Biol 1989; 9:2424-2430.
113. Ghosh S, Baltimore D. Activation in vitro of NF-κB by phosphorylation of its inhibitor IκB. Nature 1990; 344:678-682.
114. Kumar A, Haque J, Lacoste J et al. Double-stranded RNA-dependent protein kinase activates transcription factor NF-κB by phosphorylating IκB. Proc Natl Acad Sci USA 1994; 91:6288-6292.
115. Maran A, Maitra RK, Kumar A et al. Blockage of NF-κB signaling by selective ablation of an mRNA target by 2-5A antisense chimeras. Science 1994; 265:789-792.
116. Berra E, Diaz-Meco MT, Lozano J et al. Evidence for a role of MEK and MAPK during signal transduction by protein kinase C ζ. EMBO J 1995; 14:6157-6163.
117. Diaz-Meco MT, Dominguez I, Sanz L et al. ζPKC induces phosphorylation and inactivation of IκB-α in vitro. EMBO J 1994; 13:2842-2848.
118. Li S, Sedivy LM. Raf-1 protein kinase activates the NF-κB transcription factor by dissociating the cytoplasmic NF-κB-IκB complex. Proc Natl Acad Sci USA 1993; 90:9247-9251.
119. Finco TS, Baldwin AS. κB site-dependent induction of gene expression by diverse inducers of Nuclear Factor κB requires Raf-1. J Biol Chem 1993; 268: 17676-17679.
120. Baker SJ, Reddy EP. Transducers of life and death: TNF receptor family and associated proteins. Oncogene 1996; 12:1-9.
121. Bakalkin GY, Yakovleva T, Terenius L. NF-κB-like factors in the murine brain. Brain Res Mol Brain Res 1993; 20: 137-146.
122. Koong AC, Chen EY, Giaccia AJ. Hypoxia causes the activation of nuclear factor κB through the phosphorylation of IκB on tyrosine residues. Cancer Res 1994; 54:1425-1430.
123. Belka C, Wiegmann K, Adam D et al. Tumor necrosis factor (TNF)-α activates c-raf-1 kinase via the p55 TNF receptor engaging neutral sphingomyelinase. EMBO J 1995; 14:1156-1165.
124. Bomsztyk K, Rooney JW, Iwasaki T et al. Evidence that interleukin-1 and phorbolesters activate NF-κB by different pathways:role of protein kinase C. Cell Regul 1991; 2:329-335.
125. Su B, Karin M. Mitogen-activated protein kinase cascades and regulation of gene expression. Curr Opin Immunol 1996; 8:402-411.
126. Hirano M, Osada S-i, Aoki T et al. MEK kinase is involved in tumor necrosis fac-

tor-induced NF-κB activation and degradation of IκB-α. J Biol Chem 1996; 271:13234 -13238.
127. Meyer CF, Wang X, Chang C et al. Interaction between c-Rel and the mitogen-activated protein kinase kinase kinase 1 signaling cascade in mediating κB enhancer activation. J Biol Chem 1996; 271:8971-8976.
128. Beyaert R, Cuenda A, Berghe WV et al. The p38/RK mitogen-activated protein kinase pathway regulates interleukin-6 synthesis response to tumor necrosis factor. EMBO J 1996; 15:1914-1923.
129. Vandenabeele P, Declercq W, Beyaert R et al. Two tumour necrosis factor receptors: structure and function. Trends Cell Biol 1995; 5:392-399.
130. Hsu H, Xiong J, Goeddel DV. The TNF receptor 1-associated protein TRADD signals cell death and NF-κB activation. Cell 1995; 81:495-504.
131. Hsu H, Huang J, Shu H-B et al. TNF-dependent recruitment of the protein kinase RIP to the TNF receptor-1 signaling. Immunity 1996; 4:387-396.
132. Rothe M, Sarma V, Dixit VM et al. TRAF2-mediated activation of NF-κB by TNF receptor 2 and CD40. Science 1995; 269:1424-1427.
133. Cheng G, Baltimore D. TANK, a co-inducer with TRAF2 of TNF- and CD40L-mediated NF-κB activation. Genes & Dev 1996; 10:963-973.
134. Hsu H, Shu H-B, Pan M-G et al. TRAAD-TRAF2 and TRAAD-FADD interactions define two distinct TNF receptor 1 signal transduction pathways. Cell 1996; 84:299-308.
135. Staal FJ, Roederer M, Herzenberg LA. Intracellular thiols regulate activation of nuclear factor κB and transcription of human immunodeficiency virus. Proc Natl Acad Sci USA 1990; 87:9943-9947.
136. Schreck R, Rieber P, Baeuerle PA. Reactive oxygen intermediates as apparently widely used messengers in the activation of the NF-κB transcription factor and HIV-1. EMBO J 1991; 10:2247-2258.
137. Schreck R, Meier B, Maennel DN et al. Dithiocarbamates as potent inhibitors of nuclear factor κB activation in intact cells. J Exp Med 1992; 175:1181-1194.
138. Suzuki YJ, Biochem LP. Inhibition of NF-κB DNA binding activity by α–tocopheryl succinate. Mol Biol Int 1993; 31:693-700.
139. Meier B, Radeke HH, Selle S et al. Human fibroblasts release reactive oxygen species in response to interleukin-1 or tumour necrosis factor-α. Biochemical J 1989; 295:539-545.
140. Schmidt KN, Traenckner EB, Meier B et al. Induction of oxidative stress by okadaic acid is required for activation of transcription factor NF-κB. J Biol Chem 1995; 270:27136-27142.
141. Los M, Schenk H, Hexel K et al. IL-2 gene expression and NF-κB activation through CD28 requires reactive oxygen production by 5-lipoxygenase. EMBO J 1995; 14:3731-3740.
142. Schreck R, Albermann K, Baeuerle PA. Nuclear factor κB: An oxidative stress-responsive transcription factor of eukaryotic cells. Free Rad Res Comms 1992; 17:221-237.
143. Meyer M, Schreck R, Baeuerle PA. H_2O_2 and antioxidants have opposite effects on activation of NF-κB and AP-1 in intact cells: AP-1 as secondary antioxidant-responsive factor. EMBO J 1993; 12: 2005-2015.
144. Schmidt KN, Amstad P, Cerutti P et al. The roles of hydrogen peroxide and superoxide as messengers in the activation of transcription factor NF-κB. Chemistry & Biology 1995; 2:13-22.
145. Kretz-Remy C, Mehlen P, Mirault ME et al. Inhibition of IκB-α phosphorylation and degradation and subsequent NF-κB activation by glutathione peroxidase overexpression. J Cell Biol 1996; 133: 1083-1093.
146. Pahl HL, Baeuerle PA. A novel signal transduction pathway from the endoplasmic reticulum to the nucleus is mediated by transcription factor NF-κB. EMBO J 1995; 14:2580-2588.
147. Pahl HL, Baeuerle PA. Expression of influenza virus hemagglutinin activates transcription factor NF-κB. J Virol 1995; 69:1480-1484.
147a.Pahl HL, Baeuerle PA. Activation of NF-κB by ER stress requires both Ca2+ and reactive oxygen intermediates as messengers. FEBS Lett 1996; 392:129-136.

148. Naumann M, Scheidereit C. Activation of NF-κB in vivo is regulated by multiple phosphorylations. EMBO J 1994; 13: 4597-4607.
149. Israel A. A role for phosphorylation and degradation in the control of NF-κB activity. Trends Gen 1995; 11:203-205.
150. Finco TS, Baldwin AS. Mechanistic aspects of NF-κB regulation: the emerging role of phosphorylation and proteolysis. Immunity 1995; 3:263-272.
151. Beg AA, Finco TS, Nantermet PV et al. Tumor necrosis factor and interleukin-1 lead to phosphorylation and loss of IκBα: a mechanism for NF-κB activation. Mol Cell Biol 1993; 13:3301-3310.
152. Cordle SR, Donald R, Read MA et al. Lipopolysaccharide induces phosphorylation of MAD3 and activation of c-Rel and related NF-κB proteins in human monocytic THP-1 cells. J Biol Chem 1993; 268:11803-11810.
153. Mellits KH, Hay RT, Goodbourn S. Proteolytic degradation of MAD3 (IκB–α) and enhanced processing of the NF-κB precursor p105 are obligatory steps in the activation of NF-κB. Nucl Acids Res 1993; 21:5059-5066.
154. Frantz B, Nordby EC, Bren G et al. Calcineurin acts in synergy with PMA to inactivate IκB/MAD3, an inhibitor of NF-κB. EMBO J 1994; 13:861-870.
155. Rechsteiner M, Rogers S. PEST sequences and regulation by proteolysis. Trends Biochem Sci 1996; 21:267-271.
156. Barroga CF, Stevenson JK, Schwarz EM et al. Constitutive phosphorylation of IκB-α by casein kinase II. Proc Natl Acad Sci USA 1995; 92:7637-7641.
157. McElhinny JA, Trushin SA, Bren GD et al. Casein kinase II phosphorylates IκB-α at S-283, S-289, S-293 and T- 291 and is required for its degradation. Mol Cell Biol 1996; 16:899-906.
158. Lin R, Beauparlant P, Makris C et al. Phosphorylation of IκBa in the C-terminal PEST domain by casein kinase II affects intrinsic protein stability. Mol Cell Biol 1996; 16:1401-1409.
159. Schwarz E, Antwerp D, Verma I. Constitutive phosphorylation of IκBα by casein kinase II occurs preferentially at serine 293: requirement for degradation of free IκBα. Mol Cell Biol 1996; 16: 3554-3559.
160. Brown K, Gerstberger S, Carlson L et al. Control of IκB-α proteolysis by site-specific, signal-induced phosphorylation. Science 1995; 267:1485-1488.
161. Brockman J, Scherer D, McKinsey T et al. Coupling of a signal response domain in IκB-α to multiple pathways for NF-κB activation. Mol Cell Biol 1995; 15: 2809-2918.
162. Traenckner EB-M, Pahl HL, Henkel T et al. Phosphorylation of human IκB-α on serines 32 and 36 controls IκB-α proteolysis and NF-κB activation in response to diverse stimuli. EMBO J 1995; 14: 2876-2883.
163. Whiteside ST, Ernst MK, LeBail O et al. N- and C-terminal sequences control degradation of MAD3/IκB-α in response to inducers of NF-κB activity. Mol Cell Biol 1995; 15:5339-5345.
164. Sun S, Elwood J, Greene W. Both amino- and carboyl-terminal sequences within IκBa regulate its inducible degradation. Mol Cell Biol 1996; 16:1058-1065.
165. Miyamoto S, Maki M, Schmitt MJ et al. Tumor necrosis factor α-induced phosphorylation of IκBα is a signal for its degradation but not dissociation from NF-κB. Proc Natl Acad Sci USA 1994; 91:12740-12744.
166. Finco TS, Beg AA, Baldwin AS. Inducible phosphorylation of IκB-α is not sufficient for its dissociation from NF-κB and is inhibited by protease inhibitors. Proc Natl Acad Sci USA 1994; 91: 11884-11888.
167. DiDonato JA, Mercurio F, Karin M. Phosphorylation of IκB-α precedes but is not sufficient for its dissociation from NF-κB. Mol Cell Biol 1995; 15: 1302-1311.
168. Alkalay I, Yaron A, Hatzubai A et al. Stimulation-dependent IκBα phosphorylation marks the NF-κB inhibitor for degradation via the ubiquitin-proteasome pathway. Proc Natl Acad Sci USA 1995; 92:10599-10603.
169. Palombella VJ, Rando OJ, Goldberg AL et al. The ubiquitin-proteasome pathway is required for processing the NF-κB1 precursor protein and the activation of NF-κB. Cell 1994; 78:773-785.
170. Traenckner EB, Wilk S, Baeuerle PA. A proteasome inhibitor prevents activation

of NF-κB and stabilizes a newly phosphorylated form of IκB-α that is still bound to NF-κB. EMBO J 1994; 13: 5433-5441.
171. Lin YC, Brown K, Siebenlist U. Activation of NF-κB requires proteolysis of the inhibitor IκB-α: signal-induced phosphorylation of IκB-α alone does not release active NF-κB. Proc Natl Acad Sci USA 1995; 92:552-556.
172. Deshaies RJ. Make it or break it: The role of ubiquitin-dependent proteolysis in cellular regulation. Trends Cell Biol 1995; 5:428-434.
173. Peters J-M. Proteasomes: Protein degradation machines of the cell. Trends Biochem Sci 1994; 19:377-382.
174. Chen Z, Hagler J, Palombella VJ et al. Signal-induced site-specific phosphorylation targets IκB-α to the ubiquitin-proteasome pathway. Genes Dev 1995; 9:1586-1597.
175. Baldi L, Brown K, Franzoso G et al. Critical role for lysines 21 and 22 in signal-induced, ubiquitin-mediated proteolysis of IκB-α. J Biol Chem 1996; 271:376-379.
176. Scherer DC, Brockman JA, Chen Z et al. Signal-induced degradation of IκB-α requires site-specific ubiquitination. Proc Natl Acad Sci USA 1995; 92: 11259-11263.
177. Traenckner EB, Baeuerle PA. Appearance of apparently ubiquitin-conjugated IκB-α during its phosphorylation-induced degradation in intact cells. J Cell Sci Suppl 1995; 19:79-84.
178. Rodriguez MS, Wright J, Thompson J et al. Identification of lysine residues required for signal-induced ubiquitination and degradation of IκB-α. Oncogene 1996; 12:2425-2435.
179. Chen ZJ, Parent L, Maniatis T. Site-specific phosphorylation of IκBα by a novel ubiquitination-dependent protein kinase activity. Cell 1996; 84:853-862.
180. Imbert V, Peyron JF, Farahi D et al. Induction of tyrosine phosphorylation and T-cell activation by vanadate peroxide, an inhibitor of protein tyrosine phosphatases. Biochem J 1994; 297:163-173.
181. Menon SD, Guy GR, Tan YH. Involvement of a putative protein-tyrosine phosphatase and IκB-α serine phosphorylation in nuclear factor κB activation by tumor necrosis factor. J Biol Chem 1995; 270:18881-18887.
182. Singh S, Aggarwal BB. Protein-tyrosine phosphatase inhibitors block tumor necrosis factor-dependent activation of the nuclear transcription factor NF-κB. J Biol Chem 1995; 270:10631-10639.
182a. Imbert V, Rupec RA, Livolsi A, Pahl HL, Traenckner EB, Mueller-Dieckmann C, Farahifar D, Rossi B, Auberger P, Baeuerle PA, Peyron JF. Tyrosine phosphorylation of IκB-α activates NF-κB without proteolytic degradation of IκB-α. Cell 1996; 86:787-798.
183. Riviere Y, Blank V, Kourilsky P et al. Processing of the precursor of NF-κB by the HIV-1 protease during acute infection. Nature 1991; 350:625-626.
184. Fang CM, Maniatis T. Generation of p50 subunit of NF-κB by processing of p105 through an ATP dependent pathway. Nature 1991; 354:395-398.
185. Donald R, Ballard DW, Hawiger J. Proteolytic processing of NF-κB/IκB in human monocytes: ATP-dependent induction by pro-inflammatory mediators. J Biol Chem 1995; 270:9-12.
186. Kichan MLM, Logeat F, Israel A. Phosphorylation of p105 PEST sequence via a redox-insensitive pathway up-regulates processing of p50 NF-κB. J Biol Chem 1996; 271:6084-6091.
187. Harhaj EW, Maggirwar SB, Sun SC. Inhibition of p105 processing by NF-κB proteins in transiently transfected cells. Oncogene 1996; 12:2385-2392.
188. Orian A, Whiteside S, Israel A et al. Ubiquitin-mediated processing of NF-κB transcriptional activator precursor p105. J Biol Chem 1995; 270:21707-21714.
189. Rousset R, Desbois C, Bantignies F et al. Effects on NFKB1/p105 processing of the interaction between the HTLV-1 transactivator Tax and the proteasome. Nature 1996; 381:328-331.
190. Lernbecher T, Müller U, Wirth T. Distinct NF-κB/Rel transcription factors are responsible for tissue-specific and inducible gene activation. Nature 1993; 365:767-770.
191. Schmitt-Ulrich R, Memet S, Lilienbaum A et al. NF-κB activity in transgenic mice: developmental regulation and tissue specifity. Devolopment 1996; 122: 2117-2128.

192. Sha WC, Liou HC, Tuomanen E et al. Targeted disruption of the p50 subunit of NF-kB leads to multifocal defects in immune responses. Cell 1995; 80: 321-330.
193. Snapper CM, Zelazowski P, Rosas FR et al. B cells from p50/NF-kappa B knockout mice have selective defects in proliferation, differentiation, germ-line CH transcription, and Ig class switching. J Immunol 1996; 156:183-191.
194. Beg AA, Sha WC, Bronson RT et al. Embryonic lethality and liver degeneration in mice lacking the RelA component of NF-κB. Nature 1995; 376:167-170.
195. Weih F, Carrasco D, Durham SK et al. Multiorgan inflammation and hematopoietic abnormalities in mice with a targeted disruption of RelB, a member of the NF-κB/Rel family. Cell 1995; 80: 331-340.
196. Burkly L, Hession C, Ogata L et al. Expression of relB is required for the development of thymic medulla and dendritic cells. Nature 1995; 373:531 - 536.
197. Kontgen F, Grumont RJ, Strasser A et al. Mice lacking the c-rel proto-oncogene exhibit defects in lymphocyte proliferation, humoral immunity, and interleukin-2 expression. Genes Dev 1995; 9:1965-1977.
198. Gerondakis S, Strasser A, Metcalf D et al. Rel-deficient T cells exhibit defects in production of interleukin 3 and granulocyte-macrophage colony-stimulating factor. Proc Natl Acad Sci USA 1996; 93:3405-3409.
199. Klement JF, Rice NR, Car BD et al. IκB-α deficiency results in a sustained NF-κB response and severe widespread dermatitis in mice. Mol Cell Biol 1996; 16:2341-2349.
200. Beg AA, Sha WC, Bronson RT et al. Constitutive NF-κB activation, enhanced granulopoiesis, and neonatal lethality in IκB-α-deficient mice. Genes Dev 1995; 9:2736-2746.
201. Baeuerle PA, Henkel T. Function and activation of NF-κB in the immune system. Annu Rev Immunol 1994; 12: 141-179.
202. Kabrun N, Enrietto PJ. The Rel family of proteins in oncogenesis and differentiation. Semin Cancer Biol 1994; 5: 103-112.
203. Carrasco D, Rizzo CA, Dorfman D et al. The v-rel oncogene promotes malignant T-cell leukemia/lymphoma in transgenic mice. EMBO J 1996; 15:3640-3650.
204. Gilmore TD. Role of rel family genes in normal and malignant lymphoid cell growth. Cancer Surveys 1992; 15:69-87.
205. Ballard DW, Walker WH, Doerre S et al. The v-Rel oncogene encodes a κB enhancer binding protein that inhibits NF-κB function. Cell 1990; 63:803-814.
206. Houldsworth J, Mathew S, Rao PH et al. REL prot-oncogene is frequently amplified in extranodal diffuse large cell lymphoma. Blood 1996; 87:25-29.
207. Ueshima Y, Bird ML, Vardiman JW et al. A 14:19 translocation in B-cell chronic leukemia: a new recurring chromosome aberration. Int J Cancer 1985; 36: 287-290.
208. Liptay S, Schmid RM, Perkins ND et al. Related subunits of NF-κB map to two distinct loci associated with translocations in leukemia, NKFB1 and NFKB2. Genomics 1992; 13:287-292.
209. Fracchiolla NS, Lombardi L, Salina M et al. Structural alterations of the NF-κB transcription factor *lyt-10* in lymphoid malignancies. Oncogene 1993; 8: 2839-2845.
210. Chang C-C, Zhang J, Lombardi L et al. Rearranged NFKB-2 genes in lymphoid neoplasms code for constitutively active nuclear transactivators. Mol Cell Biol 1995; 15:5180-5187.
211. Migliazza A, Lombardi L, Rocchi M et al. Heterogeneous chromosomal aberrations generate 3′ truncations of the NFKB2/lyt-10 gene in lymphoid malignancies. Blood 1994; 84:3850-3860.
212. Zhang J, Chang CC, Lombardi L et al. Rearranged NFKB2 gene in the HUT78 T-lymphoma cell line codes for a constitutively nuclear factor lacking transcriptional repressor functions. Oncogene 1994; 9:1931-1937.
213. Ellisen LW, Bird J, West DC et al. TAN-1, the human homolog of the *Drosophila* notch gene, is broken by chromosomal translocations in T lymphoblastic neoplasms. Cell 1991; 66:649-661.
214. Bargou RC, Leng C, Krappmann D et al. High-level nuclear-factor κB and Oct-2 is a common feature of cultured

Hodgkin/Reed-Sternberg cells. Blood 1996; 87:4340-4347.
215. Dejardin E, Bonizzi G, Bellahcene A et al. Highly-expressed p100/p52 (NFKB2) sequesters other NF-κB-related proteins in the cytoplasm of human breast cancer cells. Oncogene 1995; 11:1835-1841.
216. Mukhopadhyay T, Roth JA, Maxwell SA. Altered expression of the p50 subunit of the NF-κB transcription factor complex in non-small cell lung carcinoma. Oncogene 1995; 11:999-1003.
217. Beauparlant P, Kwan I, Bitar R et al. Disruption of IκB-α regulation by antisense RNA expression leads to malignant transformation. Oncogene 1994; 9:3189-3197.
218. Sodroski J. The human T-cell leukemia virus (HTLV) transactivator (tax) protein. Biochem Biophys Acta 1992; 1114:19-29.
219. Lindholm PF, Marriot SJ, Gittlin SD et al. Induction of nuclear factor NF-κB DNA binding activity after exposure of lymphoid cells to soluble Tax 1 protein. New Biol 1990; 2:1034-1043.
220. Schreck R, Grassmann R, Fleckenstein B et al. Antioxidants selectively suppress activation of NF-κB by human T-cell leukemia virus type I Tax protein. J Virol 1992; 66:6288-6293.
221. Munoz E, Israel A. Activation of NF-κB by the Tax protein of HTLV-1. Immunobiology 1995; 193:128-136.
222. Hiscott J, Petropoulos L, Lacoste J. Molecular interactions between HTLV-1 Tax protein and the NF-κB/IκB-α complex. Virology 1995; 214:3-11.
223. Siddiqui A, Gaynor R, Srinivasan A et al. Trans-activation of viral enhancers including long terminal repeat of the human immunodeficiency virus by the hepatitis B virus X protein. Virology 1989; 169:479-484.
224. Meyer M, Caselmann WH, Schluter V et al. Hepatitis B virus transactivator MHBst: activation of NF-κB, selective inhibition by antioxidants and integral membrane localization. EMBO J 1992; 11:2991-3001.
225. Twu JS, Chu K, Robinson WS. Hepatitis B virus X gene activates κB-like enhancer sequences in the long terminal repeat of human immunodeficiency virus 1. Proc Natl Acad Sci USA 1989; 86:5168-5172.
226. Kekule AS, Lauer U, Weiss L et al. Hepatitis B virus transactivator HBx uses a tumour promoter signalling pathway. Nature 1993; 361:742-745.
227. Benn J, Schneider RJ. Hepatitis B virus HBx protein deregulates cell cycle checkpoint controls. Proc Natl Acad USA 1995; 92:11215-11219.
228. Hammarskjold ML, Simurda MC. Epstein-Barr virus latent membrane protein transactivates the human immunodeficiency virus type 1 long terminal repeat through induction of NF-κB activity. J Virol 1992; 66:6496-6501.
229. Scala G, Quinto I, Ruocco MR et al. Epstein-Barr virus nuclear antigen 2 transactivates the long terminal repeat of human immunodeficiency virus type 1. J Virol 1993; 67:2853-2861.
230. Herrero JA, Mathew P, Paya CV. LMP-1 activates NF-κB by targeting the inhibitory molecule IκB-α. JVirol 1995; 69: 2168-2174.
231. Baldassarre F, Mallardo M, Mezza E et al. Regulation of NF-κB through the nuclear processing of p105 (NF-κB1) in Epstein-Barr virus-immortalized B cell lines. J Biol Chem 1995; 270: 31244-31248.
232. Fauci AS. Multifactorial nature of human immunodeficiency virus disease: Implications for therapy. Science 1993; 262: 1011-1018.
233. Nabel G, Baltimore D. An inducible transcription factor activates expression of human immunodeficiency virus in T cells. Nature 1987; 326:711-713.
234. Roulston A, Lin R, Beauparlant P et al. Regulation of human immunodeficiency virus type 1 and cytokine gene expression in myeloid cells by NF-κB/Rel transcription factors. Microbiol Rev 1995; 59:481-505.
235. Moses AV, Ibanez C, Gaynor R et al. Differential role of long terminal repeat control elements for the regulation of basal and Tat-mediated transcription of the human immunodeficiency virus in stimulated and unstimulated primary human macrophages. J Virol 1994; 68: 298-307.
236. Alcami J, Lera TLd, Folgueira L et al. Absolute dependence on κB-responsive elements for initiation and Tat-mediated amplification of HIV transcription in

blood CD4 T lymphocytes. EMBO J 1995; 14:1552-1560.
237. Bachelerie F, Alcami J, Arenzana-Seisdedos F et al. HIV enhancer activity perpetuated by NF-κB induction on infection of monocytes. Nature 1991; 350 :709-712.
238. Paya CV, Ten RM, Bessia C et al. NF-κB-dependent induction of the NF-κB p50 subunit gene promoter underlies self-perpetuation of human immunodeficiency virus transcription in monocytic cells. Proc Natl Acad Sci USA 1992; 89:7826-7830.
239. Roulston A, Beauparlant P, Rice N et al. Chronic human immunodeficiency virus tpye I infection stimulates distinct NF-κB/rel DNA binding activities in myelomonblastic cells. J Virol 1993; 67: 5235-5246.
240. Jacque JM, Fernandez B, Arenzana-Seisdedos F et al. Permanent occupancy of the human immunodeficiency virus type 1 enhancer by NF-κB is needed for persistent viral replication in monocytes. J Virol 1996; 70:2930-2938.
241. Mihm S, Ennen J, Pessara U et al. Inhibition of HIV-1 replication and NF-κB activity by cysteine and cysteine derivatives. AIDS 1991; 5:497-503.
242. Eck HP, Gmunder H, Hartmann M et al. Low concentrations of acid-soluble thiol (cysteine) in the blood plasma of HIV-1-infected patients. Biol Chem Hoppe Seyler 1989; 370:101-108.
243. Westendorp MO, Shatrov VA, Schulze-Osthoff K et al. HIV-1 Tat potentiates TNF-induced NF-κB activation and cytotoxicity by altering the cellular redox state. EMBO J 1995; 14:546-554.
244. Mallery SR, Bailer RT, Hohl CM et al. Cultured AIDS-related Kaposi's sarcoma (AIDS-KS) cells demonstrate impaired bioenergetic adaption to oxidant challenge: Implication for oxidant stress in AIDS-KS pathogenesis. Cell Biochem 1995; 59:317-328.
245. Droege W. Cysteine and glutathione deficiency in AIDS patients: A rationale for the treatment with N-acetyl-cysteine. Pharmacology 1993; 46:61-65.
246. Aillet F, Gougerot-Pocidalo FM, Virelizier J et al. Appraisal of potential therapeutic index of antioxidants on the basis of their in vitro effects on HIV replication in monocytes and interleukin 2-induced lymphocyte proliferation. AIDS Res Hum Retroviruses 1994; 10:405-411.
247. Kaltschmidt B, Baeuerle PA, Kaltschmidt C. Potential involvement of the transcription factor NF-κB in neurological disorders. Mol Aspects Med 1993; 14: 171-190.
248. Kaltschmidt C, Kaltschmidt B, Neumann H et al. Constitutive NF-κB Activity in Neurons. Mol Cell Biol 1994; 14: 3981-3992.
249. Lukasiuk K, Kaczmarek L, Condorelli DF. Inducible and constitutive factor NF-κB-like DNA binding activies in rat brain cells cultured in vitro. Neurochem Int 1995; 26:173-178.
250. Kaltschmidt C, Kaltschmidt B, Baeuerle PA. Brain synapses contain inducible forms of the transcription factor NF-κB. Mech Develop 1993; 43:135-147.
251. Carter BD, Kaltschmidt C, Kaltschmidt B et al. Selective activation of NF-κB by nerve growth factor through the neurotrophin receptor p75. Science 1996; 272:542-545.
252. Guerrini L, Blasi F, Denis-Donini S. Synaptic activation of NF-κB by glutamate in cerebellar granule neurons in vitro. Proc Natl Acad Sci USA 1995; 92: 9077-9081.
253. Kaltschmidt C, Kaltschmidt B, Baeuerle PA. Stimulation of ionotropic glutamate receptors activates transcription factor NF-κB in primary neurons. Proc Natl Acad Sci USA 1995; 92:9618-9622.
254. Yang K, Mu XS, Hayes RL. Increased cortical nuclear factor-κB (NF-κB) DNA binding activity after traumatic brain injury in rats. Neurosci Lett 1995; 197: 101-104.
255. Salminen A, Liu PK, Hsu CY. Alteration of transcription factor binding activities in the ischemic rat brain. Biochem Biophys Res Commun 1995; 212: 939-944.
256. Prasad AV, Pilcher WH, Joseph SA. Nuclear factor-κB in rat brain: Enhanced DNA-binding activity following convulsant-induced seizures. Neurosci Lett 1994; 170:145-148.
257. Behl C, Davis JB, Lesley R et al. Hydrogen peroxide mediates amyloid β protein toxicity. Cell 1994; 77:817-827.
258. Barger SW, Horster D, Furukawa K et al.

Tumor necrosis factors α and β protect neurons against amyloid β-peptide toxicity: Evidence for involvement of a κB-binding factor and attenuation of peroxide and Ca2+ accumulation. Proc Natl Acad Sci USA 1995; 92:9328-9332.

259. Grilli M, Ribola M, Alberici A et al. Identification and characterization of a κB/Rel binding site in the regulatory region of the amyloid precursor protein gene. J Biol Chem 1995; 270:26774-26777.

260. Handel ML, Mcmorrow LB, Gravallese EM. Nuclear factor κB in rheumatoid synovium-localization of p50 and p65. Arthritis & Rheumatism 1995; 38: 1762-1770.

261. Marok R, Winyard PG, Coumbe A et al. Activation of the transcription factor nuclear factor-κB in human inflamed synovial tissue. Arthritis & Rheumatism 1996; 39:583-591.

262. Williams DH, Jeffery LJ, Murray EJ. Aurothioglucose inhibits induced NF-κB and AP-1 activity by acting as an IL-1 functional antagonist. Biochem Biophys Acta 1992; 1180:9-14.

263. Kopp E, Ghosh S. Inhibition of NF-κB by sodium salicylate and aspirin. Science 1994; 265:956-959.

264. Liao F, Andalibi A, Qiao JH et al. Genetic evidence for a common pathway mediating oxidative stress, inflammatory gene induction, and aortic fatty streak formation in mice. J Clin Invest 1994; 94:877-884.

265. Brand K, Page S, Rogler G et al. Activated transcription factor nuclear factor-κB is present in the atherosclerotic lesion. J Clin Invest 1996; 97:1715-1722.

266. Lindner V, Collins T. Expression of NF-κB and IκB-α by aortic endothelium in an arterial injury model. Am J Pathol 1996; 148:427-438.

267. Venkataraman L, Burakoff SJ, Sen R. FK506 inhibits antigen receptor-mediated induction of c-rel in B and T lymphoid cells. J Exp Med 1995; 181:1091-1099.

268. Sen J, Venkataraman L, Shinkai Y et al. Expression and induction of nuclear factor-κB-related proteins in thymocytes. J Immunol 1995; 154:3213-3221.

269. Muraoka K, Fujimoto K, Sun XG et al. Immunosuppressant FK506 induces interleukin-6 production through the activation of transcription factor Nuclear-Factor-κB-implications for FK506 nephropathy. J Clin Invest 1996; 97:2433-2439.

270. Schmidt KN, Podda M, Packer L et al. Anti-psoriatic drug anthralin activates transcription factor NF-κB in murine kerationocytes. J Immunol 1996; 156:4514-4519.

271. Perera PY, Qureshi N, Vogel SN. Paclitaxel (taxol)-induced NF-κB translocation in murine macrophages. Infection & Immunity 1996; 64:878-884.

272. Siebenlist U, Franzoso G, Brown K. Structure, regulation and function of NF-κB. Annu Rev Cell Biol 1994; 10:405-455.

273. Murphy TL, Cleveland MG, Kulesza P et al. Regulation of interleukin 12 p40 expression through an NF-κB half-site. Mol Cell Biol 1995; 15:5258-5267.

274. Toth CR, Hostutler RF, Baldwin Jr AS et al. Members of the nuclear factor kappa B family transactivate the human c-myb gene. J Biol Chem 1995; 270: 7661-7671.

275. Wright KL, White LC, Kelly A et al. Coordinate regulation of the human TAP1 and LMP2 genes from a shared bidirectional promoter. J Exp Med 1995; 181: 1459-1471.

276. Gum R, Lengyel E, Juarez J et al. Stimulation of 92 kDa gelatinase B promoter activity by ras is mitogen-activated protein kinase kinase-1 independent and requires multiple transcription factor binding sites including closely spaced PEA3/ETS and AP-1 sequences. J Biol Chem 1996; 271:10672-19680.

277. Mori N, Prager D. Transactivation of the interleukin-1α promoter by human T-cell leukemia virus type I and type II TAX proteins. Blood 1996; 87:3410-3417.

278. Michaelson JS, Singh M, Snapper CM et al. Regulation of the 3′IGH enhancers by a common set of factors, including κB-binding proteins. JImmunol 1996; 156: 2828-2839.

279. Zhao JL, Freeman GJ, Gray GS et al. A cell type-specific enhancer in the human B7.1 gene regulated by NF-κB. J Exp Med 1996; 183:777-789.

280. Yokoo T, Kitamura M. Dual regulation of Il-1β-mediated matrix metalloprotease-9 expression in mesangial cells by NF-κB and AP-1. Am J Physiol 1996; 39:123-130.

281. Sharma HW, Higginssochaski K, Perez JR et al. A DNA motif present in α-V integrin promoter exhibits dual binding preference to distinct transcription factors. Anticancer Res 1995; 15:1857-1867.
282. Gazzinelli RT, Sher A, Cheever A et al. Infection of human immunodeficiency virus 1 transgenic mice with Toxoplasma gondii stimulates proviral transcription in macrophages in vivo. J Exp Med 1996; 183:1645-1655.
283. Levitz SM, Tabuni A, Nong SH et al. Effects of Interleukin-10 on human peripheral blood mononuclear cell responses to Cryptococcus neoformans, Candida albicans, and lipopolysaccharide. Infection & Immunity 1996; 64: 945-951.
284. Schofield L, Novakovic S, Gerold P et al. Glycophosphatidylinositol toxin of plasmodium up-regulates intracellular adhesion molecule-1,vascular cell adhesion molecule-1, and E-selectin expression in vascular endothelial cells and increases leukocyte and parasite cytoadherence via tyrosine kinase-dependent signal transduction. J Immunol 1996; 156:1886-1896.
285. Pak J, Faller DV. Moloney murine leukemia virus activates NF-κB. J Virol 1996; 70:4167-4172.
286. Zhu Z, Tang WL, Ray A et al. Rhinovirus stimulation of interleukin-6 in vivo and in vitro-evidence for NF-κB-dependent transcriptional activation. J Clin Invest 1996; 97:421-430.
287. McDonald PP, Cassatella MA, Bald A et al. CD30 ligation induces NF-κB activation in human T cells. Eur J Immunol 1995; 25:2870-2876.
288. Rensing-Ehl A, Hess S, Ziegler-Heitbrock HW et al. Fas/Apo-1 activates NF-κB and induces interleukin-6 production. J Inflamm 1995; 45:161-174.
289. Haddad EB, Salmon M, Koto H et al. Ozone induction of cytokine-induced neutrophil chemoattractant (CINC) and NF-κB in rat lung-inhibition by corticosteroids. FEBS Letters 1996; 379: 265-268.
290. Rupec RA, Baeuerle PA. The genomic response of tumor cells to hypoxia and reoxygenation. Differential activation of transcription factors AP-1 and NF-κB. Eur J Biochem 1995; 243:632-640.
291. Shen YM, Rattan V, Sultana C et al. Cigarette smoke condensate-induced adhesion molecule expression and transendothelial migration of monocytes. American J Physiol 1996; 39:H1624-H1633.
292. Janssen YM, Barchowsky A, Treadwell M et al. Asbestos induces NF-κB DNA-binding activity and NF-κB-dependent gene expression in tracheal epithelial cells. Proc Natl Acad Sci USA 1995; 92:8458-8462.
293. Bereta J, Cohen MC, Bereta M. Stimulatory effect of oubain on VCAM-1 and iNos expression in murine endotheleial cells-involvement of NF-κB. FEBS Letters 1996; 377:21-25.
294. Noble PW, McKee CM, Cowman M et al. Hyaluronan fragments activate an NF-κB/IκB-a autoregulatory loop in murine macrophages. J Exp Med 1996; 183: 2373-2378.
295. Shenkar R, Schwartz MD, Terada LS et al. Hemorrhage activates NF-κB in murine lung mononuclear cells in vivo. American J Physiol 1996; 14:L729-L735.
296. Fujimoto K, Yasuda H, Sata Y et al. A role for phosphorylation in the proteolytic processing of the human NFKB1 precursor. Gene 1995; 165:183-189.
297. Kim CM, Koike K, Saito I et al. HBx gene of hepatitis B virus induces liver cancer in transgenic mice. Nature 1995; 317-320.
298. Pierce JW, Read MA, Ding H et al. Salicylates inhibit IkB-α phosphorylation, endothelial-leukocyte adhesion molecule expression, and neutrophil transmigration. J Immunol 1996; 156:3961-3969
299. Bruder JT, Heidecker G, Tan TH et al. Oncogene activation of HIV-LTR driven expression via the NF-κB binding sites. Nucl Acids Res 1993; 21:5229-5234.
300. Li CC, Korner M, Ferris DK et al. NF-kappaB/Rel family members are physically associated phosphoproteins. Biochem J 1994; 303:499-506.
301. Chirmule N, Kalyanaraman VS, Pahwa S. Signals transduced through the CD4 molecule on T lymphocytes activate NF-kB. Biochem Biophys Res Commun 1994; 203:498-505.

CHAPTER 10

The Hepatic Nuclear Factor 1

David J.D. Sourdive and Moshe Yaniv

INTRODUCTION

In higher eukaryotes, many genes are specifically expressed in a limited number of tissues, or at precise stages of development of the organism. Frequently, transcription initiation is a major regulatory step that controls such tissue-specific expression. Ex vivo and in vitro studies have shown that this regulation results from the action of neighboring cis-acting DNA sequences encompassing promoters and enhancers. Dissection of these DNA fragments by mutational analysis has led to the identification of short cis-acting elements often shared by promoters and enhancers of several genes. Biochemical studies revealed that these short DNA elements mediate the effect of trans-acting factors that specifically bind to them. The study of transcription regulation has focused in part on these proteins and the corresponding genes. Several groups have chosen to study the transcriptional regulation of liver-specific genes in vertebrates as a model system. Indeed, many genes were known to be specifically expressed in this organ, which offers several advantages: ease to work with, large size of the organ, ability to regenerate, etc. The analysis of many liver-specific promoters and enhancers led to the identification of a limited number of liver enriched transcription factors, each binding to sequences in specific genes. These include the hepatic nuclear factor 1-α[1-3] (HNF1-α, also referred to as APF, LFB1, HNF1, HP1), HNF3, HNF4, and members of two basic-zipper families, C/EBP and DBP/LAP. HNF1-α was first defined as the factor, present in liver, that specifically binds to the "B" element of α-1-antitrypsin promoter and to similar elements in the promoters of β-fibrinogen, albumin, α-fetoprotein and other liver-specific genes.

THE HNF1 FAMILY

HNF1-α protein was purified from rat livers and micro-sequenced; complementary DNAs coding for this specific binding activity were cloned by several groups.[4-6] Another factor able to bind specifically to the same DNA elements had been initially identified in dedifferentiated hepatoma cell lines[3,7-10] and in F9 embryonic carcinoma after induction of differentiation.[11,12] This factor was named HNF1-β (and is also referred to as vHNF1, vAPF or LFB3). Using the sequence coding for rat HNF1-α as a probe, clones coding for HNF1-β were isolated

Transcription Factors in Eukaryotes, edited by Athanasios G. Papavassiliou.

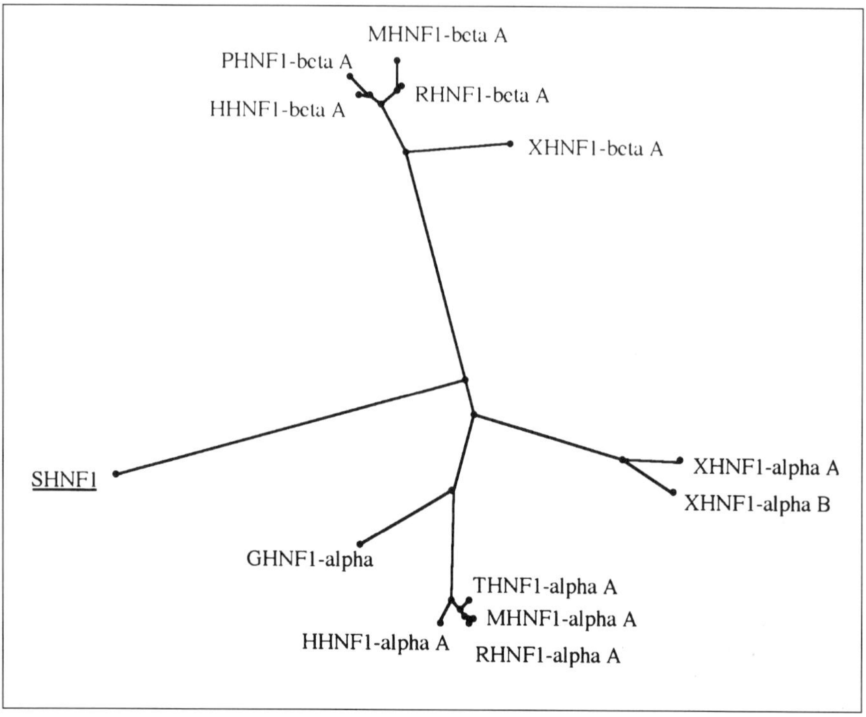

Fig. 10.1. Radial phylogenetic tree (neighbor join method) of HNF1-α and HNF1-β protein sequences in mouse (M), rat (R), hamster (T), human (H), pig (P), chicken (G), *Xenopus* (X), salmon (S). Among tetrapodians, HNF1-β is more conserved than HNF1-α. SHNF1 is closer to tetrapodian HNF1-α than to tetrapodian HNF1-β.

from rat,[13,14] mouse[12] and human.[16] A comparison of the sequences coding for HNF1-α and HNF1-β suggested that these factors were coded by different genes.[12-14] This was confirmed by localizing the two genes on different chromosomes in man and mouse (12q24.3, 17q11.2-q21.1 and 5F, 11B4-D respectively).[15,16] As of today, HNF1-α and HNF1-β are the only two genes known to code for factors able to bind to HNF1-α-specific sites.

In salmon, there seems to be only one gene for the HNF1 family.[17] Sequence analysis shows that it is closer to tetrapodian HNF1-α than to HNF1-β (see Fig. 10.1).[18] Although actively sought, no gene bearing sequence similarities with HNF1-α, or coding for a DNA-binding protein with the same sequence specificity as HNF1-α, has ever been found in invertebrates. The sequence similarity between HNF1-α and HNF1-β suggests that these two genes emerged after the duplication of a common ancestor gene. Since salmon HNF1 is the only nontetrapodian HNF1 sequence available, dating this duplication event is not straightforward. Indeed, one hypothesis would be that all actinopterigyans have only HNF1-α, which would place the duplication event in the sarcopterigyan branch, after divergence from the actinopterigyans (about 420 million years ago).[18] The emergence of a second gene could then be associated with the appearance of new physiological functions or organs (such as aerial respiration with lungs, see below). However, it cannot be ruled out that some ancestor of salmon had both HNF1-α and HNF1-β, and lost one gene during evolution (teleosteans are considered to be recent fish). In this case, the duplication would have occurred, at the latest, in a gnathostome, but could even be more ancient (see Fig. 10.2).[18]

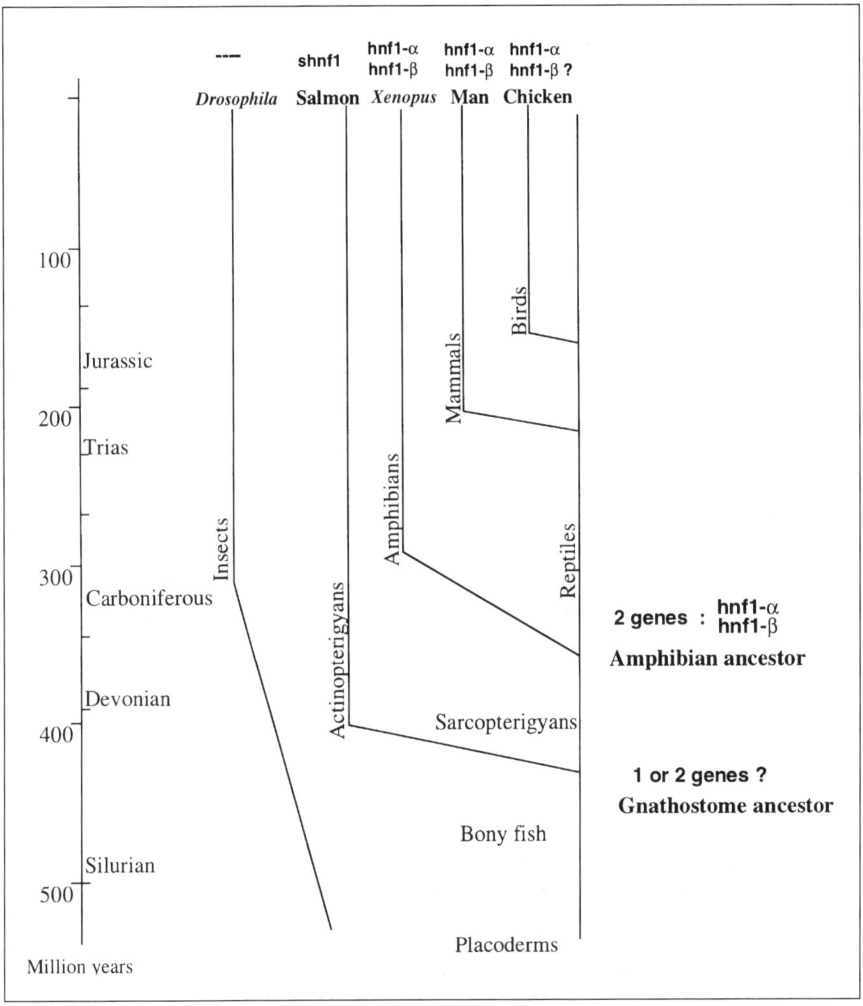

Fig. 10.2. Evolution of HNF1 family genes in the same organism.

The genomic structure of HNF1-α and part of the structure of HNF1-β was determined in rodents and man,[19,20] and in chicken[21] and *Xenopus*.[22] The exon-intron pattern of HNF1-α is largely conserved between these tetrapods. Alternative splicing of premessenger RNAs coding for both factor leads to one major and two minor isoforms of HNF1-α and HNF1-β (see Fig. 10.3).[13,20] All isoforms of HNF1-α and HNF1-β spontaneously form dimers in vitro.[5,12-14] They can all form homodimers or heterodimers without any apparent preference for a given HNF1 family member. All observations made in vitro,[12-14,16,23] ex vivo[7,12-14,23-25] or in vivo[13,26] show a distribution of homodimers and heterodimers reflecting the stochiometry of the different isoforms present. Furthermore, all dimers formed bind the same sites on DNA with comparable affinity. Thus, altogether two genes can theoretically code for 21 variant transcription factors all binding to the same target sites.

EXPRESSION OF THE HNF1 FAMILY

The expression patterns of HNF1-β and HNF1-α are best characterized in rodents. Expression of HNF1-α has also been studied

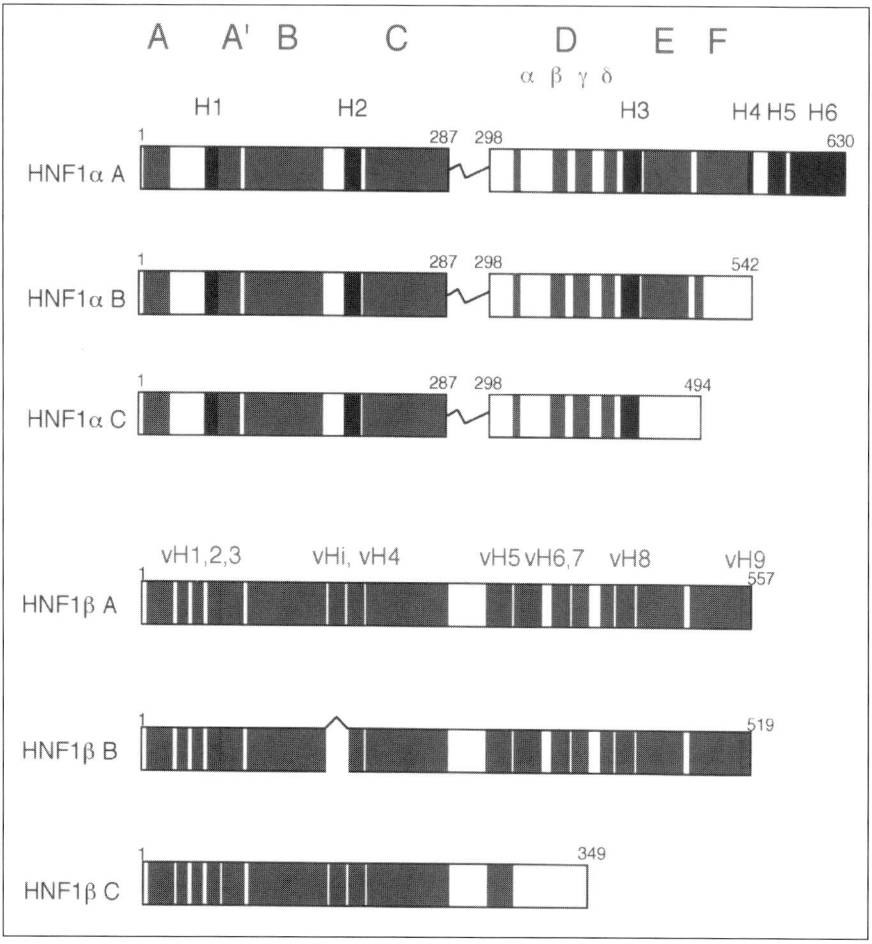

Fig. 10.3. Alternatively spliced forms of HNF1-α and HNF1-β in tetrapodians. Regions conserved in both HNF1-α and HNF1-β members are in red. Regions only conserved among HNF1-α members are in blue. Regions only conserved among HNF1-β members are in green. The nomenclature of these regions is that of refs. 18, 38. (Please see color insert.)

in *Xenopus*[27] and salmon.[17] Both genes are expressed at early stages of development, and the expression of HNF1-β always precedes the appearance of HNF1-α. In mouse, HNF1-β messenger RNA can be detected in the visceral endoderm of the yolk sac as early as day 6.5 postcoitum, while HNF1-β DNA-binding activity can be detected in the same tissue at a constant level until day 9, when it starts decreasing until birth.[28,29] Both HNF1-α and its messenger RNA are detected later, at day 8.5 in the yolk sac, and its expression increases until birth.[26,28,29]

Both HNF1-α and HNF1-β are expressed in the primitive gut and the embryonic liver and kidney. At day 9 in mouse, the HNF1-β messenger RNA is detectable near the area of the gut that will become the hepatic bud.[28] The HNF1-α mRNA appears one day later in the same area.[26,28] Then both genes are expressed in fetal and adult liver. At day 10.5 in mouse HNF1-β is expressed in the metanephros, while HNF1-α is only detectable in that tissue the next day.[26,28]

At the adult stage, the liver is the organ of highest HNF1-α expression (one report[30] gave an estimate of 22,000 HNF1-α molecules per diploid nucleus in the rat liver). HNF1-β is also present in liver, but to a much lesser extent (about 10 to 20 times less than HNF1-α). Interestingly, transcription of HNF1-α seems not to be affected during liver regeneration after surgical partial hepatectomy,[31] while the surgical shock, also observable in sham operated animals, results in a transient decrease in HNF1-α messenger RNA, suggesting an effect of the acute phase on HNF1-α RNA stability.

In the kidney, HNF1-α and HNF1-β are expressed at roughly similar levels, but HNF1-β is detectable in a wider spectrum of cell types within this organ.[32] The gut is the third major adult tissue where the HNF1-α protein[27] or both HNF1-α and HNF1-β messenger RNAs[6,11-13,17,26,33] are detected (HNF1-α is predominant over HNF1-β in this organ). Some other tissues express lower amounts of both HNF1-α and HNF1-β at the adult stage, such as pancreas,[12] thymus,[6,20,25] spleen[6,8,12,17] (especially in salmon), chick oviduct,[34] and uterus epithelium.[29]

Not only does HNF1-β systematically precede HNF1-α during development, but it is also expressed in a wider spectrum of cell types. The lung[6,11,12,27,33] does not express HNF1-α, while HNF1-β is clearly detectable in that organ. Some other tissues express lower, yet detectable, amounts of HNF1-β and no detectable HNF1-α, such as the thyroid gland,[4,12] the esophagus[12] and the ovary.[11,33]

With the exception of the kidney (especially at the fetal stages), the alternatively spliced minor isoforms HNF1-α C and HNF1-β C represent just a few percent of total HNF1-α or HNF1-β present.[20,25,29] HNF1-α B is even less expressed (3 to 20 times less than HNF1-α C), while the ratio of HNF1-β A to HNF1-β B seems to vary little (between 3 and 5.)[25,29]

Thus, HNF1-α is preferentially expressed in epithelia involved in extensive trans-membranne solute or protein transit, and is not limited to tissues derived from one single embryonic layer.[35] HNF1-β has a both temporally and spatially wider expression. For a compilation of reports on HNF1-α and HNF1-β expression in tissues see Table 10.1 (for cell lines see ref. 18).

STRUCTURE OF THE HNF1-ALPHA PROTEIN

Analysis of the HNF1-α nucleic acid and amino acid sequence showed that it comprises two major regions bearing different structures. The codon usage differs between the amino-terminal and the carboxy-terminal halves of the nucleotide sequence coding for HNF1-α.[4-6] Secondary structure predictions suggest that the former codes for a well structured essentially α-helical polypeptide (residues 1 to 284 in rat HNF1-α), while the C-terminal half of the protein is poorly structured.[18] A short flexible "hinge" region, made of glycines and prolines, links these two "domains". The N-terminal halves of HNF1-α and HNF1-β are very similar and are conserved in the minor isoforms of both factors, while the C-terminal part varies much more from one HNF1 family member to another (see Fig. 10.3).

Experimental data support these predictions made from sequence analysis. While the carboxy-terminal half of HNF1-α is rapidly degraded in vitro at room temperature,[4,5,36] its amino-terminus is stable (even at 30°C) whether produced in vitro or purified from tissues or recombinant bacteria. The abnormally reduced electrophoretic mobility of HNF1-α often observed in denaturing gels[4,5,37,38] is attributed to its N-terminal domain. The extensive O-glycosylation of HNF1-α[5,30] cannot explain this phenomenon, since it remains visible when HNF1-α is produced in vitro or in *E. coli*.[5,37] It has been suggested that detergent-resistant "kinked" structures (possibly due to prolines) could persist within the first 200 amino-acids of HNF1-α.[4]

The 284 amino-acid long amino-terminal domain of HNF1-α includes the DNA-binding domain of this factor. When expressed alone,[4,5,36-40] or fused to other polypeptides,[4,23,36,37,39] this domain binds specific sites on DNA as well as wild type HNF1-α. Deletion of either the amino or carboxy terminal ends of this domain leads

Table 10.1. Non-exhaustive compilation of reports of detection of HNF1 family members (taken from ref. 18).

Tissue	Factor	mRNA detection	Protein detection
Liver	HNF1-α++	T 5; N 3,5,9,23 R 7,10,11,18 ; I 8,13; P 18	F 1,2; B 2,9; W 10, 21
	HNF1-β+	N 13; R 9,10,11,18; P 18; I 13	B 9 ; W 9,10
Kidney	HNF1-α++	N 9,23; R 4,7,10,11,18 F 4; B 9; P 18; I 8,13	W 10,21
	HNF1-β++	N 9; R 7,10,11,18 ; P 18; I 13	B 9; W 21
Intestine	HNF1-α+	N 9,23; R 4,7,10,11; P 18;	W 21
	HNF1-β+	N 9 ; R 10,11 ; P 18	
Stomach	HNF1-α+	R 7,10,11	W 21
	HNF1-α+	R 10,11	
Pancreas	HNF1-α+	R 10	
	HNF1-β+	R 10	
Thymus	HNF1-α±	R 4; P 18	
	HNF1-β±	P 18	
Spleen	HNF1-α-	N 3,9; R 10,11	W 10,21
	HNF1-α+	N 23; R 4	B 2
	HNF1-β-	N 9, R 10,11	W 10
Lung	HNF1-α-	R 4,7,10,11; P 18	W 21
	HNF1-β+	R 10,11; P 18	B 2
Thyroid gland	HNF1-α-	N 3; R 10	
	HNF1-β+	R 10	
Esophagus	HNF1-α-	R 10	
	HNF1-β+	R 10	
Ovary	HNF1-α-	R 7,11	
	HNF1-β+	R 11	
Muscle	HNF1-α-	N 23; R 10	W 21
	HNF1-β-	R 10	
Heart	HNF1-α-	N 9; R 4,7,10	W 21
	HNF1-β-	N 9; R 10,11	

Table 10.1. Non-exhaustive compilation of reports of detection of HNF1 family members (taken from ref. 18). (con't)

Tissue	Factor	mRNA detection	Protein detection
Brain	HNF1-α-	R 7,10,11; P 18	B 2; W 10,21
	HNF1-β-	R 10,11; P 18	
Skin	HNF1-α-	R 4	
Sub Maxil. gland	HNF1-α-	R 10	
	HNF1-β-	R 10	
Blood	HNF1-α-		W 21
Testis	HNF1-α-	N 3 ; R 10	W 21

Abbreviations:
- T Transcription rate ("run on").
- I Detection by in situ hybridization.
- N Northern blot.
- R Ribonuclease protection.
- P Detection by RT-PCR.
- F DNAse I footprinting.
- B Gel retardation.
- W Western blot.
- \+ Factor detected
- ++ Large amounts detected
- \- No detection

Cross-references:
1 ref. 1
2 ref. 8
3 ref. 4
4 ref. 6
5 ref. 10
6 ref. 14
7 ref. 11
8 ref. 26
9 ref. 13
10 ref. 12
11 ref. 33
12 ref. 87
13 ref. 16
14 ref. 88
15 ref. 40
16 ref. 89
17 ref. 29
18 ref. 20
19 ref. 90
20 ref. 25
21 ref.27 (*Xenopus*)
22 ref. 34 (chicken)
23 ref. 17 (salmon)
24 ref. 50

to the loss of function.[5,37] Therefore, it is the smallest contiguous fragment of HNF1-α which is both necessary and sufficient for specific DNA binding. The structure of parts of this 284 amino-acid long domain have been further studied.

The amino-terminal domain of HNF1-α dimerizes in solution,[38] and is able to heterodimerize with the full length HNF1-α protein.[4,5,12,13,23,37,38,40] Segment "A", comprising the 31 extreme amino terminal residues (see Fig. 10.3), is strictly required for the dimerization of HNF1-α.[5] As a lone peptide, it can dimerize in solution,[37,41,42] or form hetero-dimers with the full length HNF1-α protein.[37] The three dimensional structure of this dimerization domain is unknown. However, from studies of its secondary structure,[41-43] it seems that it comprises two α-helical stretches (K4-E18 and A25-G32) separated by a "kinked" linker. This domain could potentially fold into a four helix bundle, stabilized by hydrophobic interactions between surfaces bearing a characteristic "LLxxLL" motif.[41-43]

The comparison of the sequence of segment "C", comprising residues 197 to 284, with protein databases led to the prediction that it would fold as a homeodomain.[4-6] "C" was also predicted to be atypical in that it bears a large loop between helix 2 and the recognition helix, where only a turn is found in classical homeodomains.[44] Furthermore, deletion of this loop did not alter the binding of HNF1-α to DNA.[37] The resolution of the three dimensional structure of this isolated domain by NMR[45] and X-ray crystallography[46] confirmed these predictions (see Fig. 10.4).

The structure of the region between segments A and C is predicted to comprise many amphipatic α-helices with a basic surface resembling that of DNA-binding domains of other transcription factors.[4-6] However, the complete three dimensional structure of the N-terminal domain of HNF1-α is unknown.

DNA BINDING OF THE HNF1-ALPHA PROTEIN

Many HNF1-α binding sites have been functionally characterized (for a compilation see refs. 18, 35, 47) and appear to be degenerate palindromes. A 15 base pair long consensus sequence has been derived for these sites (Fig. 10.5). As often seen in sequences bound by homeodomains, the center of this pseudopalindromic site is A/T rich.[35] The binding implies contacts between HNF1-α and the DNA all along the site, both in the major and the minor grooves (see Fig.10.5).[35]

Tomei et al[38] have proposed a modular dissection of the HNF1-α binding domain. Specific DNA binding is achieved by the combined action of three complementary segments, "A", "B" and "C", highly conserved among the amino-terminal halves of all members of the HNF1 family (see Fig.10.3).

Segment "C" alone is monomeric in solution. As most isolated homeodomains it has some affinity for nonspecific DNA (with an estimated[38] K_d of about 2×10^{-8} M), but a poor specificity for the palindromic HNF1-α binding sites (about 10-fold higher affinity). In addition, there is no cooperativity in the successive binding of several "C" segments onto the same DNA molecule, and no preference for palindromes versus direct repeats of HNF1-α half-sites. The latter also applies to "A-C" fusions, which are dimers in solution, have only two times more affinity for specific sites and only a fifty fold specificity.

Segment "B", spanning residues 100 to 175, plays a key role in the binding of HNF1-α to DNA. A polypeptide comprising both "B" and "C" qualitatively mimics this interaction.[38] "B" reduces the binding of "C" to nonspecific DNA resulting in a five hundred fold specificity for HNF1-α cognate sites. It also privileges the binding to palindromes versus direct repeats of half-sites. Finally, although "B-C" is monomeric in solution, two such polypeptides will cooperatively bind to each HNF1-α cognate site. Preliminary studies (T. Chouard, personal communication) demonstrate that the B domain directly contacts DNA in the complex.

Addition of segment "A" to "B-C" functionally reconstitutes the DNA-binding domain of HNF1-α, able to dimerize in solution, bind its cognate sites with a high affinity (an estimated K_d of 1.6×10^{-11} M) and a high specificity (about five thousand fold).[38]

Since no direct structural data are available on HNF1-α-DNA complexes, the relative positions of the two HNF1-α DNA-binding domains is unknown. Yet, from known structures of cocrystals of homeodomains and DNA, one can draw models for HNF1-α. Wolberger et al[48] have described a common feature between these previously characterized structures : an absolutely conserved asparagine residue within the recognition helix (N270 for HNF1-α) makes a direct contact with an adenine. There are four possible adenine residues that could be in contact with N270 of HNF1-α

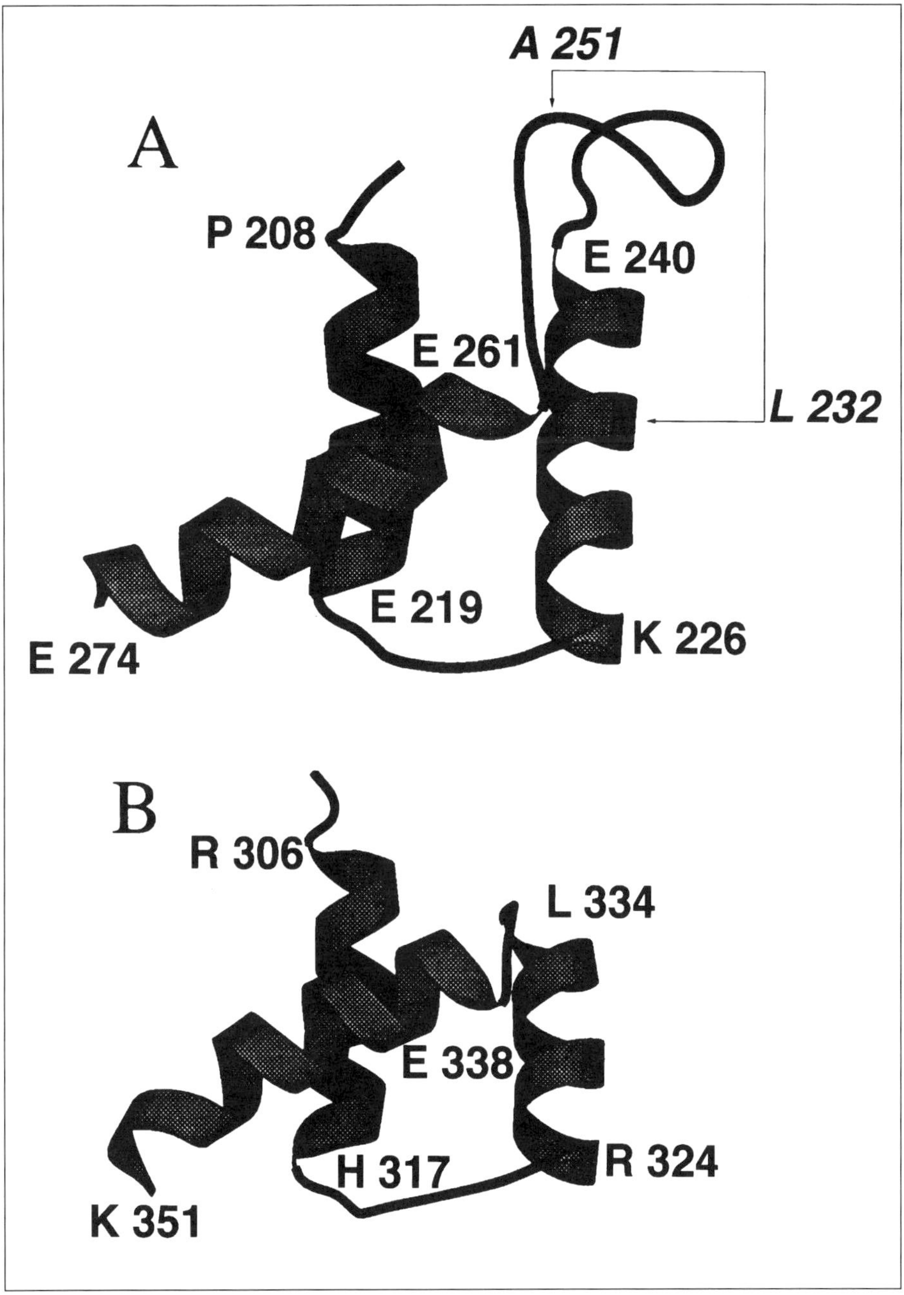

Fig. 10.4. Comparison of the three dimensional structure of homeodomains from HNF1-α in solution (A) and Antp (B) as determined by X-ray crystallography.[45] The loop ranging E240 to E261 is not structured in the isolated homeodomain. Segment L232 to A251 has been shown to be dispensable for normal HNF1-α DNA binding.[37]

A

Gene	Species	pos.	orient.	5' ---------> 3'	Characteris.
Albumin	rat, mouse	-60	->	GGTTAATGATCTACA	B,F,D,T,E, M
	man	-65	->	AGTTAATAATCTACA	B,F
		-357	->	GGTTAGTAATTACTA	B,F,D,E
		-1774	<-	TGTTACTAATTGACA	B,F,E,
	cow	-64	->	AGTTAATAATCTACA	B,F,T
	xenopus 68kD	-66	->	GGTTAATAATTTTCC	B,M,T,E
	salmon	-58	->	AGCTAATGATTGACA	B
α-Feto-Protein	rat, mouse, man	-62	->	GGTTACTAGTTAACA	B,F,T,E,M
	rat, mouse, man	-128	->	TGTTAATTATTGGCA	B,F,E,M
	man	-3445	->	GATTAATAATTACAC	B,F,E
α1-Antitrypsin	man	-76	->	GGTTAATATTCACCA	B,F,D,T,M,V
	mouse	-76	->	GGTTAATATTCTCCA	D,E
β-Fibrinogen	rat	-82	<-	AGTTAATATTTGACA	B,F,D,T,E
α-Fibrinogen	rat	-45	<-	GGTTAATCATCACCC	B,F,E
Transthyretin	mouse	-131	->	GGTTACTTATTCTCC	B,F,E,M
L-Pyruvate Kinase	rat	-90	<-	GGTTAAAGTATAACC	B,F,D,T,E
Aldolase B	rat	-119	->	AGTTATTGAATAAAC	B,F,D,T
C-reactive Proteine	man	-59	<-	TGTAAATAATTTTCC	B,E,M
		-153	<-	AGTTATTTATTACAA	B,E,M
Glut.-S-Transf. Ya	rat	-147	<-	uGGTAATGATTAATu	B,D,E
IGFBP-1	man	-60	<-	GGTTAATGATTGGCA	B,E,M
P450-CYP-2E1	rat, man	-94	<-	GGTTTATTATTAGCT	B,F,T
	man	-1035	->	TGTTAATATAAAAGT	B,F,E,V,M
2,6-Sialyltransferase	rat	-54	->	TGTTAATGTTTAACG	B,F,E
α1-Microgl./Bikunin	man	-2803	->	GGTTAATTTTTAAAA	B,E
		-2734	->	GGTTAATAATCTCAG	B,E
Apolipoprotein B	man	+845	->	TGTTTATCAGTGACT	B,E,M
Apolipoprotein AII	man	-554	->	AGTTAATGTGTAATG	B,F,E
Aminopeptidase N	pig	-78	->	GGTTAATTTTTGCCC	B,F,E
Vitellogenin A2	xenopus	-116	->	AGGTAATTGTTTACA	B,F,T,E
Insulin I	rat	-222	->	TGTTAATAATCTAAT	B,E
HBV preS1	man	-87	->	AGTTAATCATTACTT	B,F,E

B

position	-7	-6	-5	-4	-3	-2	-1	0	+1	+2	+3	+4	+5	+6	+7
VERY FAVORABLE	G	G	T	T	A	A	T	A	A	T	T	A	A	C	A
FAVORABLE	T A							T	T		C	T G	C		C
UNFAVORABLE		A	G	A	T	C T G	A	G C	G	A G	A	C	G T	A T G	T G
EXCLUDED	C	T C	A	G C	G C		G C		C	C	G				

C

```
                                    1 / 2
                                    |   |
                                    v   v
Major groove   X X ? ? ? X ? o y y X X X X X
Contacts       G G T T A A T n A T T A A C C
Minor groove   ? ? ~ ~ x x ~ ? o ~ o o ~ ? o
```

(at positions -2, -3, +4 and +5 from the center of symmetry of the HNF1-α binding site).[35] Should N270 contact an adenine 5' of the center of symmetry of the site, the two homeodomains of the HNF1-α dimer would be in close contact with each other, as is the case in the *Drosophila* transcription factor engrailed.[35,46] Should N270 contact an adenine 3' of the center of symmetry of the site, the two homeodomains of the HNF1-α dimer would be distant enough to avoid direct contact with each other, as is the case in the yeast MAT α2 protein (Fig. 10.6).[35]

Recent biochemical data suggest that the two HNF1-α homeodomains are in close contact within the dimer bound to DNA (T. Chouard, personal communication). In this configuration, the two homeodomains occupy the center of the site, and segment "B" makes the contacts observed with the 5' and 3' ends of the palindrome. DNA bending is compulsory for the two homeodomains to fit so close to one another.[35,46] The direct interaction of "B" with DNA probably provides the necessary lever to bend DNA. Thus, there is a direct functional analogy between segment "B" and the POU specific domains found in the POU homeoprotein family.[35] (The sequence of "B" is predicted to fold into amphipathic helices with a basic surface, a fact that could explain the sequence similarity between the C-terminus of "B" and the N-terminus of POU specific domains which comprise helix-turn-helix motifs.)

TRANSCRIPTION ACTIVATION BY HNF1-ALPHA

HNF1-α behaves as a transcriptional activator in vitro[30,37], ex vivo[36,39,49] and in vivo.[50] (with the exception of HNF1-β C, all the isoforms of HNF1-β and HNF1-α are transcriptional activators.[12,13,20,25]). However, the amplitude of this activation varies with the cellular context.[12,13,20,25,36,39,49] The mechanisms of transcription activation by HNF1-α are much less understood than its DNA-binding or dimerization properties. However, a modular approach has led to a gross mapping of the activation potential of HNF1-α. The carboxy terminal half of the protein is necessary for activation of transcription in vitro[37] and ex vivo[36,39]. When fused to a heterologous DNA-binding domain, it behaves as an activation element,[36] although such fusions do not activate transcription as strongly as HNF1-α. Thus, the C-terminal half of HNF1-α is necessary and partially sufficient for transcription activation. Various attempts to locate smaller activation elements within this segment have shown that, in modular terms, this potential is "distributed all along" the 332 amino acid long portion. Progressive C-terminal deletion of the protein leads to a gradual loss of activation in vitro[37] and ex vivo[36,39]. The combined use of these truncation mutants with internal deletion mutants of HNF1-α showed that some short peptides seem more crucial than others for activation[36,39,49] or proper nuclear localization of the protein.[36] When fused to the amino-terminal half of HNF1-α, two internal segments (S392-A547 and S548-Q628) are partly sufficient for activation ex vivo.[39,49] These activating portions of HNF1-α have been termed "glutamine rich" and "serine/threonine acidic"[5,18,36,39] by reference to other homeoproteins.[51]

Currently, unlike DNA binding or dimerization, transcription activation by HNF1-α has not been satisfactorily explained through this modular approach. The

Fig. 10.5. (opposite) HNF1-α binding sites (nonexhaustive compilation taken from ref. 18). (A) Wild type sites with position and orientation relative to start site, and characterization (B=gel retardation, F=DNAse I footprinting, D=DMS methylation interference, T=in vitro transcription, E=ex vivo activity, V=in vivo activity, M=characterization of mutant alleles). (B) Consensus deduced from compilation of naturally occurring sites. (C) Contacts between HNF1-α and its cognate site analyzed by methylation interference. X = strong interaction, x=weak interaction, o=absence of interaction. ~=possible interaction, y=N6 dam methylation interferes with binding (one of the two possible adenines intereferes), ?=contact not determined.

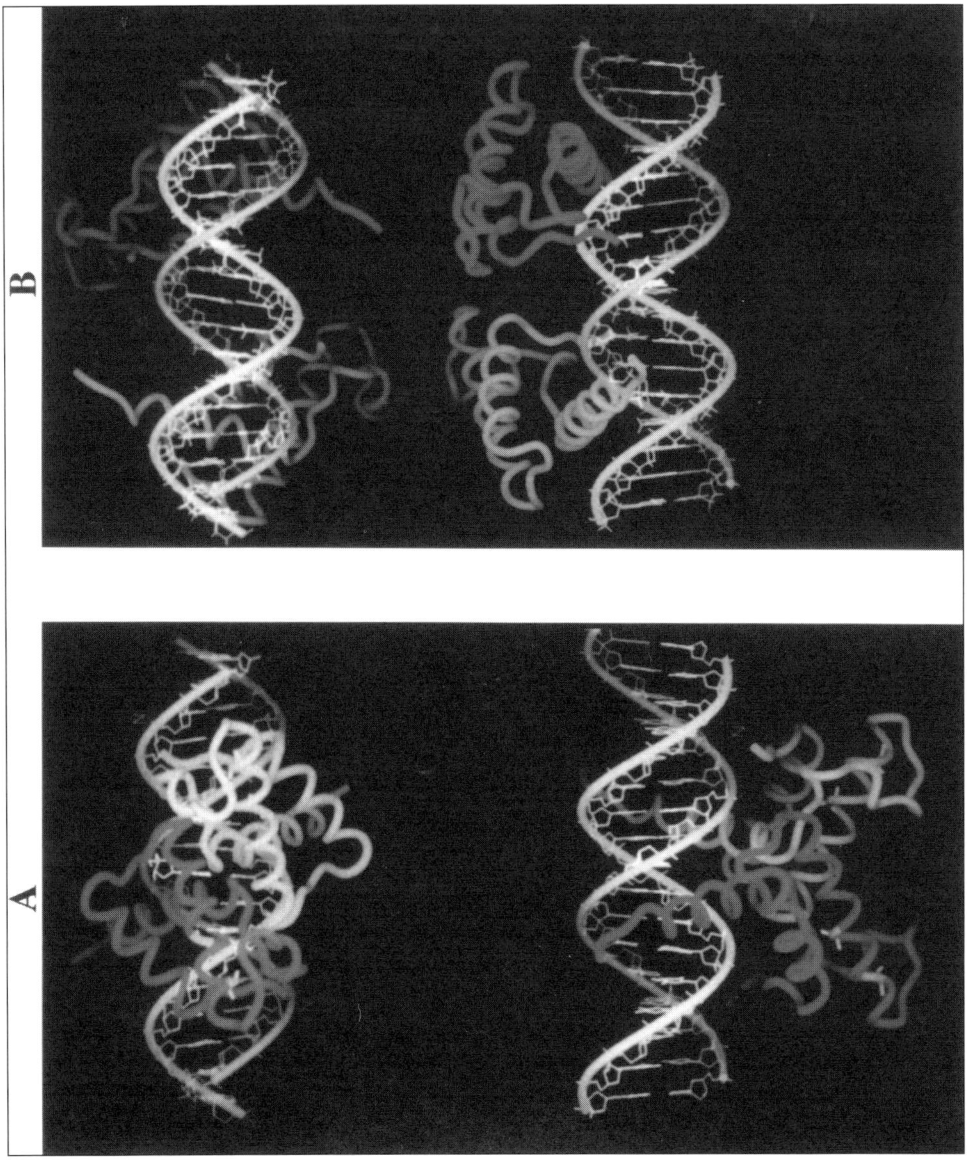

Fig. 10.6. Models of HNF1-α homeodomain/DNA interactions deduced from known three dimensional structure of complexes obtained with homeodomains of engrailed (column A) or Mat-α 2 (column B) (taken with permission from ref. 35).

amino-terminal half of HNF1-α is regarded solely as a DNA-binding domain. However, it could play a direct role in transcription activation per se. When this domain is replaced by a heterologous DNA-binding domain, the resulting chimera acts as a transcription activator ex vivo. Yet, unlike wild type HNF1-α, such fusions cannot homosynergize to activate transcription ex vivo from a promoter bearing multiple binding sites.[36] The fact that the DNA-binding domain of HNF1-α is necessary for homosynergy is reminiscent of the jun/fos transcription family members.[52] Furthermore, although the amino-terminal half of HNF1-α does not activate transcription, parts of it, when isolated and fused to heterologous DNA-binding domains, can behave as strong transcriptional activators in yeast.[18] In other homeoproteins, notably POU transcription factors sharing analogous features with HNF1-α (see above), several "positive control" mutations abolishing activation without affecting DNA binding proved to be substitutions within the homeodomain[53] or the POU specific domain.[54] Although this precedent has yet to be confirmed for HNF1-α (no naturally occurring HNF1-α mutant had been described when this manuscript was submitted; see note added in proof), a direct role of the DNA-binding domain of HNF1-α in its activation potential is plausible.

"HNF1-ALPHA KNOCK-OUT" MICE

The HNF1 gene was recently disrupted by homologous recombination in mice by Pontoglio et al[50]. In the mutant strain, the coding part of the first exon was replaced by that of lacZ, leading to a total absence of all HNF1-α isoforms. In all aspects, HNF1-$\alpha^{+/-}$ heterozygote mice are indistinguishable from their wild-type litter mates. The HNF1-$\alpha^{-/-}$ homozygotes develop normally to birth, but they fail to thrive. Starting during their second week of life, mutant homozygotes show a drastic liver enlargement, and in 20% of the cases, histological abnormalities in the gastric glandular epithelium. No macroscopic pathological signs are visible in the pancreas, the gut or the kidneys. Most mutant homozygotes die at the weaning period (around days 25 to 35). The hepatomegalia is accompanied by a severe hyperphenylalaninemia. The HNF1-$\alpha^{-/-}$ homozygotes show the typical traits of the human renal Fanconi syndrome: polyuria, glucosuria, aminoaciduria, phosphaturia and bone demineralization. Half of the mutant homozygotes have a daily diuresis of more than 40% of their body weight, with a very high glucose concentration in the urine. The renal phenotype of these mice can be attributed to dysfunctions in the transport processes which normally occur in the proximal tubules of the kidney and enable the reabsorption of glucose and other small molecules from primary urine.

Both the rate of transcription and the messenger RNA accumulation of previously suspected or known hepatic targets of HNF1-α have been examined in these mice.[50] While genes like albumin, α-1 antitrypsin or β-fibrinogen have a reduced (2- to 5-fold) transcription rate and mRNA accumulation, only one known gene, coding for phenylalanine hydroxylase (PAH), is completely down-regulated in these HNF1-$\alpha^{-/-}$ mutant mice. Apart from PAH, the reduction in the transcription rate of these known targets does not seem to lead to defects in the functions coded by these genes. However, the hepatomegalia systematically observed in the mutant homozygotes cannot be attributed to the severe hyperphenylalaninemia caused by the lack of PAH. Therefore, it is probable that other hepatic functions, yet to be identified, are dramatically reduced in these animals. The transport mechanisms that are affected in the kidneys rely on the sodium electrochemical gradient across the apical brush border of proximal tubular cells. Pontoglio et al[50] suggest that mutant homozygotes have a reduced (2-fold) number of Na^+/glucose cotransporter molecules in that membrane. However, the renal targets of HNF1-α have yet to be identified. Interestingly, HNF1-β transcription is upregulated (two fold) in the liver of these mice, but the final DNA-binding activity does not exceed 10%

to 15% of HNF1-α activity found in normal mouse liver. Although HNF1-β homodimers bind to the same DNA sequences and can also act as a transcription activator, they fail to complement the lack of HNF1-α. This may be explained by the low level of HNF1-β activity or by the specific requirement for HNF1 in the activation of liver genes.

The phenotype of HNF1-α mutant homozygotic mice confirms that HNF1-α is not critical for development, but is important for maintenance of hepatic and renal differentiation. The relatively mild reduction in transcription observed for many known targets of HNF1-α suggests that this protein contributes to a concerted action of several ubiquitous and tissue-specific transcription factors, an observation made in earlier ex vivo studies.[55-57] The mouse *pah* gene regulatory sequences seem to be atypical[58] and not well conserved among other mammals such as humans.[50,58] Therefore, it is not clear to what extent the phenotype observed in HNF1-α mutant homozygotes can be transposed to other species. It is also important that even though the global level of DNA-binding activity in kidney extracts decreases only 2-fold (HNF1-β is as abundant as HNF1-α in this tissue), HNF1-β homodimers fail to activate the transcription of a transporter in the proximal tubules.

THE DIMERIZATION COFACTOR OF HNF1-ALPHA

One of the groups that purified HNF1-α from rat liver in order to clone a cDNA coding for this transcription factor found it to be in a heterotetrameric complex comprising two HNF1-α molecules and a dimer of an 11 kDa protein.[59] Since no other function for this protein was known at the time, it was named DCoH (for "dimerization cofactor of HNF1-α"). Later, groups studying the recycling of tetrahydrobiopterin isolated the pterin-4-α carbinolamine dehydratase (PCD) which proved to have a sequence identical to that of DCoH.[60,61] Tetrahydrobiopterin (BH4) acts as a compulsory electron donor in the hydroxylation reactions of phenylalanine,[62] tyrosine[63] and tryptophan.[63] It is also required for the oxidation of the alkyl-1-mono ethers of glycerol[64,65] and is involved in the last step of NO synthesis.[66] When oxidized, BH4 becomes an unstable 4-α carbinolamine which can spontaneously either isomerize into an unrecyclable primapterin[67-74] (C7-substituted pterin) or lose a water molecule.[75] PCD favors this dehydration and prevents the formation of primapterin.

PCD/DCoH is predominantly found in liver and kidney.[59,76] It is also expressed in the stomach (D. Sourdive, unpublished results), the gut,[59] and to a lesser extent in the heart, the ovary, the lung,[59]the eye (D. Sourdive, unpublished results) and parts of the brain.[76] In *Xenopus*, it proved to be a maternal protein[77] localized in the cytoplasm of the egg, and translocated into the nucleus at the start of zygotic transcription. In the larva or the adult, wherever HNF1-α of HNF1-β is expressed, PCD/DCoH can be found both in the cytoplasm and the nucleus,[77] while in tissues only expressing PCD/DCOH, such as the pigmented epithelium of the eye, this protein seems to be strictly cytoplasmic.[77]

In solution, PCD/DCoH forms homotetramers whose three dimensional structure was determined by X-ray crystallography.[78,79] Two PCD/DCoH dimers are held together by a four helix bundle buried in the heart of the complex, and two large "saddle" shaped surfaces are exposed to the solvent. Interestingly, these homotetramers have no apparent affinity for HNF1-α dimers (Ralf Ficner, Beat Thöny unpublished data). However, if HNF1-α dimers are present while PCD/DCoH is being synthesized, PCD/DCoH-HNF1-α heterotetramers will form (the same applies to all HNF1-α or HNF1-β isoforms).[18,40,59,77] When pure homodimers of members of the HNF1 family are mixed in vitro they can exchange their monomers, leading to the formation of heterodimers.[12-14,40] However, PCD/DCoH-HNF1-α complexes lose this ability[40,59] as if PCD/DCoH "locks" the HNF1-α and HNF1-β dimer population. The dimerization domain of HNF1-α is both necessary[59] in vitro and sufficient[18,91] ex vivo for the binding of

HNF1-α to PCD/DCoH. Therefore, it has been proposed that the α helices of the dimerization domain of HNF1-α would form a four helix bundle with one dimer of PCD/DCoH preventing the interaction with another PCD/DCoH dimer.[18,91]

Human pterin-4-α carbinolamine dehydratase mutants account for 4% of total cases of BH4 deficiency.[80] The phenotype is milder than that observed with mutants of other BH4 synthesis enzymes.[81] A mild hyperphenylalaninemia and accumulation of C7-substituted pterins are the major traits of this phenotype. When PCD deficiency somatically occurs in the skin, it also results in accumulation of C7-substituted pterins and vitiligo.[82] In all cases PAH expression seems unaffected and it is still not clear whether the mild hyperphenylalaninemia results from a lack of tetrahydrobiopterin or the C7-pterins accumulated act as inhibitors of PAH.[67-74] Thus, there is little in common between the phenotypes observed in HNF1-α knock-out mice and in patients with mutant PCD/DCoH leading to a loss of PCD activity.

Human PCD point mutants cloned from such patients[83] are still able to bind ex vivo to HNF1-α or the HNF1-α dimerization domain.[18,91] Since the catalytic pocket of pterin-4-α carbinolamine dehydratase is not part of the PCD/DCoH surface interacting with HNF1-α it has been predicted that PCD/DCoH retains its enzymatic activity in PCD/DCoH-HNF1-α heterotetramers.[18,91]

The possible transcriptional role of PCD/DCoH is a complicated issue. When fused to a heterologous DNA-binding domain, PCD/DCoH does not seem to behave as a transcriptional activator ex vivo.[59,84] Transcriptional activation by HNF1-α ex vivo can be enhanced by coexpression of PCD/DCoH (3-fold[77] to 200-fold in one report,[59] under certain conditions). However, it seems that this effect observed in transient transfection assays is independent of the PCD/DCoH-HNF1-α interactions,[84] and would therefore be accounted for by a "global" or indirect activation effect of PCD/DCoH in these experimental conditions.

Phenylalanine hydroxylase is the major consumer of tetrahydrobiopterin in the organism,[85] and is the only gene known to be under the strict transcriptional control of HNF1-α in vivo.[50] DCoH messenger RNA accumulation is only mildly affected by disruption of the HNF1-α gene. Wherever PAH is expressed, so are HNF1-α and PCD/DCoH. It has been proposed that the regulation of the phenylalanine hydroxylase gene transcription may involve PCD/DCoH, possibly via PCD/DCoH-HNF1-α complexes.[79,84,86] *Pseudomonas aeruginosa*, one of the few bacteria reported to synthesize tetrahydrobiopterin, bears on one single operon both the phhA and phhB genes coding for the homologues of PAH and PCD/DCoH.[86] Deletion of phhB seems to lead to the loss of transcription of the whole operon.[86] Both the enzymatic activity of phhB and the transcription of the operon are complemented by ectopic expression of wild type human PCD/DCoH, but not by an enzymatically deficient mutant (C81→R) of PCD/DCoH isolated from a patient.[84] The mechanisms of the transcriptional effect of PHHB (or ectopically expressed DCoH) in *P. aeruginosa* are still unknown, yet the observations made in this bacterium seem difficult to transpose to mammals. Indeed, patients bearing enzymatically deficient mutants of DCoH all have a normally transcribed *pah* gene.[81,85] In fact, there are no reports of any transcriptional defects in these patients.

The possible transcriptional role of PCD/DCoH in mammals still remains to be elucidated; however, the tripartite relationship between HNF1-α, PAH and DCoH is intriguing.

PERSPECTIVES

Since its discovery,[1] a great deal has been learnt about Hepatic Nuclear Factor 1. This transcription factor has been extensively studied from many perspectives including genetics, three-dimensional structure, DNA binding, biochemistry, protein-protein interactions, evolution and transcription activation. However, many questions regarding HNF1-α remain to be answered.

HNF1-α was first identified and cloned as a DNA-binding protein. The definition of the "DNA-binding domain" by a modular dissection of HNF1-α still raises questions. The role and structure of segment "B", a possible analog of POU specific domains, are still not fully understood. The large loop in the core of the homeodomain is a feature unique to the HNF1 family and highly conserved through evolution. Since it does not seem to play a role in DNA binding, its function remains to be understood. Likewise, the mechanisms of transcription activation by HNF1-α are hardly understood at all.

Much of what was discovered about HNF1-α has been transposed to HNF1-β, the other member of the HNF1 family. However, many functional aspects of HNF1-β and HNF1-β-HNF1-α heterodimers remain to be understood. Since HNF1-β-HNF1-α heterodimers purified from tissues are bound by PCD/DCoH (a reaction which requires the heterodimer to be present at the time of PCD/DCoH synthesis), these complexes probably exist in vivo, especially in kidney. A specific role for these dimers has yet to be found, all the more as HNF1-β cannot complement HNF1-α in vivo although it binds the same cognate sites on DNA.

HNF1-α is required for survival in mice, but not for initial differentiation and organogenesis. Since HNF1-β is expressed in a broader range of tissues, and always appears earlier than HNF1-α, the inactivation of the HNF1-β gene is an obvious step in the study of the HNF1 family during the development. Besides, it is also the compulsory step towards double mutant mice (HNF1-$\alpha^{-/-}$ and HNF1-$\beta^{-/-}$) expressing no member of the HNF1 family.

The fact that DCoH was first discovered as an HNF1-α binding protein has contributed to a predominant "transcripto-centric" bias in the quest for a biological role of PCD/DCoH-HNF1-α heterotetramers. While it is far from being demonstrated, a direct role of PCD/DCoH in transcription regulation, especially in transcription activation, is still being investigated, although unseen in vivo. However, this puzzle presents an opportunity to broaden the general perspective of HNF1-α functions. Indeed, since PCD/DCoH is predicted to be enzymatically active when complexed to HNF1-α and bound to DNA, the role of this enzyme in the nucleus and the way HNF1-α contributes to its function could lead to new insights regarding this protein as more than a simple transcription activator.

Note Added in Proof

Autosomal dominant mutations in the human HNF1-α gene were recently found to be associated with MODY3 (Maturity Offest of Diabetes of the Young) form on non-insulin dependent diabetes mellitus (NIDDM).[92] Re-examination of the few hnf1$\alpha^{-/-}$ mice that survive beyond weaning, showed that the homozygous null but not heterozygous animals are diabetic (M. Pontoglio, personal communication). These studies confirm that HNF1-α also plays a major role in pancreatic β-islet function.

References

1. Courtois G, Morgan JG, Campbell LA, Fourel G, Crabtree GR. Interaction of a liver-specific nuclear factor with the fibrinogen and α 1-antitrypsin promoters. Science 1987; 238(4827):688-92.
2. Gorski K, Carneiro M, Schibler U. Tissue-specific in vitro transcription from the mouse albumin promoter. Cell 1986; 47(5):767-76.
3. Cereghini S, Raymondjean M, Carranca AG, Herbomel P, Yaniv M. Factors involved in control of tissue-specific expression of albumin gene. Cell 1987; 50(4):627-38.
4. Frain M, Swart G, Monaci P et al. The liver-specific transcription factor LF-B1 contains a highly diverged homeobox DNA binding domain. Cell 1989; 59(1): 145-57.
5. Chouard T, Blumenfeld M, Bach I, Vandekerckhove J, Cereghini S, Yaniv M. A distal dimerization domain is essential for DNA-binding by the atypical HNF1 homeodomain. Nucleic Acids Res 1990; 18(19):5853-63.
6. Baumhueter S, Mendel DB, Conley PB et al. HNF-1 shares three sequence mo-

tifs with the POU domain proteins and is identical to LF-B1 and APF. Genes Dev 1990; 4(3):372-9.
7. Cereghini S, Blumenfeld M, Yaniv M. A liver-specific factor essential for albumin transcription differs between differentiated and dedifferentiated rat hepatoma cells. Genes Dev 1988; 2(8):957-74.
8. Baumhueter S, Courtois G, Crabtree GR. A variant nuclear protein in dedifferentiated hepatoma cells binds to the same functional sequences in the β fibrinogen gene promoter as HNF-1. Embo J 1988; 7(8):2485-93.
9. Hardon EM, Frain M, Paonessa G, Cortese R. Two distinct factors interact with the promoter regions of several liver-specific genes. Embo J 1988; 7(6): 1711-9.
10. Cereghini S, Yaniv M, Cortese R. Hepatocyte dedifferentiation and extinction is accompanied by a block in the synthesis of mRNA coding for the transcription factor HNF1/LFB1. Embo J 1990; 9(7): 2257-63.
11. Kuo CJ, Conley PB, Hsieh CL, Francke U, Crabtree GR. Molecular cloning, functional expression, and chromosomal localization of mouse hepatocyte nuclear factor 1. Proc Natl Acad Sci USA 1990; 87(24):9838-42.
12. De Simone V, De Magistris L, Lazzaro D et al. LFB3, a heterodimer-forming homeoprotein of the LFB1 family, is expressed in specialized epithelia. Embo J 1991; 10(6):1435-43.
13. Rey-Campos J, Chouard T, Yaniv M, Cereghini S. vHNF1 is a homeoprotein that activates transcription and forms heterodimers with HNF1. Embo J 1991; 10(6):1445-57.
14. Mendel DB, Hansen LP, Graves MK, Conley PB, Crabtree GR. HNF-1 α and HNF-1 β (vHNF-1) share dimerization and homeo domains, but not activation domains, and form heterodimers in vitro. Genes Dev 1991; 5(6):1042-56.
15. Bach I, Galcheva GZ, Mattei MG et al. Cloning of human hepatic nuclear factor 1 (HNF1) and chromosomal localization of its gene in man and mouse. Genomics 1990; 8(1):155-64.
16. Bach I, Mattei MG, Cereghini S, Yaniv M. Two members of an HNF1 homeoprotein family are expressed in human liver. Nucleic Acids Res 1991; 19(13): 3553-9.
17. Deryckere F, Byrnes L, Wagner A, McMorrow T, Gannon F. Salmon HNF1: cDNA sequence, evolution, tissue specificity and binding to the salmon serum albumin promoter. J Mol Biol 1995; 247(1):1-10.
18. Sourdive DJD. Interactions moléculaires et activation transcriptionnelle: l'exemple du facteur HNF1 [Ph. D.]. Université de Paris 7 - Denis Diderot, 1995.
19. Bach I, Pontoglio M, Yaniv M. Structure of the gene encoding hepatocyte nuclear factor 1 (HNF1). Nucleic Acids Res 1992; 20(16):4199-204.
20. Bach I, Yaniv M. More potent transcriptional activators or a transdominant inhibitor of the HNF1 homeoprotein family are generated by alternative RNA processing. EMBO J 1993; 12(11):4229-4242.
21. Hörlein A, Grajer KH, Igo KT. Genomic structure of the POU-related hepatic transcription factor HNF-1 α. Biol Chem Hoppe Seyler 1993; 374(7):419-25.
22. Zapp D, Bartkowski S, Zoidl C, Klein HL, Ryffel GU. Genomic structure of the Xenopus laevis liver transcription factor LFB1. Gene 1993; 134(2):251-6.
23. Denecke B, Bartkowski S, Senkel S, Klein HL, Ryffel GU. Chimeric liver transcription factors LFB1 (HNF1) containing the acidic activation domain of VP16 act as positive dominant interfering mutants. J Biol Chem 1993; 268(24):18076-82.
24. Angrand PO, Rousset JP, Weiss MC. Cell phenotype, binding affinity and promoter structure modulate transactivation by HNF1 and LAP. J Cell Sci 1992; 103(4):1083-92.
25. Ringeisen F, Rey-Campos J, Yaniv M. The transactivation potential of variant hepatocyte nuclear factor 1 is modified by alternative splicing. J Biol Chem 1993; 268(34):25706-11.
26. Blumenfeld M, Maury M, Chouard T, Yaniv M, Condamine H. Hepatic nuclear factor 1 (HNF1) shows a wider distribution than products of its known target genes in developing mouse. Development 1991; 113(2):589-99.
27. Bartkowski S, Zapp D, Weber H et al. Developmental regulation and tissue dis-

tribution of the liver transcription factor LFB1 (HNF1) in Xenopus laevis. Mol Cell Biol 1993; 13(1):421-31.
28. Ott MO, Rey CJ, Cereghini S, Yaniv M. vHNF1 is expressed in epithelial cells of distinct embryonic origin during development and precedes HNF1 expression. Mech Dev 1991; 36(1-2):47-58.
29. Cereghini S, Ott MO, Power S, Maury M. Expression patterns of vHNF1 and HNF1 homeoproteins in early postimplantation embryos suggest distinct and sequential developmental roles. Development 1992; 116(3):783-97.
30. Lichtsteiner S, Schibler U. A glycosylated liver-specific transcription factor stimulates transcription of the albumin gene. Cell 1989; 57(7):1179-87.
31. Flodby P, Antonson P, Barlow C, Blanck A, Porsch HI, Xanthopoulos KG. Differential patterns of expression of three C/EBP isoforms, HNF-1, and HNF-4 after partial hepatectomy in rats. Exp Cell Res 1993; 208(1):248-56.
32. Lazzaro D, De Simone V, De Magistris L, Lehtonen E, Cortese R. LFB1 and LFB3 homeoproteins are sequentially expressed during kidney development. Development 1992; 114(2):469-79.
33. Mendel DB, Crabtree GR. HNF-1, a member of a novel class of dimerizing homeodomain proteins. J Biol Chem 1991; 266(2):677-80.
34. Grajer KH, Horlein A, Igo Kemenes T. Hepatic nuclear factor 1 α (HNF-1 α) is expressed in the oviduct of hens and interacts with regulatory elements of the lysozyme gene. Biol Chem Hoppe Seyler 1993; 374(5):319-26.
35. Chouard T. Ontogenèse précoce, différenciation terminale et homéodomaine Le cas du facteur nucléaire hépatique HNF1 [Thèse de l'Université Paris VII]. Université Paris VII, 1993.
36. Sourdive DJ, Chouard T, Yaniv M. The HNF1 C-terminal domain contributes to transcriptional activity and modulates nuclear localisation. C R Acad Sci Iii 1993; 316(4):385-94.
37. Nicosia A, Monaci P, Tomei L et al. A myosin-like dimerization helix and an extra-large homeodomain are essential elements of the tripartite DNA binding structure of LFB1. Cell 1990; 61(7):1225-36.
38. Tomei L, Cortese R, De Francesco R. A POU-A related region dictates DNA binding specificity of LFB1/HNF1 by orienting the two XL-homeodomains in the dimer. Embo J 1992; 11(11):4119-29.
39. Raney AK, Easton AJ, Milich DR, McLachlan A. Promoter-specific transactivation of hepatitis B virus transcription by a glutamine- and proline-rich domain of hepatocyte nuclear factor 1. J Virol 1991; 65(11):5774-81.
40. Nicosia A, Tafi R, Monaci P. Transdominant inhibition of transcription activator LFB1. Nucleic Acids Res 1992; 20(20):5321-8.
41. De Francesco R, Pastore A, Vecchio G, Cortese R. Circular dichroism study on the conformational stability of the dimerization domain of transcription factor LFB1. Biochemistry 1991; 30(1):143-7.
42. Pastore A, De Francesco R, Castiglione MM, Nalis D, Cortese R. The dimerization domain of LFB1/HNF1 related transcription factors: a hidden four helix bundle? Protein Eng 1992; 5(8):749-57.
43. Pastore A, De Francesco R, Barbato G, Castiglione MM, Motta A, Cortese R. 1H resonance assignment and secondary structure determination of the dimerization domain of transcription factor LFB1. Biochemistry 1991; 30(1):148-53.
44. Finney M. The homeodomain of the transcription factor LF-B1 has a 21 amino acid loop between helix 2 and helix 3 [letter]. Cell 1990; 60(1):5-6.
45. Leiting B, De FR, Tomei L, Cortese R, Otting G, Wuthrich K. The three-dimensional NMR-solution structure of the polypeptide fragment 195-286 of the LFB1/HNF1 transcription factor from rat liver comprises a nonclassical homeodomain. Embo J 1993; 12(5):1797-803.
46. Ceska TA, Lamers M, Monaci P, Nicosia A, Cortese R, Suck D. The X-ray structure of an atypical homeodomain present in the rat liver transcription factor LFB1/HNF1 and implications for DNA binding. Embo J 1993; 12(5):1805-10.
47. Tronche F, Yaniv M. HNF1, a homeoprotein member of the hepatic transcription regulatory network. [Review]. Bioessays 1992; 14(9):579-87.

48. Wolberger C, Vershon AK, Liu B, Johnson AD, Pabo CO. Crystal structure of a MAT α 2 homeodomain-operator complex suggests a general model for homeodomain-DNA interactions. Cell 1991; 67(3):517-28.
49. Wu KJ, Wilson DR, Shih C, Darlington GJ. The transcription factor HNF1 acts with C/EBP α to synergistically activate the human albumin promoter through a novel domain. J Biol Chem 1994; 269(2): 1177-82.
50. Pontoglio M, Barra J, Hadchouel M et al. Hepatocyte nuclear factor 1 inactivation results in hepatic dysfunction, phenylketonuria, and renal Fanconi syndrome. Cell 1996; 84(4):575-85.
51. Theill LE, Castrillo JL, Wu D, Karin M. Dissection of functional domains of the pituitary-specific transcription factor GHF-1. Nature 1989; 342(6252):945-8.
52. Oliviero S, Struhl K. Synergistic transcriptional enhancement does not depend on the number of acidic activation domains bound to the promoter. Proc Natl Acad Sci USA 1991; 88(1):224-8.
53. Radovick S, Nations M, Du Y, Berg LA, Weintraub BD, Wondisford FE. A mutation in the POU-homeodomain of Pit-1 responsible for combined pituitary hormone deficiency. Science 1992; 257 (5073):1115-8.
54. Pfaffle RW, DiMattia GE, Parks JS et al. Mutation of the POU-specific domain of Pit-1 and hypopituitarism without pituitary hypoplasia. Science 1992; 257 (5073):1118-21.
55. Tronche F, Rollier A, Bach I, Weiss MC, Yaniv M. The rat albumin promoter: cooperation with upstream elements is required when binding of APF/HNF1 to the proximal element is partially impaired by mutation or bacterial methylation. Mol Cell Biol 1989; 9(11):4759-66.
56. Tronche F, Rollier A, Herbomel P et al. Anatomy of the rat albumin promoter. Mol Biol Med 1990; 7(2):173-85.
57. Tronche F, Rollier A, Sourdive D, Cereghini S, Yaniv M. NFY or a related CCAAT binding factor can be replaced by other transcriptional activators for cooperation with HNF1 in driving the rat albumin promoter in vivo. Journal of Molecular Biology 1991; 222(1):31-43.
58. Faust DM, Catherin AM, Barbaux S, Belkadi L, Imaizumischerrer T, Weiss MC. The activity of the highly inducible mouse phenylalanine hydroxylase gene promoter is dependent upon a tissue-specific, hormone-inducible enhancer. Molecular & Cellular Biology 1996; 16(6):3125-3137.
59. Mendel DB, Khavari PA, Conley PB et al. Characterization of a cofactor that regulates dimerization of a mammalian homeodomain protein. Science 1991; 254(5039):1762-7.
60. Citron BA, Davis MD, Milstien S et al. Identity of 4a-carbinolamine dehydratase, a component of the phenylalanine hydroxylation system, and DCoH, a transregulator of homeodomain proteins. Proc Natl Acad Sci USA 1992; 89(24): 11891-4.
61. Hauer CR, Rebrin I, Thony B et al. Phenylalanine hydroxylase-stimulating protein/pterin-4 α-carbinolamine dehydratase from rat and human liver. Purification, characterization, and complete amino acid sequence. J Biol Chem 1993; 268(7):4828-31.
62. Kaufman S. The structure of the phenylalanine-hydroxylation cofactor. Proc Natl Acad Sci USA 1963; 50:1085.
63. Kaufman S, Fisher DB. Hepatic phenylalanine hydroxylase and PKU. In: Hayaishi O, ed. Molecular Mechanism of Oxygen Activation. New York: Academic Press, 1974: 285-365. vol 1975.
64. Tietz A, Lindberg M, Kennedy EP. A new pteridine-requiring enzyme system for the oxydation of glyceryl ethers. J Biol Chem 1964; 239(12):4081-90.
65. Kaufman S, Pollock RJ, Summer GK, Das AK, Hajra AK. Dependence of an alkyl glycol-ether monooxygenase activity upon tetrahydropterins. Biochim Biophys Acta 1990; 1040(1):19-27.
66. Tayeh MA, Marletta MA. Macrophage oxidation of L-arginine to nitric oxide, nitrite, and nitrate. Tetrahydrobiopterin is required as a cofactor. J Biol Chem 1989; 264(33):19654-8.
67. Curtius HC, Adler C, Rebrin I, Heizmann C, Ghisla S. 7-Substituted pterins: formation during phenylalanine hydroxylation in the absence of dehydratase. Biochem Biophys Res Commun 1990; 172(3):1060-6.

68. Curtius HC, Matasovic A, Schoedon G et al. 7-Substituted pterins. A new class of mammalian pteridines. J Biol Chem 1990; 265(7):3923-30.
69. Curtius HC, Adler C, Heizmann C, Blau N, Rebrin I, Ghisla S. 7-substituted pterins: formation and occurrence. J Nutr Sci Vitaminol Tokyo 1992; 5173:501-504.
70. Curtius HC, Ghisla S, Hasegawa H, Blau N, Rebrin I. Progress in the study of biosynthesis and role of 7-substituted pterins: function of pterin-4α-carbinolamine dehydratase. Adv Exp Med Biol 1993; 338(107):107-10.
71. Davis MD, Kaufman S. 7-Tetrahydrobiopterin is an uncoupled cofactor for rat hepatic phenylalanine hydroxylase. Febs Lett 1991; 285(1):17-20.
72. Davis MD, Kaufman S, Milstien S. Conversion of 6-substituted tetrahydropterins to 7-isomers via phenylalanine hydroxylase-generated intermediates. Proc Natl Acad Sci USA 1991; 88(2):385-9.
73. Davis MD, Ribeiro P, Tipper J, Kaufman S. "7-tetrahydrobiopterin," a naturally occurring analogue of tetrahydrobiopterin, is a cofactor for and a potential inhibitor of the aromatic amino acid hydroxylases. Proc Natl Acad Sci USA 1992; 89(21):10109-13.
74. Schallreuter KU, Wood JM, Pittelkow MR et al. Regulation of melanin biosynthesis in the human epidermis by tetrahydrobiopterin. Science 1994; 263(5152): 1444-6.
75. Kaufman S. A protein that stimulates rat liver phenylalanine hydroxylase. J Biol Chem 1970; 245(18):4751-9.
76. Davis MD, Kaufman S, Milstien S. Distribution of 4α-hydroxytetrahydropterin dehydratase in rat tissues. Comparison with the aromatic amino acid hydroxylases. Febs Lett 1992; 302(1):73-6.
77. Pogge von Strandmann E, Ryffel GU. Developmental expression of the maternal protein XDCoH, the dimerization cofactor of the homeoprotein LFB1 (HNF1). Development 1995; 121(4): 1217-26.
78. Ficner R, Sauer UH, Ceska TA, Stier G, Suck D. Crystallization and preliminary crystallographic studies of recombinant dimerization cofactor of transcription factor HNF1/pterin-4 α-carbinolamine dehydratase from liver. Febs Lett 1995; 357(1):62-4.
79. Endrizzi JA, Cronk JD, Wang W, Crabtree GR, Alber T. Crystal structure of DCoH, a bifunctional, protein-binding transcriptional coactivator. Science 1995; 268(5210):556-9.
80. Thony B, Heizmann CW, Mattei MG. Chromosomal location of two human genes encoding tetrahydrobiopterin-metabolizing enzymes: 6-pyruvoyl-tetrahydropterin synthase maps to 11q22.3-q23.3, and pterin-4 α-carbinolamine dehydratase maps to 10q22. Genomics 1994; 19(2):365-8.
81. Citron BA, Kaufman S, Milstien S, Naylor EW, Greene CL, Davis MD. Mutation in the 4α-carbinolamine dehydratase gene leads to mild hyperphenylalaninemia with defective cofactor metabolism. Am J Hum Genet 1993; 53(3): 768-74.
82. Schallreuter KU, Wood JM, Ziegler I et al. Defective tetrahydrobiopterin and catecholamine biosynthesis in the depigmentation disorder vitiligo. Biochim Biophys Acta 1994; 1226(2):181-92.
83. Köster S, Thony B, Macheroux P et al. Human pterin-4 α-carbinolamine dehydratase/dimerization cofactor of hepatocyte nuclear factor-1 α. Characterization and kinetic analysis of wild-type and mutant enzymes. Eur J Biochem 1995; 231:414-23.
84. Hansen LP. HNF1-α is post-translationally regulated by heterodimerization and protein stabilization [Ph.D.]. Stanford University, 1994.
85. Scriver CR, Kaufman S, Woo SLC. The Hyperphenylalaninemias. In: Scriver CR, Beaudet AL, Sly WS, Vall D, ed. The Metabolic Basis of Inherited Disease. MacGRAW-Hill, 1989: 495-546. vol 93.
86. Zhao G, Xia T, Song J, Jensen RA. Pseudomonas aeruginosa possesses homologues of mammalian phenylalanine hydroxylase and 4 α-carbinolamine dehydratase/DCoH as part of a three-component gene cluster. Proc Natl Acad Sci USA 1994; 91(4):1366-70.

87. Kuo CJ, Mendel DB, Hansen LP, Crabtree GR. Independent regulation of HNF-1 α and HNF-1 β by retinoic acid in F9 teratocarcinoma cells. Embo J 1991;10(8):2231-6.
88. Emens LA, Landers DW, Moss LG. Hepatocyte nuclear factor 1 α is expressed in a hamster insulinoma line and transactivates the rat insulin I gene. Proc Natl Acad Sci USA 1992; 89(16):7300-4.
89. Kuo CJ, Conley PB, Chen L, Sladek FM, Darnell JJ, Crabtree GR. A transcriptional hierarchy involved in mammalian cell-type specification. Nature 1992; 355 (6359):457-61.
90. Rollier A, DiPersio CM, Cereghini S et al. Regulation of albumin gene expression in hepatoma cells of fetal phenotype: dominant inhibition of HNF1 function and role of ubiquitous transcription factors. Mol Biol Cell 1993; 4(1):59-69.
91. Sourdive DJD, Transy C, Garbay S, Yaniv M. The bifunctional DCoH protein binds to HNF1 independently of its 4-α-carbinolamine dehydratase activity. Nucl Acids Res 1997; 25(8):1476-84.
92. Yamagata K, Oda N, Kaisaki PJ et al. Mutations in the hepatocyte nuclear factor-1α gene in maturity-onset diabetes of the young (MODY3). Nature 1996; 384: 455-458.

CHAPTER 11

Regulation of Gene Expression by POU Family Transcription Factors

David S. Latchman

INTRODUCTION

The Pou (Pit, Oct, Unc) family of transcription factors was originally identified on the basis of a common region which was found in the otherwise unrelated factors Pit-1, Oct-1, Oct-2 and the Unc-86 regulatory protein (for a review see refs. 1, 2). This common region consisted of approximately 150-160 amino acids and was made up of a POU homeodomain which was related to that found in the homeobox proteins and another region N-terminal to it which was called the POU specific domain. The two components of the POU domain together constitute the DNA-binding region of these regulatory factors.

Each of the original POU factors had been identified on the basis of its important role in a specific pattern of gene expression. Thus inactivation of the gene encoding the Unc-86 protein in the nematode results in the failure to form specific neuronal cell types including sensory neurons.[3,4] Similarly the mutation of the gene encoding the Pit-1 factor results in a failure to develop specific cell types within the pituitary gland, resulting in abnormal growth and development in mammals (for a review see ref. 5). Hence these factors play a critical role in the regulation of gene expression in individual neuronal cell types such that severe defects occur when the genes encoding them are inactivated by mutation.

This association of the POU family factors with gene regulation, particularly in neuronal cell types, was strengthened by the isolation of a number of novel POU domain factors by He et al.[6] These authors used a degenerate PCR approach based on the existence of conserved regions within the original POU family transcription factors. They isolated a number of cDNA clones derived from novel, previously uncharacterized POU factors and were able to show that many of these were specifically expressed in individual neuronal cell types within the central and peripheral nervous systems. Thus the POU family transcription factors appear to play a particularly critical role in the regulation of gene expression in neuronal cells.

Studies in our laboratory have concentrated on two groups of POU family transcription factors, namely the Oct-2 factor and the closely related factors Brn-3a, Brn-3b and Brn-3c.

Transcription Factors in Eukaryotes, edited by Athanasios G. Papavassiliou.

In initial studies we used the same degenerate PCR approach to isolate all the POU factors which were expressed in the immortalized ND7 cell line derived from sensory neurons.[7] In these experiments in addition to cDNA clones encoding the ubiquitously expressed POU factor Oct-1, we also isolated cDNA clones derived from the Oct-2 factor and several derived from different forms of Brn-3. These factors have therefore constituted the primary focus of our investigations.

OCT-2

The Oct-2 factor was originally identified on the basis that unlike Oct-1, it was expressed only in B lymphocytes and not in a variety of other cell types.[8,9] Subsequently however, He et al[6] showed that Oct-2 was also expressed in neuronal cells and this has since been confirmed in a number of laboratories including our own.[10-13] Hence this factor appears to be specific to B lymphocytes and neuronal cells whilst being absent in most other types. Interestingly, although Oct-2 is encoded by a single gene,[8,14,15] the protein exists in a number of different forms which are produced by alternative splicing of the primary RNA transcript.[13,16,17]

This finding is of particular interest since the Oct-2 molecule appears to contain specific regions which can activate gene expression (activation domains) and other regions which can inhibit gene expression (inhibitory domains). Thus initial studies[18,19] defined one activation domain at the N-terminus of the molecule and a second strong domain at the C-terminus of the molecule. Subsequent studies have shown, however, that the N-terminal region of the molecule also contains an inhibitory domain capable of repressing transcription although the precise boundaries of this domain have been mapped differently by different groups.[20-22]

Because of the different locations of the activation and inhibitory domains, the various forms produced by alternative splicing have different effects on gene expression. Thus the Oct-2.1 form of Oct-2, which has both activating and inhibitory domains, has a net stimulatory effect on gene expression indicating that the activation domains can overcome the inhibitory effects of the N-terminal inhibitor domain, resulting in activation of transcription. In contrast, in the Oct-2.4 and Oct-2.5 forms alternative splicing has removed the C-terminal activation domain.[16] In this situation therefore the inhibitory domain effect predominates and these isoforms have a net inhibitory effect on transcription.[17]

This difference in activity between Oct-2.1 on the one hand and Oct-2.4 and 2.5 on the other hand is especially important since these factors are produced in different relative proportions in neuronal cells and in B cells.[17] Thus in B lymphocytes, the predominant form of Oct-2 is the activating Oct-2.1 form, accounting for the initial observation that Oct-2 stimulates gene expression in this cell type. In contrast, in neuronal cells Oct-2.4 and 2.5 predominate, accounting for the fact that the Oct-2 expressed in neuronal cells is known to have a generally inhibitory effect on gene expression, repressing for example the herpes simplex virus immediate early genes[11] and the cellular tyrosine hydroxylase gene.[23] Hence the alternative splicing of Oct-2 to produce different forms with different activities in different cell types explains the paradox in early observations, which suggested that an Oct-2-binding octamer motif stimulated gene expression in B cells[24] but inhibited it in neuronal cells.[11,25]

These findings render of interest the mechanism by which the inhibitory domain in the Oct-2 transcription factor actually represses gene expression. Four mechanisms for such repression have been described (Fig. 11.1) (for a review see refs. 26-28).

In the first of these mechanisms (Fig. 11.1a) the negatively acting factor acts by binding to the DNA and blocking the binding of a positively acting factor. In the case of Oct-2, we have shown that a region of forty amino acids from 142-181 of the Oct-2 molecule can act as a separate inhibitory domain in mammalian cells when linked to the DNA-binding domain of the GAL-4 yeast transcription factor and delivered to a promoter bearing GAL-4 sites.[20,29]

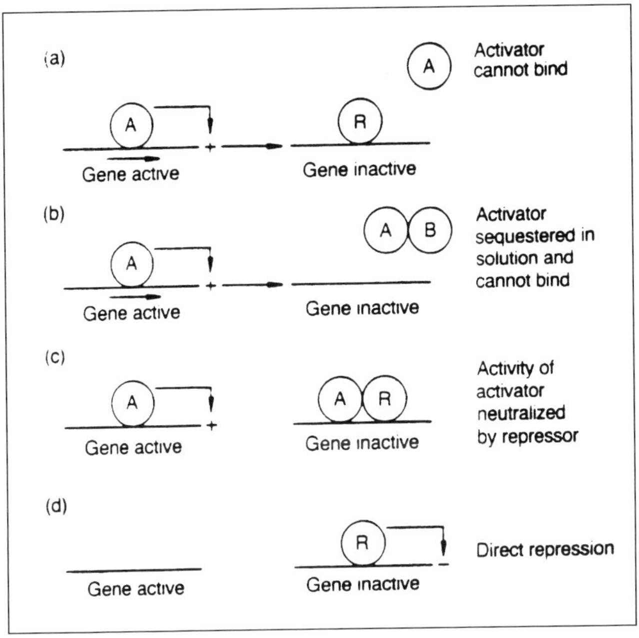

Fig. 11.1. Mechanisms by which different factors can repress gene expression. Note that mechanisms A-C constitute indirect repression by which the negatively acting factor inhibits gene expression by interfering with the activity of a positively acting factor. In contrast mechanism D involves direct repression in which the inhibitor directly reduces the activity of the basal transcriptional complex of RNA polymerase and associated factors. With permission from Latchman D. Gene Regulation: A Eukaryotic Perspective, 2nd edition. Chapman and Hall: New York, 1995.

Since no positively acting transcription factors binding to GAL-4 sites exist in mammalian cells, this eliminates passive blockage of DNA binding as the mechanism by which Oct-2 inhibits transcription. Similarly since repression by this Oct-2 - GAL-4 hybrid factor cannot be observed on a promoter which lacks GAL-4 sites,[20] this eliminates the second mechanism of repression (Fig. 11.1b), in which the inhibitory factor removes the positively acting factor from DNA by formation of a protein-protein complex.

The third mechanism of transcriptional repression involves the inhibitory factor binding adjacent to the positively acting factor and inhibiting its ability to activate transcription by interfering with its activation domain. This mechanism is known as quenching (Fig. 11.1c). To test this possibility we investigated the ability of the Oct-2 inhibitory domain (delivered to DNA by its own DNA-binding domain) to inhibit activation by a number of different types of activation domain including acidic, glutamine-rich and proline-rich domains (for a review of activation domains see ref. 30). In these experiments, the Oct-2 inhibitory domain was able to inhibit activation by all the classes of activation domain tested. It is unlikely that a forty amino acid region can interact with all these different types of activation domain directly to quench their activity. It thus appears likely that the Oct-2 inhibitory domain acts downstream of the activators by interfering with the activity of the basal transcriptional complex. This would allow it to inhibit the activity of a wide variety of activation domains since it would act downstream of these by interfering with their common target, the basal transcriptional complex. The Oct-2 factor thus appears to be a direct repressor which interferes directly with the process of transcription (Fig. 11.1d) rather than an indirect repressor which acts by blocking the action of a positively acting factor (Fig. 11.1a-c). In agreement with this idea we have recently defined an interaction between the Oct-2 inhibitory domain and the TFIIEβ component of the basal transcriptional complex.

BRN-3

As described above, in the case of Oct-2 functional diversity is generated by alternative splicing which produces different forms of Oct-2 with either activating or inhibitory actions on gene expression. In the case of Brn-3, such diversity is generated primarily by the existence of three different but closely related factors which are encoded by different genes. Thus the original Brn-3 factor was isolated in the degenerate PCR experiments of He et al[6] and was shown to be expressed in the brain. During the course of our similar experiments using the ND7 cell line, however, we generated both clones which were identical to the original Brn-3 factor and others which showed seven amino acid differences in the predicted protein within the POU domain. We thus renamed the original factor Brn-3a and named the novel but closely related factor Brn-3b.[7] Subsequently it has become clear from the work of several laboratories that three closely related Brn-3 factors exist. These factors are known as Brn-3a (also known Brn-3.0[7,31,33]). Brn-3b (also known as Brn-3.2[7,32-34]) and Brn-3c (also known as Brn-3.1[31,35]). These factors are each encoded by distinct genes but are closely related to one another.

Together with the *Drosophila* factors I-POU and tI-POU and the nematode factor Unc-86 they constitute a separate subfamily within the POU domain family.[2] The fact that the Brn-3 factors constitute the most similar mammalian factor to the critical nematode regulatory protein Unc-86 has focused attention on the role of these factors in neuronal cells. Thus each of these factors is expressed in a distinct but overlapping pattern in the developing and adult nervous systems.[31,34,35,36] In addition, when ND7 cells are induced to cease dividing and differentiate to a neuronal-like phenotype bearing neuritic processes,[37,38] the levels of Brn-3a expression increase greatly whilst Brn-3b falls and Brn-3c remains unchanged.[7]

Most interestingly this opposite expression pattern of Brn-3a and Brn-3b is paralleled by their antagonistic effects on promoter activity. Thus Brn-3a can activate both an artificial promoter containing its binding site[39,40] and the promoter of the gene encoding α-internexin, an intermediate filament protein expressed in neuronal cells.[41] In contrast Brn-3b can both inhibit the basal activity of these promoters and abolish their stimulation by Brn-3a, whilst Brn-3c has a weak stimulatory effect on both promoters. The net result of ND7 cell differentiation is therefore to increase the level of the activating Brn-3a factor whilst decreasing the level of the inhibitory Brn-3b factor. In agreement with this, the activity of both test promoters rises when they are transfected into ND7 cells and the cells induced to differentiate. This confirms that the observed changes in endogenous levels of the different forms of Brn-3 can regulate target gene activity.[39]

In view of the different effects of Brn-3a and Brn-3b on these promoters, it is possible to use chimeric constructs to map the regions of Brn-3 which are required for activation of the different target promoters. In these experiments, we showed that the two promoters differed in their requirements for activation. Thus the test promoter containing a synthetic Brn-3 binding site required the C-terminal DNA-binding POU domain for activation.[40] In contrast the α-internexin promoter required an N-terminal activation domain (Fig. 11.2).[41] Clearly such differences could arise from a difference in the sequence of the Brn-3 target sites in each promoter or from a difference in its context relative to other binding sites in the promoter. By deletion analysis of the α-internexin promoter we were able to define the region which was responsible for its stimulation by Brn-3a. Insertion of this region into the same context in the test promoter

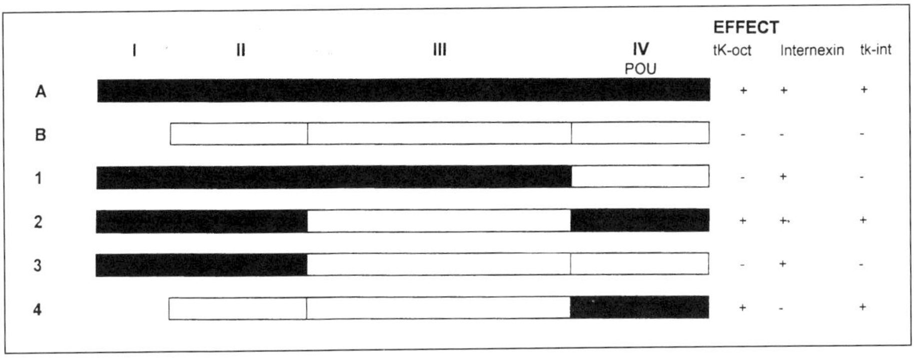

Fig. 11.2. Summary of gene activation data obtained using either the tk promoter carrying a synthetic-octamer motif (tk-Oct) (data from ref. 40) or carrying the response element from the α-internexin promoter (tk-int)[42] or with the intact α-internexin promoter[41] together with Brn-3a (A) or b (B) or with constructs encoding chimeric proteins with different regions derived from Brn-3a or Brn-3b. Note that region IV contains the POU domain of the factors. With permission from Budhram-Mahadeo V et al. Journal of Biological Chemistry 1996; 271:9108-9113.

resulted in its activation becoming dependent upon the POU region activation domain.[42] It thus appears that the context of the Brn-3 binding site in the different promoters, rather than its precise sequence, controls the activity of the different activation domains.

Interestingly, where such a sequence context renders the promoter responsive to chimeras containing the Brn-3a POU activation domain, the promoter can also be stimulated by the Brn-3a POU activation domain when it is expressed in isolation in the absence of any other region of the molecule. In contrast the corresponding POU domain of Brn-3b has no effect.[41] Since the Brn-3a and Brn-3b POU domain regions exhibit only seven amino acid differences, one or more of these differences must be responsible for this difference in activity. Six of these differences are located in the linker region between the POU-specific domain and the POU homeodomain, which is generally divergent between different factors and is believed simply to allow the two other domains to move against one another. In contrast the POU-specific domains of Brn-3a and Brn-3b are identical whilst the homeodomain contains only a single difference. However, this single difference in the POU homeodomain is located at a position which appears to be critical for interaction of POU proteins with other regulatory proteins. Thus, in the structure of the molecule the region containing this difference is located on the surface away from the DNA.[43]

Moreover alteration in the amino acid at this position within the Oct-2 transcription factor to that normally found at the corresponding position in Oct-1, allows Oct-2 to interact with the herpes simplex virus protein VP16, which is normally a property of Oct-1 alone.[44] This suggests that the difference in the amino acids at this position may allow Brn-3a to recruit a second molecule to the promoter which actually activates gene expression. In contrast Brn-3b cannot do this and binds DNA preventing activation by Brn-3a. This thus constitutes an example of the first mechanism of transcriptional repression illustrated in Figure 11.1a, whereby the inhibitory factor prevents the activating factor from binding to the DNA.

Thus, in the case of the Brn-3 factors, the production of three closely related factors encoded by different genes and with different activities results in the functional diversity which in the case of Oct-2 is obtained by alternative splicing of the transcript of a single gene. The existence of such closely related but functionally different forms of Brn-3 with distinct expression patterns obviously raises the question of their functional role. In particular we have been interested in the reasons why the levels of Brn-3a rise and those of Brn-3b fall when ND7 cells are induced to undergo a growth arrest and extend neurite processes. By using an antisense approach we have been able to inhibit the increased Brn-3a expression which occurs during the differentiation of ND7 cells.[45]

Such inhibition did not have any effect on the normal growth arrest of ND7 cells, indicating that Brn-3a does not play a critical role in this process. In contrast inhibition of increased Brn-3a expression prevents ND7 cells from extending neurite processes, suggesting that Brn-3a may play a critical role in this aspect of differentiation.[45] In agreement with this, we have demonstrated that Brn-3a is capable of transactivating the promoter of the gene encoding the SNAP-25 protein which is a component of the synaptic plasma membrane and which plays a critical role in neurite outgrowth by a number of different types of neuronal cell.[46] ND7 cell which have been treated with antisense Brn-3a do not show the rise in SNAP-25 expression which normally occurs upon differentiation.

Hence inhibition of Brn-3a expression results in a failure to activate the expression of at least one gene whose protein product is critical for neurite outgrowth as well as a failure of neurite outgrowth itself. This suggests that the rise in Brn-3a expression which normally occurs during the differentiation of ND7 cells may play a critical role in stimulating the expression of genes whose protein products are required for this process.

Interestingly our recent finding that Brn-3a can also activate the promoter of the synapsin I gene, a synaptic vesicle protein, suggests that Brn-3a may be able to induce the expression of a number of genes whose protein products play a role in synapse formation. Thus Brn-3a may play a critical role in the induction of the neuronal phenotype during development. In agreement with this a recent study[36] has demonstrated that Brn-3a expression is restricted to postmitotic CNS neurons being expressed very early in the development of such neurons indicating a role in the specification of the mature neuronal phenotype.

Overall, therefore, it is clear that the differently alternatively spliced forms of the Oct-2 factor and the different forms of Brn-3 encoded by different genes together play a critical role in the regulation of gene expression in neuronal cells, acting both by inhibiting and activating the expression of specific genes. As these factors are the only POU proteins which appear to be expressed in mature sensory neurons they are likely to play an especially critical role in the development and differentiation of these particular neurons. This role represents one aspect of the important role of the numerous different POU factors in the specification and differentiation of the many different neuronal cell types.

References

1. Verrijzer CP, van der Vliet PC. POU domain transcription factors. Biochimica et Biophysica Acta 1993; 1173:1-21.
2. Wegner M, Drolet DW, Rosenfeld MG. POU-domain proteins. Structure and function of developmental regulators. Current Opinion in Developmental Biology 1993; 5:488-498.
3. Desai C, Garriga G, McIntire SL et al. A genetic pathway for the development of *Caenorhabditis elegans* HSN motor neurons. Nature 1988; 336:638-646.
4. Finney M, Ruvkin G, Horvitz HR. The *C. elegans* cell lineage and differentiation gene *unc-86* encodes a protein with a homeodomain and extended similarity to transcription factors. Cell 1988; 55:757-769.
5. Radovick S, Nations M, Du Y et al. A mutation in the POU-homeodomain of Pit-1 responsible for combined pituitary hormone deficency. Science 1992; 257:1115-1118.
6. He X, Treacy MN, Simmons DM et al. Expression of a large family of POU-domain regulatory genes in mammalian brain development. Nature 1989; 340:35-42.
7. Lillycrop KA, Budhram-Mahadeo VS, Lakin ND et al. A novel POU family

transcription factor is closely related to Brn-3 but has a distinct expression pattern in neuronal cells. Nucl Acids Res 1992; 20:5093-5096.

8. Singh H, Sen R, Baltimore D et al. A nuclear factor that binds to a conserved sequence motif in transcriptional control elements of immunoglobulin genes. Nature 1980; 319:154-158.
9. Staudt LH, Singh H, Sen R et al. A lymphoid-specific protein binding to the octamer motif of immunoglobulin genes. Nature 1986; 323:640-643.
10. Kemp LM, Dent CL, Latchman DS. Transcriptional repression of herpes simplex virus immediate early genes and of octamer-containing cellular promoters in neuronal cells is mediated by the octamer DNA element. Neuron 1990; 4:215-222.
11. Lillycrop KA, Dent CL, Wheatley SC et al. The octamer binding protein Oct-2 represses HSV immediate early genes in cell lines derived from latently infectable sensory neurons. Neuron 1991; 7:381-390.
12. Scholer HR, Hatzopoulos AK, Balling R. A family of octamer-specific proteins present during mouse embryogenesis: evidence for germline-specific expression of an Oct factor. EMBO J 1989; 8:2543-2550.
13. Stoykova AS, Steiner S, Erselius JR et al. Mini-Oct and Oct-2c: Two novel functionally diverse murine Oct-2 products are differentialy expressed in the CNS. Neuron 1992; 841:541-558.
14. Clerc RG, Corcoran LM, LeBowitz JH et al. The B-cell-specific Oct-2 protein contains POU box and homeo box type domains. Genes and Development 1988; 2:1570-1581.
15. Scheidereit C, Cromlish JA, Gerster T et al. A human lymphoid-specific transcription factor that activates immunoglobulin genes is a homeobox protein. Nature 1988; 336:551-557.
16. Wirth T, Priess A, Annweiler A et al. Multiple Oct-2 isoforms are generated by alternative splicing. Nucl Acids Res 1991; 19:43-51.
17. Lillycrop KA, Latchman DS. Alternative splicing of the Oct-2 transcription factor is differentialy regulated in B cells and neuronal cells and results in protein isoforms with opposite effects on the activity of octamer/TAATGARAT-containing promoters. Journal of Biological Chemistry 1992; 267:24960-24966.
18. Gerster T, Balmaceda C-G, Roeder RG. The cell type-specific octamer transcription factor OTF-2 has two domains required for the activation of transcription. EMBO J 1990; 9:1635-1643.
19. Muller-Immergluck AM, Schaffner W, Matthias P. Transcription factor Oct-2A contains functionally redundant activation domains and works selectively from a remote enhancer protein in non-lymphoid (HeLa) cells. EMBO J 1990; 9:1625-1634.
20. Lillycrop KA, Dawson SJ, Estridge JK et al. Repression of a herpes simplex virus immediate-early promoter by the Oct-2 transcription factor is dependant upon an inhibitory region at the N-terminus of the protein. Mol Cell Biol 1994; 14:7633-7642.
21. Annweiler A, Zwilling S, Wirth T. Functional differences between the Oct-2 transactivation domains determine the transactivation potential of individual Oct-2 isoforms. Nucl Acids Res 1994; 22:4250-4258.
22. Friedl EM, Matthias P. Transcription activation and repression, two properties of the lymphoid-specific transcription factor Oct-2a. European Journal of Biochemistry 1995; 234:308-316.
23. Dawson SJ, Yoon SO, Chikaraishi DM et al. The Oct-2 transcription factor represses tyrosine hydroxylase expression via the heptamer motif in the promoter. Nucl Acids Res 1994; 22:1023-1028.
24. Wirth T, Staudt L, Baltimore D. An octamer oligonucleotide upstream of a TATA motif is sufficient for lymphoid specific promoter activity. Nature 1987; 329:176-178.
25. Dent CL, Lillycrop KA, Estridge JK et al. The B cell and neuronal forms of the octamer binding protein Oct-2 differ in DNA binding specificity and functional activity. Molecular and Cellular Biology 1991; 1:3925-3930.
26. Levine M, Manley JL. Transcriptional repression of eukaryotic promoters. Cell 1989; 59:405-408.

27. Goodboun S. Negative regulation of transcriptional initiation in eukaryotes. Biochimica et Biophysica Acta 1990; 1032:53-77.
28. Clark AR, Docherty K. Negative regulation of transcription in eukaryotes. Biochemical Journal 1993; 296:521-541.
29. Lillycrop KA, Latchman DS. The inhibitory domain in the Oct-2 transcription factor represses gene activity in a cell type-specific and promoter independent manner. Molecular Biology Reports 1995; 21:87-94.
30. Latchman DS. *Eukaryotic Transcription Factors*, London, San Diego:Academic Press, 1995. Ed. 2
31. Gerrero MR, McEvilly RJ, Turner E et al. Brn 3.0: A POU domain protein expressed in the sensory immune and endocrine systems that functions on elements distinct from known octamer motifs. Proc Natl Acad Sci USA 1993; 90:10841-10845.
32. Theil T, McLean-Hunter S, Zornig M et al. Mouse Brn-3 family of POU transcription factors: A new amino terminal domain is crucial for the oncogenic activity of Brn-3A. Nucl Acids Res 1993; 21:5921-5929.
33. Theil T, Zechner U, Klett C et al. Chromosomal localization and cDNA sequences of the murine Brn-3 family of developmental control genes. Cytogenetics and Cell Genetics 1994; 66:267-271.
34. Turner EE, Jenne KJ, Rosenfeld MG. Brn-3.2: A Brn-3-related transcription factor with distinctive central nervous system expression and regulation by retinoic acid. Neuron 1994; 12:205-218.
35. Ninkina NN, Stevens GEM, Wood JN et al. A novel Brn-3 like POU transcription factor expressed in subsets of rat sensory and spinal cord neurons. Nucl Acids Res 1993; 21:3175-3182.
36. Fedtsova NG, Turner EE. Brn-3.0 expression identifies early post-mitotic CNS neurons and sensory neural precursors. Mechanisms of Development 1996; 53:291-304.
37. Wood JN, Bevan SJ, Coote P et al. Novel cell lines display the properties of nociceptive sensory neurons. Proceedings of the Royal Society of London Series B 1990; 241:187-194.
38. Suburo AM, Wheatley SC, Horn D et al. Intracellular redistribution of neuropeptides and secretory proteins during differentation of neuronal cell lines. Neuroscience 1992; 46:881-889.
39. Budhram-Mahadeo VS, Theil T, Morris PJ, Lillycrop KA, Möröy T, Latchman DS. The DNA target site for the Brn-3 POU family transcription factors can confer responsiveness to cyclic AMP and removal of serum in neuronal cells. Nucl Acids Res 1994; 22:3092-3098.
40. Morris PJ, Theil T, Ring CJA et al. The opposite and antagonistic effects of the closely related POU family transcription factors on the activity of a target promoter are dependent upon differences in the POU domain. Molecular and Cellular Biology 1994; 14:6907-6914.
41. Budhram-Mahadeo VS, Morris PJ, Lakin ND et al. Activation of the α-internexin promoter by the Brn-3a transcription factor is dependent on the N-terminal region of the protein. Journal of Biological Chemistry 1995; 270:2853-2858.
42. Budhram-Mahadeo V, Morris PJ, Lakin ND, Dawson SJ, Latchman DS. The different activities of the two activation domains of the Brn-3a transcription factor are dependent on the context of the binding site. Journal of Biological Chemistry 1996; 271:9108-9113.
43. Kissinger CR, Liu B, Martin-Bianco E et al. Crystal structure of an engrailed homeodomain-DNA complex at 2.8 Å resolution: a framework for understanding homeodomain-DNA interactions. Cell 1990; 63:579-590.
44. Lai J-S, Cleary MA, Herr W. A single amino acid exhange transfers VP-16 induced positive control from the Oct-1 to the Oct-2 POU homeodomain. Genes and Development 1992; 6:2058-2065.
45. Lakin ND, Morris PJ, Theil T et al. Regulation of neurite outgrowth and SNAP-25 gene expression by the Brn-3a transcription factor. Journal of Biological Chemistry 1995; 270:15858-15863.
46. Osen-Sand A, Catsicas M, Staple JK et al. Inhibition of axonal growth by SNAP-25 antisense oligonucleotides in vitro and in vivo. Nature 1993; 364:445-448.

CHAPTER 12

Oncogenic Transcription Factors

Ronald M. Wisdom

INTRODUCTION

Understanding the mechanisms by which normal cells undergo conversion to malignant growth is an important problem in biology. Many different approaches have led to the identification of genes that, when mutated, have the potential to contribute to tumor formation. Biochemically, the proteins encoded by these genes execute diverse molecular functions. A large group of oncogenes encode transcription factors; these oncogenic transcription factors are important in the study of cancer for two reasons. First, many of the oncogenes that are not transcription factors appear to regulate transcription factors, and so there is every reason to believe that what is learned about cancer from the study of oncogenic transcription factors will be generalizable. Second, the cellular and molecular basis of the malignant phenotype is very complex, involving changes in cell morphology, nutritional requirements, cell cycle controls, extracellular matrix interactions, cell-cell interactions, and susceptibility to apoptosis. It is likely that the identification of target genes for oncogenic transcription factors will reveal a great deal about the molecular control of these processes, and will help to explain how they are coordinately regulated in the generation of a tumor. In this chapter, the emphasis is on mechanisms of transformation; for a discussion of the mechanisms of transcriptional control, the reader is referred to other chapters in this book.

IDENTIFICATION OF ONCOGENIC TRANSCRIPTION FACTORS

Oncogenic transcription factors have been brought into view by several different routes. For some, the evidence of their participation in tumor formation is compelling, but the evidence that they are transcriptional regulators is inferred from homology arguments; for others, the evidence that they function as transcriptional regulators is decisive, but their role in tumor formation is less clear. The most common means of identifying oncogenic transcription factors has been by virtue of their presence in the genomes of oncogenic retroviruses, by their activation by retroviral insertion or chromosomal translocation, and by their association with the oncoproteins of DNA tumor viruses.

Many oncogenic transcription factors were first identified by virtue of their presence in the genomes of avian or murine retroviruses. These include v-Fos, v-Jun, v-Myc, v-Rel, v-ErbA,

Transcription Factors in Eukaryotes, edited by Athanasios G. Papavassiliou.

v-Ski, v-Myb, v-Ets, and v-Qin. Most of these molecules were identified before the molecular characterization of transcription factors was detailed enough to permit their identification as transcription factors. For example, the proteins of the AP-1 complex have been identified twice: first, as the products of oncogenic retroviruses (FBJ-MuSV for Fos, ASV-17 for Jun), and later, as components of a transcription factor complex able to bind to DNA sequences that mediate transcriptional induction in response to phorbol esters[1,5,36,39] (see also chapter 5). By definition, all of the transcription factors identified by this approach show dominant genetics; that is, they exert their transforming activity in the presence of the wild type genes.

A second method of identifying oncogenic transcription factors has been by study of the transforming proteins of the DNA tumor viruses. These have most prominently included transformation by SV40 and adenovirus. The large T Ag of SV40 is a multifunctional protein that associates with several cellular proteins, including the transcriptional regulator p53. As with the AP-1 complex, p53 has in fact been identified twice; first as a cellular protein associated with large T Ag, and second as the gene that is mutated in the Li-Fraumeni syndrome, which involves an inherited predisposition to cancer[32,40] (see also chapter 16). As with many transcription factors involved in retrovirally induced tumors, the ability of p53 to function as a transcription factor became apparent long after its identification. A second oncogenic transcription factor identified by the study of DNA tumor viruses is the transcription factor E2F. The regulation of E2F activity is complex, but involves the formation of a multiprotein complex that includes pRb, the product of the retinoblastoma susceptibility gene.[13,19] pRb itself binds to E1A, a transforming gene from adenovirus; the interaction of E1A with pRb disrupts the pRb-E2F complex, resulting in the release of free, active E2F[47,70] (see also chapter 16). pRb also interacts with the large T Ag of SV40, and this interaction activates E2F in a manner similar to that observed in adenovirus infected cells; conversely, the adenovirus E1B protein binds to and inactivates p53.[6,55] Thus, both SV40 and adenovirus target p53 for inactivation and E2F for activation (via inactivation of pRb). This combinatorial action is also displayed by the human papilloma virus variants that are associated with human genital tract cancers.[46,68] The selective targeting of these transcription factors by DNA tumor viruses must reflect underlying regulatory strategies used by the host cells.

A third method of identifying oncogenic transcription factors has come from the study of genetic rearrangements induced by retroviral insertions, chromosomal translocations and gene amplifications. The *c-myc* gene was initially identified as the cellular homolog of *v-myc*, the transforming gene in the MC29 virus. It was later identified as the site of retroviral insertion in bursal lymphomas in chickens induced by ALV (avian leukosis virus), as a site of chromosomal translocation in plasmacytomas of mice and Burkitt's lymphomas in humans, and as an amplified locus in some human tumors[18,34] (see also chapter 13). The *c-myc* gene serves as a paradigm for the activation of an oncogenic transcription factor by chromosomal rearrangement. This model has since been successfully applied to the study of genes that are associated with chromosomal translocations in human leukemias in lymphomas. The characterization of genes that are rearranged in chormosomal translocations includes *scl/tal-1*, *lyl*, the retinoic acid receptor α gene, *rbtn1*, *rbtn2*, *E2A*, *Pbx*, *Tel*, *AML-1*, and many others. These genes are of special interest because of their direct association with specific human tumors.

Finally, a fourth method of identification of oncogenic transcription factors should be mentioned. Many transcription factors belong to gene families, and the identification of family members by a variety of approaches, most prominently low stringency hybridization, has led to the identification of family members with transforming activity. This method is of importance not only for it's ability to identify new family members. The sequence alignments that are possible within gene families can rapidly identify conserved and divergent regions that can be specifically correlated with different biochemical activities.

ONCOGENIC TRANSCRIPTION FACTORS AND INTRACELLULAR SIGNALING

Activation of AP-1 Signaling by Mitogens

Many forms of cell signaling function to couple the detection of a specific environmental stimulus with changes in cell behavior that are governed, at least in part, by changes in transcription. This has motivated attempts to understand proliferative and anti-proliferative stimuli in terms of their effects on individual specific transcription factors.

The mechanisms of transcription factor activation by growth factors are best understood in the case of AP-1. AP-1 activity is regulated by many different extracellular stimuli, but it is the regulation by mitogens, including serum, peptide growth factors, and phorbol esters, that seems especially relevant for the control of the proliferative response. This form of regulation is complex, and occurs at several different levels. First, an important level of control is at the level of the abundance of the various DNA-binding subunits. AP-1 is composed of a mixture of dimeric complexes derived from the Fos and Jun gene families. Many of the individual subunits are strongly, but transiently, induced by mitogenic stimulation. Second, the individual Fos and Jun proteins do not activate transcription identically, making AP-1 subunit composition a determinant of activity. Third, individual subunits are subject to regulation by phosphorylation in response to the activation of specific signaling pathways (see also chapter 5).

Increased production of AP-1 proteins in response to growth factor signaling is best understood in the case of the increased production of c-Fos mRNA and protein in response to serum. Many different components in whole serum appear to be able to mediate induction of the *c-fos* gene, including peptide growth factors that bind to receptors with tyrosine kinase activity, and lysophosphatidic acid (LPA), which signals through a G-protein coupled receptor. Activation of the *c-fos* gene can be carried out through several distinct cis-acting elements, including the cis-inducible element (SIE), the serum response element (SRE), and the cyclic AMP response element (CRE) (see also chapter 5). For many responses in vitro, individual elements appear to be able to mediate activation, but analysis of the *c-fos* promoter in vivo suggests that all of these elements are essential for *c-fos* expression in the animal. The SIE appears to be able to mediate activation by some peptide growth factors, including PDGF and EGF. SIE activity is mediated by binding proteins derived from the STAT family of transcription factors, which contain SH2 domains and are directly activated by tyrosine phosphorylation. Whether phosphorylation of the STAT proteins is mediated directly by the growth factor receptors is not clear. The receptors can directly phosphorylate STAT proteins in vitro; however, the STAT proteins are also clearly activated by nonreceptor protein tyrosine kinases in vivo in response to other stimuli, such as interferons. The role of the CRE, which is also responsive to increases in cytoplasmic calcium, is most clearly documented in the case of c-Fos induction by neuronal depolarization.

The SRE is a promoter element that is responsive to many different stimuli. The SRE is a complex element that binds several different transcription factors, of which the most important are serum response factor (SRF) and the Ets family members Elk-1 and Sap1, which function as ternary complex factors (TCF) (see also chapter 14). Both the TCF and SRF binding sites within the SRE are required for transcriptional activation by mitogens.[65] One mechanism of SRE activation is by the phosphorylation of Elk-1, an Ets family transcription factor, within its C-terminal activation domain.[21,41] In vitro, Elk-1 is phosphorylated within the activation domain by MAP kinase and JNK/SAP kinase; in vivo, it is likely that both kinases are able to mediate SRE activation.[41,69] In the case of peptide growth factors, such as PDGF and EGF, it is likely that MAP kinase activation mediates SRE activation; JNK/SAP kinase is likely to play a role in activation of the c-Fos promoter by TNF and stress-stimuli, such as UV light and heat shock. The activation of Elk-1 and the SRE by MAP kinase is also likely to play a major role in the induction of the c-Fos promoter by many stimuli that activate MAP kinase; these include activation of the Ras GTPases, the raf protein kinase, and the tumor promoting phorbol esters.[15] Although activation of the SRE by other stimuli is not as well understood, there appears to be a method of activating the SRE that is independent of TCF/Elk-1/Sap1 activation. This is best characterized in the case of activation of the SRE in response to LPA, in which case SRF appears to mediate activation in the absence of TCF proteins.[22]

The induction of other AP-1 component proteins by mitogenic signaling pathways is not as well understood. The *FosB* gene is regulated in a manner very similar to the *c-fos* gene, and the *FosB* promoter contains a similar array of cis-acting sites, including the SRE.[23] Therefore, it is very likely that *FosB* regulation will be mediated by many of the same signaling pathways. The other Fos family genes, *Fra-1* and *Fra-2*, are induced by stimuli with different kinetics than *c-fos* and *FosB*, implying that their transcriptional control may be different. The regulation of the Jun family of genes is less well characterized; it is clear, however, that the Jun family of genes are induced at the level of the respective mRNAs by many of the stimuli that activate c-Fos. There is data to suggest that c-Jun regulation in response to serum or phorbol esters is mediated, at least in part, by positive autoregulation.

An additional level of control of AP-1 activity is by posttranslational regulation. This is best documented in the case of c-Jun. The activation domain of c-Jun is phosphorylated by JNK/SAP kinase on serines 63 and 73, which results in a marked increase in activity[2,51,62] (see also chapter 5). The regulation of JNK/SAP kinase is controlled by many stimuli, including activation of the cytokine receptors and several forms of stress, such as inhibition of protein synthesis, UV light, and heat shock.[9,31] All of the AP-1 component proteins are phosphoproteins, and it seems likely that these phosphorylations regulate the activity of the complexes, although this has not been systematically addressed for most of the family members.[8]

The role of AP-1 in mediating the cellular proliferative response to growth factors has guided much of the mechanistic work on the regulation of different AP-1 promoters. There is compelling evidence that AP-1 is an effector of the mitogenic response to peptide growth and the transforming response to Ras proteins.[20,27,28,37,54] However, some AP-1 components, such as c-Fos, are induced by many stimuli that are not mitogenic. For example, UV light, a potent inducer of AP-1 activity, results in cell cycle arrest, not proliferation.[1,56] Furthermore, genetic evidence suggests that AP-1 activation is critical both for cell cycle arrest and survival after UV exposure and for proliferation in response to growth factors[27,28,54,56] (see also chapter 5). The mechanisms by which AP-1 is apparently able to participate in both cell cycle arrest and proliferation under different conditions is not clear, but a resolution to this paradox is certain to be informative with regard to the ways in which information from multiple intracellular signaling pathways is integrated.

AP-1 serves as the best understood example of the investigation of cell signaling pathways by studying the activation of specific transcription factors. The mRNAs for many other oncogenic transcription factors, including c-Rel, c-Myc, and c-Myb are induced by mitogenic stimuli. The regulation of these transcription factors by extracellular changes and the participation of these proteins in deregulated growth implies that they function as effectors in intracellular signaling (see also chapters 9, 13, and 15, respectively).

In summary, the characterization of AP-1 activation has helped to reveal the biochemical and genetic basis of signaling from the cell membrane in response to many different agents. The data support the notion that AP-1 is an effector of many cytoplasmic signaling molecules. The complexity of the signaling is great, and further investigation is likely to reveal even greater complexity. Much of the complexity may be used to allow the integration of multiple stimuli at the level of a single transcription factor. Further study is also likely to reveal a role for other oncogenic transcription factors as effectors of proliferative stimuli.

Activation of p53 by DNA Damage

How cells detect and respond to DNA damage is an interesting and important problem in cancer biology. The p53 tumor suppressor protein is induced in response to DNA damage by many agents, including chemicals that alkylate DNA, ionizing irradiation, UV irradiation, and chemotherapeutic agents. Just as AP-1 has served as an entry point into the study of growth factor induced signaling to the nucleus, p53 activation by DNA damaging agents has become a model system for the response to genetic damage (see also chapter 16).

p53 appears to be able to respond to several different forms of DNA damage, including the formation of DNA adducts, the formation of intra- and inter-strand crosslinks, the formation of broken strands, and the presence of mismatched DNA in the form of bubbles.[24,35] The response to all of these forms of DNA damage would initially seem to suggest that p53 is at the end of a signaling pathway in response to DNA damage. However, recent evidence suggests that p53 is actually able to directly bind, in a sequence nonspecific manner, to DNA lesions. This binding is mediated by the C-terminal domain of p53, and is not competitively inhibited by the sequence specific interaction with DNA that is mediated by the central domain. Furthermore, transcriptional activation by p53 is markedly enhanced by pretreatment with DNA damaging agents, and this is accompanied by an increase in p53 sequence specific DNA binding. The data suggest the existence of an intramolecular signaling pathway from the p53 C-terminal domain (which recognizes DNA damage) to the central domain (which mediates the sequence specific DNA binding that is a prerequisite for transcriptional activation). The details of this possible allosteric interaction are unclear, but structural studies are likely to be informative.

It seems likely that p53 interacts with other intracellular signaling pathways to mediate its activities. In particular, there is evidence that p53 can mediate either apoptosis or growth arrest, depending on the cellular context. The determinants of this decision (apoptosis vs. growth arrest) are not entirely clear; there is evidence to support the idea that activation of growth factor signaling pathways, and the status of other genetic loci, in particular Rb and $p21^{CIP1/WAF1}$, are important in determining the cellular response to p53 activation[3,7] (see also chapter 16).

ONCOGENIC TRANSCRIPTION FACTORS AND CELL CYCLE CONTROL

The malignant phenotype is complex and pleiotropic, but deregulated control of cellular proliferation is a hallmark. Recent progress in understanding of cell cycle control has made it possible to attempt to determine the basis by which oncogenes in general, and oncogenic transcription factors in particular, control cell growth. It is likely that these investigations will reveal the existence of novel cell cycle control mechanisms.

The Cell Cycle Machinery

Cell cycle progression in mammals is regulated by a family of protein kinases composed of regulatory subunits, the cyclins, and catalytic subunits, the cyclin-dependent kinases (cdks). Different kinases govern different cell cycle transitions, presumably by phosphorylating distinct substrates.[66] The prototypic cdk, cdc2, governs the G2 to M transition, while other cyclin-cdk complexes govern other cell cycle transitions. Many oncogenic transcription factors appear to be under the control of peptide growth factors. Several observations suggest that transcriptional mechanisms are likely to be especially important in the control of the G1/S transition. First, the progression of cells through G1, but not other phases of the mammalian cell cycle, is sensitive to the presence of protein synthesis inhibitors.[49] Second, the induction of growth factor responsive transcriptional activators, such as AP-1, occurs within a time window that corresponds to G1, suggesting that this is when these transcription factors are likely to operate. Finally, the cell cycle arrest that occurs in response to growth factor deprivation is a G1 arrest. Together, these observations suggest that many oncogenic transcription factors are likely to exert their proliferative effects by mediating G1 progression. Preliminary observations suggest that this is true, and it is seems likely that one mode of action of many oncogenic transcription factors is to couple changes in the extracellular environment to changes in cell cycle regulatory molecules.

The regulation of cyclin-cdk complexes occurs at many levels.[44] First, catalytic activity is dependent on the presence of the cyclin regulatory subunit. As the abundance of the cyclin regulatory subunits is intricately controlled by both regulated synthesis and regulated proteolysis, this is a major mechanism of control of overall kinase activity. There is a requirement for phosphorylation of a threonine residue near the carboxyl-terminus by the enzyme CAK (cdk activating kinase), itself a cyclin-cdk complex.[12,38] In addition, kinase activity is negatively regulated by phosphorylation of a tyrosine residue near the amino-terminus; activation can occur by decreased kinase activity or increased phosphatase activity. Both the N-terminal negative regulatory site and the C-terminal positive regulatory site are subject to regulation in vivo. Third, several small cdk inhibitor proteins (ckis, such as $p21^{CIP1/WAF-1}$, $p16^{INK4}$, $p27^{KIP1}$, etc.) exist that can bind to and inhibit the activity of otherwise fully active cyclin-cdk complexes. The abundance of these cdk inhibitors, or ckis, is controlled by different stimuli, and they each preferentially inhibit different subsets of cyclin-cdk enzymes. For example, p21 binds to and inactivates many cyclin-cdk complexes, while p16 is a relatively specific inhibitor of cyclin D-cdk4.[16,17,58,72] Cell cycle progression is potentially regulated by transcriptional controls at several different steps:

1) The abundance of different cyclins is regulated throughout the cell cycle, in part at the level of transcription.
2) The kinases and phosphatases that mediate the phosphorylation and dephosphorylation of the cdks are subject to transcriptional control.
3) The abundance of different inhibitor ckis is under transcriptional control.
4) Transcription factors might function as effectors of the cyclin-cdk complexes. There is now evidence to suggest that each of these forms of transcriptional control of the cell cycle is effected by oncogenic transcription factors.

The G1 to S transition occurs by the sequential activation of a series of cdks. A schematic diagram of critical events in cell cycle progression is shown in Figure 12.1. The first change to occur in the cell cycle machinery is the activation of cyclin D-cdk4 in response to growth factor treatment of cells.[42] The D-type cyclins are unusual in that they do not show cell cycle dependent changes, but instead are subject to growth factor regulation, such that they are maximally expressed following growth factor treatment.[42] The cyclin Dcdk4 complex is unusual in one other regard; cyclin D cdk is a poor histone H1 kinase, and an efficient pRb kinase, and it is believed that activation of cyclin D-cdk is responsible for the phosphorylation and inactivation of pRb that occurs in mid to late G1[30,43] (see also chapter 16). This event is important for several reasons. First, the phosphorylation inhibits the interaction of pRb with E2F, and therefore, this phosphorylation is coincident with E2F activation.[4] E2F binds to the promoter regions of several different enzymes involved in DNA replication, including thymidylate synthestase and dihydrofolate reductase, and helps to mediate the transition into S phase.[23,61] Second, classical studies have defined a point in late G1 beyond

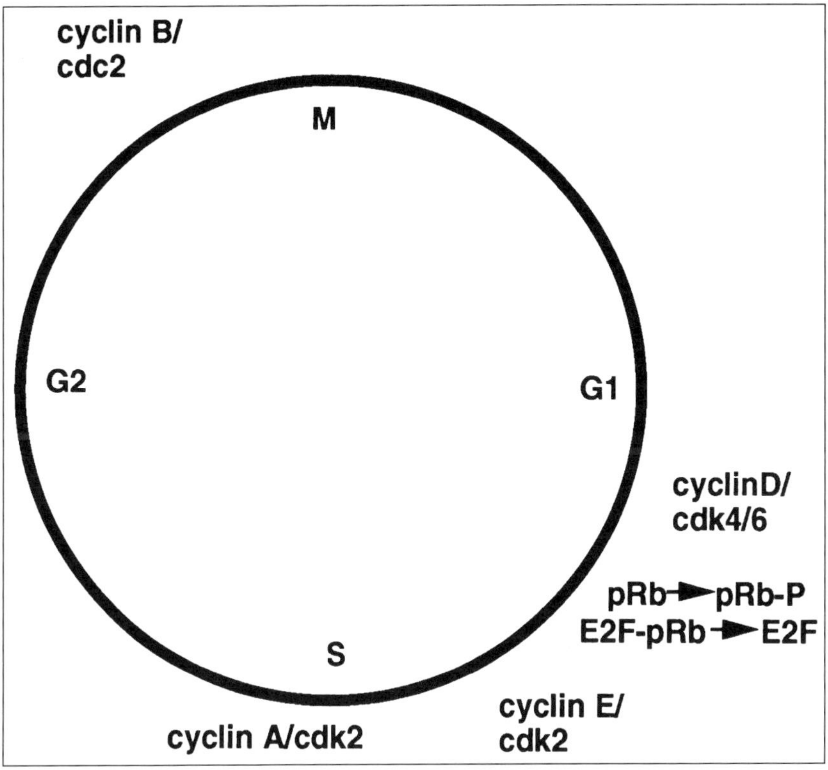

Fig. 12.1. Schematic diagram of critical events in cell cycle progression.

which de novo protein synthesis is not required for continued cell cycle progression, and this point is temporally coincident with pRb phosphorylation.[49] Finally, a number of studies suggest that pRb must be inactivated, either functionally or by mutation, for G1 progression. As growth factor deprivation results in G1 arrest that occurs before pRb hyperphosphorylation, growth factor signaling must, either directly or indirectly, affect the phosphorylation of pRb (see also chapter 16).

Activation of cyclin D-cdk4 is followed by cyclin E-cdk 6 activation, which is then followed by activation of cyclin E-cdk2 and cyclin A-cdk2. The sequential activation not only controls progress through G1 and S but also generates a set of biochemical markers for cell cycle progression. Classical studies show that progression from S through M in mammalian cells does not require new protein synthesis, suggesting that transcriptional induction is less likely to regulate progression through these phases of the cell cycle.

Transcriptional Control of Cyclin Abundance

Cell cycle arrest due to growth factor or serum deprivation occurs in early G1, prior to the phosphorylation and inactivation of pRb. Upon treatment of serum starved cells with peptide growth factors, the expression of many oncogenic transcription factors is induced. The most rapid induction is of the transcription factor AP-1, where changes in the level of c-Fos mRNA can be detected within five minutes. The induction of immediate early genes, many of which are transcription factors, mediates the activation of late genes. Among the late genes induced in response to treatment of cells with growth factors are the D-type cyclins. Cyclin D1 was initially identified as an mRNA induced by the macrophage growth factor CSF-1; its transcriptional induction is dependent on prior protein synthesis, presumably reflecting the requirement for de novo synthesis of a transcription factor.[42] It is an attractive hypothesis that the induction of the D-type cyclins by growth factors is a method of coupling growth factor treatment with cell cycle progression. However, several experiments argue that this model is too simplistic. First, cyclin D mRNA is not induced in some cell types after growth factor treatment. Second, the synthesis of new cyclin D does not result in the immediate appearance of cyclin D-cdk4 kinase activity; instead a lag period follows. The nature of the lag phase is not clear, but may reflect the need to biochemically activate the enzyme. The mechanisms by which cyclin D mRNA is induced in response to growth factor treatment are not clear, but AP-1 may play a role. It should be noted that cyclin D-cdk4 activity is regulated at many points that are not under direct transcriptional control, but the abundance of the D-type cyclins can be limiting under some circumstances.[43,52,53] The cyclin D-cdk complex is responsible for the inactivation of pRb, and the transcriptional control of cyclin D suggests a mechanism whereby transcription might play a role in regulating G1 progression by phosphorylating pRb.

Transcriptional Control of Cyclin-cdk Activity

The expression of *c-myc* mRNA is induced after treatment of quiescent cells with growth factors, and abundant evidence indicates that ectopic expression of c-Myc mRNA can induce G1 progression in some cells. The points in cell cycle progression that are influenced by expression of the c-Myc transcription factor have been analyzed using conditional Myc alleles which are estrogen dependent.[63] The decreased c-Myc expression in density arrested cells is accompanied by decreased cyclin E-cdk2 activity.[63] Expression of c-Myc in this setting activates cyclin E-cdk2 kinase activity, without altering the level of either cyclin E or cdk2, consistent with activation by a post-translational mechanism. The cdc25A gene, which encodes a dual specificity phosphatase capable of activating cyclin-cdk complexes, appears to be a transcriptional target of the c-Myc protein.[14] It is not clear if the cyclin E-cdk2 complex, which is present in an inactive form, is activated by Myc-dependent cdc25A activation, but this provides a model for Myc-induced G1 progression. Likewise, it is not clear if c-Myc activation alters the expression of cdk complexes exclusively by activation of cdc25A, or if other mechanisms are involved. Finally, the relationship of cdc25A activation to activation of the ornithine decarboxylase (ODC) gene, another putative c-Myc target gene, is not clear.

Transcriptional Control of CDK Inhibitors

There is a growing family of small proteins that function as inhibitors of cyclin-cdk complexes. Some of the ckis function as tumor suppressors, and their regulation is critical to normal cell cycle control. The p21 protein is an inhibitor of many cyclin-cdk complexes, and is able to mediate the cell cycle arrest that is induced by several forms of DNA damage.[10,17,72] The p21 gene is subject to transcriptional control by the p53 tumor suppressor protein, such that p21 induction by DNA damage is p53 dependent.[10]

The $p16^{INK4A}$ protein is a second cyclin-cdk inhibitor. Unlike p21, which inhibits many different types of cyclin-cdk complexes, p16 specifically targets complexes containing cdk4, which phosphorylates pRb.[58] A form of feedback regulation was suggested by the marked increase in abundance of p16 mRNA and protein in tumors that have suffered mutations in pRb. This feedback loop is due to the constitutive E2F activation that occurs in the presence of Rb mutations; consistent with this interpretation, functional E2F binding sites have been identified in the promoter of the p16 gene.

The family of ckis includes several other members, including the p27 protein, which is believed to play a role in the density-dependent G1 arrest, the cell cycle arrest that occurs in response to serum starvation, and the anti-proliferative effects of the TGF-β family of proteins.[50,64] p27 expression, and the expression of other ckis, is affected by many different stimuli that regulate cell cycle progression, but the molecular basis of this regulation is not understood in detail. However, the expression of these proteins is likely to be regulated at the level of transcription.

Transcription Factor Regulation by cdk Activity

The cdks regulate cell cycle progression by phosphorylating a series of substrates, which then execute the events required for cell cycle progression. The phosphorylation and inactivation of the pRb protein, which is mediated by cyclin D-cdk4 complexes, results in the release of active E2F; in this manner, E2F activity is regulated by cyclin D-cdk4[13,19,43] (see also chapter 16). E2F is a sequence specific DNA-binding activity composed of heterodimeric complexes of proteins derived from the E2F and DP family of genes, which encode a series of bZIP transcription factors. Originally identified on the basis of it's role in transcriptional control of the adenovirus life cycle, E2F is believed be a transcription factor that directly regulates the G1 to S phase transition.[47] Putative E2F target genes include many genes expressed selectively during S phase, including the genes that encode dihydrofolate reductase, thymidylate synthetase, and DNA polymerase α.[23,47,61] In addition, E2F activates the p16 gene in a manner dependent on prior phosphorylation of pRb. The evidence that E2F is oncogenic is conflicting; over-expression of E2F1 induces abnormal cell growth in cell culture, but targeted disruption of the E2F1 gene suggests it may function as a tumor suppressor in the animal.[11,25,60,73] Reconciliation of these apparently discrepant results awaits a more complete analysis of the E2F and DP family proteins, both in the animal and in cell culture. Nonetheless, the data strongly suggest that E2F is one effector of cyclin D-cdk4 activity.

The results outlined above demonstrate that many of the molecular events of cell cycle control are regulated at the level of transcription. In many cases, the relevant transcription factors have previously been identified by their oncogenic potential. Furthermore, there are hints in the literature of additional potential sites of transcriptional control of cell cycle. For example, the *S. pombe* cdc5 gene, which is essential for the G2/M transition, encodes a transcription factor related to the *c-myb* protein, demonstrating a crucial role for transcriptional control of the G2/M transition. Transcriptional control of the cyclin and cki genes is poorly understood, and it seems likely that further investigation into the regulation of these critical cell cycle regulators will reveal additional mechanisms by which the deregulation of these transcription factors results in cancer. Conversely, the continued characterization of oncogenic transcription factors is likely to reveal novel regulatory controls of the cell cycle.

CONTROL OF DEVELOPMENT AND DIFFERENTIATION BY ONCOGENIC TRANSCRIPTION FACTORS

The correct development of specialized mammalian tissues is dependent on the execution of a coordinated program, and many steps in the developmental program are regulated at the level of transcription. A number of developmental decisions are under the control of oncogenic transcription factors. In addition, the regulated program of differentiation of different specialized cell types is also regulated by transcriptional control. There is now good evidence that aberrant transcriptional control of differentiation and development can serve as a cellular mechanism contributing to malignancy. The role of oncogenic transcription factors in contributing to malignancy by altering programs of differentiation and development are most clearly demonstrated in the case of normal hematopoiesis and leukemia, its malignant counterpart.

Normal Hematopoiesis

Normal blood is formed from self-renewing hematopoietic stem cells, which generate many specialized cell types by

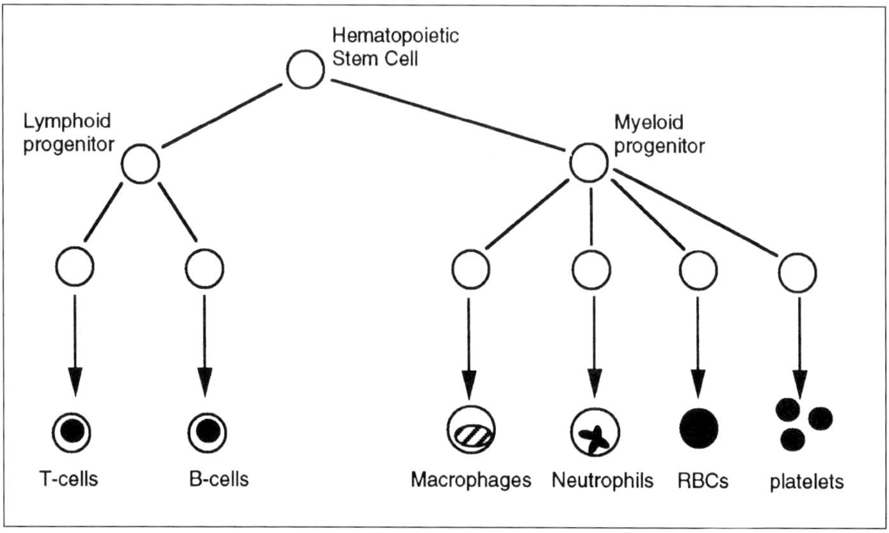

Fig. 12.2. Generation of specialized cell types by differentiation of hematopoietic stem cells.

differentiation along a hierarchical program (Fig. 12.2). In the adult organism, hematopoiesis is normally confined to the bone marrow and spleen, but during development several different sites function as blood-forming tissues, including the yolk sac and the liver. The analysis of different knockout strains of mice has revealed the roles of some oncogenic transcription factors in orchestrating this complex developmental process. The ways in which hematopoiesis is impaired in these mutant animals is described below.

Control of Hematopoiesis by Transcription Factors

The scl/tal1 gene, initially isolated on the basis of its involvement in chromosomal translocations in acute lymphoblastic leukemia (ALL), encodes a basic-helix-loop-helix (bHLH) transcription factor. The lim domain-containing protein rbtn2 interacts physically with scl/tal1 in vivo, generating a protein complex that functions in transcriptional activation. scl/tal1 is most abundantly expressed in erythroid cells within the bone marrow, but is also expressed to a lower level in other hematopoietic cells and in the developing vasculature. Mice deficient in either scl/tal1 or rbtn2 exhibit severe hematopoietic defects, with no evidence of primitive yolk-sac derived erythropoiesis.[59,67] In addition, embryonic stem cells homozygous for disruption of the scl/tal1 gene are unable to differentiate to erythroid cells in vitro, and yolk sac cells from rbtn2 mutant embryos form macrophages, but not erythroid cells. The correspondence of the phenotypes between scl/tal1 and rbtn2 deficient mice implies that both components of this protein complex are required to generate an active transcription factor complex. This result is also intriguing in view of recent data showing that in some cases of ALL there are simultaneous mutations in the scl/tal1 gene and the rbtn2 gene. scl/tal1 and rbtn2 are absolutely required for erythropoiesis.

Several other targeted disruptions in transcription factors identified on the basis of their oncogenic potential have resulted in severe defects in hematopoiesis, but with the development of some hematopoiesis in the developing embryo. The *c-myb* gene was initially identified as the cellular homolog of *v-myb*, the transforming gene of avian myeloblastosis virus (AMV). c-Myb expression in the animal is widespread, but highest levels are observed in hematopoietic cells.

Mice deficient for *c-myb* die in utero of severe anemia; histologic analysis reveals that normal yolk-sac derived hematopoiesis occurs, but there is a severe defect in fetal liver hematopoiesis.[45] In addition, the erythroid lineage is most severely affected, and megakaryocyte development is normal (see also chapter 15).

In contrast to the strong erythroid phenotype of *c-myb* deficient mice, PU.1 deficient mice are most severely affected in the development of the myeloid lineage. The PU.1 transcription factor is an Ets family member that is expressed primarily in myeloid cells during development (see also chapter 14). Mice deficient for PU.1 show severely impaired myeloid development; in addition, there are prominent defects in erythropoiesis and megakaryocytic development, and lymphoid development is also impaired.[57]

The AML1 gene was initially isolated on the basis of its involvement in a chromosomal translocation present in a large fraction of cases of acute myelogenous leukemia (AML). The AML1 protein is a mammalian homolog of the *Drosophila* runt protein. AML1 forms a DNA-binding complex with the protein product of the CBF-β gene, which is also involved in a large fraction of cases of AML. Although the AML1 gene is widely expressed in the animal, targeted disruption of AML1 in mice results in a selective defect in hematopoiesis.[48] This defect is not complete, and AML1$^{-/-}$ ES cells can differentiate into hematopoietic cells in vitro, but the defect is severe enough to result in embryonic death.[48] In this regard, it is interesting to note that transcriptional control by AML1-CBF-β complexes is implicated in the expression of many hematopoietic growth factors, including IL-3 and GM-CSF. The phenotype of CBF-β null animals is not yet known, but one might expect a defect in hematopoiesis similar to the defect observed in AML1 null mice.

The cases described above have demonstrated the role of oncogenic transcription factors in the development of multiple hematopoietic cell types. A more restricted role is shown by analysis of the *c-fos* knockout animal. The *c-fos* gene is expressed at very high levels in macrophages in the animal; in addition, it is expressed in osteoclasts, bone resorbing cells that are derived from macrophages. The only hematopoietic abnormality in *c-fos* mutant mice is a defect in osteoclast formation, which results in thickened bones; otherwise, the formation of blood forming cells appears normal.[26] This represents, then, an example of an extremely restricted developmental defect in hematopoiesis.

Several other oncogenic transcription factors are implicated in lymphoid development. For example, the products of the E2A and *c-rel* genes are both required for normal B lymphocyte development. In addition, several other transcription factors are known to be required for normal hematopoiesis (for example the GATA1, GATA2, and GATA3 proteins), but are not implicated in the development of malignancy. Based on the precedents described above, it seems possible that some of these factors may participate in tumor formation. Likewise, several oncogenic transcription factors have been suggested to play a role in normal hematopoiesis, but decisive evidence for their involvement has been missing because early embryonic death for other reasons precludes analysis of hematopoiesis in vivo.

This section has focused on the role of transcription factors in orchestrating events in the development and differentiation of hematopoietic cells; by implication, the oncogenic forms of these transcription factors contribute to leukemia by deregulating developmental decisions during hematopoiesis. This paradigm is not restricted to the hematopoietic system; there is good evidence that abnormal transcriptional control of other developmental processes can participate in the generation of other tumor types. For example, the Pax-3 transcription factor participates in dermomyotome formation during development and is implicated in the formation of rhabdomyosarcomas.

SUMMARY

Oncogenic transcription factors contribute to malignancy by several different cellular mechanisms. Transcriptional control of intracellular signaling pathways, of the cell cycle, and of cell differentiation and developmental programs is likely to underlie many forms of cancer. The investigation of oncogenic transcription factors has not only increased our understanding of cancer; it has also contributed to our understanding of basic biologic processes, such as intracellular signaling, the response to external stimuli, control of cell growth, and the development of complex organisms. There is every reason to believe that this intersection of biology and disease will continue to be informative for the understanding of each.

References

1. Angel P, Imagawa M, Chiu R, Stein B, Imbra RJ, Rahmsdorf HJ, Jonat C, Herrlich P, Karin M. Phorbol-ester inducible genes contain a common cis element recognized by a TPA-modulated trans-acting factor. Cell 1987l; 49: 729-739.
2. Binetruy B, Smeal T, Karin M. Ha-ras augments c-jun activity and stimulates phosphorylation of its transactivation domain. Nature 1991; 351:122-127.
3. Brugarolis J, Chandrasekaran C, Gordon JI, Beach D, Jacks T, Hannon GJ. Radiation-induced cell cycle arrest compromised by p21 deficiency. Nature 1995; 377:552-557.
4. Chellappan SP, Hiebert S, Mudryj M, Horowitz JM, Nevins JR. The E2F transcription factor is a cellular target for the RB protein. Cell 1991; 65:1053-1061.
5. Curran T, Peters G, Van Beveran C, Teich N, Verma IM. FBJ-murine osteosarcoma virus: Identification and molecular cloning of biologically active proviral DNA. J Virol 1982; 44:674-682.
6. DeCaprio JA, Ludlow JW, Figge J, Shew J-Y, Huang C-M, Lee W-H, Marsilio E, Paucha E, Livingston DM. SV40 large tumor antigen forms a specific complex with the product of the retinoblastoma susceptibility gene. Cell 1988; 54:275-283.
7. Deng C, Zhang P, Harper JW, Elledge SJ, Leder P. Mice lacking p21CIP1/WAF1 undergo normal development, but are defective in G1 checkpoint control. Cell 1995; 82:675-684.
8. Deng T, Karin M. c-Fos transcriptional activity stimulated by H-ras activated protein kinase distinct from JNK and ERK. Nature 1994; 371:171-175.
9. Derijard B, Hibi M, Wu I, Barrett T, Su B, Deng T, Karin M, Davis RJ. JNK-1 : A protein kinase stimulated by UV light and Ha-ras that binds and phosphorylates the c-Jun activation domain. Cell 1994; 76:1025-1037.
10. El-Deiry W, Tokino T, Velculescu V, et al. WAF1, a Potential mediator of p53 tumor suppression. Cell 75:817-825.
11. Field SJ, Tsai F-Y, Kuo F, Zubiaga AM, Kaelin WG Jr, Livingston DM, Orkin SH, Greenburg ME. E2F-1 functions in mice to promote apoptosis and suppress proliferation. Cell 85:549-561.
12. Fisher RP, Morgan DO. A novel cyclin associates with MO15/CDK7 to form the CDK-activating kinase. Cell 1994; 78: 713-724.
13. Flemington EK, Speck SH, Kaelin WG. E2F-1-mediated transactivation is inhibited by complex formation with the retinoblastoma susceptibility gene product. Proc Natl Acad Sci 1993; 90:6914-6918.
14. Galaktionov K, Chen X, Beach D. Cdc25 cell-cycle phosphatase as a target of c-myc. Nature 1996; 382:511-517.
15. Graham R, Gilman M. Distinct protein targets for signals acting at the c-fos serum response element. Science 1991; 251:189-192
16. Gu Y, Turck CW, Morgan DO. Inhibition of CDK2 activity in vivo by an associated 20K regulatory subunit. Nature 1993; 366:707-710.
17. Harper JW, Adami GR, Wei N, Keyomarsi K, Elledge SJ. The p21 cdk-interacting protein Cip1 is a potent inhibitor of G1 cyclin-dependent kinases. Cell 1993; 75:805-816.
18. Hayward WS, Neel BG, Astrin SM. Activation of a cellular oncogene by promoter insertion in ALV-induced lymphoid leukosis. Nature 1981; 290: 475-480.
19. Helin K, Harlow E, Fattaey A. Inhibition of E2F-1 transactivation by direct binding of the retinoblastoma protein. Mol. Cell Biol 1993; 13:6501-6508.

20. Hilberg F, Aguzzi A, Howells N, Wagner EF. c-jun is essential for normal mouse development and hepatogenesis. Nature 1993; 365, 179-181.
21. Hill CS, Marais R, John S, Wynne J, Dalton S, Treisman R. Functional analysis of a growth factor-responsive transcription factor complex. Cell 1993; 73: 395-406.
22. Hill CS, Wynne J, Treisman R. The Rho family GTPases RhoA, Rac1, and CDC42Hs regulate transcriptional activation by SRF. Cell 1995; 81:1159-1170.
23. Hsiao D, McMahon SL, Farnham PJ. Multiple DNA elements are required for the growth regulation of the mouse E2F1 promoter. Genes and Dev 1994; 8: 1526-1537.
24. Jayaraman L, Prives C. Activation of p53 sequence-specific DNA binding by short single strands of DNA requires the p53 C-terminus. Cell 1995; 81:1021-1029.
25. Johnson DG, Cress WD, Jakoi L, Nevins JR. Oncogenic capacity of the E2F-1 gene. Proc Natl Acad Sci USA 1994; 91:12823-12827.
26. Johnson RS, Spiegelman BM, Papaioannou V. Pleiotropic effects of a null mutation in the c-fos proto-oncogene. Cell 1992; 71:577-586.
27. Johnson RS, van Lingen B, Papaioannou VE, Spiegelman BM. A null mutation at the c-jun locus causes embryonic lethality and retarded cell growth in culture. Genes Dev 1993; 7:1309-1317.
28. Johnson R, Spiegelman B, Hanahan D, Wisdom R. Cellular transformation and malignancy induced by ras require c-jun expression. Mol Cell Biol 1996; 16: 4504-4511.
29. Kastan MB, Onyekwere O, Sidransky D, Vogelstein B, Craig RW. Participation of p53 protein in the cellular response to DNA damage. Cancer Res 1991; 51: 6304-6311.
30. Kato J, Matsushime H, Hiebert SW, Ewen ME, Sherr CJ. Direct binding of cyclin D to the retinoblastoma gene product (pRb) and pRb phosphorylation by the cyclin D-dependent kinase, CDK4. Genes and Dev 1993; 7, 331-342.
31. Kyriakis JM, Banerjee P, Nikolakaki E, Dai T, Rubie E, Ahmad M, Avruch J, Woodgett JR. The stress-activated kinase subfamily of c-Jun kinases. Nature 1994; 369: 156-160.
32. Lane DP, Crawford LV. T-antigen is bound to host proteins in SV40-transformed cells. Nature 1979; 278, 261-263.
33. Lazo PS, Dorfman K, Noguchi T, Mattei MG, Bravo R. Structure and mapping of the FosB gene: FosB downregulates the FosB promoter. Nucl Acids Res 1992; 20:343-350.
34. Leder P, Battey J, Lenoir G, Moulding C, Murphy W, Potter H, Stewart T, Taub R. Translocations among antibody genes in human cancer. Science 1983; 222: 765-770.
35. Lee S, Elenbaas B, Levine A, Griffith J. p53 and its 14 kDa C-terminal domain recognize primary DNA damage in the form of insertion/deletion mismatches. Cell 1995; 81:1013-1020.
36. Lee W, Mitchell P, Tjian R. Purified transcription factor AP-1 interacts with TPA-inducible enhancer elements. Cell 1987b; 49:741-752.
37. Lloyd A, Yancheva N, Wasylyk B. Transformation suppressor activity of a Jun transcription factor lacking its activation domain. Nature 1991; 352, 635-638.
38. Makela T, Tassan JP, Nigg EA, Frutiger S, Hughes GJ, Weinberg RA. A cyclin associated with the CDK-activating kinase MO15. Nature 1994; 371:254-257.
39. Maki Y, Bos T, Davis C, Starbuck M, Vogt P. Avian sarcoma virus 17 carries the jun oncogene. Proc Natl Acad Sci 1987; 84:2848-2852.
40. Malkin D, Li FP, Strong LC, Fraumeni JF Jr, Nelson CE, Kim DH, Kassel J, Gryka MA, Bischoff FZ, Tainsky MA, Friend SH. Germ line p53 mutations in a familial syndrome of breast cancer, sarcomas, and other neoplasms. Science 1990; 250:1233-1238.
41. Marais R, Wynne J, Treisman R. The SRF accessory protein Elk-1 contains a growth factor regulated transcriptional activation domain. Cell 1993; 73: 381-393.
42. Matsushime H, Roussel MF, Ashmun RA, Sherr CJ. Colony-stimulating factor 1 regulates novel cyclins during the G1 phase of the cell cycle. Cell 1991; 65, 701-713.

43. Matsushime H, Ewen ME, Strom DK, Kato J-Y, Hanks SK, Roussel MF, Sherr CJ. D-type cyclin-dependent kinase activity in mammalian cells. Mol Cell Biol 1994; 14:2066-2076.
44. Morgan DO. Principles of CDK regulation. Nature 1995; 374:131-134.
45. Mucenski ML, McLain K, Kier AB, Swerdlow SH, Schreiner CM, Miller TA, Pietryga DW, Scott WJ, Potter S. A functional c-myb gene is required for normal fetal hepatic hematopoiesis. Cell 1991; 65:677-689.
46. Munger K, Werness BA, Dyson N, Phelps WC, Harlow E, Howley PM. Complex formation of human papillomavirus E7 proteins with the retinoblastoma tumor suppressor gene product. EMBO J 1989; 8:4099-4105.
47. Nevins JR. E2F: a link between the Rb tumor suppressor protein and viral oncoproteins. Science 1992; 259, 216-219.
48. Okuda T, van Deursen J, Hiebert SW, Grosveld G, Downing JR. AML-1, the target of multiple chromosomal translocations in human leukemia, is essential for normal fetal liver hematopoiesis. Cell 1996; 84:321-330.
49. Pardee AB. G1 events and regulation of cell proliferation. Science 1989; 246: 603-608.
50. Polyak K, Kato J-Y, Solomon MJ, Sherr CJ, Massague J, Roberts JM, Koff A. $p27^{Kip1}$, acyclin-dependent kinase inhibitor, links transforming growth factor-β and contact inhibition to cell cycle arrest. Genes Dev 8:9-22
51. Pulverer BJ, Kyriakis JM, Avruch J, Nikolakaki E, Woodgett JR. Phosphorylation of c-jun mediated by MAP kinases. Nature 1991; 353:670-674.
52. Quelle DE, Ashmun RA, Shurtleff SA, Kato J-Y, Bar-Sagi D, Roussel MF, Sherr CJ. Overexpression of mouse D-type cyclins accelerates G1 phase in rodent fibroblasts. Genes and Dev. 1993; 7: 1559-1571.
53. Resnitzky D, Gossen M, Bujard H, Reed SI. Acceleration of the G1/S phase transition by expression of cyclins D1 and E with an inducible system. Mol Cell Biol 1994; 14:1669-1679.
54. Saez E, Rutberg SE, Mueller E, Oppenheim H, Smoluk J, Yuspa SH, Spiegelman BM. c-fos is required for malignant progression of skin tumors. Cell 1995; 82, 721-732.
55. Sarnow P, Ho YS, Williams J, Levine AJ. Adenovirus E1b-58kd tumor antigen and SV40 large tumor antigen are physically associated with the same 54kd cellular protein in transformed cells. Cell 1982; 28:387-394.
56. Schreiber M, Baumann B, Cotten M, Angel P, Wagner E. Fos is an essential component of the mammalian UV response. EMBO J 1995; 14:5338-5349.
57. Scott EW, Simon MC, Anastasi J, Singh H. Requirement of transcription factor PU.1 in the development of multiple hematopoietic lineages. Science 1994; 265: 1573-1577.
58. Serrano M, Hannon G, Beach D. A new regulatory motif in cell-cycle control causing specific inhibition of cyclin D/CDK4. Nature 1993; 366:704-707.
59. Shivdasani RA, Mayer EL, Orkin SH. Absence of blood formation in mice lacking the T-cell leukemia oncoprotein tal-1/scl. Nature 1995; 373:432-434.
60. Singh P, Wong SH, Hong W. Over-expression of E2F-1 in rat embryos fibroblasts leads to neoplastic transformation. EMBO J 1994; 13:3329-3338.
61. Slansky JE, Li Y, Kaelin WG, Farnham PJ. A protein synthesis-dependent increase in E2F1 mRNA correlates with growth regulation of the dihydrofolate reductase promoter. Mol Cell Biol 1993; 13:1610-1618.
62. Smeal T, Binetruy B, Mercola D, Brover BA, Heidecker G, Rapp UR, Karin M. Oncogenic and transcriptional cooperation with Ha-ras requires phosphorylation of c-jun on serines 63 and 73. Nature 1991; 354: 494-496.
63. Steiner P, Phillip A, Lukas J, Godden-Kent D, Pagano M, Mittnacht S, Bartek J, Eilers M. Identification of a Myc-dependent step during the formation of active G1 cyclin-cdk complexes. EMBO J 1995; 14:4814-4826.
64. Toyoshima H, Hunter T. p27, a novel inhibitor of G1 cyclin-cdk protein kinase activity, is related to p21. Cell 1994; 78:67-74.

65. Treisman R. Ternary complex factors: Growth factor regulated transcriptional activators. Curr Opin Genet Dev 1994; 4:96-101.
66. van den Heuvel S, Harlow E. Distinct roles for cyclin-dependent kinases in cell cycle control. Science 1993; 262: 2050-2054.
67. Warren AJ, Colledge WH, Carlton MBL, Evans MJ, Smith AJH, Rabbits TH. The oncogenic cysteine-rich LIM domain protein Rbtn2 is essential for erythroid development. Cell 1994; 78:45-57.
68. Werness BA, Levine AJ, Howley PM. Association of human papillomavirus types 16 and 18 E6 proteins with p53. Science 1990; 248:76-79.
69. Whitmarsh AJ, Shore P, Sharrocks AD, Davis RJ. Integration of MAP kinase signal transduction pathways at the serum response elements. Science 1995; 269: 403-407.
70. Whyte P, Buchkovich K, Hoprowitz JM, Friend SH, Raybuck M, Weinberg RA, Harlow E. Association between and oncogene and an anti-oncogene: the adenovirus E1A proteins bind to the retinoblastoma gene product. Nature 1988; 334:124-129.
71. omitted in proof.
72. Xiong Y, Hannon GJ, Zhang H, Casso D, Kobayashi R, Beach D. p21 is a universal inhibitor of cyclin kinases. Nature 1993; 366:701-704.
73. Yamasaki L, Jacks T, Bronson R, Goillot E, Harlow E, Dyson N. Tumor induction and tissue atrophy in mice lacking E2F-1. Cell 1996; 85:537-548.

CHAPTER 13

Transcriptional Regulation by the Myc-Max-Mad Network

Bernhard Lüscher, Matthias Austen, Anette Sommer, Stefanie Hilfenhaus and Marie Henriksson

INTRODUCTION

The Myc network of proteins plays a pivotal role in the regulation of cell growth. This network consists of the Myc proteins, including c-Myc, N-Myc, and L-Myc, of Max and of the Mad proteins, including Mad1, Mxi1 (also referred to as Mad2), Mad3, and Mad4. The members of this network function as heterodimers. Max is the common dimerization partner for all Myc and Mad proteins analyzed to date and is the only member known to homodimerize. In addition to these proteins several factors have been identified which associate with one or more individual members within this network (for a review see ref. 1).

The analyses of chicken retroviruses carrying *myc* sequences and subsequently of tumor cells of both human and animal origin have provided strong evidence that Myc proteins play important roles in the development of many different types of neoplasias. These findings have been supported experimentally by showing that Myc has tumorigenic potential in tissue culture as well as in animal models (for a review see refs. 1, 2). In this respect the cotransformation assay in primary rat embryo fibroblasts provides a useful model. These primary rat cells can be transformed by c-Myc and with reduced potency by N- and L-Myc in cooperation with activated Ras. Such studies have permitted the definition of two regions in c-Myc essential for transformation, the transactivation domain at the N-terminus and the DNA-binding/dimerization domains at the C-terminus of the protein (Fig. 13.1).[3] Furthermore c-Myc and to a lesser extent N-Myc and L-Myc were found to facilitate tumor development in transgenic mouse models.[4-8] These *myc* transgenic mice have been used for the identification of a number of novel cooperating oncogenes.[9,10]

From the studies described above it has become apparent that constitutive (over-) expression of Myc is not sufficient for neoplastic growth but additional genetic events are required. The primary effect of constitutive Myc expression in vivo seems to be the inhibition of differentiation. This is probably due to a reduced ability of the cells to exit the cell

Transcription Factors in Eukaryotes, edited by Athanasios G. Papavassiliou.

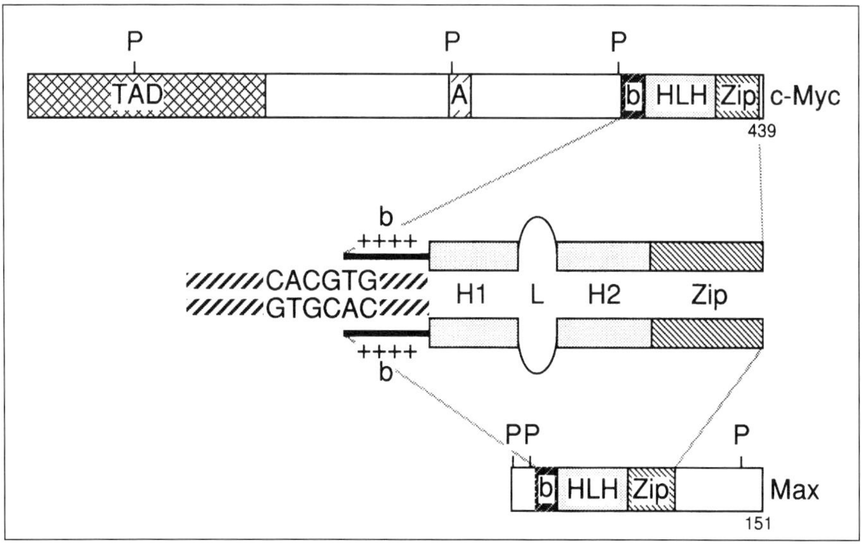

Fig. 13.1. Schematic representation of a Myc/Max heterodimer. The structural elements in c-Myc and in Max are indicated: transactivation domain (TAD), acidic region (A), basic region (b), helix-loop-helix domain (HLH), and leucine zipper (ZIP). P indicates the location of mapped phosphorylation sites (for details see Fig. 13.4). The bHLH/ZIP regions are enlarged to indicate that the basic region and helix 1 as well as helix 2 and the leucine zipper form two α-helices connected by the loop region.[38] The interaction of these four α-helices allows the precise positioning of the basic region to interact with the E-box sequence 5'-CACGTG. E-boxes have been defined as sequences with the core 5'-CANNGT.[107]

cycle and therefore undifferentiated cells accumulate. It has been hypothesized that this enlarged pool of precursor/undifferentiated cells increases the probability that some cells will acquire additional mutations, thereby leading in a stepwise process to malignant transformation.[11,12] The reduced efficiency to differentiate upon deregulated Myc expression has also been observed in tissue culture.[13-17]

In addition to the described effects on growth and differentiation it has become apparent in the last few years that Myc can also induce apoptosis (for a review see refs. 18, 19). In tissue culture this is particularly obvious in situations where Myc is overexpressed and serum has been withdrawn. The analysis of serum components revealed the presence of survival factors which have potent apoptosis inhibitory functions but not necessarily growth stimulating activities.[20] Thus separable signals may modulate Myc function regarding the control of growth and apoptosis.

As has been shown for transformation, the Myc regions relevant for both inhibition of differentiation and induction of apoptosis comprise the transactivation and the DNA-binding/dimerization domains, the hallmarks of transcription factors (for a review see ref. 1). Myc can function as a transcriptional regulator and, according to the prevailing view, Myc influences the biological processes described above by regulating gene transcription. Several potential Myc target genes have been identified (see below) but their products fall short of explaining the effects of Myc on cell growth control. Thus the genes whose products are responsible for mediating Myc function have yet to be identified.

In contrast to Myc, its dimerization partner Max is not able to transactivate.[21] In addition Mad proteins have been suggested to repress gene transcription.[22,23] Thus Myc and Mad proteins may functionally antagonize each other. In this respect it is worthwhile to consider the expression pattern of the different members of the Myc/Max/Mad network. Whereas the expression of Myc proteins is mainly confined to cycling cells,[24-27] Mad proteins are predominantly expressed in differentiating, noncycling cells.[22,23,28,29] In contrast to the very precise regulation of the expression of Myc and Mad proteins, Max expression is rather constitutive.[30,31] Together the different expression patterns of Myc network proteins result in an array of combinations of various dimeric complexes which ultimately may define the expression of target genes and therefore influence cell fate.

In this review we summarize the current knowledge of the role of the Myc network in the regulation of gene transcription. We will refer to c-, N-, and L-Myc and to Mad1, Mxi1, Mad3, and Mad4 as Myc and Mad, respectively, when general statements can be made regarding the function and the properties of these two protein families. It should be kept in mind, however, that for the Myc family the most thorough analyses have been performed with c-Myc. The assumption that N- and L-Myc have similar functional properties compared to c-Myc is in many aspects certainly a simplification. A limited number of studies have been published on the Mad proteins to date which on first sight suggest that the four members are functionally related. Future work will most likely reveal differences, not least because of the expression patterns observed for Mad family members.

MYC IS A SEQUENCE SPECIFIC DNA-BINDING PROTEIN

With the realization that the basic DNA-binding region (b), the helix-loop-helix (HLH) and the leucine zipper (ZIP) motifs in Myc proteins are related to structural elements found in a diverse group of transcription factors (for a review see ref. 32), the assumption that Myc may encode a transcriptional regulator was tested experimentally. Several studies have demonstrated that the basic region contacts DNA whereas the HLH/ZIP domain mediates protein dimerization (Fig. 13.1). Myc/Max, Max/Max, and Mad/Max protein complexes have been identified in vivo and shown to bind the E-box consensus sequence 5'-CACGTG (for a review see ref. 1). This information has then been used to study the molecular details of DNA binding (for a review see ref. 33).

The dimerization of proteins of the Myc/Max/Mad network involving the HLH/ZIP domains has been proposed, based on model building and on mutational analysis, to generate a novel protein folding motif, the parallel, four-helix bundle.[34-37] The solution of the three-dimensional structure of a Max/Max-cognate DNA cocrystal confirmed the structural predictions made previously.[38] The crystal structure revealed that the basic region and helix 1 form a contiguous α-helix as does helix 2 together with the leucine zipper. The two α-helices are connected by the loop. In addition the structure unveiled three amino acids in the basic region which make direct contacts to specific bases within the E-box core sequence.[38] All three amino acids are conserved among the Myc/Max/Mad family proteins identified to date, including the newly discovered Mad3 and Mad4 (Fig. 13.2, residues 5, 9, 13). His-28 in Max9 (Fig. 13.2, residue 5), which is encoded by one of the alternatively spliced max mRNAs,[39] contacts a G on the lower strand (Fig. 13.2, base 6') whereas Glu-32 (Fig. 13.2, residue 9) contacts a C and the A on the upper strand (Fig. 13.2, bases 1, 2). Therefore these two amino acids, His-28 and Glu-32, in the context of the bHLH domain determine that Max homodimers, and in analogy also the other dimers of the Myc/Max/Mad family, bind to E-box sequences.

The last residue of the basic region (Arg-36 in Max9; Fig. 13.2, residue 13) is of particular interest. This amino acid contacts the G of the core CG (Fig. 13.2, base 4') of the Myc recognition site and is therefore of crucial importance for discriminating 5'-CA*CG*TG against other E-box sequences, in particular 5'-CA*GC*TG. The former sequence is recognized by a group of bHLH

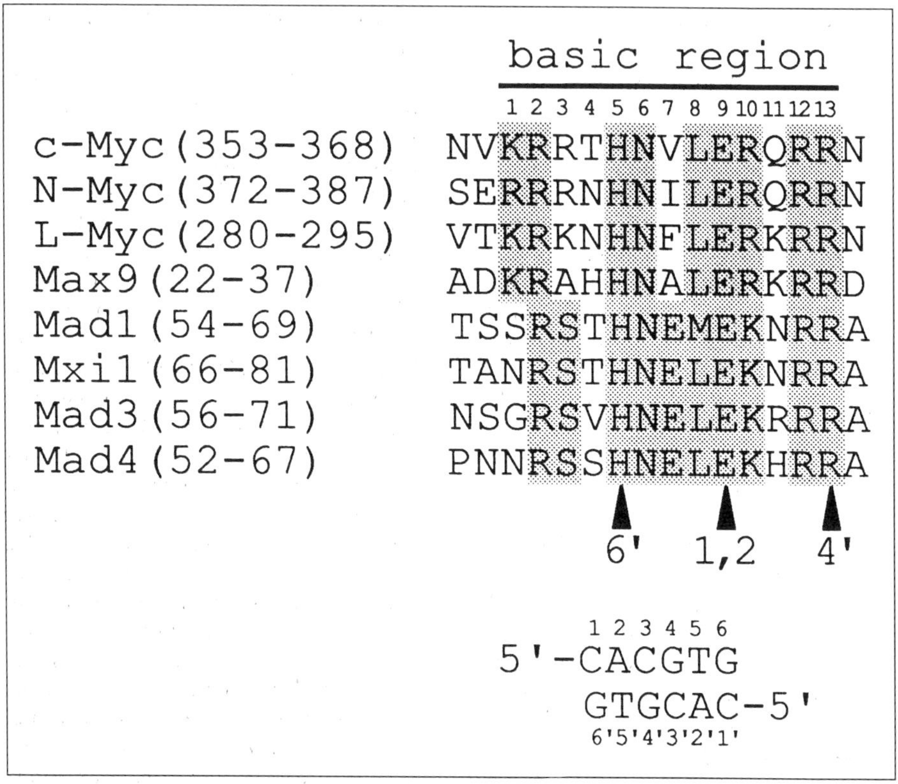

Fig. 13.2. The basic region of the members of the Myc-Max-Mad family. The basic regions of the indicated proteins are displayed. Identical/conserved amino acids within the basic region are highlighted. Amino acids 5, 9, and 13, which in the Max/Max-cognate DNA cocrystal were shown to contact specific bases, are marked with arrow heads.[38] Max9 refers to the product of one of the alternatively spliced mRNAs of max.[39] The Max/Max consensus DNA-binding site is given below. The specific bases contacted are number 1, 2, 4' and 6'. Only the contacts of one molecule of the protein dimer with DNA are considered. Additional contacts of the Max basic region to the phosphate backbone are not shown.[38] In analogy to the Max data it seems likely that the corresponding amino acids in the basic regions of the other members of the Myc network are also responsible for specific contacts to the DNA. The sequences of the indicated proteins were derived from the following references: c-Myc,[108] N-Myc,[109] L-Myc,[110] Max9,[39] Mad1,[22] Mxi1,[29] Mad3 and Mad4.[23]

proteins (class B), all containing an Arg at the relevant position, whereas the latter sequence is bound by bHLH proteins (class A) without a basic amino acid at this position.[40] The class B includes, besides the Myc/Max/Mad family, USF and TFE3 among others, whereas MyoD, E12, and E47 are members of class A. Thus this Arg (Fig. 13.2, residue 13) explains why all Myc/Max/Mad family complexes prefer 5'-CA*CG*TG but cannot bind to 5'-CA*GC*TG. In addition to 5'-CA*CG*TG, Myc/Max dimers have also been shown to bind to 5'-CA*TG*TG as well as to other sequences, albeit with lower apparent affinity.[30,41-43] The binding to 5'-CA*TG*TG suggests that one specific contact to the two nucleotides in the center is sufficient for protein-DNA interaction. However the nucleotide at the third position (Fig. 13.2, base 3) cannot be a G or an A, implying that T may be able to make a specific contact to the protein although with lower affinity. Since the palindromic sequence 5'-CA*TA*TG has not been found in binding site selection assays, it appears that two weak contacts to the E-box core sequence are not sufficient for binding.

The relevance of an Arg at position 13 (Fig. 13.2) is further supported by mutational analysis. Mutation of the Met at this position in a CBF-AP4 chimeric molecule to Arg was sufficient to change the specificity from 5'-CAGCTG to 5'-CACGTG.[40] A somewhat different result was obtained using a mutant Myc fusion protein. Mutation of the Arg corresponding to amino acid 367 in c-Myc (Fig. 13.2, residue 13) to Val, a switch from a class B bHLH to a class A bHLH amino acid, abolished binding to 5'-CA*CG*TG as well as to 5'-CA*GC*TG.[35] However weak binding to the latter sequence typical for class A bHLH proteins could be restored by mutating the Leu corresponding to amino acid 362 in Myc to Arg (as in E47; Fig. 13.2, residue 8). These findings indicate that the last Arg of the basic region is an important determinant for specifying the inner two bases but it does not seem to be the only amino acid relevant. In addition to the specific interactions described above, a number of amino acids in the basic region make contacts to the phosphate backbone, including Lys-24 and Arg-25 in Max9 (Fig. 13.2, residues 1, 2).[38] Additional amino acids in the loop region also contact the phosphate backbone.

It should be noted however that the sequence 5'-CACGTG does not appear to be sufficient for high affinity binding by dimers of the Myc/Max/Mad family. In addition to this core sequence up to three bases on either side are relevant for high affinity binding and were enriched in binding site selection assays (for a summary see Fig. 13.3A). Interestingly, differences regarding these flanking sequences were found when Myc/Max dimers were compared to Max/Max dimers.[30,44,45] A priori this is surprising since no specific amino acid-base contacts in addition to the ones described above, in particular to bases flanking the 5'-CACGTG sequence, could be observed in the Max/Max-cognate DNA cocrystal.[38] The structural basis for the differences regarding flanking sequences is unclear at present. One possibility is that amino acids located in the loop or in as yet unidentified regions outside the bHLH/ZIP domains are involved in contacting certain bases outside the E-box. The former possibility seems unlikely since the loops in Myc network proteins are too short to penetrate the major groove;[38] the latter cannot be answered to date since structural data on full length proteins are not available.

DNA-binding site selection protocols have been used to systematically search for optimal E-box flanking sequences. In different studies either bacterially expressed proteins,[30] in vitro translated proteins,[45] or immunoprecipitated Max (IP Max) proteins[42] were employed. Similar conclusions can be drawn from these three studies regarding sequence requirements. Myc/Max complexes show a more restricted specificity than Max/Max complexes, observed by comparing optimal as well as disfavored sequences (Fig. 13.3A). Some of the sequences which disfavor Myc/Max binding were also shown to be less potent in Myc-dependent reporter gene assays.[45] For Myc/Max binding the disfavoring T at position -1 respective to the E-box (Fig. 13.3A) was also observed in a yeast system.[44] It should be noted that the extended Myc/Max and Max/Max recognition sequences determined experimentally were in most cases not palindromic. The binding sites outlined in Fig. 13.3A are the interpretation of the analysis of half sites. It is unclear why the analysis using IP Max did not reveal any disfavoring bases. One possible explanation is that the material obtained by immunoprecipitating Max is heterogeneous in regard to the dimeric partner proteins. Such immunoprecipitates are expected to contain Myc/Max and Max/Max dimers but potentially also Mad/Max dimers. Thus the different complexes in the IP Max material may even out disfavored bases of individual dimers.

The findings derived from the in vitro binding site selections are in good

A

			-3 -2 -1 ... 1 2 3
Myc/Max	(bacterial)	optimal	AACCACGTGGTT
		disfavored	YYT ARR
	(in vitro transl.)	optimal	GACCACGTGGTC
		disfavored	CYT ARG
Max/Max	(bacterial)	optimal	AANCACGTGNTT
		disfavored	-C- -G-
	(in vitro transl.)	optimal	RANCACGTGNTY
		disfavored	CY- -RG
IP Max		optimal	TANCACGTGNTA
		disfavored	--- ---

B

ODC (intron)	MB1	GGCCACGTGTCG
	MB2	GGACACGTGGCC
(5')		GCCCACGTGGCC
prothymosin		GGCCACGTGCTC
ECA39		CGCCACGTGTCA
p53 (human)		TCCCATGTGCTC
(murine)		TCCCACGTGCTC
CAD		AGCCACGTGGAC

Fig. 13.3. Extended binding sites for Myc/Max and Max/Max dimeric complexes. A. Binding site selections were performed using either bacterially produced Myc (C-terminal region, including bHLH/ZIP) and Max,[30] in vitro translated proteins,[45] or immunoprecipitated Max (IP Max) proteins.[42] Optimal binding sites and disfavored bases are indicated. N: any base; R: purine base; Y: pyrimidine base. It should be noted that in the selection with bacterially expressed proteins 40% of the E-box sequences were 5'-CATGTG. This sequence was also identified in the IP Max experiments.
B. Binding sites of several Myc regulated genes are summarized. The sequences were derived from the following references: mouse ODC,[81] both Myc/Max binding sites (MB1 and MB2, respectively) in the first intron are conserved between mammals; human ODC,[89] the Myc/Max binding site 5' of the promoter is not conserved; rat prothymosin-α;[83] ECA39;[86] human p53;[84] murine p53;[85] and hamster CAD.[87]

agreement with the sequences of Myc/Max binding sites obtained from the analysis of Myc regulated genes (see below). The extended Myc binding sites of several of these genes, consisting of 12 base pairs with the E-box sequence at the center, are close to optimal sites and disfavored bases are observed only in a few cases (for a summary see Fig. 13.3B).

At present little information regarding the specificity of DNA binding of Mad/Max complexes has been obtained. However bacterially synthesized protein as well as in vivo derived Mad/Max can bind to a Myc/Max consensus sequence.[22,46] It will now be of interest to determine the relative affinity of Mad/Max complexes to the sequences described above in comparison to Myc/Max or Max/Max dimers. This information will be important in order to analyze the role of Mad proteins in the regulation of Myc function.

THE MYC TRANSACTIVATION DOMAIN

The transactivation domain (TAD) of Myc proteins is localized at the N-terminus (see Fig. 13.1). This region of the protein is essential for Myc function in transformation, inhibition of differentiation, and in-

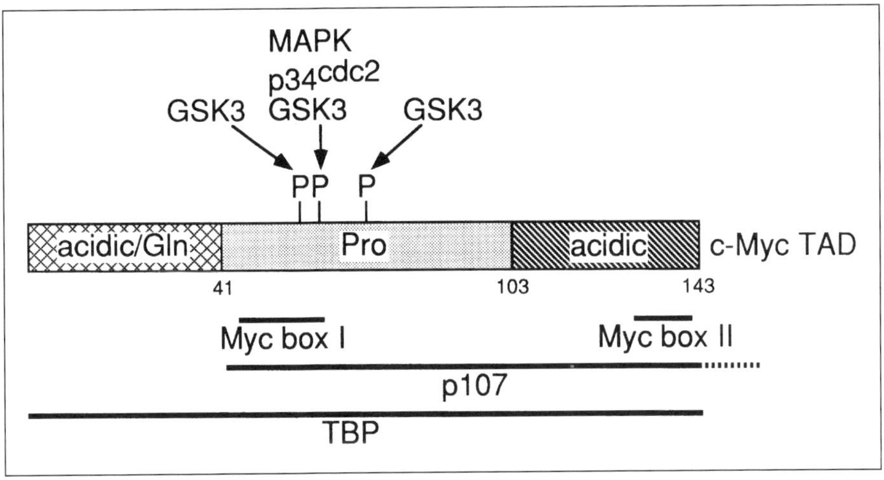

Fig. 13.4. The transactivation domain of c-Myc. The TAD of c-Myc consists of the first 143 amino acids. It is composed of three subdomains. Three phosphorylation sites have been mapped in the TAD: Thr-58, Ser-62, and Ser-71 in human c-Myc. These sites can be phosphorylated in vitro by several different kinases as indicated. Myc box I (amino acids 45-63) and Myc box II (amino acids 129-141) are highly conserved regions in all known Myc proteins. In addition these regions are conserved during evolution. The regions important for the interaction with the pocket protein p107 and the TATA box binding protein (TBP) are indicated. The p107 interaction domain extends to about amino acid 160.

duction of apoptosis (for a review see ref. 1). As for the analysis of many other transactivation domains the first detailed characterization was performed in the context of GAL4 DNA-binding domain (GAL4) fusion proteins.[21] This study established that the first 143 amino acids of c-Myc possess transactivating function and suggested the presence of at least three subdomains. The first (amino acid 1-41) contains acidic as well as glutamine residues, the second (amino acid 42-103) is rich in proline, and the third (amino acid 104-143) is slightly acidic (Fig. 13.4). A fusion of the second and third subdomain in the context of a GAL4 fusion protein had the highest transactivating activity.[21] Thus the TAD of Myc cannot easily be classified into one of several subgroups defined by amino acid composition[47,48] but rather appears to represent a composite domain.

It was furthermore demonstrated that Myc can transactivate synthetic reporter gene constructs containing Myc/Max consensus binding sites.[49-52] However, whereas GAL4-MycTAD fusion proteins resulted in a robust transactivation, wild type c-Myc is a rather poor transactivator. This suggested the presence of negative regulatory elements and removal or suppression of these may provide an opportunity to stimulate transactivation. In the context of GAL4 fusion proteins the region between the TAD and the bHLH/ZIP region inhibits transactivation and may therefore contain the potential negative regulatory elements.[21] Myc has to dimerize with Max to exert its transactivating activity, but overexpression of Max and under certain circumstances of Mad proteins results in repression of reporter gene expression.[22,23,49-52] This is consistent with the apparent lack of transactivation domains in Max and Mad proteins. Together these findings support a model in which the relative level of expression of the different Myc/Max/Mad family members determines the rate of transcription from a responsive promoter. In situations with excess Myc/Max complexes, target genes will be activated, whereas inhibition of transcription should be observed when Mad/Max and Max/Max complexes predominate.

This model is most likely an oversimplification of events that occur within a cell. Additional interacting proteins and cooperating factors, as well as distinct binding site specificities will undoubtedly affect the function of the different dimeric complexes and contribute to gene regulation.

Several proteins have been shown to interact with the TAD of c-Myc and are therefore candidates for regulation of Myc specific gene expression. These proteins include p107,[53-55] TBP,[56,57] and a newly identified protein binding to Myc box II.[58] Whereas the interaction with TBP, which has been shown to bind to a growing number of different transcription factors, may be relevant for stimulating transactivation, binding of p107 appears to be involved in the inhibition of Myc specific gene expression. Whether the Myc box II binding protein can influence Myc transcription is presently unclear, but since it has been implicated in Myc transformation it will likely affect transactivating activity.

Transactivation domains cannot only be classified according to their amino acid composition but also by whether or not they can activate from distal positions in promoters.[59] With the identification of Myc/Max consensus binding sites in the first intron of several genes (see below), the activity of Myc on proximal and distal E-box sites was compared.[60] c-Myc was capable of activating efficiently from both positions, in line with the findings for acidic transactivation domains.[59] These observations were then extended by showing that USF, which is much more abundant than Myc and also binds 5'-CACGTG, can only activate from proximal sites but not from a distance.[61] Deletion of Myc box I or II abolishes the ability of Myc to activate from distal sites but does not affect transactivation from proximal positions.[61] Together these studies suggest that the position of a binding site relative to the promoter can determine transcription factor specificity and that Myc may communicate differently with the basal transcriptional machinery from proximal or distal sites. Furthermore the reduced ability of Myc box I and II mutants to transactivate a Myc target gene can now be correlated with their impaired transforming activity.[3] Thus it is probable that transformation by Myc, which in most cases is associated with deregulated and enhanced expression of the protein, is the result of forced expression of Myc target genes.

The experiments published so far regarding position dependent transactivation by c-Myc are providing a nice framework in order to understand aspects of specificity of different 5'-CACGTG binding proteins. However these findings should also be viewed cautiously, since most were derived from studies on reporter gene constructs. The spacing of a particular binding site to the promoter in an endogenous cellular gene is almost certainly different from what the linear configuration suggests, due to chromatin organization, e.g., the positioning of specific nucleosomes.[62]

Over the last years at least three phosphorylation sites have been mapped within the transactivation domain of c-Myc.[63-67] These could be endpoints of incoming signals and modulate transcriptional activity. One of these sites (Thr-58 in human c-Myc) is of particular interest since it or amino acids nearby are frequently mutated in tumor associated c-Myc and in v-Myc proteins.[68-72] Several kinases, including MAP kinases, glycogen synthase kinase 3, and $p34^{cdc2}$, have been suggested to phosphorylate these sites. In addition it has been shown that the phosphorylation of Thr-58 and Ser-62 is regulated by mitogens.[66] However, the effect of these phosphorylation sites on transactivation is controversial. Whereas some reports suggested alterations in the transactivating potential by mutating the phosphorylation sites[63,72,73] no difference could be detected in other studies.[65,66] Even so it is compelling to argue that the N-terminal phosphorylation sites in c-Myc possess regulatory function; this issue has not been resolved yet.

Recently it has been reported that Ras can stimulate the transactivating activity of the TAD of c-Myc in the context of GAL4 or glucocorticoid receptor fusion proteins.[74] The Ras responsive region was mapped to amino acids 92-143 which includes Myc box II; however, this part of the TAD does not contain any known phosphorylation sites. Therefore it seems likely that the Ras effect does not target Myc itself but may regulate the interaction with another protein, e.g., a coactivator, a TBP associated factor, or another molecule involved in the interplay between Myc and the basal transcriptional machinery. In parallel we have found that c-Myc specific transactivation of reporter genes can be stimulated by Ras (C. Seelos, M. Henriksson, C. Cerni and B. Lüscher, unpublished results). Thus signaling by growth factor receptors which activate Ras may target Myc transactivation. These findings are also of interest in light of the synergism of Ras and Myc in transformation, providing a potential explanation.

The C-terminal half of c-Myc has recently been shown to interact directly with YY1,[75] a multifunctional transcriptional regulator which can repress as well as activate transcription depending on the context (for a review see refs. 76, 77). It was shown that c-Myc can inhibit YY1-mediated activation of the *c-myc* promoter as well as alleviate repression mediated by a GAL4-YY1 fusion protein.[75] We have found that YY1 is an efficient negative regulator of c-Myc function when assayed in a Myc/Ras cotransformation assay. However, the regions in YY1 which are necessary for this effect are different from those needed for direct binding to c-Myc. Instead YY1 appears to target the transactivation domain of c-Myc by an indirect mechanism (M. Austen, C. Cerni and B. Lüscher, unpublished results).

Whereas the repression of Myc specific transactivation by Max is most likely the result of competition by Max/Max homodimers for binding sites, the effect of Mad is more complex. At present it cannot be excluded that part of the repression observed could be due to Mad/Max heterodimers occupying binding sites or due to competition for a potentially limiting supply of Max. Nevertheless the latter scenario seems unlikely since Max is in excess in most cells analyzed and a comparatively low amount of Mad is expressed, e.g., during differentiation. The finding that Mad proteins associate via their N-terminal homology region with mSin3A and mSin3B is of particular importance.[23,78,79] mSin3A and mSin3B show homology to the yeast Sin3 protein, which can repress a number of genes and has been proposed to function as a transcriptional regulator (cited in refs. 78, 80). Recently the mSin3 interacting domain in Mxi1 has been replaced by mSin3A itself. This fusion protein has comparable activity to wild type Mxi1 in transformation assays, further underscoring that the Mxi1/mSin3A, and by inference also the Mad/mSin3A complex, is functionally relevant.[80] These studies suggest that Mad proteins recruit mSin3 proteins, resulting in a complex which inhibits gene transcription. It will be important to define whether the Mad/mSin3 complexes have a dominant repressing activity.

MYC REGULATED GENES

Myc proteins have been implicated in positive as well as negative regulation of a number of genes; among these are many indirect target genes (for a review see refs. 1, 19). Here we restrict ourselves to the discussion of two direct target genes: ornithine decarboxylase (ODC)[81] because it is potentially relevant to explain Myc biology, and prothymosin-α[82,83] because its regulation by Myc is best understood in molecular terms. Both the ODC and the prothymosin-α gene have Myc/Max binding sites localized in the first intron. Interestingly this is also true for the p53 and the ECA39 genes, two other Myc responsive genes.[84-86] In the carbamoyl-phosphate synthase gene (CAD), which is also regulated by Myc, the Myc/Max consensus binding site is localized in the 5' untranslated region of the gene.[87] Thus in all instances the E-boxes are positioned 3' relative to the start of transcription. These findings suggest that the position of the Myc/Max response element relative to the core promoter region is of physiological relevance; even so no definitive answer has emerged.

It was originally observed that enforced Myc expression results in constitutive expression of the ODC gene.[88] Later it was shown that ODC reporters are Myc responsive through two E-box sequences located in the first intron.[81] These Myc/Max DNA-binding sites are conserved in mammals, suggesting their functional relevance.[81,89] The ODC gene is likely a direct target of Myc since it can be induced by c-MycER, a fusion of c-Myc with the hormone binding domain of the human estrogen receptor,[90] in the presence of protein synthesis inhibitors.[91] Moreover a number of properties are shared between ODC and Myc: ODC, the rate-limiting enzyme in the biosynthesis of polyamines, is essential for entry into S-phase, and its expression is closely associated with cell growth.[92] In addition ODC can transform established fibroblasts.[93,94] A general role for ODC in transformation is also supported by the finding that deregulated expression of this gene is commonly found in different human and animal malignancies.[95-97] Furthermore enforced expression of ODC in 32D cells increases the sensitivity of these cells to undergo apoptosis in response to IL-3 withdrawal.[98] Together these observations remind us of several functional properties described for Myc. It is therefore very attractive to speculate that ODC is an important or even critical direct downstream effector of Myc. An obvious question arising is whether ODC can explain Myc biology or whether additional important target genes exist.

The molecular analysis of the regulation of prothymosin-α by Myc has been well advanced in the last few years. Originally a conditional MycER fusion protein was used to identify this gene as a potential Myc target in a differential screening approach.[82] Induction of the prothymosin-α gene occurred in the presence of protein synthesis inhibitors, suggesting that it is directly regulated by Myc. The Myc responsive element was mapped to an E-box sequence in the first intron.[83] Several aspects of Myc function were unraveled by studying prothymosin-α reporter gene constructs. What makes this E-box specific for Myc? This is an important question to address since a number of other proteins can bind to Myc/Max consensus sites, at least in vitro. It has been shown that USF can bind to the E-box in the first intron of prothymosin-α but that it cannot activate from this site. This seems to reflect a property of the TAD of USF which does not activate transcription from distal sites.[61] In contrast the TAD of Myc can activate from both proximal and distal sites, as discussed above. Thus the location of a given binding site may be an important determinant in discriminating between competing transcription factors. Another bHLH protein, TFE3, also binds to a 5'-CACGTG sequence and is capable of activating from a distance. Selection against TFE3 appears to require a mechanism involving a repressor protein binding to a second E-box sequence in the first intron of prothymosin-α.[61,99,100] The Myc/Max consensus binding site in the prothymosin-α gene overlaps with a high affinity binding site for AP-2 and it has been shown that binding of AP-2 competes with binding of bHLH proteins to the E-box.[101] This provides a further regulatory mechanism in that binding affinities to the two overlapping sites and availability of the corresponding binding proteins will specify the transcriptional activity associated with this distal enhancer element.

The regulation of prothymosin-α by Myc proteins has been questioned in a recent report.[102] Using human prothymosin-α sequences these authors could not detect significant changes in the expression of their reporter gene constructs in response to c-Myc, dominant negative Myc, or dominant negative Max. In addition no effects were measured by mutating either the 5' E-box sequences, the one in the intron, or both. At present the discrepancies between these different reports cannot be explained. However aspects of cell type specificity and cell density could be responsible for at least part of the contrasting findings. One way to resolve this issue and to provide evidence beyond transfection data would be to determine whether the E-boxes in prothymosin-α as well as in the ODC gene are occupied by proteins at the relevant time points in vivo and whether the protection obtained is comparable to that seen with Myc/Max complexes in vitro.

Interestingly the expression of prothymosin-α has been correlated with the expression of Myc in human colorectal cancer as well as during differentiation of HL-60 cells.[103-105] Furthermore it has been shown recently using a tetracycline regulated N-Myc expression system that the induction of N-Myc resulted in the activation of expression of both the prothymosin-α and the ODC gene.[106] However it is unclear whether this effect is direct or reflects association of prothymosin-α expression with the induced growth rate observed in response to the activation of N-Myc expression. In summary both ODC and prothymosin-α are good candidates for being directly regulated by Myc.

OUTLOOK

The Myc/Max/Mad network of proteins is essential in many aspects of cell growth control, including cell cycle progression, transformation, apoptosis, and inhibition of differentiation. A large body of evidence has shown that Myc, as a heterodimer with Max, can function as a transcriptional regulator. This and all evidence available suggest that the biological effects of Myc are mediated by the regulation of gene transcription. Furthermore mutational analysis revealed that the TAD of Myc as well as the bHLH/ZIP domains are essential for transactivation and that this activity is also critical for all biological activities of Myc.

Mad/Max hetero- and Max/Max homodimeric complexes seem to repress gene transcription whereas Myc/Max complexes have the ability to activate. In addition Myc may also be involved in directly repressing the expression of specific genes. One important aim is therefore the identification of target genes relevant in mediating the biological effects of Myc and Mad proteins, respectively. It seems likely that target genes which are activated by Myc/Max dimers might be repressed by Mad/Max dimers. In this setting Myc and Mad would functionally antagonize each other. However, it is also possible that genes exist which are targeted by either Myc/Max or Mad/Max complexes. In such situations the functions of Myc and Mad proteins would be uncoupled with Mad obtaining a role beyond its proposed antagonism to Myc. With the identification of more direct target genes of Myc/Max and/or Mad/Max we will learn how these proteins exert the observed biological effects and how they are functionally related. In addition, the identification of genes which are involved in specific aspects of the Myc response rather than in all may also be of therapeutic interest in order to specifically target the pathways leading to Myc dependent transformation.

The relative expression level of the different Myc/Max/Mad family proteins probably also has an important influence on the transcriptional regulation of specific target genes. In light of the recently discovered Mad3 and Mad4 proteins it will be interesting to learn more about the prevailing complexes in cells during proliferation and differentiation. Further aspects which contribute to regulation of the Myc/Max/Mad proteins include interaction with other proteins and signaling onto the network members. The search for additional factors interacting with Myc, Max, or Mad proteins is ongoing and will certainly shed light on mechanisms regulating the activities of these proteins. So far little is known about signaling to Myc and Mad proteins. Clearly with the continual identification of novel pathways transmitting signals from the cell membrane to the nucleus, new possibilities emerge which could be relevant for targeting the network members.

REFERENCES

1. Henriksson M, Lüscher B. Proteins of the Myc network: Essential regulators of cell growth and differentiation. Adv. Cancer Res. 1996; 68:109-182.
2. Marcu KB, Bossone SA, Patel AJ. *myc* function and regulation. Annu Rev Biochem 1992; 61:809-860.
3. Stone J, de Lange T, Ramsay G, Jakobovits E, Bishop JM, Varmus H, Lee W. Definition of regions in human c-*myc* that are involved in transformation and

nuclear localization. Mol Cell Biol 1987; 7:1697-1709.

4. Stewart TA, Pattengale P, Leder P. Spontaneous mammary adenocarcinomas in transgenic mice that carry and express MTV/myc fusion genes. Cell 1984; 38: 627-637.
5. Adams JM, Harris AW, Pinkert CA, Corcoran LM, Alexander WS, Cory S, Palmiter RD, Brinster RL. The c-myc oncogene driven by immunoglobulin enhancers induces lymphoid malignancy in transgenic mice. Nature 1985; 318: 533-538.
6. Dildrop R, Ma A, Zimmerman K, Hsu E, Tesfaye A, DePinho R, Alt FW. IgH enhancer-mediated deregulation of N-*myc* gene expression in transgenic mice: generation of lymphoid neoplasias that lack c-*myc* expression. EMBO J 1989; 8:1121-1128.
7. Rosenbaum H, Webb E, Adams J, Cory S, Harris A. N-*myc* transgene promotes B lymphoid proliferation, elicits lymphomas and reveals cross-regulation with c-*myc*. EMBO J 1989; 8:749-755.
8. Möröy T, Fisher P, Guidos C, Ma A, Zimmerman K, Tesfaye A, DePinho R, Weissman I, Alt FW. IgH enhancer deregulated expression of L-*myc*: abnormal T lymphocyte development and T cell lymphomagenesis. EMBO J 1990; 9: 3659-3666.
9. Haupt Y, Alexander W.S, Barri G, Klinken SP, Adams JM. Novel zinc finger gene implicated as *myc* collaborator by retrovirally accelerated lymphomagenesis in Eμ-*myc* transgenic mice. Cell 1991; 65:753-763.
10. van Lohuizen M, Verbeek S, Scheijen B, Wientjens E, van der Gulden H, Berns A. Identification of cooperating oncogenes in Eμ-myc transgenic mice by provirus tagging. Cell 1991; 65:737-752.
11. Harris AW, Pinkert CA, Crawford M, Langdon WY, Brinster RL, Adams JM. The Eμ-myc transgenic mouse. A model for high-incidence spontaneous lymphoma and leukemia of early B cells. J Exp Med 1988; 167:353-371.
12. Nussenzweig MC, Schmidt EV, Shaw AC, Sinn E, Campos-Torres J, Mathey-Prevot B, Pattengale PK, Leder P. A human immunoglobulin gene reduces the incidence of lymphomas in c-Myc bearing transgenic mice. Nature 1988; 336:446-450.
13. Coppola JA, Cole MD. Constitutive c-*myc* oncogene expression blocks murine erythroleukemia cell differentiation but not commitment. Nature 1986; 320: 760-763.
14. Dmitrovsky E, Kuehl WM, Hollis GF, Kirsh IR, Bender TP, Segal S. Expression of a transfected human c-*myc* oncogene inhibits differentiation of a mouse erythroleukemia cell line. Nature 1988; 322: 748-750.
15. Onclercq R, Babinet C, Crémisi C. Exogenous c-*myc* gene overexpression interferes with early events in F9 cell differentiation. Oncogene Res 1989; 4:293-302.
16. Freytag S. Enforced expression of the c-*myc* oncogene inhibits cell differentiation by precluding entry into a distinct predifferentiation state in GO/G1. Mol Cell Biol 1988; 8:1614-1624.
17. Larsson L-G, Ivhed I, Gidlund M, Pettersson U, Vennström B, Nilsson K. Phorbol ester-induced terminal differentiation is inhibited in human U-937 monoblastic cells expressing a v-*myc* oncogene. Proc Natl Acad Sci USA 1988; 85:2638-2642.
18. Harrington EA, Fanidi A, Evan GI. Oncogenes and cell death. Curr Opin Gen Dev 1994; 4:120-129.
19. Packham G, Cleveland JL. c-Myc and apoptosis. Biochem Biophys Acta 1995; 1242:11-28.
20. Harrington EA, Bennett MR, Fanidi A, Evan GI. c-Myc-induced apoptosis in fibroblasts is inhibited by specific cytokines. EMBO J 1994; 13:3286-3295.
21. Kato GJ, Lee WMF, Chen L, Dang CV. Max: functional domains and interaction with c-Myc. Genes Dev 1992; 6:81-92.
22. Ayer DE, Kretzner L, Eisenman RN. Mad: a heterodimeric partner for Max that antagonizes Myc transcriptional activity. Cell 1993; 72:211-222.
23. Hurlin PJ, Quéva C, Koskinen PJ, Steingrímsson E, Ayer DE, Copeland NG, Jenkins NA, Eisenman RN. Mad3 and Mad4: Novel Max-interacting transcriptional repressors that suppress c-Myc-dependent transformation and are ex-

pressed during neural and epidermal differentiation. EMBO J 1995; 14: 5646-5659.
24. Greenberg ME, Ziff EB. Stimulation of mouse 3T3 cells induces transcription of the c-*fos* oncogene. Nature 1984; 311: 433-438.
25. Hann SR, Thompson CB, Eisenman RN. c-*myc* oncogene protein synthesis is independent of the cell cycle in human and avian cells. Nature 1985; 314:366-369.
26. Thompson CB, Challoner PB, Neiman PE, Groudine M. Levels of c-*myc* oncogene mRNA are invariant throughout the cell cycle. Nature 1985; 314:363-366.
27. Waters CM, Littlewood TD, Hancock DC, Moore JP, Evan GI. c-myc protein expression in untransformed fibroblasts. Oncogene 1991; 6:797-805.
28. Larsson L-G, Pettersson M, Öberg F, Nilson K, Lüscher B. Expression of mad, mxi1, max and c-myc during differentiation of hematopoetic cells: opposite regulation of mad and c-myc. Oncogene 1994; 9:1247-1252.
29. Zervos AS, Gyuris J, Brent R. Mxi1, a protein that specifically interacts with Max to bind Myc-Max recognition sites. Cell 1993; 72:223-232.
30. Berberich S, Hyde-DeRuyscher N, Espenshade P, Cole M. *max* encodes a sequence-specific DNA-binding protein and is not regulated by serum growth factors. Oncogene 1992; 7:775-779.
31. Blackwood EM, Lüscher B, Eisenman RN. Myc and Max associate in vivo. Genes Dev 1992; 6:71-80.
32. Lüscher B, Eisenman RN. New light on Myc and Myb, Part I. Myc. Genes Dev 1990; 4:2025-2035.
33. Ferré-D'Amaré AR, Burley SK. DNA recognition by helix-loop-helix proteins. Nucl Acids Mol Biol 1995; 9:285-298.
34. Anthony-Cahill SJ, Benfield PA, Wasserman ZR, Brenner SL, Stafford WF, Altenbach C, Hubbell W, DeGrado WF. Molecular characterization of helix-loop-helix peptides. Science 1992; 255: 979-983.
35. Halazonetis TD, Kandil A. Predicted structural similarities of the DNA-binding domains of c-Myc and EcoRI. Science 1992; 255:464-466.
36. Vinson CR, Garcia KC. Molecular model for DNA recognition by the family of basic-helix-loop-helix-zipper proteins. New Biol 1992; 4:396-403.
37. Davis LJ, Halazonetis TD. Both the helix-loop-helix and the leucine zipper motifs of c-Myc contribute to its dimerization specificity with Max. Oncogene 1993; 8:125-132.
38. Ferré-D'Amaré AR, Prendergast GC, Ziff EB, Burley SK. Recognition by Max of its cognate DNA through a dimeric b/HLH/Z domain. Nature 1993; 363:38-45.
39. Blackwood EM, Eisenman RN. Max: A helix-loop-helix zipper protein that forms a sequence-specific DNA-binding complex with Myc. Science 1991; 251: 1211-1217.
40. Dang CV, Dolde C, Gillison ML, Kato GJ. Discrimination between related DNA sites by a single amino acid residue of Myc-related basic-helix-loop-helix proteins. Proc Natl Acad Sci USA 1992; 89:599-602.
41. Prendergast GC, Ziff EB. Methylation-sensitive sequence-specific DNA binding by the c-Myc basic region. Science 1991; 251:186-189.
42. Blackwell TK, Huang J, Ma A, Kretzner L, Alt FW, Eisenman RN, Weintraub H. Binding of Myc proteins to canonical and noncanonical DNA sequences. Mol Cell Biol 1993; 13:5216-5224.
43. Ma A, Möröy T, Collum R, Weintraub H, Alt FW, Blackwell TK. DNA binding by N- and L-Myc proteins. Oncogene 1993; 8:1093-1098.
44. Fisher F, Crouch DH, Jayaraman P-S, Clark W, Gillespie DAF, Goding CR. Transcription activation by Myc and Max: flanking sequences target activation to a subset of CACGTG motifs in vivo. EMBO J 1993; 12:5075-5082.
45. Solomon DLC, Amati B, Land H. Distinct DNA binding preferences for the c-Myc/Max and Max/Max dimers. Nucl Acids Res 1993; 21:5372-5376.
46. Cerni C, Bousset K, Seelos C, Burkhardt H, Henriksson M, Lüscher B. Differential effects by Mad and Max on transformation by cellular and viral oncoproteins. Oncogene 1995; 11: 587-596.

47. Hahn S. Structure(?) and function of acidic transcription activators. Cell 1993; 72:481-483.
48. Triezenberg SJ. Structure and function of transcriptional activation domains. Curr Opin Genet Dev 1995; 5:190-196.
49. Amati B, Dalton S, Brooks MW, Littlewood TD, Evan GI, Land H. Transcriptional activation by the human c-Myc oncoprotein in yeast requires interaction with Max. Nature 1992; 359: 423-426.
50. Kretzner L, Blackwood EM, Eisenman RN. Myc and Max proteins possess distinct transcriptional activities. Nature 1992; 359:426-429.
51. Amin C, Wagner AJ, Hay N. Sequence-specific transcriptional activation by Myc and repression by Max. Mol Cell Biol 1993; 13:383-390.
52. Gu W, Cechova K, Tassi V, Dalla-Favera R. Opposite regulation of gene transcription and cell proliferation by c-Myc and Max. Proc Natl Acad Sci USA 1993; 90:2935-2939.
53. Beijersbergen RL, Hijmans EM, Zhu L, Bernards R. Interaction of c-Myc with the pRb-related protein p107 results in inhibition of c-Myc-mediated transactivation. EMBO J 1994; 13: 4080-4086.
54. Gu W, Bhatia K, Magrath IT, Dang CV, Dalla-Favera R. Binding and suppression of the Myc transcriptional activation domain by p107. Science 1994; 264251-254.
55. Hoang AT, Lutterbach B, Lewis BC, Yano T, Chou T-Y, Barrett JF, Raffeld M, Hann SR, Dang CV. A link between increased transforming activity of lymphoma-derived MYC mutant alleles, their defective regulation by p107 and altered phosphorylation of the c-Myc transactivation domain. Mol Cell Biol 1995; 15:4031-4042.
56. Hateboer G, Timmers HTM, Rustgi AK, Billaud M, van't Veer LJ, Bernards R. TATA-binding protein and the retinoblastoma gene product bind to overlapping epitopes on c-Myc and adenovirus E1A protein. Proc Natl Acad Sci USA 1993; 90:8489-8493.
57. Maheswaran S, Lee H, Sonenshein GE. Intracellular association of the protein product of the c-myc oncogene with the TATA-binding protein. Mol Cell Biol 1994; 14:1147-1152.
58. Brough DE, Hofmann TJ, Ellwood KB, Townley RA, Cole MD. An essential domain of the c-Myc protein interacts with a nuclear factor that is also required for E1A-mediated transformation. Mol Cell Biol 1995; 15:1536-1544.
59. Seipel K, Georgiev O, Schaffner W. Different activation domains stimulate transcription from remote ('enhancer') and proximal ('promoter') positions. EMBO J 1992; 11:4961-4968.
60. Packham G, Bello-Fernandez C, Cleveland JL. Position and orientation independent transactivation by c-Myc. Cell Mol Biol Res 1994; 40:699-706.
61. Desbarats L, Gaubatz S, Eilers M. Discrimination between different E-box binding proteins at an endogenous target gene of *c-myc*. Genes Dev 1996; 10:447-460.
62. Paranjape SM, Kamakaka RT, Kadonaga JT. Role of chromatin structure in the regulation of transcription by RNA polymerase II. Annu Rev Biochem 1994; 63:265-297.
63. Seth A, Alvarez E, Gupta S, Davis RJ. A phosphorylation site located in the NH_2-terminal domain of c-Myc increases transactivation of gene expression. J Biol Chem 1991; 266:23521-23524.
64. Seth A, Gonzalez FA, Gupta S, Raden DL, Davis RJ. Signal transduction within the nucleus by mitogen-activated protein kinase. J Biol Chem 1992; 267:24796-24804.
65. Henriksson M, Bakardjiev A, Klein G, Lüscher B. Phosphorylation sites mapping in the N-terminal domain of c-Myc modulate its transforming potential. Oncogene 1993; 8:3199-3209.
66. Lutterbach B, Hann SR. Hierarchical phosphorylation at N-terminal transformation-sensitive sites in c-Myc protein is regulated by mitogens and in mitosis. Mol Cell Biol 1994; 14:5510-5522.
67. Pulverer BJ, Fisher C, Vousden K, Littlewood T, Evan G, Woodgett JR. Site-specific modulation of c-Myc cotransformation by residues phosphorylated in vivo. Oncogene 1994; 9:59-70.
68. Frykberg L, Graf T, Vennström B. The transforming activity of the chicken c-

myc gene can be potentiated by mutations. Oncogene 1987; 1:415-421.
69. Symonds G, Hartshorn A, Kennewell A, O'Mara M-A, Bruskin A, Bishop JM. Transformation of murine myelomonocytic cells by *myc*: point mutations in v-*myc* contribute synergistically to transforming potential. Oncogene 1989; 3: 285-294.
70. Bhatia K, Huppi K, Spangler G, Siwarski D, Iyer R, Magrath I. Point mutations in the c-Myc transactivation domain are common in Burkitt's lymphomas and mouse plasmacytomas. Nature Genetics 1993; 5:56-61.
71. Yano T, Sander CA, Clark HM, Dolezal MV, Jaffe ES, Raffeld M. Clustered mutations in the second exon of the *myc* gene in sporadic Burkitt's lymphomas. Oncogene 1993; 8:2741-2748.
72. Albert T, Urlbauer B, Kohlhuber F, Hammersen B, Eick D. Ongoing mutations in the N-terminal domain of c-Myc affect transactivation in Burkitt's lymphoma cell lines. Oncogene 1994; 9: 759-763.
73. Gupta S, Seth A, Davis RJ. Transactivation of gene expression by Myc is inhibited by mutation at the phosphorylation sites Thr-58 and Ser-62. Proc Natl Acad Sci 1994; 90:3216-3220.
74. Colman MS, Ostrowski MC. The transactivation potential of a c-Myc N-terminal region (residues 92-143) is regulated by growth factor/Ras signaling. Nucl Acids Res 1996; 24:1971-1979.
75. Shrivastava A, Saleque S, Kalpana GV, Artandi S, Goff SP, Calame K. Inhibition of transcriptional regulator Yin-Yang-1 by association with c-Myc. Science 1993; 262:1889-1892.
76. Hahn S. The ying and yang of mammalian transcription. Curr Biol 1992; 2: 152-154.
77. Shrivastava A, Calame K. An analysis of genes regultaed by the multi-functional transcriptional regulator Yin Yang-1. Nucl Acids Res 1994; 22:5151-5155.
78. Ayer DE, Lawrence QA, Eisenman RN. Mad.Max transcriptional repression is mediated by ternary complex formation with mammalian homologs of the yeast repressor Sin3. Cell 1995; 80:767-776.
79. Schreiber-Agus N, Chin L, Chen K, Torres R, Rao G, Guida P, Skoultchi AI, DePinho RA. An amino-terminal domain of Mxi1 mediates anti-Myc oncogenic activity and interacts with a homolog of the yeast transcriptional repressor SIN3. Cell 1995; 80:777-786.
80. Rao G, Alland L, Guida P, Schreiber-Agus N, Chen K, Chin L, Rochelle JM, Seldin MF, Skoultchi AI, DePinho RA. Mouse Sin3A interacts with and can functionally substitute for the amino-terminal repression domain of the Myc antagonist Mxi1. Oncogene 1996; 12: 1165-1172.
81. Bello-Fernandez C, Packham G, Cleveland JL. The ornithine decarboxylase gene is a transcriptional target of c-Myc. Proc Natl Acad Sci USA 1993; 90: 7804-7808.
82. Eilers M, Schirm S, Bishop JM. The Myc protein activates transcription of the α-prothymosin gene. EMBO J 1991; 10: 133-141.
83. Gaubatz S, Meichle A, Eilers M. An E-box element localized in the first intron mediates regulation of the prothymosin α gene by c-*myc*. Mol Cell Biol 1994; 14:3853-3862.
84. Roy B, Beamon J, Balint E, Reisman D. Transactivation of the human p53 tumor suppressor gene by c-Myc/Max contributes to elevated mutant p53 expression in some tumors. Mol Cell Biol 1994; 14:7805-7815.
85. Reisman D, Elkind NB, Roy B, Beamon J, Rotter V. c-Myc trans-activates the p53 promoter through a required downstream CACGTG motif. Cell Growth Diff 1993; 4:57-65.
86. Benvenisty N, Leder A, Kuo A, Leder P. An embryonically expressed gene is a target for c-Myc regulation via the c-Myc-binding sequence. Genes Dev. 1992; 6:2513-2523.
87. Miltenberger RJ, Sukow KA, Farnham PJ. An E-box-mediated increase in *cad* transcription at the G1/S-phase boundary is suppressed by inhibitory c-Myc mutants. Mol Cell Biol 1995; 15:2527-2535.
88. Askew D, Ashum R, Simmons B, Cleveland J. Constitutive c-*myc* expression in IL-3-dependent myeloid cell line sup-

presses cycle arrest and accelerates apoptosis. Oncogene 1991; 6:1915-1922.
89. Peña A, Reddy CD, Wu S, Hickok NJ, Reddy EP, Yumet G, Soprano DR, Soprano KJ. Regulation of human ornithine decarboxylase expression by the c-Myc/Max protein complex. J Biol Chem 1993; 268:27277-27285.
90. Eilers M, Picard D, Yamamoto KR, Bishop JM. Chimaeras of Myc onco-protein and steroid receptors cause hormone dependent transformation of cells. Nature 1989; 340:66-68.
91. Wagner AJ, Meyers C, Laimins LA, Hay N. c-Myc induces the expression and activity of ornithine decarboxylase. Cell Growth Diff 1993; 4:879-883.
92. Bowlin TL, McKnown BJ, Sunkara PS. Ornithine decarboxylase induction and polyamine biosynthesis are required for the growth of interleukin-2- and interleukin-3-dependent cell lines. Cell Immunol 1986; 98:341-350.
93. Auvinen M, Paasinen A, Andersson LC, Hölttä E. Ornithine decarboxylase activity is critical for cell transformation. Nature 1992; 360:355-358.
94. Moshier JA, Dosecu J, Skunca M, Luk GD. Transformation of NIH/3T3 cells by ornithine decarboxylase overexpression. Cancer Res 1993; 53:2618-2622.
95. Pegg AE. Polyamine metabolism and its importance in neoplastic growth and as a target for chemotherapy. Cancer Res 1988; 48:759-774.
96. Katz A, Kahana C. Rearrangement between ornithine decarboxylase and the switch region of the γ1 immunoglobulin gene in α-difluoromethylornithine resistant mouse myeloma cells. EMBO J 1989; 8:1163-1167.
97. Tonin PN, Yeger H, Stallings RL, Srinivasan PR, Lewis WH. Amplification of N-Myc and ornithine decarboxylase genes in human neuroblastoma and hydroxyurea-sensitive cell lines. Oncogene 1989; 4:1117-1121.
98. Packham G, Cleveland JL. Ornithine decarboxylase is a mediator of c-Myc-induced apoptosis. Mol Cell Biol 1994; 14:5741-5747.
99. Genetta T, Ruezinsky D, Kadesch T. Displacement of an E-box-binding repressor by basic helix-loop-helix proteins: Implications for B-cell specificity of the immunoglobulin heavy-chain enhancer. Mol Cell Biol 1994; 14:6153-6163.
100. Weintraub H, Genetta T, Kadesch T. Tissue-specific gene activation by MyoD: Determination of specificity by *cis*-acting repression elements. Genes Dev 1994; 8:2203-2211.
101. Gaubatz S, Imhof A, Dosch R, Werner O, Mitchell P, Buettner R, Eilers M. Transcriptional activation by Myc is under negative control by the transcription factor AP-2. EMBO J 1995; 14:1508-1519.
102. Mol PC, Wang R-H, Batey DW, Lee LA, Dang CV, Berger SL. Do products of the myc proto-oncogene play a role in transcriptional regulation of the prothymosin α gene? Mol Cell Biol 1995; 15:6999-7009.
103. Dosil M, Alvarez-Fernandez L, Gomez-Márquez J. Differentiation-linked expression of prothymosin α gene in human myeloid leukemia cells. Exp Cell Res 1993; 204:94-101.
104. Mori M, Barnard GF, Staniumas RJ, Jessup JM, Steele GD Jr, Chen LB. Prothymosin-α mRNA expression correlates with that of c-*myc* in human colon cancer. Oncogene 1993; 8:2821-2826.
105. Smith MR, Al-Katib A, Mohammad R, Silverman A, Szabo P, Khilnani S, Kohler W, Nath R, Mutchnik MG. Prothymosin α gene expression correlates with proliferation, not differentiation, of HL-60 cells. Blood 1993; 82:1127-1132.
106. Lutz W, Stöhr M, Schürmann J, Wenzel A, Löhr A, Schwab M. Conditional expression of N-myc in human neuroblastoma cells increases expression on α-prothymosin and ornithine decarboxylase and accelerates progression into S-phase early after mitogenic stimulation of quiescent cells. Oncogene 1996; 13:803-812.
107. Leonardo M, Pierce JW, Baltimore D. Protein-binding sites in Ig enhancers determine transcriptional activity and inducibility. Science 1987; 236:1573-1577.

108. Stanton LW, Watt R, Marcu KB. Translocation breakage and truncated transcripts of c-myc oncogene in murine plasmacytomas. Nature 1983; 303:401-406.
109. DePinho RA, Leguoy E, Feldman LB, Kohl NE, Yancopoulos GD, Alt FW. Structure and expression of the murine N-myc gene. Proc Natl Acad Sci USA 1986; 83:1827-1831.
110. DePinho RA, Hatton KS, Tesfaye A. Yancopoulos GD, Alt FW. The human myc gene family: structure and activity of L-myc and an L-myc pseudogene. Genes Dev 1987; 1:1311-1326.

CHAPTER 14

Ets Transcription Factors: Partners in the Integration of Signal Responses

Bohdan Wasylyk and Alfred Nordheim

INTRODUCTION: ETS PROTEINS

The first Ets protein was reported about thirteen years ago.[1,2] Since that time a family of proteins has been identified that are related at the sequence level in the Ets domain (reviewed in refs. 3-8). However, since the structure of the Ets domain was established, an extended family of potential distant structural relatives have been identified that have a similar fold.[9-12] Over 40 Ets protein sequences have been reported, in diverse species from man to the annelid Nereis,[13] representing about 25 different proteins. The Ets-domain interacts specifically and as a monomer with DNA sequences related to 5'-C/AGGAA/T-3'. Comparisons of the structure of the most divergent Ets domains (Ets1 and PU.1/Spi1, about 35% homology) show that they form a variant subclass of the helix-turn-helix DNA-binding proteins. There are strictly conserved amino acids in the Ets domain that are important for DNA binding.

The Ets proteins are modulators of transcription that have been implicated in many biological processes. They have multiple domains involved in physical and functional interactions in both cis and trans. Some of these domains are found in more than one Ets factor, which is a hallmark of subfamilies of Ets proteins. Subfamily members have presumably arisen during evolution by gene duplication followed by divergence.[8,14,15] Examples of domains that will be considered later include the Pointed domain, found in about eight members of the Ets1 subfamily, and the B and C-domains of the TCF subclass of Ets proteins. The various domains as well as the Ets domain mediate interactions with other modulators to form multiprotein complexes on promoters that finally result in either transcription activation or repression. Some domains appear to be mainly involved in cis regulation, affecting DNA binding or transcription activation by the Ets protein.

Transcription Factors in Eukaryotes, edited by Athanasios G. Papavassiliou.

The activities of Ets proteins are regulated at different levels, most notably by protein phosphorylation by a variety of kinases, but also mRNA stability.[16] Genetic studies in *Drosophila* and *Caenorhabditis* and reverse genetics in mammalian cells have implicated at least three of the Ets subfamilies in the growth factor-Ras-Map kinase cascades, including the Ets1 subfamily, the TCFs and ERF. The ubiquitous growth factor-Ras-Map kinase signal is converted into specific regulation by "anchor" proteins, such as Pit1/GHF1 for Ets1 and the prolactin promoter in pituitary cells and SRF for the TCFs and the fos promoter in fibroblasts.

Ets proteins are involved in differentiation, development and tumorigenesis. The genetic studies in *Drosophila* and *Caenorhabditis* point to roles of Ets proteins in the differentiation of the R7 photoreceptor and the vulva. Gene inactivation by homologous recombination in mice of PU.1, Fli1 and Ets1[17-20] indicate that these factors are particularly important in hematopoiesis. There are various links to tumorigenesis. The Ras component of the growth factor-Ras-Map kinase pathway is mutated in about 35% of human tumors.[21] The first member of the Ets family, v-Ets, was identified as a fusion protein with gag and myb in the E26 acute avian leukemia virus.[1,2] The oncogenic properties of Ets proteins have been studied in cells by both overexpression and by inhibition of Ras transformation with transdominant mutants. Erythroleukemia induced by Friend virus insertion led to the identification of the *fli-1* and *spi-1*(PU.1) *ets* genes.[5,22] The *erg*[23] and *ets1*[24] genes are also found at proviral insertion sites. A variety of human tumors have translocations that produce altered Ets proteins. These include the fusions of the EWS gene with Fli1, Erg and Etv-1, TLS/FUS-Erg and Tel with the PDGF receptor, Abl and MN1. Ets1 expression is induced by tumor invasion in surrounding stromal fibroblasts and in capillaries during angiogenesis.[25,26] *ets-1* and *erg-b* gene loci have been implicated in other translocations, but the consequences at the protein level have not been established.[3] Ets proteins could mediate transcription deregulation in human tumors, such as breast cancer. The Ets2 gene is localized on chromosome 21 which is triplicated in Downs syndrome. Transgenic mice that overexpress Ets2 have some of the features of Downs syndrome, suggesting that Ets2 overexpression may generate some of the phenotypic abnormalities.[27]

In this review we will discuss some highlight of the functions of Ets proteins, to illustrate how they are regulated and how they participate in normal and pathological biological processes. Special emphasis will be given to the Ets1 and TCF subclasses of Ets proteins.

THE ETS 1 SUBFAMILY OF ETS TRANSCRIPTION FACTORS

Multiple Functional Domains Built Around the Ets Domain

The Ets1 factor will be described in some detail to illustrate how, around the common theme of a conserved DNA-binding domain, the distinct functions of the Ets proteins are composed. The DNA-binding domains of Ets proteins are subtly different, in ways that affect both interactions with DNA and other factors. The additional conserved domains help to define the Ets subfamilies and confer both common and distinct features. The Ets1 subfamily is composed of eight factors (Fig. 14.1) that all contain an additional conserved domain, the Pointed or B domain. The founding Ets factor, the vEts fusion protein encoded by the E26 virus, is most closely related to the p68 isoform of Ets1. This isoform is found only in birds and reptiles whereas p54 is conserved in vertebrates, from amphibians to mammals.[28,29] Ets1/p68 is closely related to Ets2. Both proteins have seven domains that are similar in sequence and function (A-F, Fig. 14.1). Studies from a number of laboratories show that A and C are transcription activation domains (ACT1 and 2), B is a regulator (REG) of the activation domains that probably mediates protein-protein interactions, D and F are regulators of DNA binding. The A domains of Ets1/p68 and Ets2, unlike that of Ets1/

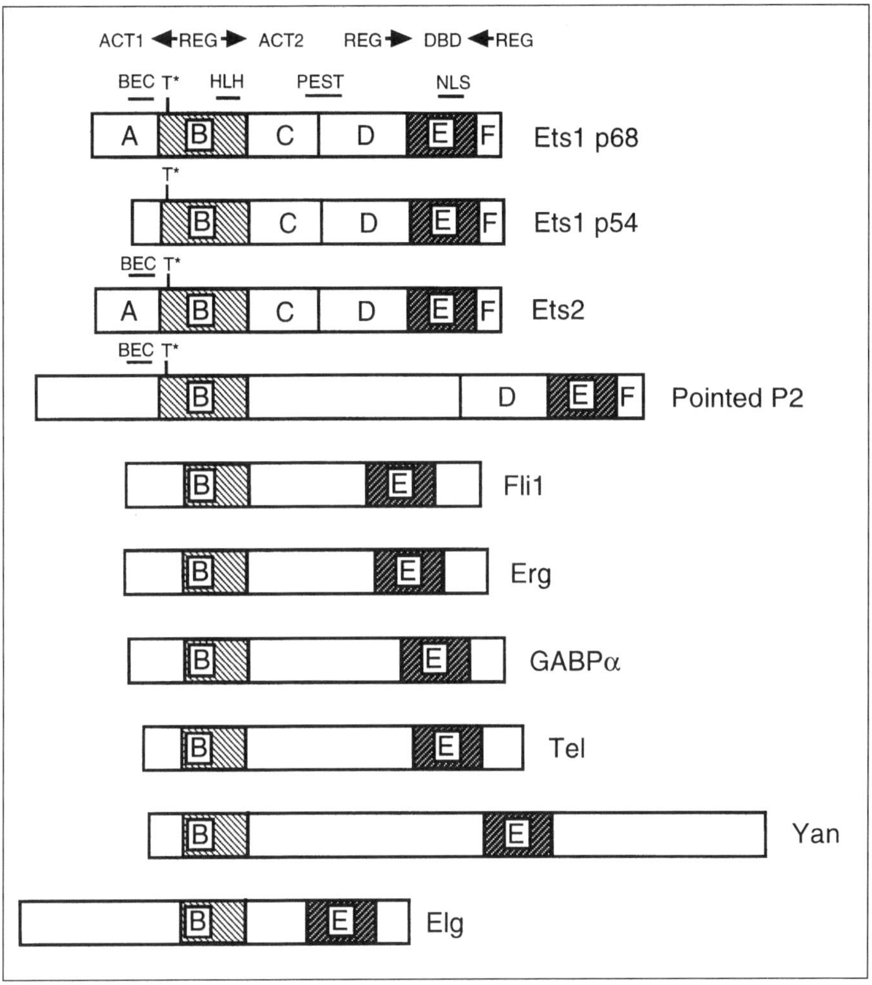

Fig. 14.1. The Ets1 subfamily. The members of the Ets1 subfamily of proteins are represented. Pointed P2 and Yan are *Drosophila* proteins, the others vertebrate. The subfamily is characterized by two similar sequences, the Ets (E) and the B domains (also called Pointed or RIII). The sequences that are similar in B are longer in Ets1, Ets2 and Pointed P2 (see Fig. 14.5) and contain the MAP kinase phosphorylation site that is conserved from *Drosophila* to man (T*) . BEC is found in Ets1/p68, Ets2 and Pointed P2. D and F domains are found in Ets1, Ets2 and Pointed P2. Also indicated are the helix-loop-helix sequence of the B domain, the PEST sequence and the nuclear localization sequence of Ets1/p68. ACT1 and ACT2 are transcription activation domains found in Ets1/p68 and Ets2. REG are regulatory domains. DBD is the Ets DNA-binding domain.

p54, act as independent activation domains.[30] They have a highly related sequence, BEC, that is required for transcription activation.[29] Interestingly, BEC is also found in a *Drosophila* protein, Pointed P2,[29] that in addition has the adjacent conserved Pointed domain[31] and a D domain related in sequence and function.[32] Ets1 and 2 presumably arose by duplication of an ancestral gene, that also evolved towards Pointed.[14,29] The more distantly related proteins, Erg and Fli1 (Fig. 14.1), are proposed to have evolved from a gene duplication that preceded that of Ets1 and 2.[14]

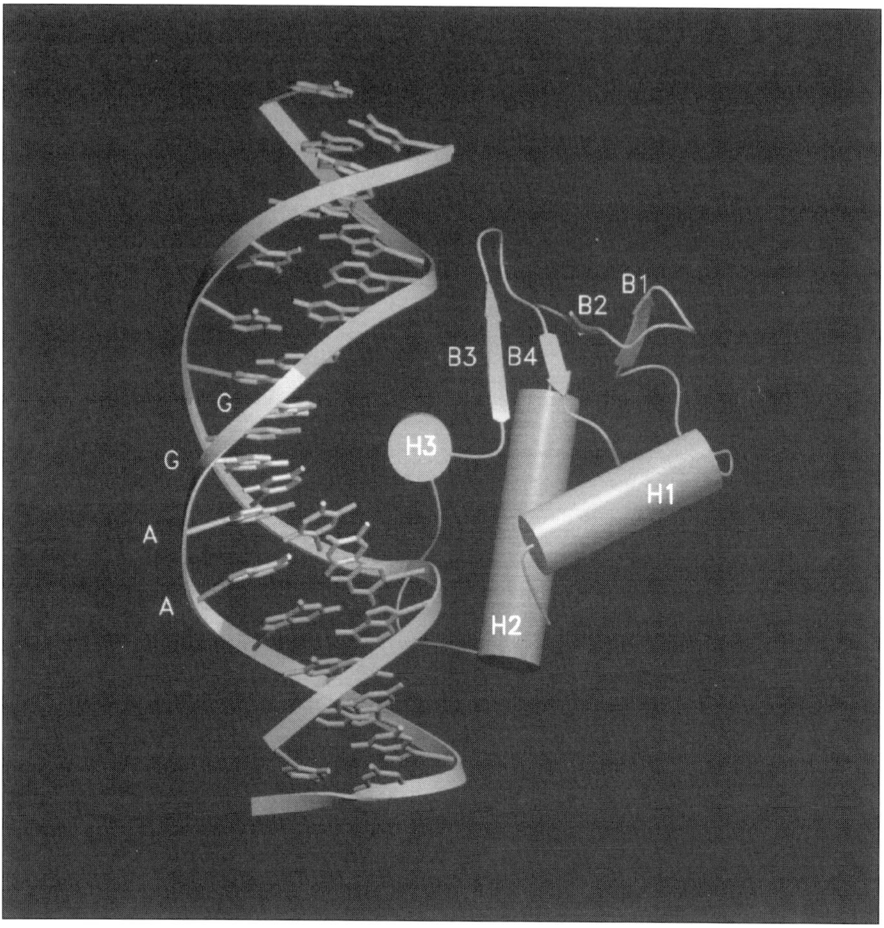

Fig. 14.2. Three dimensional structure of the Ets domain. The picture was generated with Molscript from the structure determined by Kodandapani et al.[10] The core recognition sequence (5'-GGAA-3') is labeled. The three α–helices, H1-3 pack on the β sheet formed from the β-strands B1-4. H3 is the recognition helix of the H2-T-H3 helix-turn-helix motif. The N-terminus is oriented towards the bottom.

The Ets Domain and Surrounding Regulatory Elements

Ets domains are composed of about 85 amino acids. Structures have been determined by NMR for Ets1[9,11,33] and Fli1[34,35] and X-ray crystallography for PU.1.[10,10a] The original NMR structure of the Ets1 Ets domain bound to DNA is wrong, due to an unfortunate misassignment, and the structure has now been retracted.[11] The structure of the PU.1 Ets domain bound to DNA is very similar to those of both Fli1 and revised Ets1, and it correlates with the mutational analyses and chemical shift data on other Ets proteins, and thus appears to be a paradigm for Ets domain structures.[10a] The Ets domain free in solution and bound to DNA is composed of a three helix bundle (H1 to 3, Fig. 14.2) packed on a scaffold formed by a four stranded β sheet (B1 to 4). Helix 3 (H3) of the helix-turn-helix motif interacts with DNA in the major groove. The structure is called a winged helix-turn-helix to highlight the role of the β sheet and associated wing like loops in the structure and to distinguish it from other helix-turn-helix motifs.[9-10a,33-35] An extended family of proteins have winged helix-turn-helix folds, including *E. coli* CAP,[36] heat shock factor,[37-40] hepatocyte nuclear factor HNF-3γ[41] and others.[42-47] Although the sequences of these proteins are not significantly similar, a related pattern of hydrophobic aminoacids that form the core of the related structures is detected. The Ets domain is a distinct group that differs in the number and lengths of the β-strands, the lengths of the helices and loops and their relative disposition. The most divergent Ets domains are those of Ets1 and PU.1, which are about 35% similar at the sequence level (Fig. 14.3). The conserved amino acids are clustered in the secondary structures and form the hydrophobic core, including in Ets1 the W and FLxxxL in helix 1, the IxW in β1, the F in β2, W in helix 2, YxxL in helix 3 and YxF in β4 (Fig. 14.3, see arrows). Divergent amino acids are mainly found in the loops and turns and may confer distinct properties such as DNA-binding specificities.[9]

The PU.1 Ets domain bound to DNA has a distinct pattern of DNA recognition from the novel loop-helix-loop architecture (Fig. 14.2). Helix 3 lies deep in the major groove and makes contacts through two conserved arginines with the core sequence (5'-GGA-3') in both strands. All Ets domains have these arginines (Fig. 14.3) and require the core sequence for binding. The two loops contact the phosphate backbone, contributing to extensive DNA contacts over 30 Å. They neutralize the charge of the phosphate backbone on one side, apparently thereby causing the DNA to uniformly curve towards the protein, forming an 8° bend. Strictly conserved lysines, in the loop between H2 and H3 and in the wing (B3), contact the backbone. The contacts made by the loop are unique to Ets proteins amongst the helix-tern-helix superfamily. The length and sequence of both contact loops differs amongst members of the family, and may be important for specific recognition by the different members.[48]

The Ets1 Ets domain bound to DNA was erroneously thought[11] to have a distinct feature, true and extreme intercalation by tryptophan.[49] Before the retraction,[11] Kopandapani et al[10] suggested from the PU.1 structure that all Ets domains interact in similar ways with DNA, and that tryptophan intercalation may have occurred during renaturation.[50] Werner et al[11] argued for intercalation from the slow kinetics of Ets1 domain DNA interactions and the generation of a DNAase I hypersensitive site upon DNA binding.[51] There are intercalating factors: TBP, PurR and SRY, which penetrate to progressively lesser degrees.[49] The H1 trypthophan of Ets proteins, and the corresponding tyrosine of PU.1, turn out to be located in the hydrophobic core and do not intercalate.[10,10a]

The DNA-binding domain of Ets1 is subject to cis regulation by surrounding adjacent amino acid sequences (D, F in Fig. 14.1) and to trans regulation by other factors. D and F-like domains, related in sequence and/or function, are found in Ets2 and Pointed[32] and possibly Erg[52] and Fli1.[53] Functionally equivalent elements are present in other Ets factors. For example, Net has two domains that inhibit DNA binding (see below).[54] Understanding the molecular mechanisms of intramolecular regulation between domains of a protein is of growing interest for many factors (transcription factors, kinases, etc.), one example being the tumor suppressor p53.[55]

The Ets domain appears to be constrained from DNA binding by interactions with sequences in D and F (Fig. 14.1), which drive dissociation from the DNA. Ets domains lacking either element form stable complexes with DNA. The change in sequence of the F domain of vEts enhances complex stability with DNA and alters its biological properties.[56] A natural isoform of Ets1, produced by alternative splicing, lacks the D domain. Structural and biochemical studies have led to an allosteric model of inhibition of DNA binding which does not involve direct masking of the Ets domain from the DNA.[57] The first helix (H1) of the Ets domain is proposed to form a helix bundle with two helices from domain D and a helix from F (I-1, I-2 and H4, respectively, see inhibitor Fig. 14.4). H1 thus couples the

```
                                        strand     strand                          strand  strand
                     helix 1              ß1         ß2    helix 2      helix 3      ß3      ß4
Net    hum.  ITLWQFLLQLLLDQK..HEHLICWTSNDG.EFKLLKAEEVAKL.WGLRKNKTN.MNYDKLSRALRYYYDKNIIKKVI.GQKFVYKFV
Sap1   hum.  ITLWQFLLQLLQKPQ..NKHMICWTSNDG.QFKLLQAEEVARL.WGIRKNKPN.MNYDKLSRALRYYYVKNIIKKVN.GQKFVYKFV
Lin-1  elg.  ITLWQFLLELLQQDQ..NGDIIDWTSRDG.EFRLIDAEAVARK.WGQRKAKPH.MNYDKLSRALRYYYEKNIIKKVI.GKKFVYRFV
Elk1   hum.  VTLWQFLLQLLREQG..NGHIISWTSRDGGEFKLVDAEEVARL.WGLRKNKTN.MNYDKLSRALRYYYDKNIIRKVS.GQKFVYKFV
ERF    hum.  IQLWHFILELLRKEE..YQGVIAWQG.DYGEFVIKDPDEVARL.WGVRKCKPQ.MNYDKLSRALRYYYNKRILHKTK.GKRFTYKFN
Pok    dro.  RLLWDFLQQLLNDRNQKYSDLIAWKCRDTGVFKIVDPAGLAKL.WGIQKNHLS.MNYDKMSRALRYYYRVNILRKVQ.GERHCYQFL
Tel    hum.  RLLWDYVYQLLSDS..RYENFIRWEDKESKIFRIVDPNGLARL.WGNHKNRTN.MTYEKMSRALRHYYKLNIIRKEP.GQRLLFRFM
Elf1   hum.  IYLWEFLLALLQDKAT.CPKYIKWTQREKGIFKLVDSKAVSRL.WGKHKNKPD.MNYETMGRALRYYYQRGILAKVE.GQRLVYQFK
Ets-4  dro.  IHLWQFLKELLASPQV.NGTAIRWIDRSKGIFKIEDSVRVAKL.WGRRKNRPA.MNYDKLSRSIRQYYKKGIMKKTERSQRLVYQFC
E74A   dro.  TYLWEFLLKLLQDREY.CPRFIKWTNREKGVFKLVDSKAVSRL.WGMHKNKPD.MNYETMGRALRYYYQRGILAKVD.GQRLVYHFV
ER81   mur.  LQLWQFLVALLDDPS..NSHFIAWTGRGME.FKLIEPEEVARR.WGIQKNRPA.MNYDKLSRSLRYYYEKGIMQKVA.GERYVYKFV
ERM    hum.  LQLWQFLVTLLDDPA..NAHFIAWTGRGME.FKLIEPEEVARR.WGIQKNRPA.MNYDKLSRSLRYYYEKGIMQKVA.GERYVYKFV
PEA3   mur.  LQLWQFLVALLDDPT..NAHFIAWTGRGME.FKLIEPEEVARL.WGIQKNRPA.MNYDKLSRSLRYYYEKGIMQKVA.GERYVYKFV
Ets-1  mur.  IQLWQFLLELLTDKS..CQSFISWTGDGWE.FKLSDPDEVARR.WGKRKNKPK.MNYEKLSRGLRYYYDKNIIHKTA.GKRYVYRFV
v-Ets  vir.  IQLWQFLLELLTDKS..CQSFISWTGDGWE.FKLSDPDEVARR.WGKRKNKPK.MDYEKLSRGLRYYYDKNVIHKTA.GKRYVYRFV
Ets-2  avi.  IQLWQFLLELLTDKS..CQSFISWTGDGWE.FKLADPDEVARR.WGRRKNKPK.MNYEKLSRGLRYYYDKNIIHKTS.GKRYVYRFV
Erg    hum.  IQLWQFLLELLSDSS..NSSCITWEGTNGE.FKMTDPDEVARR.WGERKSKPN.MNYDKLSRALRYYYDKNIMTKVH.GKRYAYKFD
Ets-6  dro.  IQLWQFLLELLADSS..NANAISWEGQSGE.FRLIDPDEVARR.WGERKAKPN.MNYDKLSRALRYYYDKNIMTKVH.GKRYAYKFD
Fli-1  hum.  IQLWQFLLELLSDSA..NASCITWEGTNGE.FKMTDPDEVARR.WGERKSKPN.MNYDKLSRALRYYYDKNIMTKVH.GKRYAYKFD
Elg    dro.  VQLWQFLLEILTDCE..HTDVIEWVGTEGE.FKLTDPDRVARL.WGEKKNKPA.MNYEKLSRALRYYYDGDMISKVS.GKRFAYKFD
GABPα  mur.  IQLWQFLLELLTDKD..ARDCISWVGDEGE.FKLNQPELVAQK.WGQRKNKPT.MNYEKLSRALRYYYDGDMICKVQ.GKRFVYKFV
ER71   mur.  IQLWQFLLELLQDGA..RSSCIRWTGNSRE.FQLCDPKEVARL.WGERKRKPG.MNYEKLSRGLRYYYRRDIVLKSG.GRKYTYRFG
Pu1    hum.  IRLYQFLLDLLRSGDMK..DSIWWVDKDKGTFQSSKHKEALAHRWGIQKGNRKKMTYQKMARALRNYGKTGEVKKVK..KKLTYQFS
```

(hum: Human, mur: Murine, elg: C. Elegans dros: Drosophila, vir: Virus, avi: Avian).

Fig. 14.3. Alignment of Ets domain sequences. Twenty-three Ets domain sequences were aligned using programs from the GCG package. Secondary structure is indicated on the sequence, conserved sequences are highlighted and hydrophobic amino acids from the core of the Ets1 structure are indicated by arrows.

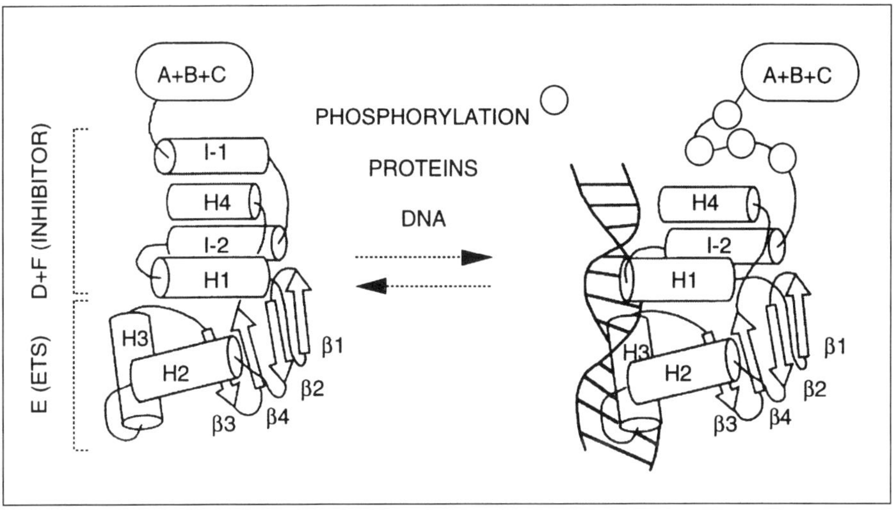

Fig. 14.4. Allosteric model of regulation of Ets1 DNA-binding activity. The model is adapted mainly from the work of B. Graves and collaborators (see text). A+B+C: activation domains and regulator (see Fig. 14.1). H1-3 and β1-4: Ets domain (see Figs. 14.2 and 14.3). H4: α helix C-terminal to Ets domain in region F. I-1, I-2: α helices in domain D. The equilibrium between bound and free protein is proposed to be regulated by phosphorylation, protein interactions and the sequence of the Ets motif (DNA). Note that phosphorylation of the D domain inhibits DNA binding (see text).

inhibitory sequences with the DNA-binding part of the Ets domain. In the presence of DNA the structure is found to be rearranged, most notably by unfolding of the N-terminal helix (I-1). This model explains both biochemical studies and structural analysis of a minimal Ets1 protein that is refractory to DNA binding.[9,57-59] The model has interesting implications for regulation of DNA binding. Ets1 cooperates functionally with a number of transcription factors (e.g., Sp1, Myb, AP1, PU.1, CBF/PEBP2, ATF/CREB, GHF1/Pit1). Amongst the potential mechanisms of cooperation is direct interactions with the D and F domains that induce the relaxed conformation.[60] This mechanism may be mimicked by antibodies that interact with the Ets1 D domain and stimulate DNA binding. Antibodies directed against either inhibitory domain of Net also stabilize DNA binding.[54] There may be additional inhibitory sequences in Ets1. Cooperative DNA binding by Ets1 with PEBP2 to the TCRα enhancer requires additional sequences N-terminal from the minimum inhibition domain[61] that have also been shown to inhibit DNA binding by others.[62] However, the level of DNA binding by Ets1 lacking these additional sequences can still be enhanced.[61] Phosphorylation is proposed to either stimulate or inhibit DNA binding.[57] Ets1 is phosphorylated in vivo,[63-67] and at least four of the phosphorylation sites in the D domain are in the region that changes conformation upon DNA binding (see Fig. 14.4).[65] Phosphorylation of the D domain has been shown to inhibit DNA binding[65,68] and transactivation by Ets1.[69]

The Ets domain of Ets1 complexes with other proteins, either directly (AP1, MafB) or in the presence of DNA (Pax-5). Ets1 cooperates functionally with AP1 in transfection assays and in whole organisms during development[70] and combined Ets1 and AP1 binding sites have been identified in a large number of promoters (see list in ref. 71 and also refs. 72-75). Ets and AP1 form a complex in vitro.[71,74] Bassuk and Lieden[71] have shown that Ets and the Jun and Fos components of AP1 form a heterotrimer in which

Ets interacts directly with Jun and Jun recruits Fos. Ets and Jun interact through their DNA-binding domains. The Ets domains of Ets1, Elf1, Fli1 and PU.1 can form the complex, suggesting that the interaction is mediated by a conserved feature of the Ets domain. The basic region of Jun mediates both the Ets interaction and DNA binding, whereas the leucine zipper forms the complex with Fos. In cell extracts Elf-1 is found to be associated with the three Jun factors (cJun, JunB and JunD) and four Fos proteins (cFos, *FosB*, Fra1 and Fra2). The promiscuity of the interaction is compatible with the large number of promoters that contain both elements, but raises the challenge of identifying the mechanisms that confer specificity. Transcription is usually increased by Ets combined with AP1 (see, however, ref. 76). Interestingly, activation by Ets1 is repressed by another AP1 like factor, MafB.[77] As with AP1, the interactions are mediated by the DNA-binding domains of Ets1 and MafB (basic region and/or the leucine zipper). MafB inhibits Ets1 mediated transactivation of the transferrin receptor, which is essential for erythroid differentiation, and blocks differentiation of erythroblasts. MafB is proposed to maintain the myelomonocytic cell lineage by repressing the activity of erythroid specific promoters despite the presence of Ets1.

Ets family proteins are recruited to the B cell specific promoter, mb-1, by Pax-5.[78] Pax-5 is an important regulator of B cell specific transcription. It acts as a docking protein for Ets factors, somewhat reminiscent of SRF and the *c-fos* promoter (see below). However, the Ets factor can also act as the docking factor, if a single base pair is altered in the overlap between the two motifs to more closely resemble an Ets motif. The interactions involve the DNA-binding domains of both proteins and DNA. Pax-5 can recruit a number of Ets proteins, including Ets1, Elk1 and Net, but, interestingly, not SAP1a. The determinant of specificity has been narrowed down to a single amino acid, valine 68, that lies just C-terminal to H3 (Figs. 14.2 and 14.3). Mutation of valine 68 to aspartic acid, found in Elk1, Net and Ets1, permits recruitment, perhaps by direct interaction between the factors bound to adjacent sites on the DNA. Intriguingly, valine 68 of SAP1a and aspartic acid 68 of Elk1 are critical determinants of their different DNA-binding specificities.[48] Structure predictions for the TCFs suggest that the critical amino acids face away from the DNA, towards the Pax5 binding site (towards the bottom in Fig. 14.2).

The Pointed or B Domain and Adjacent Activation Domains

Regions A-C of Ets1 and Ets2 are involved in transcription activation. The A regions of Ets1/p68 and Ets2 have a common function, despite overall sequence differences, whereas that of Ets1/p54 has no known function. The C (or RIII) regions are also activation domains in both Ets1 and Ets2. The B regions resemble the "Pointed" domain of Pointed P2, and are postulated to be both regulatory domains that control the activities of the surrounding activation domains and protein-protein interaction domains.[79] Interestingly, this domain is the target for regulation of Ets activity by the Ras signaling cascade, by a mechanism that has been conserved from *Drosophila* to man.

Ras controls several signaling pathways (see below). The best established downstream effector is Raf kinase, which initiates a series of phosphorylations through the MEK and MAPK kinases, leading to the modification of transcription factors.[80,81] Another downstream effector is MEKK1, which is at the beginning of the JNKK/SEK/MKK4 and JNK/SAPK cascade. There is at least one more similar cascade involving p38.[81,82] However, there are many other potential downstream targets of Ras, including, PI(3)K, p120GAP, ralGDS, PKCζ and KSR, although their precise roles still need to be established.[83-85] The Ras/MAP kinase pathway has been studied by biochemical approaches in mammalian cells and genetic studies in *Caenorhabditis*, *Drosophila* and yeast. The emerging patterns indicate that Ets factors are evolutionary conserved effectors of the pathways.

In *Caenorhabditis* the Ets gene product, Lin-1, has been shown genetically to act after the Ras-Raf-MEK-MAPK cascade that mediates development of the vulva. Lin-1 is a negative regulator with multiple potential MAPK phosphorylation sites. The Ets domain is most similar to the Elk/Sap1/Net subfamily of Ets proteins in mammals, both in sequence and N-terminal position, but it lacks the other similarities outside the Ets domain.[86] In *Drosophila*, the two Ets proteins Pointed P2 and Yan have been shown genetically and biochemically to be targets of the Ras/rolled MAPK pathway.[87,88] Alternative splicing produces two Pointed proteins, P1 and P2, with a common carboxy-terminal Ets domain. Both Pointed P2 and Yan have consensus MAPK phosphorylation sites. In transfection assays in *Drosophila* cells P2 is a weak activator that is stimulated by expression of activated forms of Ras or rolled MAP kinase. Yan is a negative regulator of the constitutively active Pointed P1 isoform, and Ras down-regulates Yan.[87,89] P2 is phosphorylated in vitro by rolled MAP kinase on a single site.[88]

In mammalian cells the Ras responsive factors include the Elk1/Sap1/Net subfamily (see below), ERF and Ets1/Ets2.[90] ERF is a distinct factor with an N-terminal Ets domain and a C-terminal repressor domain. It inhibits transcription and tumorigenesis, and is regulated by Ras, possibly through phosphorylation of the repressor domain by MAP kinase.[91] The Ets1/Ets2 subfamily proteins are regulated by Ras.[92-94] Ets1 and 2 and the Ras signaling cascade are found in many cell types, raising the question of how the cell specific responses are produced. In the case of the pituitary specific prolactin promoter, Ras activation of promoter activity is mediated by functional interactions between Ets1 and GHF1/Pit1.[93] The Ras/Raf/MAP kinase signaling pathway activates the prolactin promoter activity via a composite Ras response element (RRE) comprised of an Ets binding site adjacent to a Pit-1/GHF-1 binding site.[93,95] GHF1 or Ets1/p68 alone enhance the Ras response and together they synergise.[93] The growth hormone gene promoter, that is also tissue specific and has both Ets and GHF1 binding sites, fails to respond to Ras, GHF-1 and Ets1, since its proximal promoter lacks the composite organization of Ets and GHF sites.[93] Ets1-GHF1 interaction is also critical for the lactotroph-specific basal activity of the rat PRL promoter.[93a] In reconstitution transfection assays, rPRL promoter activity is low in nonpituitary cells. Expression of either GHF-1 or Ets1 activates the rPRL promoter to a low level, whereas together they activate the rPRL promoter to very high levels. Activation may involve specific interactions between GHF1 and Ets1 but not Ets2.[93a] It appears that Ets1 and GHF-1 physically interact in a DNA-independent manner to establish optimal lactotroph-specific expression of the rPRL gene. Ras mediates growth factor signals that target the Ets1 factor, whereas GHF1 acts as the cell specific anchor factor.

Ets1/p68 is has two Ras regulated activation domains, A+B and C.[32,93] A MAP kinase phosphorylation site, required for Ras activation of A+B, is conserved in the *Drosophila* Pointed P2 protein and in vertebrate Ets1/p54 and Ets2 (Fig. 14.5).[94] The conserved BEC sequence (Fig. 14.1) contributes mainly to basal activity and may affect Ras activation of Ets1. There is a remarkable conservation of the Ras/MAP kinase signaling cascade in evolution. The Ras signaling pathway converges on an evolutionarily conserved site in vertebrate Ets1 and Ets2 and *Drosophila* Pointed P2.[96] There are other similarities in regulation by Ras. In mammalian cells Ets1 and Jun cooperate transcriptionally and they interact physically (see above). In *Drosophila* Jun cooperates with Pointed to induce photoreceptor R7 fate.[70] Therefore, the classical Ras/MAP kinase targets evolutionarily related nuclear factors, such as Ets and Jun, that may in turn orchestrate conserved and coordinate responses of gene expression in recognizable patterns.

The second Ras responsive element of Ets1 (C and part of D) does not have an apparent homologue in Pointed P2. This domain has been shown to be phosphorylated by other signaling pathways. Valentine et al[66]

p68	IQEVPTGLEH	YSTDMECADV	PLLTPSSKEM	MSQALKATFS	GFAKEQQRLG	IPKDPQQWTE
ets2	LQEVPTGFDS	TSYESNNCEL	PLLTPCSKAV	MSQALKDTFS	GFTKEQCRLG	IPNNPWLWTE
p2a	LCSLSVYPDH	EASLEVPTAL	PPLTPGTNRK	VNEVLKASFA	SWEKEVQKCN	ITKDPREWTE
fli1	SRESPVDCSV	SKCNKLVGGG	EANPMNYNSY	MDEKNGPPPP	NMTTNERRVI	VPADPTLWTQ
erg	SRNSPDECSV	AKGGKMVGSP	DTVGMNYGSY	MEEKHMPPP.	NMTTNERRVI	WPADPTLWST
gah	ESEAHLVEEA	QVITLDGTKH	ITTISDETSE	QVTRWAAALE	GYRKEQERLG	IPYDPIQWST
yan	....MSKMKM	LPVQLSLNSL	NPGIWSDVLW	RCPPAPSSQL	AELKTQLPPS	LPSDPRLWSR
tel	TPAQCSIKQE	RISYTPPESP	VPSYASSTPL	HVPVPRALRM	EEDSIRLPAH	LRLQPIYWSR
elg	SSSSESPIKT	PLKRMHKEDS	EEESVEGKDV	KPVLNWVLDS	KFKREQIRLK	IPEAANEWTH

p68	THVRDWVMWA	VNEFSLKGVD	FQKFCMNGAA	LCALGKECFL	ELAPDFVGDI	LWEHLEILQK
ets2	QHVCQWLAWA	TNEFSLANVN	IHQFLMSGQD	LCNLGKERFL	ELAPDYVGDI	LWEHLEQMIK
p2a	EHVIYWLNWA	KNEFSLVSMN	LDPFYKMKGR	AMVDLGKEKF	LAITPPFTGD	IWEHLDILQK
fli1	EHVRQWLEWA	IKEYGLMEID	TSFFQNMDGK	ELCKMNKEDF	LRATSAYNTE	VLLSHLSYLR
erg	DHVRQWLEWA	VKEYGLPDVN	ILLFQNIDGK	ELCKMTKDDF	QRLTPSYNAD	ILLSHLHYLR
gah	DQVLHWVVWV	MKEFSMTDID	LTT.NISGRE	LCSLNQEDFF	QRVPR..GEI	LWSHLELLRK
yan	EDVLVHLREC	VREFDLPKLD	FDLFQMNGKA	LCLLTRADFG	HRCPGA.GDV	LHNVLQMLII
tel	DDVAQWLKWA	ENEFSLRPID	SNTFEMNGKA	LLLLTKEDFR	YRSPHSGDVL	YELLQHILKQ
elg	AHVTYWLEWA	VKQFELVGIN	MSDWQMNGQE	LCAMTHEEFN	QKLPRDPGNI	FWTHLQLLKE

Fig. 14.5. The B or Pointed domain. The sequences of the B domains of the Ets1 subfamily were aligned with programs from the GCG Package. Identical and similar sequences are highlighted. The homology is more extensive for Ets1/p68, Ets2 and Pointed P2 and covers the MAP kinase phosphorylation site that is regulated by the Ras signaling pathway (PLLT*P).

showed that in the Ets1/p54 isoform Ser-156 as well as Thr-73 is phosphorylated in response to the increased intracellular calcium following ligation of IgM on B lymphocytes. Using T cells, Rabault and Ghysdael[65] showed that calcium and anti-CD3 induce phosphorylation of four sites, two of which are in the Ras responsive domain (Ser-251 and Ser-257, equivalent to Ser-295 and Ser-301). Ets1 has also been reported to be hyperphosphorylated during mitosis in T cells[63] and in response to neuro-transmitters in astrocytes,[67] although the phosphorylation sites were not mapped. It remains to be seen if there are similarities between these and the Ras signaling cascade.

Sequences similar to the Pointed domain are found in other members of the Ets1 subfamily. The Pointed domain may in fact comprise two sub-domains. The MAP kinase phosphorylation site and flanking sequences (approximately 75-100 of Ets1/p68) are only found in Ets1, Ets2 and Pointed P2 (Figs. 14.1 and 14.5, whereas the second part of the domain (approximately amino acids 100-175) is conserved in all the proteins with a Pointed domain. The second part contains the helix-loop-helix that Seth and Papas[79] have proposed may mediate interactions with other proteins (Fig. 14.5). The role of the B or Pointed domain is particularly interesting because it is conserved in evolution and is critical for the oncogenic activation of Tel in human tumors. The Pointed domain of Tel is retained in fusion proteins produced by chromosomal translocations to various loci.[97-100]

In the Tel-AML1 fusion protein, the Pointed region of Tel is linked to the DNA-binding and activation domains of AML1.[98] The Tel-PDGFβ receptor and Tel-cAbl fusion proteins are constitutively active kinases.[97,100] The Pointed domain of the Tel-cAbl fusion protein mediates oligomerization of the protein, which is required for transformation, aberrant cytoskeletal localization and constitutive tyrosine kinase activity.[101] In the Tel-PDGFβ fusion a homo-oligomerization domain has been narrowed down to a 65 amino acid region of Tel corresponding to the widely conserved region of the Pointed domain (Fig. 14.5). The equivalent regions of other Ets proteins do not appear to mediate homodimerization, suggesting that they mediate heterotypic protein-protein interactions.[102] Interestingly, the A+B region of Ets1/p54, containing the Pointed domain, is sufficient for direct complex formation with several proteins that were identified in a yeast dual hybrid screen (Hahn SL, Wasylyk B, Hahn SL, Filipe P and Wasylyk B; manuscripts in revision).

Ets1 in Development

Ets1 is expressed in a tissue and temporal specific manner, and differentially during embryogenesis. In mouse and chicken embryos, Ets1 expression is widespread in a variety of mesodermal cells. It is first detected during gastrulation, in ventral mesoderm and then hemangioblasts, the common progenitor of the primitive erythroid cell lineage and blood vessel endothelial cells. Interestingly, the Ets1/p68 isoform is specific to hemangioblasts, whereas the Ets1/p54 isoform is expressed in lymphoid cells as well. The E26 virus encoding a derivative of Ets1/p68 maintains the multipotency of hematopoietic progenitors. Throughout development, Ets1 expression is associated with endothelial cells involved in vascularization, the formation of new blood vessels, and angiogenesis, the sprouting of vessels from existing ones. Expression in lost in large vessels and adult capillaries.[25,103-105] Ets1 is found in the brain at early stages (8 days), later on in most tissues (15 days), followed by predominant expression in lymphoid tissues, brain and organs undergoing branching morphogenesis, such as the lung. In neonatal development it is expressed only in lymphoid organs and the brain.[106,107] In the adult, Ets1 is expressed in lymphoid cells of the B and T cell lineage.[6,108,109] Ets1 expression begins in the mouse thymus two days before birth, coinciding with the appearance of mature single positive thymocytes, particularly $CD4^+CD8^-$. Expression is high in quiescent T cells, decreases upon activation and re-augments upon prolonged activation.[110] The role of Ets1 in lymphoid cell activation and development has been studied by antisense and gene knock outs.

Inhibition of Ets1 in T cells with antisense blocks IL2 induction by activators of T cells, such as phorbol esters and ionomycin.[111] Embryonic stem cells with homozygous inactive Ets genes and normal RAG2 genes have been used to reconstitute the lymphoid system of RAG2$^{-/-}$ deficient mice. These mice generated mature B and T cells, showing that Ets1 is not required for normal B and T cell development. However, there were dramatically lower numbers of mature T cells due to an increased susceptibility to cell death. The number of B cells was normal and they still responded to inducers of proliferation, but they had a larger proportion of IgM positive plasma cells.[17,18] Ets1 may be required to maintain T cells in a quiescent state and prevent inappropriate cell-cycle progression, or it may be required for the production of survival factors. In B cells it appears to prevent terminal differentiation. The knock out of two other Ets family genes has been reported, Fli1 and PU.I (Spi1), both of which are involved in the late stages of erythroleukemia induced by Friend murine leukemia virus. Fli1 inactivation leads to a defect in thymus development and hypocellularity, which is not associated with a loss of a specific sub-population of thymocytes or increased apoptosis, which suggests that Fli1 somehow affects an early progenitor. The mice expressed a truncated Fli1 protein, which may have contributed to the phenotype.[20] PU.1 is highly expressed in the monocytic and B lineages. The PU.1 knock-out caused late embryonic lethality and a defect in generation of B cells, T cells, monocytic and granulocytic progenitors without affecting the numbers of megakaryocytes and erythrocytes. PU.1 appears to be required for the generation of a multipotential lymphoid-myeloid progenitor.[19]

Ets1 in Cancer

Ras oncogene transformation of NIH3T3 mouse fibroblasts is reverted by transdominant mutants of Ets proteins, suggesting that members of the Ets family are mediators of transformation.[112,113] NIH3T3 cells express a number of Ets proteins and their individual contribution to Ras transformation remains to be established. Ets1 is not expressed in these cells, but Ets1 overexpression increases their growth and tumorigenic properties.[79] Ets1 overexpression does not immortalize primary rat embryo fibroblasts but increases the tumorigenic properties of E1A immortalized derivatives.[114] Ets1 appears to have a weak transforming activity or may require other cooperating oncogenes for transformation of primary cells. Ets1 may rather be associated with maintenance of either the quiescent or differentiated state. This is compatible with the effects of *ets1* knock-out in the mouse genome (see above). Furthermore, Ets1 levels increase during differentiation of P19 embryonal carcinoma cells[106] and Ets1 expression in human colon carcinoma cell lines reduces tumourigenesis.[115] Ets1 is not expressed in most epithelial tumors and cell lines.[108]

The *ets1* locus is the target of proviral integration in rat T cell lymphomas induced by Moloney Murine Leukemia virus.[24] The insertion site is immediately 5' to the first exon and results in only a modest increase in RNA expression, suggesting that there is perhaps a subtle effect on fine regulation of the gene. There have been no reports of chromosomal translocations that disrupt the Ets1 gene. Other genes are rearranged in human malignancies, including Erg in myeloid leukaemia,[116] and Fli1, Erg and ETV1 in Ewings sarcoma.[117-119] The Ets domain is retained in the fusion proteins produced by these translocation. There appears to be no correlation between Ets1 expression and the severity of human leukaemias.[120] Ets1 may be required for the survival of leukemic cells.[3] In invasive carcinomas, Ets1 is found to be highly expressed in the stromal fibroblasts surrounding invasive tumors and in the endothelial cells of stromal capillaries but not in the epithelial tumor cells.[25,26] Invasive processes and angiogenesis requires degradation of extracellular remodeling by proteases and their inhibitors (for example collagenase-1, stromelysin-1, urokinase plasminogen activator and TIMP). They have Ets binding sites in their promoters, suggesting that there could be a link between cytokines and angiogenic factors released from tumors, Ets1 and promoter activity. In a variety of human tumors, the expression patterns of Ets1, collagenase and urokinase suggest that they are involved in lung cancer invasion and that Ets1 may transactivate the collagenase promoter during tumor invasion.[121] Ets1 may also be involved in responses to tissue injury. Arterial cell wall injury activates smooth muscle cells to divide, migrate and secrete extracellular matrix. Balloon injury of rat arteries has been shown to increase Ets1 expression in medial smooth muscle cells, suggesting a role in vascular remodelling.[122] The correlative studies in vivo suggest important roles of Ets1 in regulation of tumor invasion, angiogenesis and tissue remodeling. However, Ets1 forms a part of regulatory circuits involving other factors and further studies will be required to establish its precise roles.

THE SERUM RESPONSE ELEMENT (SRE) RECRUITS ETS PROTEINS

Ets transcription factors mediate the nuclear uptake of intracellular signaling processes and convert them into a transcriptional response. In this scenario, activation of Ets proteins by specific, signal induced phosphorylation events leads to the transcriptional activation of promoters recognized by the modified Ets factors and their cooperating partners. The following discussion will concentrate on the serum response element (SRE),[123] one important and well characterized transcriptional control element which functions as a signal response element largely through the TCF subfamily of Ets proteins.

SREs Mediate Signal-Induced "Immediate Early Gene" Activation

The SRE was first identified as the cis-element that rendered the c-*fos* promoter responsive to stimulation of resting cells by serum.[124,125] Since then the SRE has been shown to confer rapid c-*fos* induction in response to a wide variety of extracellular stimuli, including mitogens, individual growth or differentiation factors (PDGF, EGF, or NGF), cytokines, the phorbol ester TPA, LPA, UV-light, oxidants and antioxidants.[90,126] Signaling through cascades of the MAP kinase network, induced by these stimuli, also targets the SRE.[90,127-131] The SRE is also critical for basal repression of c-*fos* transcription (see below).[132-135] In the c-*fos* promoter the SRE is positioned at coordinate -300 with regard to the start site of transcription and is situated between two other response elements, the SIE and the CRE, which mediate promoter activation by cytokines and cAMP/Ca^{2+}, respectively. In addition to c-*fos*, SREs regulate activation of many other "immediate early" genes (IEGs), whose transient transcription is induced rapidly in a cycloheximide insensitive manner. SRE-containing members of the IEG family include *egr-1, egr-2, pip 92, nur77, β-actin, vinculin* and *junB* (for reviews see 136,137). IEG induction upon extracellular signal activation is known to be crucial for cell proliferation (G0-G1 transition), cell differentiation,[138] and vertebrate development.[139]

The Structure of the SRE

Anatomically, the SRE is composed of two dyad symmetrically positioned CC/GG doublets that are separated by a string of six A/Ts (Fig. 14.6). This $CC(A/T)_6GG$ sequence has been referred to as the CArG-box. In the cocrystal with SRF core the SRE DNA is found in a highly distorted B-DNA conformation that is bent around the protein by 72°. In this segment the DNA displays a significantly narrowed major groove and a strongly widened minor groove.[140] The narrow groove faces the SRF protein. However, the bend is distributed nonuniformly around the dyad-axis of the CArG box such that near the two GG doublets an abrupt bend is seen in each case and the central $(A/T)_6$ segment is found smoothly curved.

Genomic footprinting of the c-*fos* SRE demonstrated a continuous occupancy of the SRE by proteins.[141,142] The observed

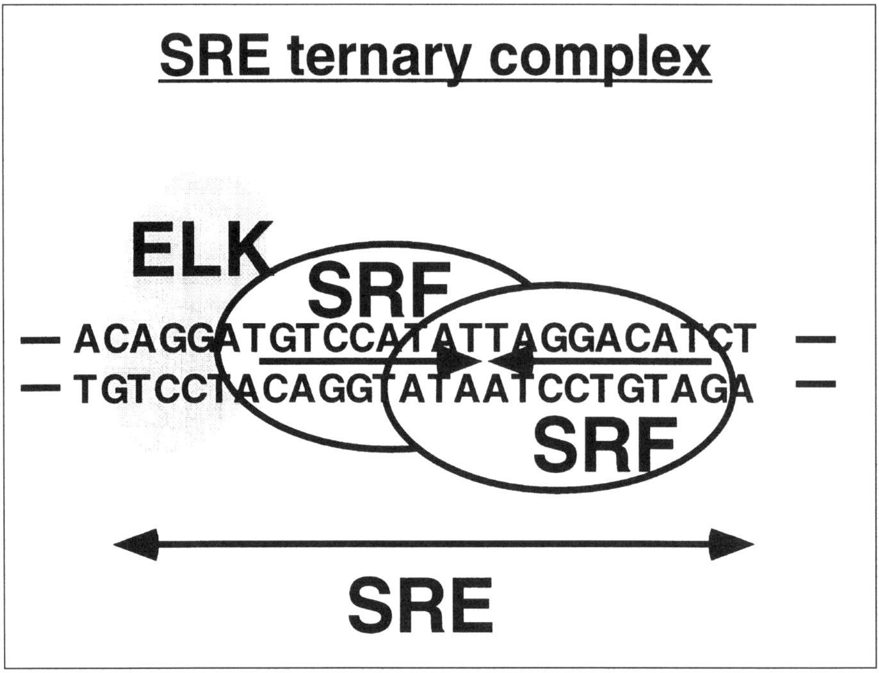

Fig. 14.6. Schematic drawing of the SRE-SRF-TCF ternary complex.

methylation protection pattern is identical to the one observed in vitro upon SRE complexation with SRF and TCF (see below). Recent evidence suggests that in vivo the c-*fos* SRE is packaged in a local chromatin structure that has a positioned nucleosome located between the SRE and sequences directly upstream of the TATA-box (Herrera R, Nordheim A, Stewart AF; submitted). This chromatin structure is postulated to permit direct or indirect protein-protein interactions between SRE-bound proteins and those interacting with the promoter-proximal CRE or with basal promoter sequences.

Proteins Interacting Directly with the SRE

In addition to SRF (see below), several other proteins were demonstrated to bind directly to the SRE in a sequence specific manner.[143] These include AP-1,[144] rNFIL-6 and rE12,[145] SRE-BP,[146] MyoD,[147] and YY1.[148-152] The contributions of these proteins to SRE activity are presently poorly understood. SRE recognition by SRF, YY1 and MyoD, has been shown to be mutually exclusive. Accordingly, YY1 and MyoD were argued to function as repressors of SRE activity.[147,150] In contrast, the demonstrated ability of YY1 to induce curvature within the murine c-*fos* promoter may aid SRE function.[151,152] It should be pointed out that none of these vitro interactions has been substantiated, by genomic footprint patterns, to occur in vivo, except for SRF.[141,142]

The SRE Binding Protein SRF

The human serum response factor (SRF),[153,154] a transcription factor of 508 amino acids in size, binds to the SRE as a homodimeric molecule. This interaction is of high affinity, displaying a K_d around 2×10^{-11}.[155,156] The central 90 amino acids of SRF, commonly referred to as the "core" region, span residues 132 to 222 and encode three important functions, namely the ability to dimerize, to bind the SRE in a sequence-specific fashion, and to interact with TCF proteins (see below).[154,157]

The structure of a cocrystal containing a dimer of SRF "core" bound to an SRE oligonucleotide was solved by X-ray crystallography.[140] It revealed a novel, striated DNA-binding structure that integrates the functions of protein dimerization and interactions with accessory proteins. The SRF "core" is built from three structural motifs (a long anti-parallel coiled-coil, a four-stranded anti-parallel β-sheet, and a dimerization surface). Sequence-specific DNA recognition is mediated through extensive minor groove interactions, contacts to the phosphate backbone, and an SRF-induced bend in the SRE double helix conformation. This arrangement might permit the formation of additional, DNA-directed protein contacts by accessory proteins. Residues important for both SRF homodimerization and additional heteromeric interactions with accessory proteins (see below) spread across two thirds of the "core" sequence.

The SRF "core" region encompasses the MADS box, a conserved 56 amino acid region present in the DNA-binding domains of many eukaryotic transcription factors (for review, see ref. 158). MADS box proteins are found throughout all eukaryotic species, ranging from the yeast *S. cerevisiae* up to plants and vertebrates, including man. Relatives of SRF, called RSRFs or MEFs, are expressed in a number of cell types, notably muscle.[159,160] MADS box proteins serve important gene regulatory functions determining cell growth during the development of multi-cellular organisms, such as floral organ development or vertebrate muscle formation. SRF is expressed fairly abundantly in many cell types. However an interesting selective expression domain was recently identified in the cardiac muscle of the developing chicken heart[161] suggesting hitherto unappreciated, important morphoregulatory functions for SRF in vertebrates.

SRF of the fruit fly *Drosophila melanogaster*, DSRF,[162] is encoded by the *pruned* allele which is expressed in the terminal branches of the tracheal system.[163] *Pruned* loss of function mutants are lethal and have terminal cells of the tracheal system that fail to extend cytoplasmic projections.[163] In *X. laevis* the XSRF protein is present in about 10^5 copies as the maternal supply to the unfertilized egg and is further synthesized after the mid-blastula transition during gastrulation.[164] This indicates an important role for SRF during early *Xenopus* development. Two MADS box proteins have been described in *S. cerevisiae*, namely Mcm1 and Arg80. Mcm1 is an essential gene product involved in the regulation of several G2-specific genes during the yeast cell cycle.[165,166] The genes encoding the B-type cyclins, *clb1* and *clb2,* fail to get activated in cells lacking functional Mcm1, leading to cell cycle arrest near the G2-M transition. In addition, Mcm1 is involved in regulating the transcriptional activity of cell-type specific genes,[167] including genes mediating the pheromone response in a- or α-cells. The signaling cascades activated by mating pheromones target nuclear transcription factors, such as STE12, which act at their target promoters through cooperative interactions with Mcm1. The yeast signaling cascades are homologous to the MAP kinase cascades in vertebrates.[168] Therefore, in both structural and functional terms Mcm1 can be considered the vertebrate SRF's homologue in the yeast.

The mammalian SRF protein has been shown to be modified post-translationally by *O*-linked GlcNAc glycosylation,[155,169] and phosphorylation.[170-172] The biological significance of either modification is presently obscure. Using the baculovirus expression system several sites of glycosylation have been mapped close to and within the transcriptional activation domain. Glycosylation has been proposed to have roles in nuclear localization, control of phosphorylation, or transcriptional activation. Phosphorylation sites have been mapped to serine residues 77, 79, 83, 85, 101, 103, 253, 435, and 446.[130,156,171,173-175] Phosphorylation at Ser-83 and 85 influences the kinetics of SRE binding, whereas phosphorylation at Ser-103 affects the overall binding affinity. Ser-103 phosphorylation occurs rapidly and transiently in growth factor stimulated cells[175] (R. Zinck and A. N., unpublished) but the use of SRF derivatives mutated at residue

103 as yet has not provided any revealing insight into the biological role of this modification event. Ser-103 of mammalian SRF is not conserved in the *X. laevis* XSRF protein,[164] raising doubts as to the biological importance of this phosphorylation.

Proteins Interacting with SRF or SRE-SRF Complexes

SRF can interact and recruit to the SRE factors, such as the TCFs, which cannot bind by themselves (see above). This recruitment is called "ternary complex formation" (for review see ref. 176).[177] SRF has been shown to establish direct protein-protein interactions with two myogenic basic-helix-loop-helix proteins, E12 and MyoD.[178] It binds E12-MyoD and E12-myogenin heterodimers but not the corresponding homodimers. The DNA-binding/dimerization domains are important for these interactions, but not contacts to DNA. E12 and MyoD have also been shown to interact with SRE DNA directly (see above).[145,147] Since SRE binding by SRF and homodimeric MyoD has been shown to be mutually exclusive it will now be interesting to measure the affinity of SRF-MyoD/E12 complexes for the SRE. Mutual modulation of the functional states of SRF and myogenic proteins may be very important in muscle development during the myoblast switch from proliferation to terminal cell differentiation. The myocardiac-specific homeobox protein Nkx-2.5 has recently been shown to interact with SRF both physically and functionally (quoted in ref. 161). SRF interacts with another homeobox protein, the Phox1 polypeptide, that helps to recruit SRF to the SRE and facilitates SRE-SRF complex formation.[179,180] Another accessory protein has recently been identified, the p65 subunit of NF-κB. It interacts both physically and functionally with SRF and stimulates its transcriptional activity.[181] SRF has been shown to interact with components of the general transcription machinery. The transcriptional activation domain of SRF (amino acid residues 412 to 508) interacts with the RAP74 subunit of TFIIF.[182,183] Finally, the viral Tax1 protein of the human T cell leukemia virus type 1 (HTLV-1), which acts as a transforming protein, has been shown to interact with SRF and stimulate SRF-mediated "immediate early" gene activation.[184]

Recruitment of Ets Proteins by the SRE-SRF Complex

The ability of SRF to interact with other proteins was first recognized in biochemical experiments, when SRF-containing HeLa cell extracts were found to form DNA complexes with slower electrophoretic migration than expected for a binary SRE-SRF complex.[177] Characterization of the additional protein component revealed a monomeric polypeptide, christened $p62^{TCF}$, which was unable to interact with the SRE alone but formed a stable SRE-SRF-$p62^{TCF}$ ternary complex upon recruitment to the SRE by SRF (Fig. 14.6). Ternary complex-forming activity[155,177] is displayed by a group of proteins[185] which are collectively referred to as the "ternary complex factors" (TCFs). TCFs have been shown to be encoded by Ets proteins.[186-189] The TCF subfamily of Ets proteins is described below. A new addition to this family is the EWS-Fli1 fusion protein, a transforming protein involved in Ewing's sarcoma. The unfused Fli1 protein is not recruited to the SRE by SRF.[190]

THE TCF SUBFAMILY OF ETS FACTORS

TCF activity is defined as the ability to be recruited to SRE sequences by SRF without being able to bind to SREs autonomously.[176,177,186,187] TCFs can bind to DNA sequences other then SREs, such as the E74 motif (5'- CGGAAG - 3').[191,192] The Ets family members Elk-1, Sap1, and Net (Erp/Sap-2), and some of their splicing variants, are TCFs. TCF recruitment to the SRE involves DNA recognition of the SRE 5' sequence motif CAGGA by the amino-terminally positioned Ets domains of the TCFs[177,187,191] and specific protein-protein contacts between the TCF B-domains (Fig. 14.7)[187,191,193,194] and the C-terminal half of the "core" of SRF.[157,195,196] The transcriptional activation potential of the TCFs is induced by multiple phosphorylations of

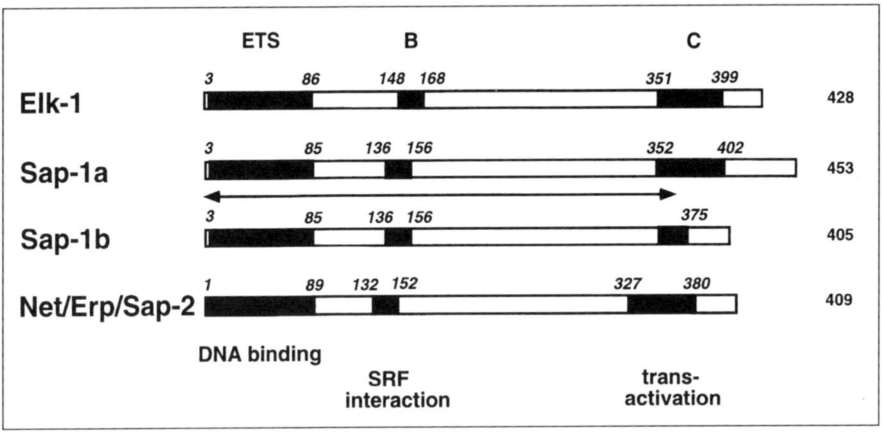

Fig. 14.7. Schematic representation of the TCFs Elk-1, Sap1, and Net/Erp/Sap2.

their conserved C-domains (Fig. 14.7) by members of the MAPK family of kinases (see below).[127,128,130,188,197] Thus, TCFs can function as signal-activated, nuclear transcription factors which control gene activity as a function of extracellular stimulation.

Elk-1

Elk-1 is the most extensively characterized TCF. Its cDNA was cloned in a homology screen for Ets domain-containing sequences[198] and was shown subsequently to encode a protein with TCF activity.[186,187] Nuclear localization of Elk-1 is determined by the Ets domain in conjunction with sequence elements located between amino acid residues 137-157.[199] Gille et al[127] were the first to report p62^{TCF}/Elk-1 phosphorylation in vitro by MAP kinase (Erk) and to correlate this modification with EGF-induced *c-fos* expression. Indeed, TCF phosphorylation and dephosphorylation closely follows the transient, signal-induced transcriptional activation profile of *c-fos*.[200] The direct stimulation of Elk-1 by the Ras/Raf/Erk pathway was subsequently demonstrated by transient transfection studies using kinase activators and TCF-dependent reporter systems.[128-130,201] Activation of TCFs via the different branches of the MAP kinase network is described in more detail below. Elk-1 is phosphorylated by the MAP kinases ERK1 and ERK2 at multiple sites, mainly Ser-324, Thr-336, Ser-383, Ser-389, and Ser-422, and, to some extent ,Thr-363, Thr-368 and Ser-417.[128,130,202] Ser-383 and Ser-389 are the preferential sites of phosphorylation, as observed upon both ERK treatment in vitro and growth factor stimulation in vivo. Mutation of these residues to alanines has the greatest detrimental effect on Elk-mediated gene activation. Phosphorylation of Ser-383 and Ser-389 facilitates ternary complex formation.[127,202] Mutations at other phosphorylation sites impair gene activation, apparently without influencing the efficiency of forming ternary complexes. Phosphorylation probably stimulates Elk-1 transcriptional activation potential by other mechanisms besides SRF-SRE complex formation.

Elk-1 is phosphorylated by other MAP kinases besides ERK1/2, namely the stress/cytokine-induced JNK/SAPK[203-206] and p38/RK.[207] These modifications also lead to activation of the Elk-1 transcriptional stimulatory function. For both ERKs and JNK/SAPKs the Elk-1 residues Ser-383 and Ser-389 have been found to be the major sites of phosphorylation. Other residues in Elk-1 may be modified by the two types of kinase with differential efficiencies.

Elk-1 is expressed at the mRNA level in most human tissues tested, with the highest level seen in brain.[208] In HeLa cells Elk-1 protein appears to be the most abundantly

Table 14.1. Genomic locations of human and murine genes encoding TCFs.

	Genes	Assignment
mouse	Net	10C-10D1
	Elk1	XA1-XA3
	Sap1	1E3-1G
human	Net	12q22-12q23
	Elk1	14q32
		Xp11.2-Xp11.1
	Sap1	1q32

expressed TCF.[185] In the mouse, highest levels of both Elk-1 mRNA and protein have been detected in the brain and thymus (S. Arsenian and A.N., unpublished).[188]

SAP1

Human Sap1 was cloned by Dalton and Treisman[187] in a functional screen that scored for SRF-recruited fusion proteins able to stimulate transcription in *S. cerevisiae*. They isolated both the full-length Sap1a clone and the splice variant Sap1b, which lacks the C domain (Fig. 14.7). In structural and functional terms Sap1 shares with Elk-1 the regions of homology A (Ets domain), B (SRF interaction domain), and C (transcriptional activation domain). Elk-1 and Sap1a form ternary complexes with SRE-SRF with comparable efficiencies.[197,208] Also, like Elk-1, Sap1 is a transcription factor whose gene activation potential requires phosphorylation by the ERK1/2 MAP kinases.[197,208] Residues Thr-361/Thr-366 and Ser-381/Ser-387 have been shown to be the targets for ERK kinases. Sap1a cooperates with the coactivator CBP.[209] The Ets domain of Sap1 suffices to mediate translocation to the nucleus.[197] In most functional assays, Elk-1 and Sap1a have behaved very similarly, but some differences are beginning to be found. Although Elk-1 and Sap1a are concomitantly induced by stimulating murine macrophage-like BAC-1 cells with CSF-1, the activated form of Sap1a persisted much longer.[210] The functional states of the two TCFs were judged by their differential migration in band-shift assays and it remains to be shown that downregulation of Sap1a and Elk-1 transcriptional activity differs kinetically. Sap1a has a different pattern of expression from Elk-1. Sap1a was found at the mRNA level in all human tissues tested.[208] Murine tissues displayed the strongest Sap1 mRNA signals in heart, liver and muscle.[188]

NET/ERP/SAP2

The Net protein (also called Erp and Sap2) was independently cloned in several laboratories.[188,189,208] It is highly related to the other TCFs, with many common features (see Fig. 14.7). However, there are a growing number of differences, which may provide insights into how common signaling pathways are resolved into specific regulation in the cell. The TCFs are encoded by different genes on syntenic regions of the human and mouse chromosomes (see Table 14.1) and they are more similar in sequence amongst themselves between mouse and man than between themselves,[188,211] suggesting that there is evolutionary pressure to maintain the three factors. Recently, some insight into distinct roles for Net have been provided by both studies of an inhibitor present in the full length protein and by the identification of isoforms with distinct properties (Giovane et al, in preparation).[54]

Net has been shown to inhibit transcription under certain basal conditions, under which Sap1a is inactive and ELK-1 stimulates.[54] Repression is mediated by an autono-

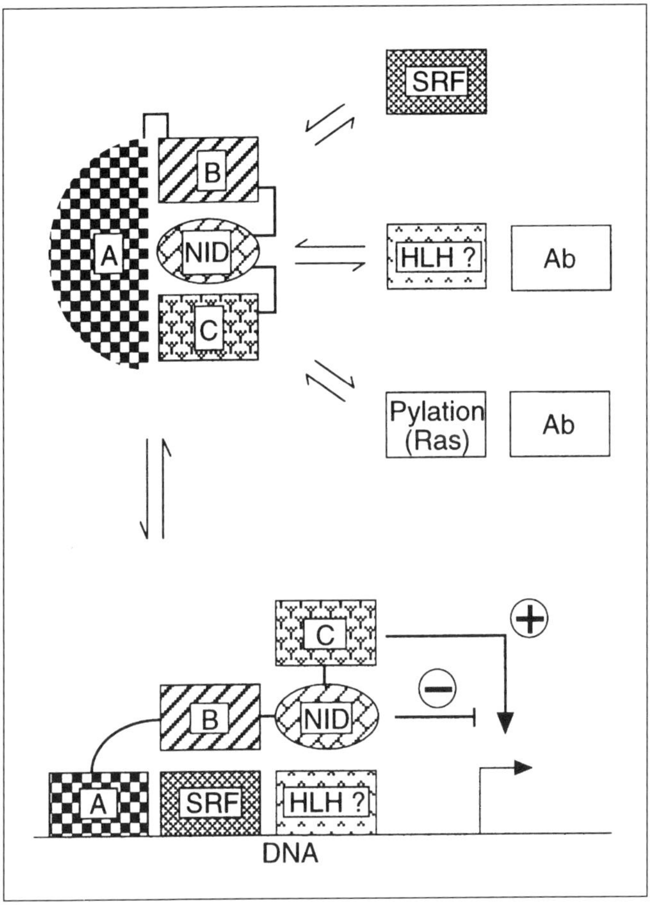

Fig. 14.8. A model for Net regulation. Net has a "closed" compact conformation resulting in a refractory DNA-binding activity (upper diagram). The conformation is altered by various mechanisms, including interaction of proteins (?) with the NID and B domain and phosphorylation (Pylation) of the C domain by components of the Ras signaling pathway. Antibodies (Ab) mimic protein-protein interactions. The DNA bound complex is proposed to be formed with the help of interactions between the B domain and SRF and/or the NID and a HLH protein. The NID inhibits basal activity (-) whereas the C domain stimulates (+) in the presence of the Ras signal. NID activity might be regulated by protein interactions or modification (not illustrated).

mous inhibitory domain, NID, comprised of about 50 amino acids. The NID is an efficient inhibitor of both DNA binding by Net and activation by Ras. It appears to adopt a helix-loop-helix (HLH) structure that resembles the bHLH domains of myogenic factors. Net interacts in vivo with a heterodimerization partner of myogenic factors, the basic-helix-loop-helix factor E47.[54]

The properties of Net are summarized in the model shown in Figure 14.8. The model proposes that Net adopts a "closed" conformation and a transition to an "open" complex is an important point of regulation by protein-protein interactions and phosphorylation. Opening is driven by protein-protein interactions between the NID and members of the HLH family, the B domain and SRF, and also the C-domain with other proteins (which are mimicked by antibodies that stimulate DNA binding, see ref. 54). The NID lies next to the B domain, suggesting that it could also affect specificity for interactions with members of the SRF family,[158] a number of which are expressed in both muscle and nonmuscle cells.[212] Phosphorylation through the Ras signaling pathway appears to be the major regulatory mechanism for the C-domain.[208] The NID has several consensus PKC phosphorylation sites, which could be targets for regulation. Net, "forced open" by mutation in the NID, appears to be more sensitive to modification,[54] suggesting that "opening" would make it more responsive to fine regulation by mechanisms such as phosphorylation. Net "opening" leads to formation of a DNA-bound complex. In the model, the DNA-protein complex is proposed to be stabilized by interactions of Net with SRF and/or bHLH proteins and to be subject to regulation. There are interesting parallels. On muscle cell differentiation, MyoD1 represses *c-fos* through an E-Box binding site for bHLH proteins in the SRE.[147] A possibility is that Net forms ternary complexes with yet to be identified bHLH factors, which are present in the cells where Net is abundant (e.g., NIH3T3 cells, see ref. 188). Myogenic bHLH factors have been shown to physically interact with SRF.[178] The NID could participate in these interactions, either with the myogenic factors, with SRF, or with both. DNA bound Net inhibits transcription by unknown mechanisms that could involve recruiting negative coactivators and nucleosomes, analogous to α2 and Mcm1,[166,213] the SRF homologue of yeast.[158] However, many mechanisms of repression have been described.[214-216] Repression by DNA bound Net could be subject to regulation, similar to the retinoic acid receptors.[217] The NID is located in sequences that differ between the ternary complex factors. An interesting possibility is that the divergent sequences of ELK-1 and Sap1a could also mediate specific functions of these otherwise highly related proteins.

The *net* gene codes for three other isoforms, Net-a, Net-b and Net-c (Fig. 14.9). Net-a and Net-b have been detected in a variety of cell lines and tissues, whereas Net-c is apparently restricted to NIH3T3 fibroblasts (Giovane et al, in preparation). Net-a is a cytoplasmic protein that does not bind to DNA in vitro, and presumably has no direct role in transcription. Net-b and Net-c lack the Ras inducible "C" activation domain and are altered in the inhibitory sequence NID. They are nuclear, bind efficiently to DNA, and inhibit transcription both under basal conditions and in response to the Ras signal (A. Giovane et al, in preparation). Net-b and Net-c could be inhibitors that bind distinct proteins through their modified NID domains. The Net isoforms are different from the Elk1 and Sap1 isoforms. Net has a quite different expression pattern compared to Elk1 and Sap1 during mouse development (H. Galinaro, in preparation). These distinct properties probably confer unique functions to the net gene that forms the basis of the selective pressure that has maintained net as a distinct gene during evolution.

SIGNAL ACTIVATION OF ETS PROTEINS BY THE MAP KINASE NETWORK

The Ets proteins are an important class of transcription factors whose activity is governed by the MAP kinase signaling network (see Fig. 14.10). The basic scaffold of this network is comprised of three major cascades that funnel into the activation of the three types of MAP kinases: ERK, JNK/SAPK, and p38/RK.[90] Our understanding of this multi-component network is rapidly expanding, mainly due to genetic analysis in the budding yeast *S. cerevisiae* .[168]

The ERK subgroup of MAP kinases consists of p44 ERK1 and p42 ERK2. MEK1 and MEK2 are the upstream activators (MAPKKs) of the ERKs and MEK activation is itself mediated by the Ras-activated MAPKKK Raf-1. The ERK pathway is activated through Ras by a variety of tyrosine kinase receptors and certain serpentine receptor-coupled G-proteins. It can also be activated directly by protein kinase C.[218,219]

The c-Jun kinases (JNKs) or stress activated protein kinases (SAPKs) represent the

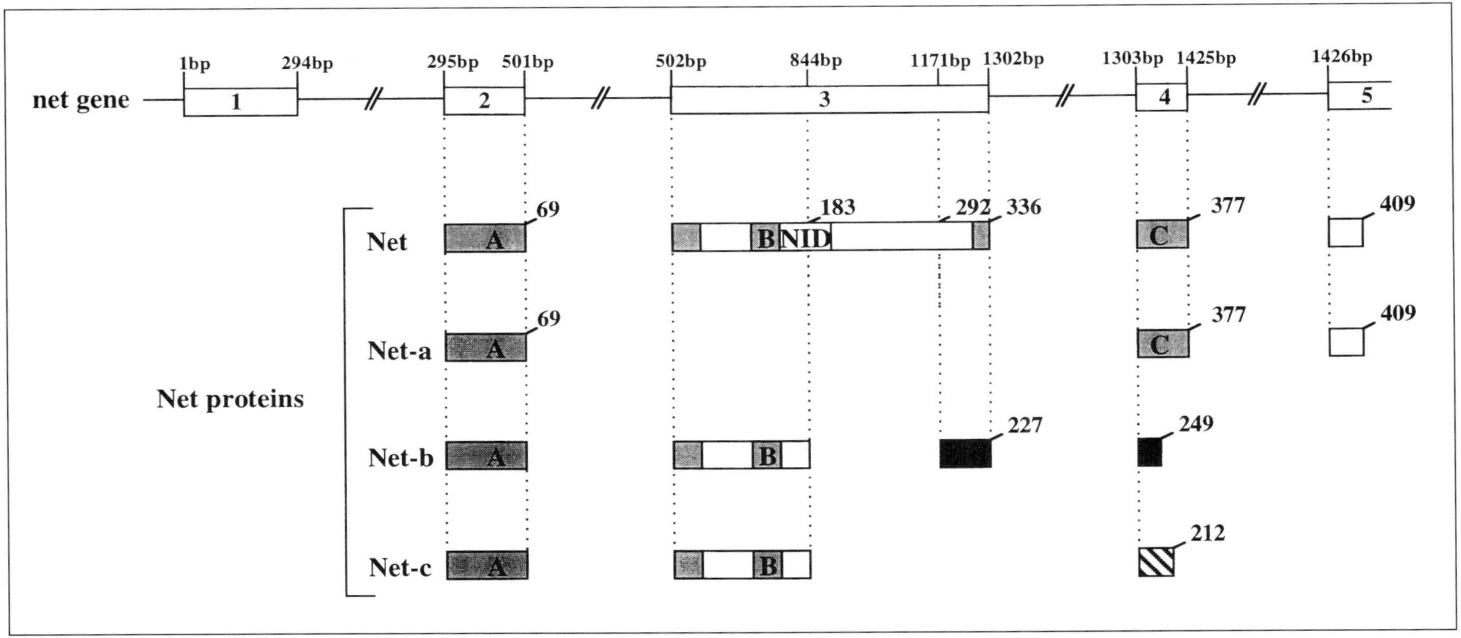

Fig. 14.9. Structure of the net gene and Net isoforms. The net gene is composed of exons 1-5 and slice acceptor and donor sites are found in exon 3. Domains A, B and C resemble elements in the other TCFs; NID is the Net inhibitor domain. The NID is absent or altered in the Net isoforms (A. Giovane et al, in preparation).

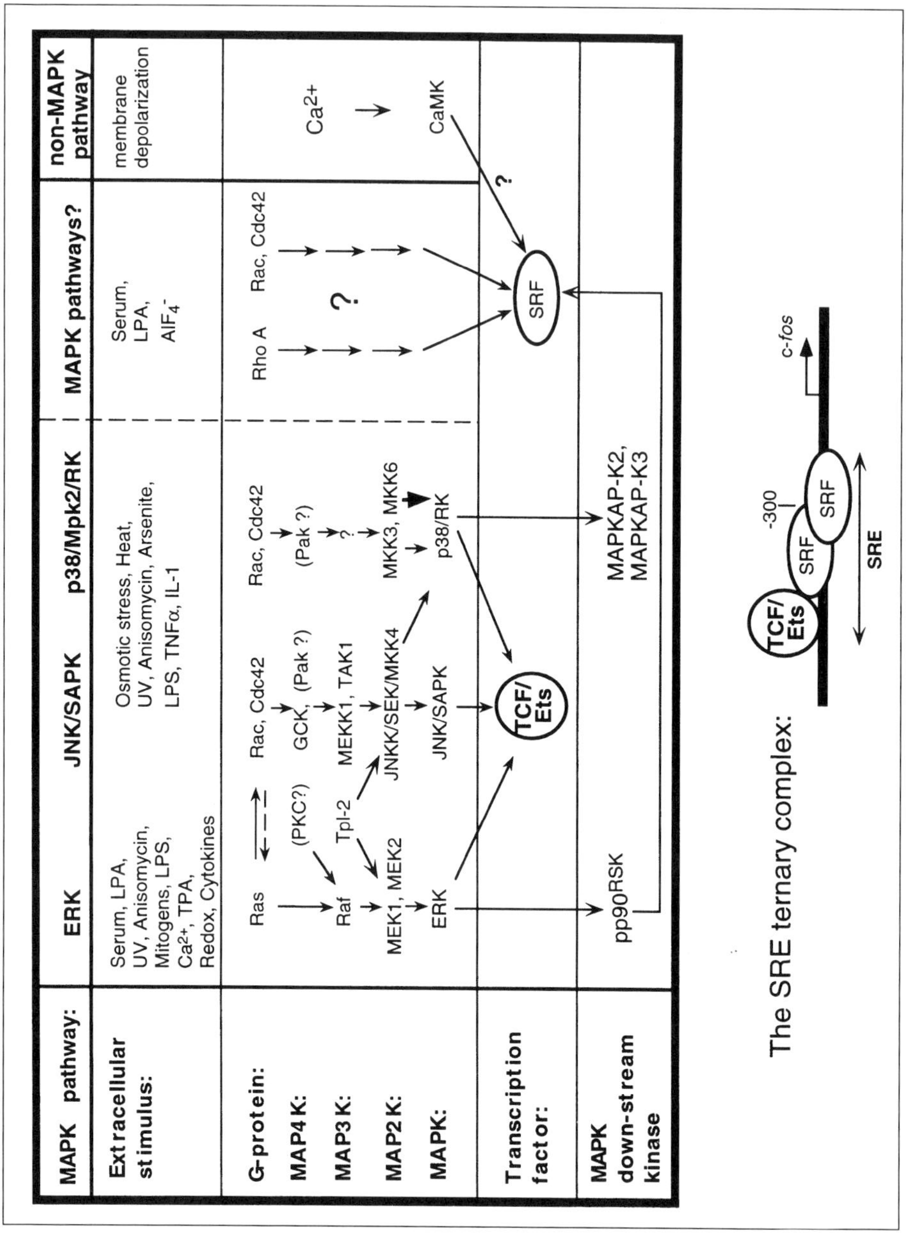

Fig. 14.10. The MAP kinase signaling network targeting the SRE.

second class of vertebrate MAPKs. They are encoded by four genes: SAPKα, SAPKβ, JNK1, and JNK2/SAPKγ; additional complexity arises through differential splicing that produces a number of protein products between 45 kDa and 54 kDa molecular weight. The role of the different JNK/SAPK isoforms is presently unclear. A MAPKK, variously known as SEK1, JNKK, or MKK4, can activate JNK/SAPKs, which in turn is activated preferentially by MEKK, a MAPKKK.

The third mammalian MAPK pathway involves a homologue of the *S. cerevisiae* protein Hog1 that responds to osmotic stress in yeast. The mammalian enzyme also functions in the osmotic stress response and thus appears to be a direct functional homologue of the yeast enzyme. It has been variously called p38, p40, Mpk2, reactivating kinase (RK), or cytokine-suppressive anti-inflammatory drug binding protein (CSBP) and can be activated by the MAPKKs called MKK3 and MKK6, as well as by SEK1/JNKK/MKK4.

Elk-1 is an efficient substrate for all three of these classes of MAP kinase, at least in vitro (see above), suggesting that Elk-1 is a point of convergence for the MAP kinase cascades. Sap1a and Net/Erp/Sap2 are similarly under Ras control and the activation of Sap1a by ERK kinases has been shown. It will be interesting to obtain a more complete understanding of the regulatory mechanisms that govern TCF activation as a consequence of the induction of the MAPK cascade.

The SRF component of SRE complexes is also subject to activation by signaling cascades and may act as a TCF-independent transcription factor. Phosphorylation of SRF by either the growth factor-induced kinase $pp90^{RSK}$ [175] or by the Ras-independent and Ca^{2+}-induced Ca^{2+}/calmodulin-dependent kinase (CaMK)[220] has been observed. Signaling induced by the LPA component of serum has been reported to activate small GTPase-dependent signaling that targets the SRF directly or indirectly via novel cascades.[221,222] This signaling apparently involves the RhoA, Rac, and Cdc42 G-proteins.

CONCLUSIONS/OPEN QUESTIONS

The functional studies of the Ets1 and TCF subfamilies of Ets proteins summarized here have uncovered two conceptually important mechanistic features underlying their activity as transcription factors. Firstly, members of these two subfamilies display the capability to utilize cooperative protein-protein interactions with other DNA-bound transcription factors to specifically recognize and control a diverse set of genes. This type of cooperation with other transcription factors expands the spectrum of potential target genes beyond the ones that can be recognized by the individual factors alone. Secondly, Ets1 and TCF proteins serve as nuclear substrates for the kinases of the MAPK signaling cascade, thereby rendering their target genes exquisitely sensitive to extracellular stimuli.

Both Ets1 and TCF subfamilies contain structurally and functionally very similar member proteins. There are some immediate questions. What are their distinct roles where they are coexpressed? Are the major differences due to their patterns of expression? How do these factors respond to temporal and spatial cues? Do they respond differently to the MAP kinases and other signaling pathways? Which are their true in vivo target genes? There are as yet few studies by homologous recombination in the murine genome and many answers to these questions will come from a more comprehensive mutational analysis. Additionally, phospho-specific antisera, able to distinguish epitopes according to their phosphorylation status, will become extremely useful for the in situ detection of the existing spatial phosphorylation/activity profiles associated with these factors.

Studies of Ets protein structure and function, such as those described in this review, are providing new and original insights into the molecular mechanisms of gene control in growth and development.

ACKNOWLEDGMENTS

We are grateful to Dr. K.R. Ely for providing the coordinates of the PU.1 DNA complex and permission to illustrate the structure (Fig. 14.2). We thank for invaluable help in preparing the figures: J-M Wurtz and L. Moulinier for Figure 14.2, A. Giovane for Figure 14.3, A. Collet and J-L Toussaint for Figure 14.4 and C. Wasylyk for Figures 14.1 and 14.5. We apologize for the many references that have not been explicitly cited, due to length limits. In general more recent references have been cited that cannot be found in many of the reviews. We apologize to those authors that have not been adequately cited. B. W. thanks for funding the Centre National de la Recherche Scientifique, the Institut National de la Santé et de la Recherche Médicale, the Centre Hospitalier Universitaire Régional, the Association pour la Recherche sur le Cancer, the Fondation pour la Recherche Médicale, the Ligue Nationale Française contre le Cancer, the Ligue Régionale (Haut-Rhin) contre le Cancer, the Ligue Régionale (Bas-Rhin) contre le Cancer (the Legs Meyer) and the Bioavenir Program (Ministère de la Recherche et Ministère de l'Industrie). A.N. was supported by the DFG, the Dr. Mildred-Scheel-Stiftung, and the Fonds der Deutschen Industrie. Both authors acknowledge financial assistance provided by PROCOPE/DAAD for French/German international collaborations.

References

1. Leprince D, Gegonne A, Coll J et al. A putative second cell-derived oncogene of the avian leukaemia retrovirus E26. Nature 1983; 306:395-397.
2. Nunn MF, Seeburg PH, Moscovici C et al. Tripartite structure of the avian erythroblastosis virus E26 transforming gene. Nature 1983; 306:391-395.
3. Bhat NK, Fischinger PJ, Seth A et al. Pleiotropic functions of ETS-1 (Review). Int J Oncol 1996; 8:841-846.
4. Crepieux P, Coll J, Stehelin D. The Ets family of proteins: weak modulators of gene expression in quest for transcriptional partners. Crit Rev Oncog 1994; 5:615-638.
5. Moreau-Gachelin F. Spi-1/PU.1: an oncogene of the Ets family. Biochim Biophys Acta 1994; 1198:149-163.
6. Tymms MJ, Kola I. Regulation of gene expression by transcription factors Ets-1 and Ets-2. Mol Reprod Dev 1994; 39:208-214.
7. Janknecht R, Nordheim A. Gene regulation by Ets proteins. Biochim Biophys Acta 1993; 1155:346-356.
8. Wasylyk B, Hahn SL, Giovane A. The Ets family of transcription factors [published erratum appears in Eur J Biochem 1993 Aug 1; 215(3):907]. Eur J Biochem 1993; 211:7-18.
9. Donaldson LW, Petersen JM, Graves BJ et al. Solution structure of the ETS domain from murine Ets-1: A winged helix-turn-helix DNA binding motif. EMBO J 1996; 15:125-134.
10. Kodandapani R, Pio F, Ni CZ et al. A new pattern for helix-turn-helix recognition revealed by the PU.1 ETS-domain-DNA complex. Nature 1996; 380 :456-460.

10a. Pio F, Kodandapani R, Ni CZ et al. New insights on DNA recognition by ets proteins from the crystal structure of the PU.1 ETS domain-DNA complex. J Biol Chem 1996; 271:23329-23337.

11. Werner MH, Clore M, Fisher CL et al. The solution structure of the human ETS1-DNA complex reveals a novel mode of binding and true side chain intercalation. Cell 1995; 83:761-771. Erratum: Cell 1996; 87.
12. Landsman D, Wolffe AP. Common sequence and structural features in the heat-shock factor and Ets families of DNA-binding domains. Trends Biochem Sci 1995; 20:225-226.
13. Lelievre-Chotteau A, Laudet V, Flourens A et al. Identification of two ets related genes in a marine worm, the polychaete annelid Nereis diversicolor. FEBS Lett 1994; 354:62-66.
14. Lautenberger JA, Burdett LA, Gunnell MA et al. Genomic dispersal of the ets gene family during metazoan evolution. Oncogene 1992; 7:1713-1719.
15. Laudet V, Niel C, Duterque-Coquillaud M et al. Evolution of the ets gene family. Biochem Biophys Res Commun 1993; 190:8-14.
16. Hensold JO, Stratton CA, Barth D et al. Expression of the transcription factor, Spi-1 (PU.1), in differentiating murine erythroleukemia cells is regulated post-

transcriptionally—Evidence for differential stability of transcription factor mrnas following inducer exposure. J Biol Chem 1996; 271:3385-3391.

17. Bories JC, Willerford DM, Grevin D et al. Increased T-cell apoptosis and terminal B-cell differentiation induced by inactivation of the Ets-1 proto-oncogene. Nature 1995; 377:Iss 6550:635-638.
18. Muthusamy N, Barton K, Leiden JM. Defective activation and survival of T cells lacking the Ets-1 transcription factor. Nature 1995; 377, Iss 6550:639-642.
19. Scott EW, Simon MC, Anastasi J et al. Requirement of transcription factor PU.1 in the development of multiple hematopoietic lineages. Science 1994; 265: 1573-1577.
20. Melet F, Motro B, Rossi DJ et al. Generation of a novel fli-1 protein by gene targeting leads to a defect in thymus development and a delay in friend virus-induced erythroleukemia. Mol Cell Biol 1996; 16:2708-2718.
21. Kiaris H, Spandidos DA. Mutations of ras genes in human tumours (Review). Int J Oncol 1995; 7:413-421.
22. Ben-David Y, Bernstein A. Friend virus-induced erythroleukemia and the multistage nature of cancer. Cell 1991; 66: 831-834.
23. Nibbs RJ, Itoh K, Ostertag W et al. Differentiation arrest and stromal cell-independent growth of murine erythroleukemia cells are associated with elevated expression of ets-related genes but not with mutation of p53. Mol Cell Biol 1993; 13:5582-5592.
24. Bellacosa A, Datta K, Bear SE et al. Effects of provirus integration in the Tpl-1/Ets-1 locus in Moloney murine leukemia virus-induced rat T-cell lymphomas: levels of expression, polyadenylation, transcriptional initiation, and differential splicing of the Ets-1 mRNA. J Virol 1994; 68:2320-2330.
25. Vandenbunder B, Queva C, Desbiens X et al. Expression of the transcription factor c-Ets1 correlates with the occurrence of invasive processes during normal and pathological development. Invasion Metastasis 1994; 14:198-209.
26. Wernert N, Gilles F, Fafeur V et al. Stromal expression of c-Ets1 transcription factor correlates with tumor invasion. Cancer Res 1994; 54:5683-5688.
27. Sumarsono SH, Wilson TJ, Tymms MJ et al. Down's syndrome-like skeletal abnormalities in Ets2 transgenic mice. Nature 1996; 379:534-537.
28. Albagli O, Flourens A, Crepieux P et al. Phylogeny of the p68c-ets-1 amino-terminal transactivating domain reveals some highly conserved structural features. Oncogene 1992; 7:1435-1439.
29. Albagli O, Soudant N, Ferreira E et al. A model for gene evolution of the ets-1/ets-2 transcription factors based on structural and functional homologies. Oncogene 1994; 9:3259-3271.
30. Schneikert J, Lutz Y, Wasylyk B. Two independent activation domains in c-Ets-1 and c-Ets-2 located in non-conserved sequences of the ets gene family. Oncogene 1992; 7:249-256.
31. Klambt C. The *Drosophila* gene pointed encodes two ETS-like proteins which are involved in the development of the midline glial cells. Development 1993; 117: 163-176.
32. Wasylyk C, Bradford AP, Gutierrez-Hartmann A et al. Conserved mechanisms of Ras regulation of evolutionary related transcription factors, Ets1 and Pointed P2. Oncogene 1996; 14:899-913.
33. Donaldson LW, Petersen JM, Graves BJ et al. Secondary structure of the ETS domain places murine Ets-1 in the superfamily of winged helix-turn-helix DNA-binding proteins. Biochemistry 1994; 33:13509-13516.
34. Liang H, Mao X, Olejniczak ET et al. Solution structure of the ets domain of Fli-1 when bound to DNA. Nat Struct Biol 1994; 1:871-875.
35. Liang H, Olejniczak ET, Mao X et al. The secondary structure of the ets domain of human Fli-1 resembles that of the helix-turn-helix DNA-binding motif of the Escherichia coli catabolite gene activator protein. Proc Natl Acad Sci USA 1994; 91:11655-11659.
36. Schultz SC, Shields GC, Steitz TA. Crystal structure of a CAP-DNA complex: the DNA is bent by 90 degrees. Science 1991; 253:1001-1007.

37. Damberger FF, Pelton JG, Harrison CJ et al. Solution structure of the DNA-binding domain of the heat shock transcription factor determined by multidimensional heteronuclear magnetic resonance spectroscopy. Protein Sci 1994; 3:1806-1821.
38. Harrison CJ, Bohm AA, Nelson HC. Crystal structure of the DNA binding domain of the heat shock transcription factor. Science 1994; 263:224-227.
39. Vuister GW, Kim SJ, Orosz A et al. Solution structure of the DNA-binding domain of *Drosophila* heat shock transcription factor [see comments]. Nat Struct Biol 1994; 1:605-614.
40. Vuister GW, Kim SJ, Wu C et al. NMR evidence for similarities between the DNA-binding regions of *Drosophila* melanogaster heat shock factor and the helix-turn-helix and HNF-3/forkhead families of transcription factors. Biochemistry 1994; 33:10-16.
41. Clark KL, Halay ED, Lai E et al. Co-crystal structure of the HNF-3/fork head DNA-recognition motif resembles histone H5. Nature 1993; 364:412-420.
42. Wilson KP, Shewchuk LM, Brennan RG et al. Escherichia coli biotin holoenzyme synthetase/bio repressor crystal structure delineates the biotin- and DNA-binding domains. Proc Natl Acad Sci USA 1992; 89:9257-9261.
43. Ramakrishnan V, Finch JT, Graziano V et al. Crystal structure of globular domain of histone H5 and its implications for nucleosome binding. Nature 1993; 362:219-223.
44. Clubb RT, Omichinski JG, Savilahti H et al. A novel class of winged helix-turn-helix protein: The DNA-binding domain of Mu transposase. Structure 1994; 2:1041-1048.
45. Fogh RH, Ottleben G, Ruterjans H et al. Solution structure of the LexA repressor DNA binding domain determined by 1H NMR spectroscopy. EMBO J 1994; 13:3936-3944.
46. Bussiere DE, Bastia D, White SW. Crystal structure of the replication terminator protein from B. subtilis at 2.6 A. Cell 1995; 80:651-660.
47. Qiu X, Verlinde CL, Zhang S et al. Three-dimensional structure of the diphtheria toxin repressor in complex with divalent cation co-repressors. Structure 1995; 3:87-100.
48. Shore P, Whitmarsh AJ, Bhaskaran R et al. Determinants of DNA-binding specificity of ETS-domain transcription factors. Mol Cell Biol 1996; 16:3338-3349.
49. Werner MH, Gronenborn AM, Clore GM. Intercalation, DNA kinking, and the control of transcription. Science 1996; 271, Iss 5250:778-784.
50. Werner MH, Clore GM, Gronenborn AM et al. Refolding proteins by gel filtration chromatography. FEBS Lett 1994; 345:125-130.
51. Nye JA, Petersen JM, Gunther CV et al. Interaction of murine ets-1 with GGA-binding sites establishes the ETS domain as a new DNA-binding motif. Genes Dev 1992; 6:975-990.
52. Siddique HR, Rao VN, Lee L et al. Characterization of the DNA binding and transcriptional activation domains of the erg protein. Oncogene 1993; 8: 1751-1755.
53. Rao VN, Ohno T, Prasad DD et al. Analysis of the DNA-binding and transcriptional activation functions of human Fli-1 protein. Oncogene 1993; 8: 2167-2173.
54. Maira S-M, Wurtz J-M, Wasylyk B. Net (ERP/SAP2), one of the Ras inducible TCFs, has a novel inhibitory domain with resemblance to the helix-loop-helix motif. EMBO J 1996; 15:5849-5865.
55. Hupp TR, Sparks A, Lane DP. Small peptides activate the latent sequence-specific DNA binding function of p53. Cell 1995; 83:237-245.
56. Garrido C, Leprince D, Plaza S et al. Back-mutation of the v-Ets to the c-Ets carboxy-terminal amino acids in the P135(gag-myb-ets) results in chicken neuroretina cells transformation and loss of basic fibroblast growth factor responsiveness. Oncogene 1996; 12:1449-1456.
57. Jonsen MD, Petersen JM, Xu QP et al. Characterization of the cooperative function of inhibitory sequences in Ets-1. Mol Cell Biol 1996; 16:2065-2073.
58. Skalicky JJ, Donaldson LW, Petersen JM et al. Structural coupling of the inhibitory regions flanking the ETS domain of

murine Ets-1. Protein Sci 1996; 5:296-309.

59. Petersen JM, Skalicky JJ, Donaldson LW et al. Modulation of transcription factor Ets-1 DNA binding: DNA- induced unfolding of an α helix. Science 1995; 269, Iss 5232:1866-1869.
60. Seth A, Robinson L, Thompson DM et al. Transactivation of GATA-1 promoter with ETS1, ETS2 and ERGB/Hu-FLI-1 proteins: stabilization of the ETS1 protein binding on GATA-1 promoter sequences by monoclonal antibody. Oncogene 1993; 8:1783-1790.
61. Giese K, Kingsley C, Kirshner JR et al. Assembly and function of a TCR α enhancer complex is dependent on LEF-1-induced DNA bending and multiple protein-protein interactions. Genes Dev 1995; 9:995-1008.
62. Lim F, Kraut N, Framptom J et al. DNA binding by c-Ets-1, but not v-Ets, is repressed by an intramolecular mechanism. EMBO J 1992; 11:643-652.
63. Fleischman LF, Pilaro AM, Murakami K et al. c-Ets-1 protein is hyperphosphorylated during mitosis. Oncogene 1993; 8:771-780.
64. Pognonec P, Boulukos KE, Gesquiere JC et al. Mitogenic stimulation of thymocytes results in the calcium-dependent phosphorylation of c-ets-1 proteins. EMBO J 1988; 7:977-983.
65. Rabault B, Ghysdael J. Calcium-induced phosphorylation of ETS1 inhibits its specific DNA binding activity. J Biol Chem 1994; 269:28143-28151.
66. Valentine MA, Czernik AJ, Rachie N et al. Anti-immunoglobulin M activates nuclear calcium/calmodulin- dependent protein kinase II in human B lymphocytes. J Exp Med 1995; 182:1943-1949.
67. Fleischman LF, Holtzclaw L, Russell JT et al. ets-1 in astrocytes: expression and transmitter-evoked phosphorylation. Mol Cell Biol 1995; 15:925-931.
68. Fisher RJ, Fivash M, Casas-Finet J et al. Real-time DNA binding measurements of the ETS1 recombinant oncoproteins reveal significant kinetic differences between the p42 and p51 isoforms. Protein Sci 1994; 3:257-266.
69. Hodge DR, Robinson L, Watson D et al. Interaction of ETS-1 and ERGB/FLI-1 proteins with DNA is modulated by spacing between multiple binding sites as well as phosphorylation. Oncogene 1996; 12:11-18.
70. Treier M, Bohmann D, Mlodzik M. JUN cooperates with the ETS domain protein pointed to induce photoreceptor R7 fate in the *Drosophila* eye. Cell 1995; 83: 753-760.
71. Bassuk AG, Leiden JM. A direct physical association between ETS and AP-1 transcription factors in normal human T cells. Immunity 1995; 3:223-237.
72. Galang CK, Garciaramirez JJ, Solski PA et al. Oncogenic Neu/ErbB-2 increases ets, AP-1, and NF-kappa B- dependent gene expression, and inhibiting ets activation blocks neu- mediated cellular transformation. J Biol Chem 1996; 271: 7992-7998.
73. Gum R, Lengyel E, Juarez J et al. Stimulation of 92-kDa gelatinase B promoter activity by ras is mitogen-activated protein kinase kinase 1-independent and requires multiple transcription factor binding sites including closely spaced PEA3/ets and AP-1 sequences. J Biol Chem 1996; 271:10672-10680.
74. Logan SK, Garabedian MJ, Campbell CE et al. Synergistic transcriptional activation of the tissue inhibitor of metalloproteinases-1 promoter via functional interaction of AP-1 and Ets-1 transcription factors. J Biol Chem 1996; 271: 774-782.
75. Noti JD, Reinemann BC, Petrus MN. Regulation of the leukocyte integrin gene CD11c is mediated by AP1 and Ets transcription factors. Mol Immunol 1996; 33:115-127.
76. Goldberg Y, Treier M, Ghysdael J et al. Repression of AP-1-stimulated transcription by c-Ets-1. J Biol Chem 1994; 269: 16566-16573.
77. Sieweke MH, Tekotte H, Frampton J et al. MafB is an interaction partner and repressor of Ets-1 that inhibits erythroid differentiation. Cell 1996; 85:49-60.
78. Fitzsimmons D, Hodsdon W, Wheat W et al. Pax-5 (BSAP) recruits Ets protooncogene family proteins to form functional ternary complexes on a B-cell-spe-

cific promoter. Genes Dev 1996; 10:2198-2211.
79. Seth A, Papas TS. The c-ets-1 proto-oncogene has oncogenic activity and is positively autoregulated. Oncogene 1990; 5:1761-1767.
80. Karin M, Hunter T. Transcriptional control by protein phosphorylation: Signal transmission from the cell surface to the nucleus. Curr Biol 1995; 5:747-757.
81. Cobb MH, Goldsmith EJ. How MAP kinases are regulated. J Biol Chem 1995; 270:14843-14846.
82. Cano E, Mahadevan LC. Parallel signal processing among mammalian MAPKs. Trends Biochem Sci 1995; 20:117-122.
83. Marshall MS. Ras target proteins in eukaryotic cells. FASEB J 1995; 9:1311-1318.
84. Downward J. KSR: A novel player in the RAS pathway. Cell 1995; 83:831-834.
85. Marshall CJ. Ras effectors. Curr Opin Cell Biol 1996; 8:197-204.
86. Beitel GJ, Tuck S, Greenwald I et al. The *Caenorhabditis* elegans gene lin-1 encodes an ETS-domain protein and defines a branch of the vulval induction pathway. Genes Dev 1995; 9:3149-3162.
87. O'Neill EM, Rebay I, Tjian R et al. The activities of two Ets-related transcription factors required for *Drosophila* eye development are modulated by the Ras/MAPK pathway. Cell 1994; 78:137-147.
88. Brunner D, Ducker K, Oellers N et al. The ETS domain protein pointed-P2 is a target of MAP kinase in the sevenless signal transduction pathway. Nature 1994; 370:386-389.
89. Rebay I, Rubin GM. Yan functions as a general inhibitor of differentiation and is negatively regulated by activation of the Ras1/MAPK pathway. Cell 1995; 81: 857-866.
90. Treisman R. Regulation of transcription by MAP kinase cascades. Curr Opin Cell Biol 1996; 8:205-215.
91. Sgouras DN, Athanasiou MA, Beal GJ Jr et al. ERF: An ETS domain protein with strong transcriptional repressor activity, can suppress ets-associated tumorigenesis and is regulated by phosphorylation during cell cycle and mitogenic stimulation. EMBO J 1995; 14:4781-4793.
92. Coffer P, de Jonge M, Mettouchi A et al. junB promoter regulation: Ras mediated transactivation by c-Ets-1 and c-Ets-2. Oncogene 1994; 9:911-921.
93. Bradford AP, Conrad KE, Wasylyk C et al. Functional interaction of c-Ets-1 and GHF-1/Pit-1 mediates Ras activation of pituitary-specific gene expression: Mapping of the essential c-Ets-1 domain. Mol Cell Biol 1995; 15:2849-2857.
93a. Bradford AD, Wasylyk C, Wasylyk B et al. Interactions of Ets1 and the Pou-homeodomain protein reconstitutes pituatory-specific gene expression. Mol Cell Biol 1997; 17:1065-1074.
94. Yang BS, Hauser CA, Henkel G et al. Ras-mediated phosphorylation of a conserved threonine residue enhances the transactivation activities of c-Ets1 and c-Ets2. Mol Cell Biol 1996; 16:538-547.
95. Conrad KE, Oberwetter JM, Vaillancourt R et al. Identification of the functional components of the Ras signaling pathway regulating pituitary cell-specific gene expression. Mol Cell Biol 1994; 14:1553-1565.
96. Dickson B. Nuclear factors in sevenless signalling. Trends Genet 1995; 11:106-111.
97. Golub TR, Barker GF, Lovett M et al. Fusion of PDGF receptor β to a novel ets-like gene, tel, in chronic myelomonocytic leukemia with t(5;12) chromosomal translocation. Cell 1994; 77:307-316.
98. Golub TR, Barker GF, Bohlander SK et al. Fusion of the TEL gene on 12p13 to the AML1 gene on 21q22 in acute lymphoblastic leukemia. Proc Natl Acad Sci USA 1995; 92:4917-4921.
99. Buijs A, Sherr S, van Baal S et al. Translocation (12;22) (p13;q11) in myeloproliferative disorders results in fusion of the ETS-like TEL gene on 12p13 to the MN1 gene on 22q11. Oncogene 1995; 10: 1511-1519.
100. Papadopoulos P, Ridge SA, Boucher CA et al. The novel activation of ABL by fusion to an ets-related gene, TEL. Cancer Res 1995; 55:34-38.
101. Golub TR, Goga A, Barker GF et al. Oligomerization of the ABL tyrosine kinase by the Ets protein TEL in human leukemia. Mol Cell Biol 1996; 16:4107-4116.

102. Jousset C, Carron C, Boureux A et al. A domain of tel conserved in a subset of ets proteins defines a specific oligomerization interface essential to the mitogenic properties of the tel-PDGFb oncoprotein. EMBO J 1997; 16:69-82.
103. Pardanaud L, Dieterlen-Lievre F. Expression of C-ETS1 in early chick embryo mesoderm: relationship to the hemangioblastic lineage. Cell Adhes Commun 1993; 1:151-160.
104. Queva C, Leprince D, Stehelin D et al. p54c-ets-1 and p68c-ets-1, the two transcription factors encoded by the c-ets-1 locus, are differentially expressed during the development of the chick embryo. Oncogene 1993; 8:2511-2520.
105. Vandenbunder B, Pardanaud L, Jaffredo T et al. Complementary patterns of expression of c-ets 1, c-myb and c-myc in the blood-forming system of the chick embryo. Development 1989; 107:265-274.
106. Kola I, Brookes S, Green AR et al. The Ets1 transcription factor is widely expressed during murine embryo development and is associated with mesodermal cells involved in morphogenetic processes such as organ formation. Proc Natl Acad Sci USA 1993; 90:7588-7592.
107. Maroulakou IG, Papas TS, Green JE. Differential expression of ets-1 and ets-2 proto-oncogenes during murine embryogenesis. Oncogene 1994; 9:1551-1565.
108. Romano-Spica V, Suzuki H, Giorgiou P et al. Expression of ets family of genes in hematopoietic cells. Int J Oncol 1994; 4:521-531.
109. Bhat NK, Komschlies KL, Fujiwara S et al. Expression of ets genes in mouse thymocyte subsets and T cells. J Immunol 1989; 142:672-678.
110. Bhat NK, Thompson CB, Lindsten T et al. Reciprocal expression of human ETS1 and ETS2 genes during T-cell activation: regulatory role for the protooncogene ETS1. Proc Natl Acad Sci USA 1990; 87:3723-3727.
111. Romano-Spica V, Georgiou P, Suzuki H et al. Role of ETS1 in IL-2 gene expression. J Immunol 1995; 154:2724-2732.
112. Langer SJ, Bortner DM, Roussel MF et al. Mitogenic signaling by colony-stimulating factor 1 and ras is suppressed by the ets-2 DNA-binding domain and restored by myc overexpression. Mol Cell Biol 1992; 12:5355-5362.
113. Wasylyk C, Maira SM, Sobieszczuk P et al. Reversion of Ras transformed cells by Ets transdominant mutants. Oncogene 1994; 9:3665-3673.
114. Topol LZ, Tatosyan AG, Ascione R et al. C-ets-1 protooncogene expression alters the growth properties of immortalized rat fibroblasts. Cancer Lett 1992; 67: 71-78.
115. Suzuki H, Romano-Spica V, Papas TS et al. ETS1 suppresses tumorigenicity of human colon cancer cells. Proc Natl Acad Sci USA 1995; 92:4442-4446.
116. Shimizu K, Ichikawa H, Tojo A et al. An ets-related gene, ERG, is rearranged in human myeloid leukemia with t(16;21) chromosomal translocation. Proc Natl Acad Sci USA 1993; 90:10280-10284.
117. Delattre O, Zucman J, Plougastel B et al. Gene fusion with an ETS DNA-binding domain caused by chromosome translocation in human tumours. Nature 1992; 359:162-165.
118. Jeon IS, Davis JN, Braun BS et al. A variant Ewing's sarcoma translocation (7;22) fuses the EWS gene to the ETS gene ETV1. Oncogene 1995; 10:1229-1234.
119. Sorensen PH, Lessnick SL, Lopez-Terrada D et al. A second Ewing's sarcoma translocation, t(21;22), fuses the EWS gene to another ETS-family transcription factor, ERG. Nat Genet 1994; 6:146-151.
120. Collyn H, Galiegue-Zouitina S, Szymiczek D et al. Quantitative and qualitative variation of ETS-1 transcripts in hematologic malignancies. Leukemia 1993; 7:1777-1785.
121. Bolon I, Gouyer V, Devouassoux M et al. Expression of c-ets-1, collagenase 1, and urokinase-type plasminogen activator genes in lung carcinomas. Am J Pathol 1995; 147:1298-1310.
122. Hultgardhnilsson A, Cercek B, Wang JW et al. Regulated expression of the ets-1 transcription factor in vascular smooth muscle cells in vivo and in vitro. Circ Res 1996; 78:589-595.
123. Treisman R. The serum response element. Trends Biochem 1992; 17:423-426.
124. Treisman R. Transient accumulation of c-fos RNA following serum stimulation

requires a conserved 5′ element and c-fos 3′ sequences. Cell 1985; 42:889-902.
125. Treisman R. Identification of a protein-binding site that mediates transcriptional response of the c-fos gene to serum factors. Cell 1986; 46:567-574.
126. Janknecht R, Cahill MA, Nordheim A. Signal integration at the c-fos promoter. Carcinogenesis 1995; 16:443-450.
127. Gille HG, Sharrocks AD, Shaw PE. Phosphorylation of transcription factor p62TCF by MAP kinase stimulates ternary complex formation at the c-fos promotor. Nature 1992; 358:414-417.
128. Marais R, Wynne J, Treisman R. The SRF accessory protein Elk-1 contains a growth factor-regulated transcriptional activation domain. Cell 1993; 73:381-393.
129. Hill CS, Marais R, John S et al. Functional analysis of a growth factor-responsive transcription factor complex. Cell 1993; 73:395-406.
130. Janknecht R, Ernst WH, Pingoud V et al. Activation of ternary complex factor Elk-1 by MAP kinases. EMBO J 1993; 12: 5097-5104.
131. Cahill MA, Janknecht R, Nordheim A. Signalling pathways: Jack of all cascades. Current Biology 1996; 6:16-19.
132. Konig H, Ponta H, Rahmsdorf U et al. Autoregulation of fos: The dyad symmetry element as the major target of repression. EMBO J 1989; 8:2559-2566.
133. Leung S, Miyamoto NG. Point mutational analysis of the human c-fos serum response factor binding site. Nucleic Acids Res 1989; 17:1177-1195.
134. Shaw PE, Frasch S, Nordheim A. Repression of c-fos transcription is mediated through p67SRF bound to the SRE. EMBO J 1989; 8:2567-2574.
135. Ernst WH, Janknecht R, Cahill MA et al. Transcriptional repression mediated by the serum response factor. FEBS Letters 1995; 357:45-49.
136. Herschman HR. Primary response genes induced by growth factors and tumor promoters. Annu Rev Biochem 1991; 60:281-319.
137. Treisman R. Journey to the surface of the cell: Fos regulation and the SRE. EMBO J 1995; 14:4905-4913.
138. Nguyen HQ, Hoffman-Liebermann B, Liebermann DA. The zinc finger transcription factor Egr-1 is essential for and restricts differentiation along the macrophage lineage. Cell 1993; 72:197-209.
139. Dawid IB. Intercellular signaling and gene regulation during early embryogenesis of Xenopus laevis. J Biol Chem 1994; 269:6259-6262.
140. Pellegrini L, Tan S, Richmond TJ. Structure of serum response factor core bound to DNA. Nature 1995; 376:490-498.
141. Herrera RE, Shaw PE, Nordheim A. Occupation of the c-fos serum response element in vivo by a multi-protein complex is unaltered by growth factor induction. Nature 1989; 340:68-70.
142. König H. Cell-type specific multiprotein complex formation over the c-fos serum response element in vivo: Ternary complex formation is not required for the induction of c-fos. Nucleic Acids Res 1991; 19:3607-3611.
143. de Belle I, Walker PR, Smith IC et al. Identification of a multiprotein complex interacting with the c-fos serum response element. Mol Cell Biol 1991; 11: 2752-2759.
144. Piette J, Yaniv M. Two different factors bind to the α-domain of the polyoma virus enhancer, one of which also interacts with the SV40 and c-fos enhancers. EMBO J 1987; 6:1331-1337.
145. Metz R, Ziff E. The helix-loop-helix protein rE12 and the C/EBP-related factor rNFIL-6 bind to neighboring sites within the c-fos serum response element. Oncogene 1991; 6:2165-2178.
146. Boulden AM, Sealy LJ. Maximal serum stimulation of the c-fos serum response element requires both the serum response factor and a novel binding factor, SRE-binding protein. Mol Cell Biol 1992; 12:4769-4783.
147. Trouche D, Grigorlev M, Lenormand J-L et al. Repression of c-fos promoter by MyoD on muscle cell differentiation. Nature 1993; 363:79-82.
148. Ryan WA Jr, Franza BR Jr, Gilman MZ. Two distinct cellular phosphoproteins bind to the c-fos serum response element. EMBO J 1989; 8:1785-1792.
149. Walsh K. Cross-binding of factors to functionally different promoter elements

in c-fos and skeletal actin genes. Mol Cell Biol 1989; 9:2191-2201.

150. Gualberto A, LePage D, Pons G et al. Functional antagonism between YY1 and the serum response factor. Mol Cell Biol 1992; 12:4209-4214.
151. Natesan S, Gilman MZ. DNA bending and orientation-dependent function of YY1 in the c-fos promoter. Genes Dev 1993; 7:2497-2509.
152. Natesan S, Gilman M. YY1 facilitates the association of serum response factor with the c-fos serum response element. Mol Cell Biol 1995; 15:5975-5982.
153. Treisman R. Identification and purification of a polypeptide that binds to the c-fos serum response element. EMBO J 1987; 6:2711-2717.
154. Norman C, Runswick M, Pollock R et al. Isolation and properties of cDNA clones encoding SRF, a transcription factor that binds to the c-fos serum response element. Cell 1988; 55:989-1003.
155. Schröter H, Mueller CGF, Meese K et al. Synergism in ternary complex formation between the dimeric glycoprotein p67SRF, polypeptide p62TCF and the c-fos serum response element. EMBO J 1990; 9:1123-1130.
156. Marais RM, Hsuan JJ, McGuigan C et al. Casein kinase II phosphorylation increases the rate of serum response factor-binding site exchange. EMBO J 1992; 11:97-105.
157. Mueller CGF, Nordheim A. A protein domain conserved between yeast MCM1 and human SRF directs ternary complex formation. EMBO J 1991; 10:4219-4229.
158. Shore P, Sharrocks AD. The MADS-box family of transcription factors. Eur J Biochem 1995; 229:1-13.
159. Pollock R, Treisman R. Human SRF-related proteins: DNA-binding properties and potential regulatory targets. Genes Dev 1991; 5:2327-2341.
160. Yu Y-Y, Breitbart RE, Smoot LB et al. Human myocyte-specific enhancer factor 2 comprises a group of tissue-restricted MADS box transcription factors. Genes Dev 1992; 6:1783-1798.
161. Croissant JD, Kim J-H, Eichele G et al. Avian serum response factor expression restricted primarily to muscle cell lineages is required for α-actin gene transcription. Developmental Biology 1996; 177:250-264.
162. Affolter M, Montagne J, Walldorf U et al. The *Drosophila* SRF homolog is expressed in a subset of tracheal cells and maps within a genomic region required for tracheal development. Development 1994; 120:743-753.
163. Guillemin K, Groppe J, Ducker K et al. The pruned gene encodes the *Drosophila* serum response factor and regulates cytoplasmic outgrowth during terminal branching of the tracheal system. Development 1996; 122:1353-1362.
164. Mohun TJ, Chambers AE, Towers N et al. Expression of genes encoding the transcription factor SRF during development of Xenopus laevis: identification of a CArG box-binding activity as SRF. EMBO J 1991; 10:933-940.
165. Maher M, Cong F, Kindelberger D et al. Cell cycle-regulated transcription of the CLB2 gene is dependent on Mcm1 and a ternary complex factor. Mol Cell Biol 1995; 15:3129-3137.
166. Althöfer H, Schleiffer A, Wassmann K et al. Mcm1 is required to coordinate G2-specific transcription in Saccharomyces cerevisiae. Mol Cell Biol 1995; 15: 5917-5928.
167. Treisman R, Marais R, Wynne J. Spatial flexibility in ternary complexes between SRF and its accessory proteins. EMBO J 1992; 11:4631-4640.
168. Ammerer G. Sex, stress and integrity: the importance of MAP kinases in yeast. Curr Opin Genet Dev 1994; 4:90-95.
169. Reason AJ, Morris HR, Panico M et al. Localization of O-GlcNac modification on the serum response transcription factor. J Biol Chem 1992; 267:16911-16921.
170. Schalasta G, Doppler C. Inhibition of c-fos transcription and phosphorylation of the serum response factor by an inhibitor of phospholipase C-type reactions. Mol Cell Biol 1990; 10:5558-5561.
171. Manak RJ, de Bisschop N, Kris RM et al. Casein kinase II enhances the DNA binding activity of serum response factor. Genes Dev 1990; 4:955-967.
172. Misra RP, Rivera VM, Wang JM et al. The serum response factor is extensively modified by phosphorylation following its synthesis in serum-stimulated fibroblasts. Mol Cell Biol 1991; 11:4545-4554.

173. Janknecht R, Hipskind RA, Houthaeve T et al. Identification of multiple SRF N-terminal phosphorylation sites affecting DNA binding properties. EMBO J 1992; 11:1045-1054.
174. Liu SH, Peng BH, Ma JT et al. Serum response element associated transcription factors in mouse embryos: Serum response factor, YY1, and PEA3 factor. Dev Genet 1995; 16:229-240.
175. Rivera VM, Miranti CK, Misra RP et al. A growth factor-induced kinase phosphorylates the serum response factor at a site that regulates its DNA-binding activity. Mol 1993; 13:6260-6273.
176. Treisman R. Ternary complex factors: Growth factor regulated transcriptional activators. Curr Opin Genet Dev 1994; 4:96-101.
177. Shaw PE, Schröter H, Nordheim A. The ability of a ternary complex to form over the serum response element correlates with serum inducibility of the human c-fos promoter. Cell 1989; 56:563-572.
178. Groisman R, Masutani H, Leibovitch MP et al. Physical interaction between the mitogen-responsive serum response factor and myogenic basic-helix-loop-helix proteins. J Biol Chem 1996; 271: 5258-5264.
179. Grueneberg DA, Natesan S, Alexandre C et al. Human and *Drosophila* homeodomain proteins that enhance the DNA-binding activity of serum response factor. Science 1992; 257:1089-1095.
180. Grueneberg DA, Simon KJ, Brennan K et al. Sequence-specific targeting of nuclear signal transduction pathways by homeodomain proteins. Mol Cell Biol 1995; 15:3318-3326.
181. Franzoso G, Carlson L, Brown K et al. Activation of the serum response factor by p65/NFκB. EMBO J 1996; 15: 3403-3412.
182. Zhu H, Joliot V, Prywes R. Role of transcription factor TFIIF in serum response factor-activated transcription. J Biol Chem 1994; 269:3489-3497.
183. Joliot V, Demma M, Prywes R. Interaction with RAP74 subunit of TFIIF is required for transcriptional activation by serum response factor. Nature 1995; 373:632-635.
184. Fujii M, Tsuchiya H, Chuhjo T et al. Interaction of HTLV-1 Tax1 with p67SRF causes the aberrant induction of cellular immediate early genes through CArG boxes. Genes Dev 1992; 6: 2066-2076.
185. Pingoud V, Zinck R, Hipskind RA et al. Heterogeneity of ternary complex factors in HeLa cell nuclear extracts. J Biol Chem 1994; 269:23310-23317.
186. Hipskind RA, Rao VN, Mueller CG et al. Ets-related protein Elk-1 is homologous to the c-fos regulatory factor p62TCF. Nature 1991; 354:531-534.
187. Dalton S, Treisman R. Characterization of SAP-1, a protein recruited by serum response factor to the c-fos serum response element. Cell 1992; 68:597-612.
188. Giovane A, Pintzas A, Maira SM et al. Net, a new ets transcription factor that is activated by Ras. Genes Dev 1994; 8:1502-1513.
189. Lopez M, Oettgen P, Akbarali Y et al. ERP, a new member of the ets transcription factor/oncoprotein family: cloning, characterization, and differential expression during B-lymphocyte development. Mol Cell Biol 1994; 14:3292-3309.
190. Magnaghi-Jaulin L, Masutani H, Robin P et al. SRE elements are binding sites for the fusion protein EWS-FLI-1. Nucleic Acids Res 1996; 24:1052-1058.
191. Janknecht R, Nordheim A. Elk-1 protein domains required for direct and SRF-assisted DNA-binding. Nucleic Acids Res 1992; 20:3317-3324.
192. Rao VN, Reddy ES. A divergent ets-related protein, elk-1, recognizes similar c-ets-1 proto-oncogene target sequences and acts as a transcriptional activator. Oncogene 1992; 7:65-70.
193. Rao VN, Reddy ES. elk-1 domains responsible for autonomous DNA binding, SRE:SRF interaction and negative regulation of DNA binding. Oncogene 1992; 7:2335-2340.
194. Treisman R, Marais R, Wynne J. Spatial flexibility in ternary complexes between SRF and its accessory proteins. EMBO J 1992; 11:4631-4640.
195. Shaw PE. Ternary complex formation over the c-fos serum response element: p62TCF exhibits dual component specificity with contacts to DNA and an ex-

tended structure in the DNA-binding domain of p67SRF. EMBO J 1992; 11: 3011-3019.
196. Shore P, Sharrocks AD. The transcription factors Elk-1 and serum response factor interact by direct protein-protein contacts mediated by a short region of Elk-1. Mol Cell Biol 1994; 14:3283-3291.
197. Janknecht R, Ernst WH, Nordheim A. SAP1a is a nuclear target of signaling cascades involving ERKs. Oncogene 1995; 10:1209-1216.
198. Rao VN, Huebner K, Isobe M et al. elk, tissue-specific ets-related genes on chromosomes X and 14 near translocation breakpoints. Science 1989; 244:66-70.
199. Janknecht R, Zinck R, Ernst WH et al. Functional dissection of the transcription factor Elk-1. Oncogene 1994; 9: 1273-1278.
200. Zinck R, Hipskind RA, Pingoud V et al. c-fos transcriptional activation and repression correlate temporally with the phosphorylation status of TCF. EMBO J 1993; 12:2377-2387.
201. Kortenjann M, Thomae O, Shaw PE. Inhibition of v-raf-dependent c-fos expression and transformation by a kinase-defective mutant of the mitogen-activated protein kinase Erk2. Mol Cell Biol 1994; 14:4815-4824.
202. Gille H, Kortenjann M, Thomae O et al. ERK phosphorylation potentiates Elk-1-mediated ternary complex formation and transactivation. EMBO J 1995; 14: 951-962.
203. Zinck R, Cahill MA, Kracht M et al. Protein synthesis inhibitors reveal differential regulation of mitogen-activated protein kinase and stress-activated protein kinase pathways that converge on Elk-1. Mol Cell Biol 1995; 15:4930-4938.
204. Cavigelli M, Dolfi F, Claret FX et al. Induction of c-fos expression through JNK-mediated TCF/Elk-1 phosphorylation. EMBO J 1995; 14:5957-5964.
205. Gille H, Strahl T, Shaw PE. Activation of ternary complex factor Elk-1 by stress-activated protein kinases. Curr Biol 1995; 5:1191-1200.
206. Whitmarsh AJ, Shore P, Sharrocks AD et al. Integration of MAP kinase signal transduction pathways at the serum response element. Science 1995; 269: 403-407.
207. Raingeaud J, Whitmarsh AJ, Barrett T et al. MKK3- and MKK6-regulated gene expression is mediated by the p38 mitogen-activated protein kinase signal transduction pathway. Mol Cell Biol 1996; 16:1247-1255.
208. Price MA, Rogers AE, Treisman R. Comparative analysis of the ternary complex factors Elk-1, SAP-1a and SAP-2 (ERP/NET). EMBO J 1995; 14:2589-2601.
209. Janknecht R, Nordheim A. Regulation of the c-fos promoter by the ternary complex factor Sap-1a and its coactivator CBP. Oncogene 1996; 12:1961-1969.
210. Hipskind RA, Buscher D, Nordheim A et al. Ras/MAP kinase-dependent and -independent signaling pathways target distinct ternary complex factors. Genes Dev 1994; 8:1803-1816.
211. Giovane A, Sobieszczuk P, Mignon C et al. Locations of the ets subfamily members net, elk1, and sap1 (ELK3, ELK1, and ELK4) on three homologous regions of the mouse and human genomes. Genomics 1995; 29:769-772.
212. Dodou E, Sparrow DB, Mohun T et al. MEF2 proteins, including MEF2A, are expressed in both muscle and non-muscle cells. Nucleic Acids Res 1995; 23:4267-4274.
213. Cooper JP, Roth SY, Simpson RT. The global transcriptional regulators, SSN6 and TUP1, play distinct roles in the establishment of a repressive chromatin structure. Genes Dev 1994; 8:1400-1410.
214. Johnson AD. The price of repression. Cell 1995; 81:655-658.
215. Kornberg RD, Lorch Y. Interplay between chromatin structure and transcription. Curr Opin Cell Biol 1995; 7: 371-375.
216. Cowell IG. Repression versus activation in the control of gene transcription. Trends Biochem Sci 1994; 19:38-42.
217. Perlmann T, Vennstrom B. The sound of silence [news; comment]. Nature 1995; 377:387-388.
218. Yamaguchi K, Ogita K, Nakamura S et al. The protein kinase C isoforms leading to MAP-kinase activation in CHO cells. Biochem Biophys Res Commun 1995; 210:639-647.

219. Nishizuka Y. Protein kinase C and lipid signaling for sustained cellular responses. FASEB J 1995; 9:484-496.
220. Miranti CK, Ginty DD, Huang G et al. Calcium activates serum response factor-dependent transcription by a Ras- and Elk-1-independent mechanism that involves a Ca2+/calmodulin-dependent kinase. Mol Cell Biol 1995; 15:3672-3684.
221. Hill CS, Wynne J, Treisman R. The Rho family GTPases RhoA, Rac1, and CDC42Hs regulate transcriptional activation by SRF. Cell 1995; 81:1159-1170.
222. Hill CS, Treisman R. Differential activation of c-fos promoter elements by serum, lysophosphatidic acid, G proteins and polypeptide growth factors. EMBO J 1995; 14:5037-5047.

CHAPTER 15

Myb Proteins: Regulators of Hemopoietic Cell Growth, Differentiation and Apoptosis

Jonathan Frampton and Thomas Graf

INTRODUCTION

As is becoming a trend for transcriptional regulators, c-Myb is the prototype for an expanding family of related factors. Originally defined as the proto-oncogenic counterpart of the v-Myb transforming protein of the avian acute leukemia viruses, it is now clear that c-Myb is a central regulator of gene expression within the hemopoietic system and perhaps in other tissues. Two homologues of c-Myb, A- and B-Myb, are expressed in distinct but partially overlapping patterns in vertebrates and a large number of more distantly related family members are found in eukaryotes from yeast to mammals. Recent reviews have extensively covered the structure, functional domains and expression of c-Myb and its homologues.[1-6] We will concentrate here more on mechanistic features, paying particular attention to known or potential target genes of c-Myb involved in growth control, differentiation and survival, with an emphasis on those regulated in hemopoietic cells.

CELL GROWTH REGULATION

Much circumstantial evidence points to a function for c-Myb in the control of proliferation, largely through its expression during specific phases of the cell cycle. Thus, in many cell types elevated expression of c-Myb is seen at the G1-S transition, and the presence of recognition elements for E2F in the c-*myb* promoter implicates this periodically active factor in its control (DeGregori et al[7] and references therein). Consistent with these observations, reductions of c-Myb function by the use of specific antisense oligonucleotides has been shown to block cells in G1 (reviewed in refs. 2-3), and c-Myb appears to be essential in late G1 for progression of fibroblasts into S phase.[8] Interestingly, B-Myb is also activated as cells enter into S phase,[9-10] but in this case it involves a repression by E2F rather than an activation.[11] More

Transcription Factors in Eukaryotes, edited by Athanasios G. Papavassiliou.

recently, in vivo footprint analyses have shown that a factor binding the E2F site in the B-*myb* promoter is lost as cells enter S phase and begin to express B-*myb* RNA. This factor has been identified as E2F4.[12]

A number of genes involved in cell cycle regulation at the G1-S boundary could in principle be regulated by Myb proteins and several candidates have been described. The timing of c-Myb expression closely correlates with that of enzymes required for progression into S phase, such as thymidine kinase and the α isoform of topoisomerase III;[8] in fact, the minimal promoter of the latter contains a perfect Myb recognition element although its role has yet to be examined.[13] Other candidate Myb targets include the cdc2 kinase gene (c-Myb and B-Myb) and cyclin D1 (B-Myb; reviewed in refs. 4-6). However, since the timing of the appearance of B-Myb does not correlate with that of cyclin D1, it is not likely that it is directly Myb-regulated. Nevertheless, a direct link between B-Myb and cell cycle control is implied by the finding that its over-expression is able to relieve a G1 arrest mediated by the p21 kinase inhibitor in glioblastoma cells.[14] Furthermore, over-expression of B-Myb in the M1 murine leukemic cell line has been shown to overcome the proliferation arrest that accompanies IL-6 induced monocytic differentiation.[15] Studies with a conditional mutant of E26 virus also support the idea that there is a direct link between Myb and cell cycle progression. Thus, inactivation of v-Myb encoded by the E26 temperature sensitive mutant *ts*21 in transformed myelomonocytic cells shifted from 37°C to 42°C leads to a rapid G1 arrest.[16-17] However, the observed effects might be secondary to maturation since under similar conditions thromboblastic cells transformed by *ts*21 differentiate without immediately withdrawing from the cell cycle.[18]

Myb may also play a part in the regulation of the growth factor IGF-1 which is required during the late G1 phase of the cell cycle in most cell types. Thus, high mRNA levels of both IGF-1 and its receptor correlate with high c-*myb* RNA expression in fibroblasts,[19] in stimulated Tlymphocytes and in HL-60 myeloblasts.[20] More direct evidence for a role of Myb in the control of IGF-1 has come from studies performed on vascular smooth muscle cells (VSMC). Here, an antisense c-*myb* oligonucleotide was shown to induce a late G1 arrest in VSMC and this correlated with suppression of the elevation in intracellular Ca^{2+} ion concentration which occurs at this point in the cell cycle.[21] From experiments involving over-expression of c-Myb, it appears that its effect in VSMC operates through an upregulation of the mRNAs for IGF-1 and its receptor.[22] It will therefore be extremely interesting to determine whether this upregulation occurs through a direct binding of c-Myb to the promoters of the corresponding genes.

Perhaps caution should be exercised in the interpretation of these results, however, because it has recently been shown that the anti-proliferative effect of the antisense c-*myb* oligonucleotides employed involves a nonantisense mechanism.[23] Indeed, this observation could well cast some doubt on the earlier results obtained with the same oligonucleotides which showed that the proliferative capacity of hemopoietic cells was diminished.[24]

Finally, a more tenuous connection between Myb and the cell cycle comes from the finding that the *S. pombe* cell cycle regulatory gene cdc5 encodes a protein which contains a region of high homology to the Myb DNA-binding domain.[25] Inactivation of this protein leads to a cell cycle arrest with cells exhibiting a 2N DNA content, suggesting an essential role in G2. Interestingly, the N-terminal portion of the protein, containing the Myb DNA-binding motif, is sufficient to rescue the defective cells.

DIFFERENTIATION

A plethora of observations in erythroid cells, monocytes, neutrophil granulocytes and lymphoid cells has shown either directly or by implication that c-Myb plays a role in the differentiation of these cell types (reviewed in refs. 5-6). A number of recent developments have shed light on the biological role performed by Myb in these cells.

Until now, it was generally accepted that transformation and the associated block to differentiation in hemopoietic cells required truncation of at least the C-terminus of c-Myb (reviewed in refs. 3, 4 and 6). However, Ferrao et al[26] have been able to transform fetal liver myelomonocytic cells with full length c-Myb by maintaining the cells at a high density. This density dependence could be partly overcome by using conditioned medium from transformed cells grown at high density, suggesting that an autocrine factor is involved.

In contrast to the original assumption that c-Myb blocks cells at early stages of differentiation, more recent results obtained in both erythroid and myeloid cells suggest that it can also block at late stages. Danish et al[27] showed that by expressing exogenous c-Myb late in erythroid differentiation by linking it to a globin gene promoter, late but not early differentiation events could be prevented with no effect on proliferation. Similarly, Bies et al[28] demonstrated a block by exogenous c-Myb of terminal granulocytic differentiation in the 32Dcl3 cell line. Since the effects of overexpression were limited to downregulation of differentiation markers these reports imply that the function of c-Myb in the control of proliferation and differentiation can be separated.

We have recently found that v-Myb affects maturation of thromboblasts into thrombocytes, as seen by a block to the differentiation of *ts*21 E26 v-Myb transformed thromboblasts.[18] Does this reflect a function of c-Myb during differentiation of this lineage? The finding that the livers of fetuses homozygous for an ablated c-*myb* allele contain megakaryocytes (the mammalian equivalents of chicken thromboblasts) while erythroid cells are absent, suggests that c-Myb is not essential for the formation of megakaryocytes.[29] However, it is not clear whether mature platelets are formed in these embryos and it is a possible that the mutated c-*myb* loci are still capable of producing a truncated c-Myb protein. It would therefore be desirable to re-examine the phenotype of mice homozygous for a completely deleted c-*myb* gene. In another approach, c-Myb overexpressed in the human K562 erythroid/ megakaryocytic cell line failed to block TPA-induced megakaryocyte formation.[30] Rather than minimizing a function of c-Myb, v-Myb might therefore interfere with A- or B-Myb function as, at least in chickens, both of these c-Myb homologues are expressed in cells of the thromboblastic lineage (unpublished data, JF and TG).

The specific ablation of cells in the thymus with a dominant negatively acting c-Myb-Engrailed chimera[31] likewise indicates a role for Myb in T cell differentiation. Correspondingly, there is an accumulating number of T cell specific genes which are regulated, at least in part, by c-Myb. The best characterized of these is the T cell receptor (TCR) δ gene (refs. 32 and 33 and references therein), but more recently the TCR γ and *lck* genes have been added to the list.[34-35] Interestingly, binding of c-Myb to the TCR δE3 enhancer is necessary for VD-J recombination[36] although it is not yet clear whether this represents a direct effect of Myb. An intriguing possibility is that c-Myb controls groups of T cell specific genes through its effect on the T cell locus control region (LCR). This locus, which is located in the first intron of the adenosine deaminase gene, contains a consensus c-Myb binding site at its core.[37]

APOPTOSIS

Research over the last few years has revealed yet another function of Myb, that of regulating components of the programmed cell death machinery. Experiments involving forced expression of Myb in hemopoietic cells have shown that the oncoprotein can either accelerate or suppress apoptosis, depending on the cell line and apoptotic stimulus used. Thus, Bies and Wolff[38] showed an acceleration of TGF-β1 induced apoptosis in M1 cells by over expression of B-Myb and Selvakumaran et al[39] obtained similar results with c-Myb. In contrast, Smarda and Lipsick[40] showed that c-Myb over expression reduces phorbol ester-induced apoptosis in the AMV transformed monoblast cell line BM2. The differences in the effects observed by forced c-Myb expression may reflect differences in the cell lines used.

More recent work with primary hemopoietic cells has shown that the Myb transcription factor, both in its oncogenic and normal form, can prevent apoptosis by directly upregulating the death preventing *bcl*-2 gene. Thus, when *ts*21 E26 mutant transformed chicken myeloblasts are shifted to the nonpermissive temperature, leading to v-Myb inactivation, a large proportion of the transformed cells undergo apoptosis,[17] while the remaining cells differentiate into macrophage-like cells.[41] This effect on cell death, which is paralleled by downregulation of *bcl*-2 expression can be reversed by ectopic expression of *bcl*-2. Similarly, transgenic mice expressing a dominantly-negatively acting Myb-Engrailed fusion protein under the control of the CD2 promoter exhibit the death of a population of $CD4^-$ $CD8^-$ T cells, suggesting that Myb normally prevents their programmed cell death.[31] Taylor et al[42] have now shown that this dominant negative fusion protein downregulates *bcl*-2 expression and causes apoptosis in the EL4 T cell line.

In both the avian myeloid and the murine T lymphoid systems Myb appears to directly activate *bcl*-2 expression through Myb recognition elements present within the downstream (P2) promoter of the gene.[17,24] These results suggest that c-Myb plays an important role in hemopoiesis by regulating the survival of certain subsets of T cells and myeloid cells. They also indicate that *myb*-expressing retroviruses have evolved to evade a cellular mechanism which prevents the elimination of cells in which *bcl*-2 is downregulated, being no longer required. Interestingly, the effect of v-Myb on *bcl*-2 regulation is lineage specific since v-Myb inactivation in thromboblastic progenitors transformed by *ts*21 neither affects *bcl*-2 expression nor apoptosis.[17] This suggests that *bcl*-2 regulation by v-Myb is important for the transforming potential of the oncoprotein for some but not all cell types. In addition, as is the case for the differentiation-related target genes of Myb (see below), the action of Myb on the *bcl*-2 promoter probably requires cooperation with other transcription factors, some of which may themselves be lineage restricted.

TARGET GENES AND COOPERATING FACTORS

Since most well characterized direct target genes of c-Myb seem to play a role in hemopoietic differentiation, it is not surprising that our knowledge of nuclear factors that cooperate with c-Myb has come exclusively from studies on lineage specific target genes. Most of these studies have centered on the TCR δ gene and on genes encoding enzymes specific to myelomonocytic cells. Thus, regulation of the TCR δ enhancer involves an interaction between c-Myb and the CBF transcription factor complex.[32-33] The CBF complex consists of the AML-1 (PEBP2αB) subunit, with a DNA-binding domain homologous to the *Drosophila* protein Runt, and a PEBP2β subunit with no DNA-binding capacity. Both of the genes encoding these polypeptides have been identified in specific translocations associated with human myeloid leukemias.[43-44] The involvement of the CBF complex is especially intriguing because it has been found that, as in c-Myb defective mice, animals homozygous for the inactivated AML-1 gene fail to develop the adult hemopoietic system.[45] This indicates that both c-Myb and CBF play a crucial role not only in differentiation but also in lineage commitment.

The TCR γ gene, which is coexpressed with TCR δ in γδ T cells, appears to have much the same requirement for a cooperation between c-Myb and CBF.[34] The TCR δ and γ enhancers might therefore represent a paradigm for regulation by Myb of genes playing an important role in developmental decisions. In contrast, the TCR α gene, which is expressed in αβ T cells, is not regulated by c-Myb, but instead by a multiprotein complex including c-Ets-1, ATF/CREB, LEF-1 and CBF.[46] Myb and Ets, in combination with other factors involved in the regulation of TCR genes, are also used in the control of additional T cell genes. For example, the genes encoding the receptor tyrosine kinase lck and the CD4 protein are activated synergistically by c-Myb and Ets family proteins.[35,47-48]

Work on the promoters of genes specifically expressed in myelomonocytic cells has revealed a number of Myb targets and is beginning to provide insights into interactions with accessory factors. In addition to the chicken *mim*-1 gene, which represents the first identified Myb target (Ness et al 1989), the list now includes the c-*fms* gene encoding the CSF-1 receptor[50] and the CD13 cell surface peptidase aminopeptidase N.[51]

The way that Myb regulates T cell specific genes appears to be mirrored by its actions on myelomonocytic genes. For example, regulation of the *mim*-1 promoter by c-Myb or v-Myb requires coordinated action with members of the C/EBP family of basic leucine zipper proteins (bZIP).[52-53] More recently, it has been shown that the CD13 gene is regulated synergistically by c-Myb in combination with an Ets-like factor, which is most likely c-Ets-2.[51] Interestingly, CD13 expression is downregulated in immature myeloid cells following their interaction with stromal cells, and this correlates with a reduction in the levels of c-Myb. Furthermore, ectopic c-Myb prevents this down-regulation of CD13 (Linda Shapiro, personal communication). The finding that c-Myb can also repress gene expression opens up a whole new aspect to the role played by Myb in gene regulation. Thus, the c-*fms* gene, which is upregulated as cells mature along the myelomonocytic pathway, is activated by Ets family proteins (specifically c-Ets-1, c-Ets-2 and PU-1) but can be repressed by c-Myb.[50] This repression appears to be active in that it is dependent upon the C-terminal portion of c-Myb. Similarly, when v-Myb is inactivated in avian myelomonocytic cells transformed by the *ts*21 E26 mutant a rapid increase in c-*fms* RNA is observed.[17] However, whether or not Myb acts directly on the c-*fms* promoter remains to be shown.

How does Myb cooperate with other transcription factors? Most studies addressing this question have been performed with Myb and C/EBPβ, a combination of factors that appears to be sufficient to activate the endogenous *mim*-1 gene in heterologous cells, such as fibroblasts.[52-53] Although some authors have been unable to show a direct interaction between c-Myb and C/EBPs in their synergistic effect on the *mim*-1 promoter,[54] Mink et al[55] have reported an interaction involving the basic region of C/EBPβ and the C-terminal end of repeat R2 of the AMV v-Myb DNA-binding domain. Curiously, however, an interaction with c-Myb or E26 v-Myb could not be observed even though the latter, rather than AMV v-Myb, upregulate *mim*-1 in myeloid cells.[49] Another accessory factor described for Myb is the CREB binding protein or CBP.[56-57] Using the *mim*-1 promoter, it was shown that CBP cooperates with both Myb and C/EBP in functional assays, but the magnitude of the effect (approximately 2- to 4-fold) is small compared to the degree of synergism seen between Myb and C/EBP.[52-53] c-Myb interacts with CBP at a site (residues 590-669) which overlaps with that occupied by phosphorylated CREB. The interaction surface on Myb has been located to its transactivation domain, although additional sequences seem also to be required.[56] Thus, to date two factors have been identified which cooperate with Myb. C/EBP is restricted within the hemopoietic system predominantly to myelomonocytic cells and is a sequence specific binding protein in its own right. In contrast, CBP is more widely expressed and it does not seem to be able to bind to DNA. From the range of lineages in which c-Myb is expressed and active, it can be predicted that there will be many more accessory factors involved in the optimum functioning and regulation of Myb. Candidates are the bZIP-type proteins p67 and p160 which were shown to interact with the leucine zipper motif of c-Myb.[58]

Constructs in which the Myb transactivation domain of the Myb-Ets oncoprotein was replaced with VP16 were found to be capable of transforming cells in vitro and of causing leukemia. However, the transformed cells exhibited a slightly different phenotype from that of wild type

E26 transformed cells, presumably reflecting differences in the pattern of target genes regulated by the Myb derivative.[59] The same study also showed that Myb does not interact with TFIID/TBP through its transactivation domain. It is therefore possible that another factor links Myb to the basal transcription machinery. Perhaps this factor is CBP.

DISTANT MYB HOMOLOGUES

A number of more distant Myb relatives have been described with interesting novel properties. Thus, a protein which recognizes the repeated TTAGGG motif in telomeres was found to contain a domain homologous to the DNA-binding domain repeat of Myb. This protein is believed to facilitate telomerase-mediated telomere elongation.[60] In addition, the DNA-binding protein mTTF-1, which specifically terminates transcription of the mouse ribosomal RNA gene by RNA pol I, has recently been shown to contain a striking homology to the Myb DNA-binding domain at its C-terminus.[61] Motifs related to the Myb DNA-binding domain also appear to be present in a wide range of proteins which are associated with transcriptional complexes. For example, the α interferon-induced multimeric transcription factor ISGF3, which is part of the JAK-STAT signaling pathway, shows significant homology to the Myb DNA-binding domain in the N terminal region of the γ subunit which has been shown to interact directly with the interferon responsive element.[62-63] Also, recent cloning of the N-CoR protein, which acts as a corepressor of nuclear hormone receptors,[64] has revealed a family of proteins with diverse roles in transcriptional regulation but which all contain a motif related to the Myb DNA-binding domain.[65]

It is even conceivable that Myb proteins are involved in translational regulation. For example, especially in light of the similarity between DNA-binding motifs of Myb and homeobox proteins,[66-67] it is interesting that the Bicoid homeodomain-containing protein of *Drosophila* can control translation of *caudal* RNA by binding to its 3' untranslated region.[68]

OUTLOOK

The identification of novel Myb target genes, beyond those involved in differentiation, has broadened our understanding of c-Myb action, and it is likely that more will be described in the near future. Besides its role as a transactivator, c-Myb has also been shown to act as a repressor and it will now be interesting to find out the mechanisms that underlie this functional dichotomy. At least on model promoters, B-Myb is also capable of both activating and repressing genes in a cell type specific manner.[69] However, the nature of the genes regulated by A-Myb and B-Myb is still a mystery, and it is not clear to what extent there is an overlap with the functions (and targets) of c-Myb. The creation of mice homozygous for defective alleles of the two close homologues of c-Myb should soon give us greater insights into the role of these proteins in the whole animal.

References

1. Shen-Ong GLC. The myb oncogene. Biochem Biophys Acta 1990; 1032:39-52.
2. Lüscher B, Eisenman RN. New light on Myc and Myb. Part II. Myb. Genes & Dev 1990; 4:2235-2241.
3. Graf T. Myb: A transcriptional activator linking proliferation and differentiation in hematopoietic cells. Curr Opin Genet Dev 1992; 2:249-255.
4. Introna M, Luchetti M, Castellano M et al. The myb oncogene family of transcription factors: Potent regulators of hematopoietic cell proliferation and differentiation. Semin Cancer Biol 1994; 5:113-124.
5. Thompson MA, Ramsay RG. Myb: An old oncoprotein with new roles. Bioessays 1995; 17:341-350.
6. Ness SA. The Myb oncoprotein: Regulating a regulator. Biochim Biophys Acta 1996; 1288:F123-139.
7. DeGregori J, Kowalik T, Nevins JR. Cellular targets for activation by the E2F1 transcription factor include DNA synthesis- and G1/S-regulatory genes. Mol Cell Biol 1995; 15:4215-4224.
8. Olson JE, Winston JT, Whitlock JA et al. Cell cycle-dependent gene expression in

V point-arrested BALB/c-3T3 cells. J Cell Physiol 1993; 154:333-342.
9. Lam EW, Robinson C, Watson RJ. Characterization and cell cycle-regulated expression of mouse B-myb. Oncogene 1992; 7:1885-1890.
10. Lam EW, Bennett JD, Watson RJ. Cell-cycle regulation of human B-myb transcription. Gene 1995; 160:277-281.
11. Lam EW, Watson RJ. An E2F-binding site mediates cell-cycle regulated repression of mouse B-myb transcription. EMBO J 1993; 12:2705-2713.
12. Zwicker J, Liu N, Engeland K et al. Cell cycle regulation of E2F site occupation in vivo. Science 1996; 271:1595-1597.
13. Fraser DJ, Brandt TL, Kroll DJ. Topoisomerase IIα promoter trans-activation early in monocytic differentiation of HL-60 human leukemia cells. Mol Pharmacol 1995; 47:696-706.
14. Lin D, Fiscella M, O'Connor PM et al. Constitutive expression of B-myb can bypass p53-induced Waf1/Cip1-mediated G1 arrest. Proc Natl Acad Sci USA 1994; 91:10079-10083.
15. Bies J, Hoffman B, Amanullah A et al. B-Myb prevents growth arrest associated with terminal differentiation of monocytic cells. Oncogene 1996; 12:355-363.
16. Beug H, Blundell PA, Graf T. Reversibility of differentiation and proliferative capacity in avian myelomonocytic cells transformed by tsE26 leukemia virus. Genes & Dev 1987; 1:277-286.
17. Frampton J, Ramqvist T, Graf T. v-Myb of E26 leukemia virus up-regulates bcl-2 and suppresses apoptosis in myeloid cells. Genes & Dev 1996; 10:2720-2731.
18. Frampton J, McNagny KM, Sieweke M et al. v-Myb DNA binding is required to block thrombocytic differentiation of Myb-Ets-transformed multipotent haematopoietic progenitors. EMBO J 1995; 14:2866-2875.
19. Reiss K, Ferber A, Travali S et al. The protooncogene c-myb increases the expression of insulin-like growth factor 1 and insulin-like growth factor 1 receptor messenger RNAs by a transcriptional mechanism. Cancer Res 1991; 51: 5997-6000.
20. Reiss K, Porcu P, Sell C et al. The insulin-like growth factor 1 receptor is required for the proliferation of hemopoietic cells. Oncogene 1992; 7:2243-2248.
21. Simons M, Morgan KG, Parker C et al. The proto-oncogene c-myb mediates an intracellular calcium rise during the late G_1 phase of the cell cycle. J Biol Chem 1993; 268:627-632.
22. Simons M, Ariyoshi H, Salzman EW et al. c-myb affects intracellular calcium handling in vascular smooth muscle cells. Am J Physiol 1995; 268:C856-C868.
23. Burgess TL, Fisher EF, Ross SL et al. The antiproliferative activity of c-myb and c-myc antisense oligonucleotides in smooth muscle cells is caused by a nonantisense mechanism. Proc Natl Acad Sci USA 1995; 92:4051-4055.
24. Gewirtz AM, Calabretta B. A c-myb antisense oligodeoxynucleotide inhibits normal human hematopoiesis in vitro. Science 1988; 242:1303-1306.
25. Ohi R, McCollum D, Hirani B et al. The Schizosaccharomyces pombe cdc5$^+$ gene encodes an essential protein with homology to c-Myb. EMBO J 1994; 13:471-483.
26. Ferrao P, Macmillan EM, Ashman LK et al. Enforced expression of full length c-Myb leads to density-dependent transformation of murine haemopoietic cells. Oncogene 1995; 11:1631-1638.
27. Danish R, el-Awar O, Weber BL et al. c-myb effects on kinetic events during MEL cell differentiation. Oncogene 1992; 7:901-907.
28. Bies J, Mukhopadhyaya R, Pierce J et al. Only late, nonmitotic stages of granulocyte differentiation in 32Dc13 cells are blocked by ectopic expression of murine c-myb and its truncated forms. Cell Growth & Diff 1995; 6:59-68.
29. Mucenski ML, McLain K, Kier AB et al. A functional c-myb gene is required for normal murine fetal hepatic hematopoiesis. Cell 1991; 65:677-689.
30. Rosson D, O'Brien TG. Constitutive c-myb expression in K562 cells inhibits induced erythroid differentation but not tetradecanoyl phorbol acetate-induced megakaryocytic differentiation. Mol Cell Biol 1995; 15:772-779.

31. Badiani P, Corbella P, Kioussis D et al. Dominant interfering alleles define a role for c-Myb in T cell development. Genes & Dev 1994; 8:770-782.
32. Hernández-Munain C, Krangel MS. Regulation of the T cell receptor δ enhancer by functional cooperation between c-Myb and core-binding factors. Mol Cell Biol 1994; 14:473-483.
33. Hernández-Munain C, Krangel MS. c-Myb and core-binding factor/PEBP2 display functional synergy but bind independently to adjacent sites in the T cell receptor delta enhancer. Mol. Cell. Biol 1995; 15:3090-3099.
34. Hsiang HH, Goldman JP, Raulet DH. The role of c-Myb or a related factor in regulating the T cell receptor gamma gene enhancer. J Immunol 1995; 154: 5195-5204.
35. McCracken S, Leung S, Bosselut R et al. Myb and Ets related transcription factors are required for activity of the human lck type I promoter. Oncogene 1994; 9: 3609-3615.
36. Hernández-Munain C, Lauzurica P, Krangel MS. Regulation of T cell receptor δ gene rearrangement by c-Myb. J Exp Med 1996; 183:289-293.
37. Ess KC, Whitaker, TL, Cost GJ et al. A central role for a single c-Myb binding site in a thymic locus control region. Mol Cell Biol 1995; 15:5707-5715.
38. Bies J, Wolff L. Accelaration of apoptosis in transforming growth factor β 1-treated M1 cells ectopically expressing B-myb. Cancer Res 1995; 55:501-504.
39. Selvakumaran M, Lin, H-K, Sjin RTT et al. The novel primary response gene MyD118 and the proto-oncogenes myb, myc, and bcl-2 modulate transforming growth factor β1-induced apoptosis of myeloid leukemia cells. Mol Cell Biol 1994; 14:2352-2360.
40. Smarda J, Lipsick JS. c-Myb prevents TPA-induced differentiation and cell death in v-Myb transformed monoblasts. Oncogene 1994; 9:237-245.
41. Beug H, Leutz A, Kahn P et al. Temperature-sensitive mutants of E26 leukemia virus allow transformed myeloblasts, but not erythroblasts or fibroblasts, to differentiate at the non-permissive temperature. Cell 1984; 39:579-588.
42. Taylor D, Badiani P and Weston K. A dominant interfering Myb mutant causes apoptosis in T cells. Genes & Dev 1996; 10:2732-2744.
43. Liu P, Tarle SA, Hajra A et al. Fusion between transcription factor CBFb/PEBP2b and a myosin heavy chain in acute myeloid leukemia. Science 1993; 261:1041-1044.
44. Miyoshi H, Shimizu K, Kozu T et al. t(8;21) breakpoints on chromosome 21 in acute myeloid leukemia are clustered within a limited region of a single gene, AML1. Proc Natl Acad Sci USA 1991; 88:10431-10434.
45. Okuda T, van-Deursen J, Hiebert SW et al. AML1, the target of multiple chromosomal translocations in human leukemia, is essential for normal fetal liver hematopoiesis. Cell 1996; 84:321-330.
46. Giese K, Kingsley C, Kirschner JR et al. Assembly and function of a TCR α enhancer complex is dependent on LEF-1-induced DNA bending and multiple protein-protein interactions. Genes & Dev 1995; 9:995-1008.
47. Salmon P, Giovane A, Wasylyk B et al. Characterization of the human CD4 gene promoter:transcription from the CD4 gene core promoter is tissue-specific and is activated by Ets proteins. Proc Natl Acad Sci USA 1993; 90:7739-7743.
48. Siu G, Wurster AL, Lipsick JS, Hendrick SM. Expression of the CD4 gene requires a Myb transcription factor. Mol Cell Biol 1992; 12:1592-1604.
49. Ness SA, Marknell Å, Graf T. The v-myb oncogene product binds to and activates the promyelocyte-specific mim-1 gene. Cell 1989; 59:1115-1125.
50. Reddy MA, Yang BS, Yue X et al. Opposing actions of c-ets/PU.1 and c-myb protoncogene products in regulating the macrophage-specific promoters of the human and mouse colony-stimulating factor-1 receptor (c-fms) genes. J Exp Med 1994; 180:2309-2319.
51. Shapiro LH. Myb and Ets proteins cooperate to transactivate an early myeloid gene. J Biol Chem 1995; 270:8763-8771.
52. Burk O, Mink S, Ringwald M et al. Synergistic activation of the chicken mim-1 gene by v-myb and C/EBP transcription factors. EMBO J 1993; 12:2027-2038.

53. Ness SA, Kowenz-Leutz E, Casini T et al. Myb and NF-M: Combinatorial activators of myeloid genes in heterologous cell types. Genes & Dev 1993 7:749-759.
54. Oelgeschläger M, Krieg J, Lüscher-Firzlaff JM et al. Casein kinase II phosphorylation site mutations in c-Myb affect DNA binding and transcriptional cooperativity with NF-M. Mol Cell Biol 1995; 15: 5966-5974.
55. Mink S, Kerber U, Klempnauer K-H. Interaction of C/EBPβ and v-Myb is required for synergistic activation of the mim-1 gene. Mol Cell Biol 1996; 16: 1316-1325.
56. Dai P, Akimaru H, Tanaka Y et al. CBP as a transcriptional coactivator of c-Myb. Genes & Dev 1996; 10:528-540.
57. Oelgeschläger M, Janknecht R, Krieg J et al. Interaction of the coactivator CBP with Myb proteins: effects on Myb-specific transactivation and on the cooperativity with NF-M. EMBO J 1996; 15: 2771-2780.
58. Favier D, Gonda TJ. Detection of proteins that bind to the leucine zipper motif of c-Myb. Oncogene 1994; 9: 305-311.
59. Frampton J, Kouzarides T, Döderlein G et al. Influence of the v-Myb transactivation domain on the oncoprotein's transformation specificity. EMBO J 1993; 12:1333-1341.
60. Chong L, van Steensel B, Broccoli D et al. A human telomeric protein. Science 1995; 270:1663-1667.
61. Evers R, Smid A, Rudloff U et al. Different domains of the murine RNA polymerase I-specific termination factor mTTF-1 serve distinct functions in transcription termination. EMBO J 1995; 14:1248-1256.
62. Veals SA, Schindler C, Leonard D et al. Subunit of an α-interferon-responsive transcription factor is related to interferon regulatory factor and Myb families of DNA-binding proteins. Mol Cell Biol 1992; 12:3315-3324.
63. Veals SA, Santa-Maria T, Levy DE. Two domains of ISGF3 gamma that mediate protein-DNA and protein-protein interactions during transcription factor assembly contribute to DNA-binding specificity. Mol Cell Biol 1993; 13:196-206.
64. Hörlein AJ, Näär AM, Heinzel T et al. Ligand-independent repression by the thyroid hormone receptor mediated by a nuclear receptor co-repressor. Nature 1995; 377:397-404.
65. Aasland R, Gibson TJ, Stewart F. The SANT domain: a putative DNA-binding domain in the SWI-SNF and ADA complexes, the transcriptional co-repressor N-CoR and TFIIIB. TIBS 1996; 21:87-88.
66. Frampton J, Gibson TJ, Ness SA et al. Proposed structure for the DNA-binding domain of the Myb oncoprotein based on model building and mutational analysis. Protein Engng 1991; 4:891-901.
67. Ogata K, Hojo H, Aimoto S, Nakai T et al. Solution structure of a DNA-binding unit of Myb: A helix-turn-helix motif with conserved tryptophans forming a hydrophobic core. Proc Natl Acad Sci USA 1992; 89:6428-6432.
68. Dubnau J, Struhl G. RNA recognition and translational regulation by a homeodomain protein. Nature 1996; 379: 694-699.
69. Tashiro S, Takemoto Y, Handa H et al. Cell type-specific trans-activation by the B-myb gene product: Requirement of the putative cofactor binding to the C-terminal conserved domain. Oncogene 1995; 10:1699-1707.

CHAPTER 16

Tumor Suppression by 'Transcription' Factors

Rachel C. Davies and Nicholas D. Hastie

INTRODUCTION

Cancer is a tremendously emotive word and a great deal of energy has gone into the study of malignant disease with the ultimate objective of developing better therapy. Some families have an increased susceptibility to particular types of cancer, strongly suggesting that there are genes which normally protect against tumor development. One might predict that these tumor suppressor genes will play a critical role in the control of cell proliferation. The ability of an individual protein to modulate function of many others by subtle changes in transcription patterns would provide a powerful way of effecting such control. The model of a 'transcription' factor which springs to mind is a molecule binding to a specific sequence of DNA upstream of a target gene and stimulating expression directly via its own transactivation domain. There are many ways transcription may be influenced, however, and the three tumor suppressor proteins RB, WT1 and p53 illustrate some of these points.

TUMOR SUPPRESSION: THE TWO-HIT HYPOTHESIS AND BEYOND

The now classical concept of tumor suppressor genes was first proposed by Knudson, following his studies of childhood retinal tumors, to explain the relative frequencies of familial, sporadic, bilateral and unilateral retinoblastoma. It was proposed that protective tumor suppressor genes normally prevent tumor development, but if both copies of the gene were defective no mechanism would be available to hold proliferation in check, and tumors would result. It has now been shown that in about 40% of cases a single defective retinoblastoma susceptibility *RB1* allele is inherited in the germline, and therefore the chances of acquiring a somatic mutation in the remaining allele and developing tumors, often in both eyes, is raised. In sporadic cases spontaneous mutations affect both alleles. The chances of acquiring a mutation in the subset of cells already carrying one defective allele are small, accounting for the observation that often only one eye is affected. The ability of tumor suppressor gene products to actively suppress growth was demonstrated subsequently when it

Transcription Factors in Eukaryotes, edited by Athanasios G. Papavassiliou.

became apparent that the reintroduction of wildtype retinoblastoma gene product, RB, into tumor cells could at least partially revert the tumor phenotype.

Although RB is the paradigm of tumor suppressors many other genes fall into this category. Wilms' tumor is another childhood tumor, predominantly affecting the kidneys. WAGR patients, who have constitutive deletions of chromosomal region 11p13, have a particularly high incidence of Wilms' tumor and this facilitated cloning of a Wilms' tumor suppressor gene, *WT1*.[1] Whilst changes at the *WT1* locus cannot account for all incidences of Wilms,' it is implicated in at least 10% of cases. *WT1* is critically affected in Denys-Drash syndrome.[2] These patients have severe kidney and gonad abnormalities, coupled to the development of Wilms' tumor. Detailed analysis has shown that around 90% of Denys-Drash patients have changes in a single *WT1* allele. Although the second allele is usually lost in the tumors of these patients the urogenital abnormalities are much more severe than WAGR patients who have a constitutive deletion of one *WT1* allele. This suggests that two hits might not always be required for abnormal WT1 function. Changes in a one allele might generate a protein able to function in a dominant or dominant-negative manner, where the wild type allele is functionally inactivated. Indeed there is some evidence that WT1 can dimerize in vitro, which might be one mechanism through which this can occur.[3]

p53 has been identified as another tumor suppressing protein and alterations within the *p53* locus are the most frequent genetic lesion associated with cancer of many different types.[4] Li-Fraumeni patients are genetically predisposed to a wide variety of tumors, and as might be predicted from the two-hit hypothesis have germline mutations within a *p53* allele, tumors arising following a second hit. This effect is also modeled in *p53* null mice which develop normally, but have an increased incidence of a tumors at an early age.[5] Among heterozygous mice lacking one *p53* allele, tumors arise in 25% of mice within 18 months; while 75% of homozygous null mice develop tumors within 6 months. These observations provide clear evidence that loss of p53 alone is not enough for abnormal growth, but is consistent with the multistep carcinogenesis model which predicts that many genetic changes contribute to the development of malignant disease.

Like Wilms' tumor alterations of both *p53* alleles are not always required and some tumors arise following mutations in a single *p53* allele, with the resulting mutant appearing to act in a dominant-negative manner. It has been demonstrated that many mutant p53 proteins are conformationally different from their wild-type counterpart. Indeed some mutant proteins can bind wild-type protein in vitro, forcing the adoption of a mutant conformation.[6] It is easy to envisage how such a mechanism would contribute to functional inactivation of wild-type protein. This ability would be dependent on p53 oligomerization, and indeed the oligomerization domain alone is able to act in a dominant negative manner to transform cells in culture and further induce tumors in transgenic mice.[7]

Significantly some mutant p53 proteins have been reported to have additional dominant 'gain of function' transforming properties, although the molecular basis of such an activity is poorly understood. Evidence for a gain of function comes largely from experiments where mutant p53 was reintroduced into null cells, conferring a growth advantage, and increased tumorigenicity.[8] More recently, however, mice with a mutant p53 transgene were crossed onto both hetero and homozygous p53 null mice.[9] The introduction of the p53 mutant accelerated the rate of tumor growth in the hetero but not the homozygous mice, more consistent with a dominant negative mode of action than a gain of function. It is clear however that the gain of function is likely to be mutation specific and the more aggressive p53 mutants remain to be tested in this type of systems. One of the most convincing arguments in favor of a dominant gain of function comes from the studies of ano-genital carcinomas.[10] In many of these tumors p53 function is ablated by the presence of a protein from human papillomavirus, HPV E6, and p53 mutation is rare in this tumor type. In some cases, however, p53 mutations have been shown to arise within HPV positive tumors and are associated with metastatic progression. In this essentially p53 null background provided by E6 there would only be selection for the acquisition of p53 mutations if the resulting proteins had some (E6 resistant) 'gain of function' actively contributing to metastatic potential. The wide range of tumor types with p53 mutations strongly suggests that p53 has a crucial role in the control of proliferation.

p53: THE HALLMARKS OF A BONA FIDE TRANSCRIPTION FACTOR

p53 has many features typical of a traditional transcription factor. It binds DNA in a sequence specific manner, and possesses a transactivation domain, which can be transferred independently to a heterologous DNA binding domain, such as GAL4. The DNA binding portion of the molecule resides in the central core, with an acidic transactivation domain at the N-terminus, and an oligomerization domain at the C-terminus. Recent structural studies have provided an insight into the functioning of the molecule.[11] The central DNA binding core is made up of series of loops suspended from a hydrophobic β-sandwich. These loops contact both the minor and major grooves of DNA. Significantly many of the residues found to be mutated in human tumors are predicted to be specifically involved in DNA recognition, consistent with the observation that most p53 mutant proteins are transcriptionally defective. Many of the other commonly mutated residues fall within regions anticipated to help to stabilize the overall structure, and indeed it is this class of defective proteins which form the so called 'mutant' conformation, where the molecule is largely denatured.

Studies of the C-terminal oligomerization domain have shown that p53 forms a tetramer which is best considered a dimer of dimers.[12] The oligomer is formed via largely electrostatic interactions across two anti-parallel β-sheets within each dimer, and held by a bundle of 4 helices (one from each monomer). Significantly the sheets are arranged in a scissors-like fashion which might facilitate subtle changes in conformation. Studies have shown that p53 can shift from a latent to an active DNA binding protein.[13] The extreme C-terminal tail, implicated in the control of sequence specific DNA binding, is located in such a position that it could influence this conformational shift.

The nature of the p53 DNA target correlates well with the tetrameric structure of p53. Two palindromic sequences based on the motif PuPuPuC(A/T) are separated by 0-13 bases, suggesting that each monomer contacts half of each palindrome. The binding site was characterized in detail by selection of genomic DNA fragments by p53 in vitro and good definition of the binding site has aided the identification of potential target genes. The most well characterized target gene is p21, which was identified by differentially screening cells containing wild-type and mutant (non-DNA binding) p53.[14] The same gene was simultaneously identified as a cyclin dependent kinase binding protein.[15] p21 has been shown to inhibit the activity of cyclin dependent kinases (cdk) in a concentration dependent manner, preventing access of the cdk-activating enzyme to the kinase. As activation of a cyclin dependent kinases is critical for cell cycle progression the ability of p53 to induce the expression of a protein with these characteristics would be consistent with the ability of p53 to arrest cells in G1 phase of the cell cycle.[16]

p53 itself is very unstable, and it has been shown that its stability dramatically increases after DNA damage. Elevated levels of p53 presumably stimulate expression of p53 target proteins, such as p21, leading to growth arrest. Such a mechanism would prevent replication of cells with damaged DNA, enabling repair to be carried out before allowing replication to proceed. Clearly, loss of p53 function would permit the accumulation of further genetic changes and ultimately this might result in the generation of cells with tumorigenic capacity. This model proposes that p53 is a 'guardian of the genome'. Indeed both p53 and p21 null cells show an inability to G1 arrest in response to some DNA insults, showing that p21 does have an effector role in this pathway.[17] Significantly p21 binds to PCNA, a component of DNA polymerase δ, (independently of cyclin dependent kinases), blocking its DNA replication but not repair functions.[18] As might be predicted p53 null cells show increased genome instability. There is an increased rate of chromosome amplification and aneuploidy, but apparently not point mutations, implying that not all DNA repair pathways are equally sensitive to p53.[19]

Other genes regulated by p53 are also consistent with its role as a tumor suppressor. pGADD45 for example contains an intronic p53 response element and is strongly expressed in response to DNA damage in a p53 dependent manner. pGADD45 has been demonstrated, like p21, to interact with PCNA, and can if overexpressed induce cell cycle arrest,[20] although its precise function at the molecular level is poorly understood. Cyclin G was isolated by differential screening of p53 induced cells and contains a p53 response element in its upstream regulatory region.[21] Although this protein has the hallmarks of a cyclin and thus is likely to be involved in cell cycle regulation no precise function has yet been identified.

The induction of p53 does not always result in growth arrest and in some cell types apoptotic cell death is triggered in response to DNA damage. For example p53 deficient T cells remain viable at much higher radiation doses than wildtype p53 expressing counterparts which rapidly undergo apoptosis at low doses of ionizing radiation. The apoptotic response of p53 null T cells to glucocorticoids remains intact, illustrating that some pathways are p53 independent.[22,23]

p53 apparently stimulates the expression of at least one protein involved in the apoptotic pathway, bax.[24] The ratio of bcl-2, which protects against cell death, and bax, a death promoting molecule, has been shown to influence cell survival. The induction of a death enhancing factor such as bax by p53 would disturb the balance, favoring apoptotic death. Surprisingly, however, bax deficient mice still show a normal p53-dependent apoptotic response.[25] This might suggest that bax is not a critical target for p53 in the induction of apoptosis, or alternatively that there may be functional redundancy within the pathway (which given the growing complexity of the bcl-2 family seems plausible). The observation that a tumor derived mutant p53 is competent for the induction of growth arrest, and yet unable to induce apoptosis perhaps suggests that the major contribution of p53 as a tumor suppressor may lie in its ability to induce apoptosis.[26] This might be further supported by the observation that p21 null mice do not show increased frequency of tumors as might otherwise be predicted.[17]

Clearly the activity of p53 has to be precisely regulated, to prevent inappropriate growth arrest or even cell death. The short half-life of p53 under normal conditions is achieved by ubiquitination, although it is not clear how turnover is prevented in response to damage.[27] Interestingly an oncogenic protein of human papillomavirus, HPV E6, denudes the cell of the tumor suppressing effects of p53 by targeting it for specific degradation. This enables virally infected cells to continue proliferation, even under conditions where p53 is induced. Additionally the DNA binding ability of the protein is controlled by an allosteric type mechanism. p53 protein initially purified from bacteria binds poorly to DNA, and yet this activity can be induced in vitro by modification of the extreme C-terminus, by binding specific antibodies for example.[13] The advantages of having a pool of 'latent' p53 in the cell would be that it is able to mount an immediate response to genetic insult.

Several kinases have been implicated in phosphorylating the C-terminal of p53, particularly CKII and cdc2, but whilst this is intriguing there is no convincing evidence that phosphorylation at these sites is required for the growth suppression by p53.[28] More recently it has been reported that short single stranded DNA oligomers can activate latent p53, by binding to the C-terminus.[29] Provocatively short oligonucleotides are released during nucleotide excision repair. Perhaps this is the activator of latent p53 in vivo. There are potential phosphorylation sites in the N-terminus of p53 for a DNA dependent protein kinase (DNA-PK), which like p53 is sensitive to double stranded DNA breaks. This raises the possibility that DNA damage might induce both p53 and a regulatory kinase. It is of significance that p53 with a specific mutation of the potential DNA PK site to a nonphosphorylatable residue has reduced ability to induce growth arrest.[30]

The activity of p53 is further controlled by mdm2, a p53 binding protein.[31] mdm2 binds to the N-terminal transactivation domain of p53, inhibiting activity independently of DNA binding, and as might be predicted mdm2 inhibits both p53 mediated growth arrest and apoptosis.[32] The mdm2 gene is amplified in some human tumors which retain wild-type p53, implying that alterations at either locus can contribute to tumor development.[33] In addition to establishing mdm2 as a downstream modulator of p53 function, this observation also re-emphasizes the central role of the p53 pathway in the control of normal proliferation. Interestingly mdm2 is itself regulated by p53 to form a negative regulatory loop, although the relatively late timing of mdm2 induction by p53 implies that this event might be more closely involved in the reinitiation of cell cycle progression than growth arrest per se.[34]

Although the ability of p53 to transactivate has been mapped to the N-terminus of the protein it is not clear exactly how activation is mediated. It has been shown that this region of p53 may physically interact with components of the basal transcription machinery such as TBP, $TAF_{II}40$, $TAF_{II}60$ and CBF.[35] These interactions might equally well contribute to the ability of p53 to repress transcription, which it appears to do in a more general manner. Indeed it has been suggested that p53 inhibits TATA-mediated transcription, which it might do by sequestration of essential RNA polymerase elements.[36] It would clearly be advantageous to switch off general transcription when mediating a growth arrest, but one particularly provocative target for transcriptional repression by p53 is bcl-2.[24] A decrease in bcl-2, coupled to a p53 dependent increase in bax would perturb the ratio between these two proteins further, presumably leading to stronger apoptotic signaling. Significantly most, but not all, tumor derived mutant p53 proteins are defective in both transactivation and suppression, suggesting that both aspects of the protein are central to normal p53 function.[28]

WT1: DNA BINDING, BUT TO WHAT END?

The sequence of WT1, with four zinc finger motifs at the C-terminus, and an acidic N-terminus immediately suggests that this protein is a transcription factor.[1] (Fig. 16.1) The zinc fingers are of the C2-H2 type and the last three show particularly strong similarity to the fingers of the early growth response protein (EGR1). Indeed WT1 and EGR1 have been shown to bind to the same DNA consensus sequence.[37] The zinc finger region of mouse EGR1 has been co-crystalized with DNA,[38] providing a model which can perhaps be extrapolated to WT1. EGR1 contacts the major groove of DNA via a semicircular 'knuckle' type arrangement of its three zinc fingers. The structure of each individual finger is maintained through the coordinated binding of zinc by the cysteine and histidine residues which are found in the center of a β-sheet and α-helix respectively. The α-helix physically contacts DNA, specifically interacting with 3 bases. Ironically the DNA is bound in the opposite direction to the protein, with the most C-terminal finger binding the most 5' DNA. One of the critical residues of each finger is an arginine at the base of the α-helix. This amino acid binds the most 3' guanine nucleotide of the triplet, an interaction which is stabilized by a neighboring aspartic acid residue. Interestingly one of the residues most commonly found to be mutated in patients is the arginine of finger 3, and as might be predicted this change ablates DNA

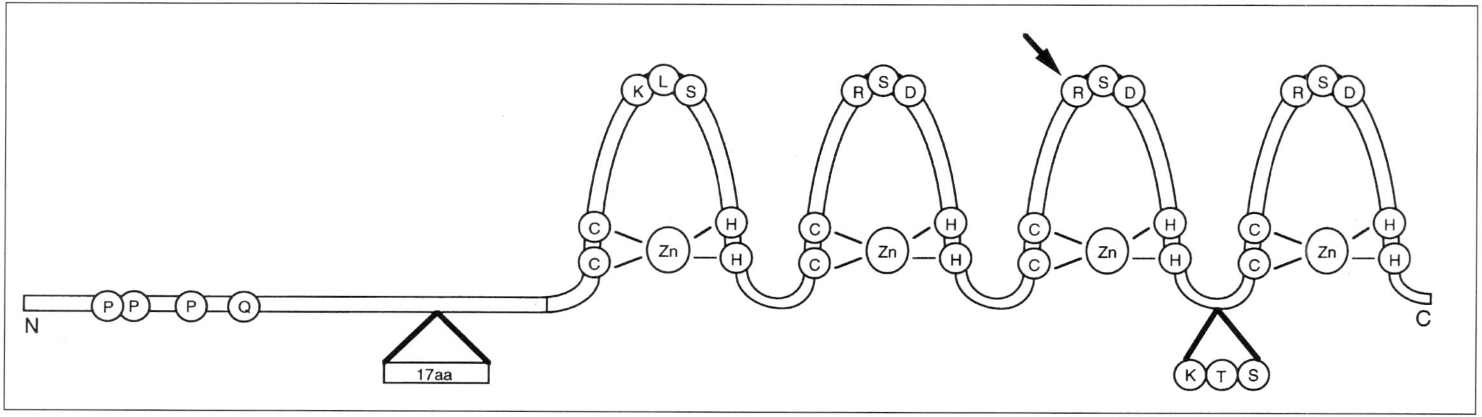

Fig. 16.1. Model of WT1. WT1 contains four zinc fingers at the C-terminus of the protein, with cysteine and histidine residues coordinately binding zinc. The residues at the end of each finger thought to be directly involved with DNA binding are shown. The most commonly mutated residue in Denys-Drash patients is indicated with an arrow. Alternative splicing can lead to the insertion of KTS between the third and fourth finger and/or 17 amino acids adjacent to the proline/glutamine rich N-terminal domain implicated in transcriptional regulation.

binding. Other frequent mutations alter the aspartic acid predicted to stabilize the arginine-DNA contact, while others affect residues involved in complexing with zinc, and would be predicted to affect the folding of the molecule.[2] This mutation profile strongly suggests that a crucial aspect of WT1 function involves DNA binding.

The first finger of WT1 displays less similarity to EGR1, and might be predicted to influence DNA binding. A variety of approaches have been used to investigate the nature of sequences bound by WT1 in vitro, and all have shown that although more extended the consensus is similar to that of EGR1, and is G rich. Significantly however WT1 is alternatively spliced (Fig. 16.1). One of the splices introduces a three amino acid motif, KTS, between the third and fourth zinc finger. From the structure of EGR1 it might be anticipated that a KTS containing protein would have impaired binding to the EGR1 consensus. Early experiments verified this hypothesis[37] and it was subsequently possible to isolate DNA sequences which +KTS forms bound more specifically.[39] The idea that alternative splicing can affect DNA specificity is not unique to WT1, but raises the possibility that WT1 proteins may target different DNA sequences. Significantly the isoform balance is critical for normal WT1 function. Denys-Drash patients have been described who have mutations at splice acceptor sites which prevent generation of the +KTS isoforms.[40] The absolute amounts of the +KTS containing protein would be equivalent to that in WAGR patients where one allele is deleted, but in these cases the relative proportions of +/-KTS forms is perturbed. The severe disease from which these patients suffer emphasizes the crucial role of the isoform ratio.

To examine what effects WT1 might be having once bound to DNA, early experiments used an EGR1 consensus coupled to a reporter gene, and found, perhaps surprisingly, that WT1 repressed, rather than activated, transcription.[41] Indeed WT1 could transfer its repressor functions when coupled to heterologous DNA binding domains independently of its zinc fingers, implying that repression does not occur by displacement of other DNA bound proteins, for example. The repression function has been mapped to the poly-proline and glutamine containing N-terminal region. It has also been reported that the efficacy of repression is increased by the presence of another alternative splice which introduces 17 amino acids into the core of the molecule.[42] Interestingly although WT1 is highly conserved in evolution the 17 amino acid splice only seems to be present in mammals.[43] Significantly two missense mutations of WT1 arising within tumors have been reported to convert the protein from a transcriptional repressor to an activator, implying that transcriptional repression may be an important part of normal WT1 function. A model consistent with WT1 as a tumor suppressing molecule might propose that WT1 is acting antagonistically to EGR1, transcriptionally repressing immediate early genes, and preventing growth stimulation. WT1 does not always function as a transcriptional repressor, however, and it has been reported for example that WT1 can both positively and negatively affect the expression of the PDGF-A chain through different DNA binding sites.[44] This raises the possibility that WT1 might be simultaneously repressing genes necessary for proliferation, while specifically activating genes necessary for urogenital differentiation.

The characterization of a DNA binding motif for WT1 has helped initiate a search for potential target genes whose transcription might be influenced by the protein. The nature of the target site, however, makes it difficult to scan regulatory regions for potential WT1 binding sites. WT1 binds GC or GT rich sequences, a component of many repeats; indeed repetitive DNA was isolated when binding sites for the -KTS form were selected from genomic DNA.[39] Despite this problem a whole array of genes have been identified which are potentially regulated by WT1 (e.g., PDGF-A, CSF-1, bcl-2, myc, nov, retinoic acid receptor-α, ornithine decarboxylase and WT-1 itself), but it is difficult to establish whether they are genuine targets for WT1 in vivo. Aberrant WT1 function can apparently give rise to three phenotypes, suggesting that WT1 has multiple roles in the development and maintenance of the urogenital tract. WT1 null mice die in utero with massive apoptosis of predominantly undifferentiated mesenchyme, implying that WT1 is essential very early in development for the mesenchyme to epithelial transition.[45] The incidence of tumors in WAGR patients, where kidney development itself is normal, emphasizes the part WT1 must play in controlling proliferation of differentiated mesenchyme. WT1 expression persists in glomerular podocytes into adulthood. Denys-Drash patients suffer from characteristic glomerular sclerosis, which may result from defects in the podocyte layer, demonstrating that WT1 is essential not only for the development but also the maintenance of kidney function. Thus WT1 might be affecting the transcription of different sets of genes to fulfill each role, which, coupled to the likelihood that different WT1 proteins might themselves have different target genes, dramatically increases the potential complexity. Nevertheless some genes are proposed to be regulated by WT1 in a manner that might make sense given the role of the WT1 in kidney development.

Insulin-like growth factor II (IGFII) would be an ideal target for WT1. IGFII is expressed in the mesenchyme, but expression drops as WT1 is induced and the cells undergo epithelial differentiation. IGFII is overexpressed in about 95% of Wilms' tumors, compared to normal tissue and, significantly, expression is maintained within some tumor derived epithelial cells not otherwise predicted to express IGFII.[46] The expression of IGFII is usually subject to imprinting, such that only one copy is expressed (usually that inherited from the father), and yet it has been found that in many cases of Wilms' tumor where the WT1 locus is apparently unaffected, there appears to be a loss of imprinting, and overexpression of IGFII.[47] Indeed this chromosomal region has been implicated in genetic predisposition to Beckwith-Wiedemann syndrome, where Wilms' tumor is associated with fetal overgrowth. The transcriptional regulation of IGFII is complex, as there are unique 5' untranslated exons expressed from different promoters. There are WT1 binding sites present in the IGFII promoter predominantly used in the fetal kidney, P3, and WT1 is able to repress transcription via these sites, at least in a transient reporter type assay.[48] As deregulated expression of IGFII can lead to fetal overgrowth it is not difficult to envisage how failure of WT1 to downregulate IGFII might contribute to tumor development. Intriguingly the IGFII receptor is also reported to be transcriptionally repressed by WT1 in vitro.

Circumstantial evidence also earmarks pax-2 as a potentially significant target for transcriptional repression by WT1.[49] The pax proteins are themselves transcription factors which have critical roles in development. pax-2 is expressed in the developing kidney and, like WT1, is implicated in the mesenchyme to epithelial transition.[50] Kidney development is very sensitive to pax-2 dosage and both under and overexpression can cause kidney abnormalities, implying that its expression must be strictly regulated.[50,51] Expression of pax-2 seems to reciprocally reflect that of WT1 (analogous to IGFII) and its maintained expression can be detected in Wilms' tumors. Significantly, aberrant expression of pax-2 can give rise to glomerular abnormalities, and it is precisely these structures where WT1 is expressed.[51] In addition pax-2 has oncogenic potential when overexpressed in some assays,[52] perhaps suggesting that an inability to suppress genes such as pax-2 might both contribute to renal problems and promote progression to tumorigenicity.

The identification of WT1 target genes is problematic as WT1 is expressed in a tissue and time specific manner. This makes it difficult to get appropriate material for differential analysis of WT1 and non-WT1 expressing cells. Recently, however, Englert et al[53] developed a system where WT1 expression could be induced in the same cell background in culture, although these were not kidney cells. Following WT1 expression apoptosis occurred (in a p53 independent manner), and close examination of this phenotype showed that synthesis of the epidermal growth factor receptor (EGFR) was down-regulated in response to WT1. Indeed the introduction of exogenous EGFR could overcome these effects, which appear to be mediated by WT1 directly repressing the EGFR promoter. Interestingly the expression of other described potential target genes was affected in this experiment. The identification of EGFR as a potential WT1 target gene is interesting, as the EGFR response pathway appears to be critical for the survival of differentiating kidney cells. Although it is not clear at present that the repression of EGFR in these cells reflects what would be occurring during development, the ability to use such a system to identify potential target genes must be a step in the right direction.

RB: PROTEIN-PROTEIN INTERACTIONS ARE THE KEY

The cloning of the *rb1* gene did not immediately give any clues about the function of the gene product. Although a nuclear protein, RB displays no hallmarks of a transcription factor, such as a DNA binding or transactivation domain. There became reason to believe that RB might affect transcription, however, following observations that RB has dramatic effects on the expression of some genes, such as myc, and that these effects were dependent on an intact 'pocket' domain, known to be disrupted in some tumors. One of the key studies used a technique similar to that employed for WT1 and p53 to select DNA sequences that RB might bind to. The experimental design differed from that of previous occasions, however, as bacterially expressed RB was incubated with a cell extract prior to DNA binding.[54] The observation that the isolation of specific DNA sequences was dependent on preincubation of RB with cell extracts clearly indicates that DNA binding is mediated by RB-binding proteins, rather than RB directly. The DNA binding site selected closely resembled TTTCGCGC, which had previously been defined as a response element for the E2F transcription factor. The ability of E2F to bind to RB was central to the cloning of E2F-1, which contains a DNA-binding region towards the N-terminus and C-terminal transactivation domain. Subsequent work has shown that 'E2F' is itself a heterodimer, E2F-1 binding through a putative amphipathic α-helix to its partner, DP-1.[55,66] Perhaps surprisingly, DP-1 lacks a transactivation domain, but is able to interact synergistically with E2F-1 to increase DNA binding and hence E2F dependent transcription. Subsequently a series of E2F and DP related molecules have been cloned, suggesting that these proteins may mix and match to differentially regulate particular target sites.

The significance of the E2F:RB interaction became apparent following identification of E2F sites in the regulatory regions of several genes essential for cell cycle progression, including c-myc, B-myb, cdc2, dihydrofolate reductase, thymidine kinase, and E2F-1 itself.[55-57] The expression of these genes is predominantly induced at the G1-S phase transition, and in many cases mutation of the E2F site abolishes cell cycle regulated expression. As the re-introduction of RB into RB deficient cells can cause reversible growth arrest in G1,[58] it was proposed that RB regulates the early stages of the cell cycle by binding E2F, thus inhibiting the expression of genes needed for progression. The interaction of RB with E2F is likely to make an important contribution to its growth suppressing properties, and RB mutants unable to bind E2F are unable to induce growth arrest.[59] As this hypothesis might predict, overexpression of E2F-1 can overcome the growth suppressive effects of RB.

Interestingly it has been demonstrated that RB can repress transcription when directed to DNA via a heterologous DNA binding domain in the absence of E2F.[60] RB shows small regions of similarity to components of the basic transcriptional machinery, TBP and TFIIB,[61] so it is possible that through binding to E2F RB becomes directed to DNA, from whence it is able to repress other transcription factors through mimicry of the basal complex. These properties might be particularly important at promoters with additional enhancer elements, and may help explain why RB can positively repress activity below basal when bound to E2F. (Fig. 16.2.) This hypothesis would imply that RB interacts with other transcription factors and indeed RB appears able to associate with a wide range of factors, including PU.1, myoD, Id-2 etc.[61,68,70] While in most cases the binding site on RB has not been rigorously assigned there is evidence that at least some of these molecules bind to RB independently of E2F, again consistent with the model.[60] Such a mechanism might shed light on the recent observations that whilst over expression of E2F-1 can be oncogenic (by overcoming the growth suppressive effects of RB),[62] E2F-1 null mice show an increased susceptibility to tumors.[63,64] If E2F functions to anchor the growth suppressive RB protein to DNA, then the absence of E2F would not permit growth suppression by RB. The observation that DNA binding by E2F-1 alone is sufficient to transform cells supports this model.[65] But what about transactivation? RB physically masks the transactivation domain of E2F-1.[66] Another property of E2F-1, the ability to induce apoptosis, is dependent on both DNA binding and transactivation,[65] perhaps suggesting that E2F may be a mediator of the apoptotic pathway if proliferative signaling is aberrant. This would tie in with the observations that RB null cells are susceptible to apoptosis, and apoptosis is decreased in E2F-1 null cells. Presumably RB is therefore functioning to suppress both proliferation and cell death.

In order for cell cycle progression to occur the repressive activity of RB must be relieved. It was demonstrated in some of the first studies that RB is differentially phosphorylated. RB is unphosphorylated in early stages of the cell cycle, and sequentially phosphorylated as division ensues to induce a characteristic shift in the mobility of the protein on SDS-PAGE gels. These studies clearly imply that the unphosphorylated form of RB is the actively growth suppressing form, and this repression is relieved once the protein is phosphorylated. The affinity of RB for binding proteins such as E2F decreases when RB is phosphorylated, leading to the simple hypothesis that once RB is phosphorylated E2F might be released, and facilitating transactivation by E2F (and adjacent transcription factors). It is now becoming clear, however, that this is a naive model, and gel retardation assays have shown that protein complexes bound to E2F sites go through a series of alterations during the cell cycle, not least because the RB related proteins p107 and p130 also bind E2F proteins, perhaps targeting different genes or differentially affecting their transcription. Surprisingly perhaps, it seems likely that these RB related proteins are the predominant regulators of E2F function in cycling cells. The expression of E2F dependent genes apparently remains normal in proliferating RB null cells, presumably regulated by these RB related proteins. It has recently been reported that following the induction of differentiation there appears to be a shift away from p107 and p130, with E2F containing complexes apparently becoming predominantly RB associated.[67] This suggests that RB may play a crucial role in complete withdrawal from the cell cycle. It is particularly significant in this context that although RB suppresses the activity of most bound transcription factors, in some cases RB potentiates the ability to stimulate transcription, e.g., NF-IL6, Id-2.[68,69] These proteins are specifically associated with the balance between differentiation and proliferation. Indeed the terminal differentiation of myoblasts appears to require cooperation between myoD and RB, and while muscle differentiation pathway is initiated in RB null mice the resulting myoblasts are not terminally differentiated and can re-enter the cell cycle.[70] Significantly RB null mice die in utero with developmental abnormalities related to aberrant terminal differentiation.[71-73] RB may therefore play a critical role in the balance between differentiation and proliferation.

Although direct association with transcription factors is central to RB function it may not be the only mode of action. Interestingly an RB binding protein, c-abl, has

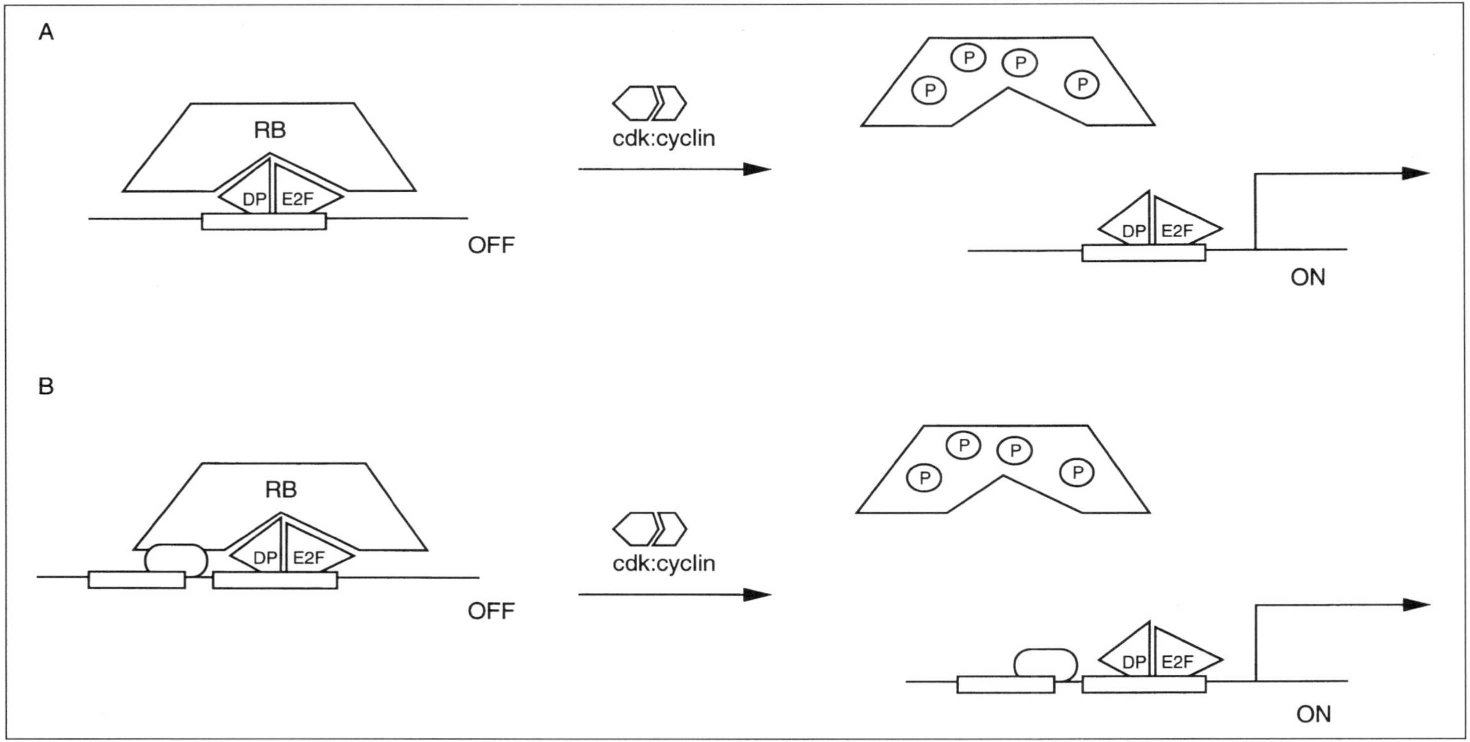

Fig. 16.2. Transcriptional regulation by RB. A. RB directly represses the transactivation potential of E2F (itself a dimer of a DP and E2F proteins), by binding the transactivation domain. RB becomes phosphorylated as cells start to move through the cell cycle, releasing E2F and stimulating E2F dependent expression. In this case mutation of the E2F site would be likely to result in an inactive gene, e.g., DHFR.[56] B. RB is anchored to the DNA via E2F and represses the activity of adjacent transcription factors. In this case mutation of the E2F site may cause an increase in expression, e.g., B-myb.[57]

been identified which binds independently of E2F to the C-terminus of RB.[74] Significantly an RB fragment containing just the C-terminus was able to inactivate the growth suppressing ability of full length RB in an apparently dominant negative manner.[75] This suggests that the ability of RB to act as an adapter molecule and bring several different proteins together is central to RB function. Indeed in this way RB might even facilitate the regulation of bound molecules. Although c-abl is competent for DNA binding its main effector region is a tyrosine kinase domain. At least one RB binding protein, cyclin D, is potentially phosphorylated on tyrosine. While RB binding would be anticipated to inhibit c-abl activity through direct binding of the kinase domain perhaps RB could be considered a scaffold, bringing enzyme and substrate close together, such that when c-abl is released from RB it is in close proximity to its substrate (or vice versa with c-abl itself displaying several potential cdk target sites). Interestingly c-abl is implicated in the phosphorylation of RNA polymerase II, stimulating the switch from initiation to elongation.[76] It is easy to envisage how regulation of such an event by RB would contribute to growth suppression.

Finally, some of the proteins to which RB binds, such as RBP-1, Rbp48, hBRM, and BRG, display hallmarks of proteins which might affect gross DNA organization and chromatin structure, in a manner similar to the SNF/SWI complex in yeast (see chapter 18).[77-80] If RB were to influence chromatin conformation this might be another way of affecting transcription.

BUILDING UP COMMUNICATION NETWORKS

Tumor suppressor genes do not function in isolation and there is increasing evidence of interactions between pathways. The convergence of the RB and p53 pathways is currently the most well established. The key factor in the regulation of RB function would appear to be phosphorylation, and it is this aspect which has provided an interesting link with p53. RB first appears to be phosphorylated towards the end of the G1 phase of the cell cycle, and many of the sites resemble those targeted by cyclin dependent kinases (Ser/Thr followed by proline). It has been demonstrated that RB can be phosphorylated by cdks in vitro. Cyclin D binds directly to RB, suggesting that the cyclin D associated kinase, probably cdk4/6, which is active early in the cell cycle, initiates RB phosphorylation. Indeed D-type cyclins can overcome an RB induced cell cycle arrest, and inappropriate expression of G1/cyclin dependent kinases can induce E2F activity, supporting this hypothesis.[81,82] This ability is antagonized by a cyclin dependent kinase inhibitor p16, which interestingly seems to be induced by RB, possibly forming a feedback loop. Significantly, analysis of tumor material has shown that there is mutually exclusive ablation of RB, an increase in cyclin D production or loss of p16.[83] All three changes would have the same net effect, withdrawing the brakes on the cell cycle and promoting cell proliferation, which strongly implies that these three proteins lie on the same pathway. It seems unlikely, however, that cyclin D/cdk complexes alone are able to fulfill the whole scheme of RB phosphorylation, and indeed they may be most important when a cell is re-entering the cell cycle from a completely quiescent state. Other cyclin dependent kinases have also been implicated in RB phosphorylation, particularly cyclin A and E associated kinases, and these are both more potent in relieving RB mediated growth arrest, and phosphorylating RB in vitro.[84] Significantly these kinases are inhibited via p21, the same p21 which is induced by p53.[15] (Fig. 16.3)

The first indications that RB and p53 might be linked came from observations that tumor virus proteins such as E1A and E7, which abrogate the growth suppressing effects of RB, could overcome p53 mediated growth arrest. The real significance of such a link became apparent following studies in mice where both p53 and RB function was abolished, either using mice with targeted null alleles, or specific viral proteins. In mice where RB function was ablated in the eye, apoptosis occurred, resulting in microphthalmia.[85-87] In the absence of both p53 and

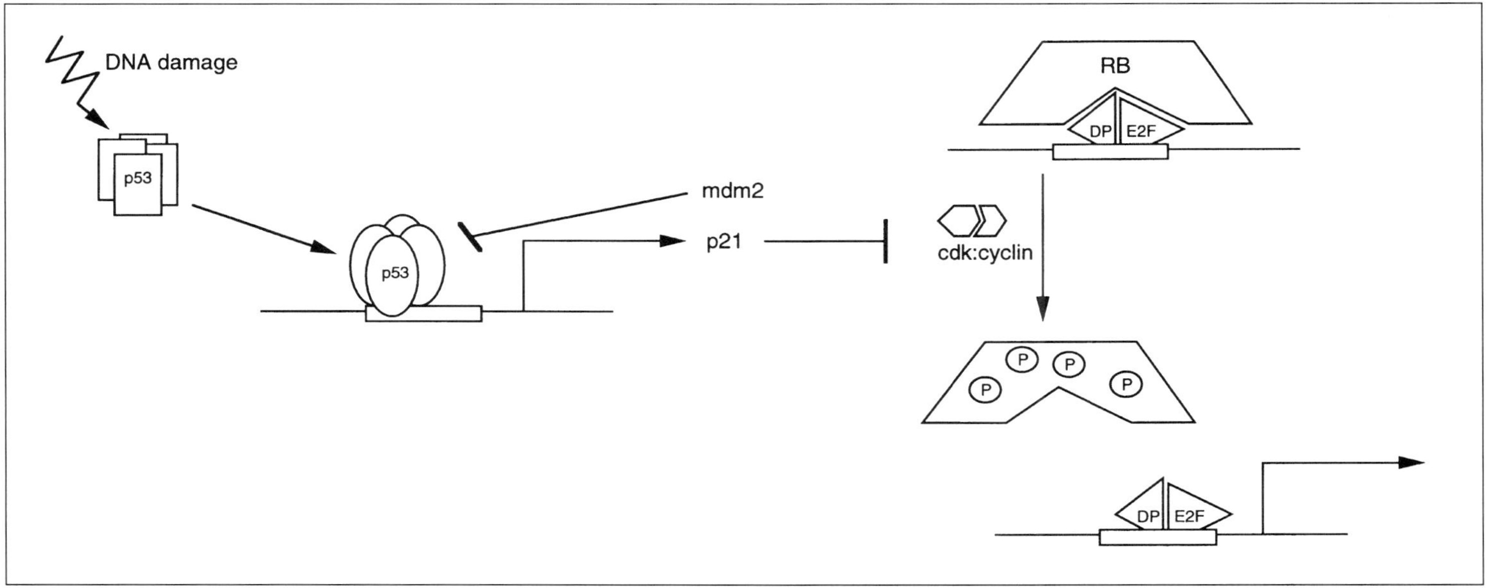

Fig 16.3. p53 and RB converge. p53 is stabilized and undergoes a conformational shift to an active DNA binding form following DNA damage. Transcription of p53 responsive genes, such as p21, is initiated. p21 inhibits the activity of cyclin dependent kinases such that transcriptional repression by RB is retained, mediating growth arrest. mdm2 is also induced in a p53 dependent manner, and represses transactivation by p53, forming a feedback loop.

RB function apoptosis was prevented, but subsequently lens tumors developed. This apparent cooperation between p53 and RB, coupled to the observation that $RB^{+/-}$ $p53^{-/-}$ mice develop a wider range of tumors than either $RB^{+/-}$ or $p53^{-/-}$ mice alone,[88] illustrates why both the RB and p53 pathways are disrupted in so many tumors.

Further evidence of communication between tumor suppressing proteins has recently accumulated. Although mdm2 was originally characterized as a p53 binding molecule, recent experiments suggest that it might also associate with RB.[89] mdm2 appears to antagonize the growth suppressive effects of both p53 and RB, apparently binding to the C-terminal domain of RB, at a site independent of E2F but in a manner that increases E2F dependent transcription. Confusingly some experiments also suggest that mdm2 might bind directly to E2F.[90] Both models could be reconciled if RB were envisaged as a scaffold, facilitating the assembly of specific complexes. Irrespective of whether mdm2 stimulates E2F activity in an RB dependent or independent manner, the net result would promote proliferation. It will be important to assess the RB/cyclin D/p21 status of tumors where mdm2 is overexpressed. If mdm2 does antagonize RB functions, in addition to those of p53, one might predict that if mdm2 were overexpressed there would be no need for alterations in the RB pathway. To complicate matters further it has also been reported that p53 might bind to E2F1 and DP1, causing a suppression of both E2F and p53 dependent transactivation.[91] Provocatively, some genes, such as myc, are responsive to both p53 and E2F, illustrating the need for effective communication.

Although tissue specific, the tumor suppressive effects of WT1 must also be coordinated with those of the ubiquitous tumor suppressing proteins. A link between WT1 and p53 function was first proposed following detection of p53:WT1 complexes in baby rat kidney cells transformed with a mutant form of WT1.[92] Cotransfection experiments have shown that although p53 suppresses the transactivation potential of WT1, that of p53 is reciprocally potentiated, while transcriptional repression by p53 is simultaneously inhibited. WT1 apparently stabilizes p53 protein and inhibits p53 mediated apoptosis.[93] Perhaps consistent with this model a high proportion of Wilms' tumors display elevated levels of p53 protein, although detailed analysis of WT1 and p53 status has not been performed. Interestingly, mice overexpressing wildtype p53 have unusually small kidneys.[94] However these abnormalities appear not to arise from a defect in WT1 expressing cells, but rather a failure of p53 expressing cells in the adjacent ureteric bud to induce appropriate differentiation signals. Significantly, p53 null mice have no urogenital abnormalities,[5] strongly suggesting that WT1:p53 interaction is not essential for normal development.

p53 mutations are rare in Wilms' tumor, perhaps unusually for a human tumor, and the small percentage of Wilms' tumors which do have p53 mutations have a particularly poor prognosis.[95] Significantly, none of the patients with p53 mutations described to date have alterations at their WT1 locus, raising the possibility that changes are mutually exclusive. Given the central role p53 occupies in the control of growth it does not seem unreasonable to predict that p53 function must somehow be compromised in every tumor, in which case it is likely that WT1 may interfere with p53 function. It has been reported, for example, that pax genes such as pax-2 might be able to inhibit p53 function by preventing transcription from the p53 promoter.[96] If pax-2 is a genuine target for WT1 then aberrant expression of pax-2 following functional compromise of WT1 would lead to downregulation of p53 and hence the promotion of growth, although it remains to be seen whether such a regulatory loop occurs in vivo. Currently there are no reports of interactions between the RB pathway and WT1. Given the increasing evidence that RB may have a role in terminal differentiation, however, one might anticipate that links will be found.

TRANSCRIPTION IS NOT THE ONLY FRUIT

There is evidence to suggest that each of these proteins may have additional roles, independently of transcription. Despite convincing evidence that p53 can induce growth arrest through a transcription dependent mechanism, the precise contribution of transactivation to the induction of apoptosis is less clear. In at least some assays apparently transcriptionally inactive p53 remains able to induce apoptosis, (although one might argue that these proteins remain able to activate as yet unidentified critical targets for p53 in apoptotic signaling).[97] More provocative evidence, however, comes from the observation that p53 dependent apoptosis can still occur in the presence of both RNA and protein synthesis inhibitors,[98] and it may transpire that cell death can be induced via transcriptionally dependent and independent mechanisms.

p53 displays other features which are apparently independent of transcription. In particular p53 can inhibit DNA replication.[99] This may be due to the ability of p53 to bind directly to RPA, a component of the DNA replication machinery that is required for the initial unwinding of DNA. This interaction appears to prevent single stranded DNA binding by RPA, and it is of further interest that p53 has anti-helicase activities, promoting the annealing of both RNA and DNA. Together these properties would constrain DNA replication, providing another mechanism through which growth might be suppressed.

The absence of p53 can lead to abnormal centrosome amplification, which might contribute to the genomic instability of cells where p53 function is compromised.[100] Significantly, p53 has been found to be associated with the centrosome, although the molecular basis for such an association is unclear.

Although the major effect of RB is documented early in the cell cycle, RB may have additional functions at mitosis. One RB binding protein, mitosin, has been documented to influence mitotic progression,[101] and intriguingly deficiencies in the yeast counterparts of other RB binding proteins, h-nuc2 and phosphatase type 1a, cause mitotic arrest.

Unexpectedly both RB and WT1 show subnuclear colocalization with splice factors.[102,103] This observation has been investigated in most detail in the case of WT1. Cells transfected with -KTS forms of WT1 display a diffuse WT1 staining pattern, resembling that of transcription factors, whereas cells transfected with the +KTS forms show a more speckled distribution, which colocalizes with splice factors. A similar distribution of WT1 is also seen under physiological conditions, in both a cell line derived from endogenous tissue and frozen sections through the kidney and gonad, implying that this relates to protein function in vivo. WT1 coimmunoprecipitates with some splice factors, suggesting that WT1 may influence pre-mRNA splicing. Interestingly fingers of the C2-H2 type have been reported to bind RNA as well as DNA,[104] and WT1 has recently been predicted from computer modeling to contain an RNA recognition motif, although ironically this does not include the fingers, but rather an N-terminal domain.[105] Whilst these observations are provocative there is currently no evidence to suggest that WT1 directly influences RNA processing.

CONCLUSIONS

It is clear from studying these tumor suppressor genes that their ability to influence transcription plays a central role in their function, but it is equally apparent that there are multiple ways of achieving this end. One area, which is perhaps the most difficult to study, but nevertheless crucial, is the relationship between proliferation and differentiation, and studies of tumor suppressing proteins might lead us to a better understanding of this balance. As p53 is the most commonly affected protein to date in human tumors it is a prime target for therapy. Transcriptional transactivation plays a central role in at least some of p53 function and this property is compromised in many tumor derived proteins. The restoration of such activity might therefore lead to the halt of tumor progression. One mechanism by which this might be achieved is through the reactivation of latent DNA binding activity, and indeed this has been possible for a subset of p53 mutants in vitro. This exciting prospect illustrates the ultimate benefit of better understanding tumor suppressor gene function.

References

1. Call KM, Glaser T, Ito CY et al. Isolation and characterisation of a zinc finger polypeptide at the human chromosome 11 Wilms' tumour locus. Cell 1990; 60:509-20.
2. Pelletier J, Bruening W, Kashtan CE et al. Germline mutations in the Wilms' tumour suppressor gene are associated with abnormal urogential development in Denys-Drash syndrome. Cell 1991; 67:437-47.
3. Reddy JC, Morris JC, Wang J et al. WT1-mediated transcriptional activation is inhibited by dominant negative mutant proteins. J Biol Chem 1995; 270: 10878-84.
4. Hollstein M, Sidransky D, Vogelstein B et al. p53 mutations in human cancers. Science 1991; 253:49-53.
5. Donehower LA, Harver M, Slagle BL et al. Mice deficient for p53 are developmentally normal but susceptible to spontaneous tumours. Nature 1992; 356: 215-21.
6. Milner J, Metcalf EA. Cotranslation of activated mutant p53 with wild type drives the wild type p53 into the mutant conformation. Cell 1991; 65:765-74.
7. Bowman T, Symmonds H, Gu L et al. Tissue-specific inactivation of p53 tumour suppression in the mouse. Genes Dev 1996; 10:826-35.
8. Wolf D, Harris N, Rotter V. Reconstitution of p53 expression in a nonproducer Ab-MuLV-transformed cell line by transfection of a functional p53 gene. Cell 1984; 38:119-26.
9. Harvey M, Vogel H, Morris D et al. A mutant p53 transgene accelerates tumour development in heterozygous but not nullizygous p53-deficient mice. Nature Genetics 1995; 9:305-11.
10. Crook T, Vousden KH. Properties of p53 mutations detected in primary and secondary cervical cancers suggest mechanisms of metastasis and involvement of environmental carcinogens. EMBO J 1992; 11:3935-40.
11. Cho Y, Gorina S, Jeffrey PD et al. Crystal structure of a p53 tumour suppressor-DNA complex: understanding tumourigenic mutations. Science 1994; 265:346-55.
12. Clore GM, Ominchinski JG, Sakaguchi K et al. High-resolution structure of the oligomerisation domain of p53 by multidimensional NMR. Science 1994; 265:386-91.
13. Hupp TR, Meek DW, Midgley CA et al. Regulation of the specific DNA binding function of p53. Cell 1992; 71:875-784.
14. El-Diery WS, Tokino T, Velescu VE et al. *WAF1*, a potential mediator of p53 tumour suppression. Cell 1993; 75: 817-25.
15. Harper JW, Adami GR, Wei N et al. The p21 cdk-interacting protein Cip1 is a potent inhibitor of G1 cyclin-depenent kinases. Cell 1993; 75:805-16.
16. Kuerbitz SJ, Plunkett BS, Walsh WV et al. Wild-type p53 is a cell cycle checkpoint determinant following irradiation. Proc Natl Acad Sci USA 1992; 89:7491-5.
17. Deng C, Zhang P, Harper JW et al. Mice lacking $p21^{CIP1/WAF1}$ undergo normal development, but are defective in G1 checkpoint control. Cell 1995; 82:675-84.
18. Li R, Waga S, Hannon GJ et al. Differential effects by the p21 CDK inhibitor on PCNA-dependent DNA replication and repair. Nature 1994; 371:534-7.
19. Sands AT, Suraokar MB, Sanchez A et al. p53 deficiency does not affect the accumulation of point mutations in a transgene target. Proc Natl Acad Sci USA 1996; 92:8517-21.
20. Smith ML, Chen I, Zhan Q et al. Interaction of the p53-regulated protein Gadd45 with proliferating nuclear antigen. Science 1995; 266:1376-80.
21. Okamoto K, Beach D. Cyclin G is a transcriptional target of the p53 tumour suppressor protein. EMBO J 1994; 13: 4816-22.

22. Lowe SW, Schmitt EM, Smith SW et al. p53 is required for radiation-induced apoptosis in mouse thymocytes. Nature 1993; 362:847-9.
23. Clarke AR, Purdie CA, Harrison DJ et al. Thymocyte apoptosis induced by p53-dependent and independent pathways. Nature 1993; 362:849-52.
24. Miyashita T, Krajewski S, Krajewski M et al. Tumour suppressor p53 is a regulator of *bcl-2* and *bax* gene expression in vitro and in vivo. Oncogene 1994; 9: 1799-805.
25. Knudson CM, Tung KSK, Tourtellotte WG et al. Bax-deficient mice with lymphoid hyperplasia and male germ cell death. Science 1995; 270:96-9.
26. Rowan S, Ludwig RL, Haupt Y et al. Specific loss of apoptotic but not cell cycle arrest function in a human tumour derived p53 mutant. EMBO J 1996; 15: 827-38.
27. Chowdary DR, Dermody JJ, Jha KK et al. Accumulation of p53 in a mutant cell line defective in the ubiquitin pathway. Mol Cell Biol 1994; 14:1997-2003.
28. Crook T, Marston NJ, Sara EA et al. Transcriptional activation by p53 correlates with suppression of growth but not transformation. Cell 1994; 79:817-27.
29. Jayaraman L, Prives C. Activation of p53 sequence-specific DNA binding by short-single strands of DNA requires the p53 C-terminus. Cell 1995; 81:1021-9.
30. Fiscella M, Ullrich SJ, Zambrano N et al. Mutation of the serine phosphorylation site of human p53 reduces the ability of p53 to inhibit cell cycle progression. Oncogene 1993; 8:1519-28.
31. Momand J, Zambetti GP, Olson DC et al. The mdm-2 oncogene product forms a complex with the p53 protein and inhibits p53-mediated trans-activation. Cell 1992; 69:1237-45.
32. Chen J, Wu X, Lin J et al. mdm-2 inhibits the G1 arrest and apoptosis functions of the p53 tumour suppressor protein. Mol Cell Biol 1996; 16:2445-52.
33. Oliner JD, Kinzler KW, Meltzer PS et al. Amplification of a gene encoding a p53-associated protein in human sarcomas. Nature 1992; 358:80-3.
34. Wu X, Bayle H, Olson D et al. The p53-mdm2 autoregulatory feedback loop. Genes Dev 1993; 7:1126-32.
35. Seto E, Usheva A, Zambetti GP et al. Wild-type p53 binds to the TATA-binding protein and represses transcription. Proc Natl Acad Sci USA 1992; 89: 12028-32.
36. Mack DH, Vartikar J, Pipas JM et al. Specific repression of TATA-mediated but not initiator-mediated transcription by wild type p53. Nature 1993; 363: 281-3.
37. Rauscher FJ, Morris JF, Tounay OE et al. Binding of the Wilms' tumour locus zinc finger protein to the EGR-1 consensus sequence. Science 1990; 250:1259-62.
38. Pavletich NP, Pabo CO. Zinc finger-DNA recognition: crystal structure of a Zif268-DNA complex at 2.1A. Science 1991; 252:809-17.
39. Bickmore WA, Oghene K, Little MH et al. Modulation of DNA binding specificity by alternative splicing of the Wilms' tumour *wt1* gene transcript. Science 1992; 257:235-7.
40. Bruening W, Bardeesy N, Silverman BL et al. Germline intronic and exonic mutations in the Wilms' tumour gene (WT1) affecting urogential development. Nature Genetics 1992; 1:144-8.
41. Madden SL, Cook DM, Morris JF et al. Transcriptional repression mediated by the WT1 Wilms' tumour gene product. Science 1991; 253:1550-3.
42. Rupprecht HD, Drummond IA, Madden SL et al. The Wilms' tumour suppressor gene is negatively autoregulated. J Biol Chem 1994; 269:6198-206.
43. Kent J, Coriat A, Sharpe PT et al. The evolution of WT1 sequence and expression pattern in the vertebrates. Oncogene 1995; 11:1781-92.
44. Wang ZY, Qiu QQ, Dueul TF. The Wilms' tumour gene product WT1 activates or represses transcription through separate functional domains. J Biol Chem 1993; 268:9172-5.
45. Kreidberg JA, Sariola H, Loring JM et al. WT-1 is required for early kidney development. Cell 1993; 74:679-91.
46. Yun K, Fidler AE, Eccles MR et al. Insulin-like growth factor II and WT1 tran-

scripts in human foetal kidney and Wilms' tumour. Cancer Res 1993; 53:5166-71.
47. Ogawa O, Eccles MR, Szeto J et al. Relaxation of insulin-like growth factor II gene imprinting implicated in Wilms' tumour. Nature 1993; 362:749-51.
48. Drummond IA, Madden SL, Rohwer-Nutter P et al. Repression of the insulin-like growth factor II gene by the Wilms' tumour suppressor, WT1. Science 1992; 257:674-8.
49. Ryan G, Steele-Perkins V, Morris JF et al. Repression of *pax-2* by WT1 during normal kidney development. Development 1995; 121:867-75.
50. Torres M, Gomez-Pardo E, Dressler GR et al. Pax-2 controls multiple steps of urogenital development. Development 1995; 121:4057-65.
51. Dressler GR, Wilkinson JE, Rothenpieler UW et al. Deregulation of Pax-2 expression in transgenic mice generates severe kidney abnormalities. Nature 1993; 362:65-7.
52. Maulbecker CC, Gruss P. The oncogenic potential of Pax genes. EMBO J 1993; 12:2361-7.
53. Englert C, Hou X, Maheswaran S et al. WT1 suppresses synthesis of the epidermal growth factor receptor and induces apoptosis. EMBO J 1995; 14:4662-75.
54. Chittenden T, Livingston DM, Kaelin WG. The T/E1A-binding domain of the retinoblastoma product can interact selectively with a sequence specific DNA-binding protein. Cell 1991; 65:1073-82.
55. La Thangue NB. DRTF/E2F: An expanding family of heterodimeric transcription factors implicated in cell-cycle control. TIBS 1994; 19:108-14.
56. Means AL, Slansky JE, McMahon SL et al. The HIP1 binding site is required for growth regulation of the dihydrofolate reductase gene promoter. Mol Cell Biol 1992; 12:1054-63.
57. Lam EW, Watson RJ. An E2F-binding site mediates cell-cycle regulated repression of mouse B-*myb* transcription. EMBO J 1993; 12:2703-13.
58. Goodrich DW, Wang NP, Qian Y et al. The retinoblastoma gene product regulates progression through the G1 phase of the cell cycle. Cell 1991; 67:293-302.
59. Qin X, Chittenden T, Livingston DM et al. Identification of a growth suppression domain within the retinoblastoma gene product. Genes Dev 1992; 6:953-64.
60. Weintraub SJ, Chow KNB, Luo RX et al. Mechanism of active transcriptional repression by the retinoblastoma protein. Nature 1995; 375:812-5.
61. Hagemeier C, Bannister AJ, Cook A et al. The activation domain of transcription factor PU.1 binds the retinoblastoma (RB) protein and the transcription factor TFIID in vitro: RB shows sequence similarity to TFIID and TFIIB. Proc Natl Acad Sci USA 1993; 90:1580-4.
62. Singh P, Wong SH, Hong W. Overexpression of E2F-1 in rat fibroblasts leads to neoplastic transformation. EMBO J 1994; 13:3329-38.
63. Yamasaki L, Jacks T, Bronson R et al. Tumour induction and tissue atrophy in mice lacking E2F-1. Cell 1996; 85:537-48.
64. Field SJ, Tsai F, Kuo F et al. E2F-1 functions in mice to promote apoptosis and suppress proliferation. Cell 1996; 85:549-61.
65. Krek W, Xu G, Livingston DM. Cyclin A-kinase regulation of E2F-1 DNA binding function underlies suppression of an S phase checkpoint. Cell 1995; 83:1149-58.
66. Helin K, Lees JA, Vidal M et al. A cDNA encoding a pRB-binding protein with properties of the transcription factor E2F. Cell 1992; 70:337-50.
67. Ikeda M, Jakoi L, Nevins JR. A unique role for the Rb protein in controlling E2F accumulation during cell growth and differentiation. Proc Natl Acad Sci USA 1996; 93:3215-20.
68. Iavarone A, Garg P, Lasorella A et al. The helix-loop-helix protein Id-2 enhances cell proliferation and binds to the retinoblastoma protein. Genes Dev 1994; 8:1270-84.
69. Chen P, Riley DJ, Chen-Kiang S et al. Retinoblastoma protein directly interacts with and activates the transcription factor NF-IL6. Proc Natl Acad Sci USA 1996; 93:465-9.
70. Scheider JW, Gu W, Zhu L et al. Reversal of terminal differentiation mediated by p107 in $Rb^{-/-}$ muscle cells. Science 1994; 264:1467-71.

71. Jacks T, Fazeli A, Schmitt EM et al. The effects of an Rb mutation in the mouse. Nature 1992; 359:295-300.
72. Clarke AR, Maandag ER, van Roon M et al. Requirement for a functional Rb-1 gene in murine development. Nature 1992; 359:328-30.
73. Lee EYP, Chang CY, Hu N et al. Mice deficient for Rb are nonviable and show defects in neurogenesis and haematopoiesis. Nature 1992; 359:288-94.
74. Welch PJ, Wang JYJ. A C-terminal protein-binding domain in the retinoblastoma protein regulates nuclear c-abl tyrosine kinase in the cell cycle. Cell 1993; 75:779-90.
75. Welch PJ, Wang JYJ. Disruption of retinoblastoma protein function by coexpression of its C pocket fragment. Genes Dev 1995; 9:31-46.
76. Baskaran R, Dahmus ME, Wang JYJ. Tyrosine phosphorylation of mammalian RNA polymerase II carboxy-terminal domain. Proc Natl Acad Sci USA 1993; 90:11167-71.
77. Taunton J, Hassig CA, Scheiber SL. A mammalian histone deacetylase related to the yeast transcriptional regulator Rpd3p. Science 1996; 272:408-11.
78. Koonin EV, Zhou S, Lucchesi JC. The chromo superfamily: new members, duplication of the chromodomain and possible role in delivering transcription regulators to chromatin. Nucleic Acids Res 1995; 23:4229-33.
79. Dunaief JL, Strober BE, Guha S et al. The retinoblastoma protein and BRG1 form a complex and cooperate to induce cell cycle arrest. Cell 1994; 79:119-30.
80. Singh P, Coe J, Hong W. A role for retinoblastoma protein in potentiating transcriptional activation by the glucocorticoid receptor. Nature 1995; 374:562-5.
81. Dowdy SF, Hinds PW, Louie K et al. Physical interaction of the retinoblastoma protein with human D cyclins. Cell 1993; 73:499-511.
82. Ewen ME, Sluss HK, Sherr CJ et al. Functional interactions of the retinoblastoma protein with mammalian D-type cyclins. Cell 1993; 73:487-97.
83. Parry D, Bates S, Mann DJ et al. Lack of cyclin D-cdk complexes in Rb-negative cells correlates with high levels of $p16^{INK4/MTS1}$ tumour suppressor gene product. EMBO J 1995; 14:503-11.
84. Hinds PW, Mittnacht S, Dulic V et al. Regulation of retinoblastoma functions by ectopic expression of human cyclins. Cell 1992; 70:993-1006.
85. Howes KA, Ransom N, Papermaster DS et al. Apoptosis or retinoblastoma: alternative fates of photoreceptors expressing the HPV-16 E7 gene in the presence or absence of p53. Genes Dev 1994; 8:1300-10.
86. Pan H, Griep AE. Altered cell cycle regulation in the lens of HPV-16 E6 or E7 transgenic mice: implications for tumour suppressor gene function in development. Genes Dev 1994; 8:1285-99.
87. Morgenbesser SD, Williams BO, Jacks T et al. p53-dependent apoptosis produced by Rb-deficiency in the developing mouse lens. Nature 1994; 371:72-4.
88. Williams BO, Remington L, Albert DM et al. Cooperative tumourigenic effects of germline mutations in Rb and p53. Nature Genetics 1994; 7:480-4.
89. Xiao Z, Chen J, Levine AJ, Modjtahedi N et al. Interaction between the retinoblastoma protein and the oncoprotein MDM2. Nature 1995; 375:694-7.
90. Martin K, Trouche D, Hagemeier C et al. Stimulation of E2F1/DP1 transcriptional activity by MDM2 oncoprotein. Nature 1995; 375:691-4.
91. O'Connor DJ, Lam EW, Griffin S et al. Physical and functional interactions between p53 and cell cycle co-operating transcription factors, E2F1 and DP1. EMBO J 1995; 14:6184-92.
92. Maheswaran S, Park S, Bernard A et al. Physical and functional interaction between WT1 and p53 proteins. Proc Natl Acad Sci USA 1993; 90:5100-4.
93. Maheswaran S, Englert C, Bennett P et al. The *WT1* gene product stabilises p53 and inhibits p53-mediated apoptosis. Genes Dev 1995; 9:2143-56.
94. Godley LA, Koop JB, Eckhaus M et al. Wild-type p53 transgenic mice exhibit altered differentiation of the ureteric bud and possess small kidneys. Genes Dev 1996; 10:836-50.

95. Bardeesy N, Falkoff D, Petruzzi MJ et al. Anaplastic Wilms' tumour, a subtype displaying poor prognosis, harbours p53 gene mutations. Nature Genetics 1994; 7:91-7.
96. Stuart ET, Haffner R, Oren M et al. Loss of p53 function through PAX-mediated transcriptional repression. EMBO J 1996; 14:5638-45.
97. Haupt Y, Rowan S, Shaulian E et al. Induction of apoptosis in HeLa cells by *trans*-activation-deficient p53. Genes Dev 1995; 9:2170-83.
98. Caelles C, Helmberg A, Karin M. p53 dependent apoptosis in the absence of transcriptional activation of p53-target genes. Nature 1994; 370:220-3.
99. Cox LS, Hupp T, Midgley CA et al. A direct effect of activated human p53 on nuclear DNA replication. EMBO J 1995; 14:2099-105.
100. Fukasawa K, Choi T, Kuriyama R et al. Abnormal centrosome amplication in the absence of p53. Science 1996; 271:1744-7.
101. Zhu X, Mancini MA, Chang K et al. Characterisation of a novel 350-kilodalton nuclear phosphoprotein that is specifically involved in mitotic-phase progression. Mol Cell Biol 1995; 15:5017-29.
102. Larsson S, Charlieu JP, Miyagawa K et al. Subnuclear localisation of WT1 in splicing or transcription factor domains is regulated by alternative splicing. Cell 1995; 81:391-401.
103. Durfee T, Mancini MA, Jones D et al. The amino-terminal region of the retinoblastoma gene product binds a novel nuclear matrix protein that co-localises to centres for RNA processing. J Cell Biol 1994; 127:609-22.
104. Clemens KR, Wolf V, McBryant SJ et al. Molecular basis for specific recognition of both RNA and DNA by a zinc finger protein. Science 1993; 260:530-3.
105. Kennedy D, Ramsdale T, Mattick J et al. An RNA recognition motif in Wilms' tumour protein (WT1) revealed by structural modelling. Nature Genetics 1996; 12:329-32.

CHAPTER 17

Transcriptional Regulation by Viral Activators

Xiao-Yong Li, Giovanni Perini and Michael R. Green

INTRODUCTION

All viruses encode one or more transcription activators, which are usually expressed immediately following viral infection or upon disruption of latency that leads to lytic growth. Synthesis of these transcription regulators is required for orderly viral gene expression and virus reproduction. Viral activators are among the first transcription factors studied, and have provided important paradigms for understanding many of the fundamental mechanisms governing the regulation of gene expression. In many respects viral activators are very similar to their cellular counterparts. At the same time, many of them also display some unusual properties. Notably, whereas some viral activators are like typical cellular activators in containing a DNA-binding domain, many others lack DNA-binding ability. It has been found that these activators usually induce a transcriptional response by interacting with cellular transcription factors.[1] Remarkably, the activators of some lentiviruses, such as the human immunodeficiency virus type 1 (HIV-1) activator Tat, function by binding to RNA elements in the nascent transcripts. Another unusual property associated with viral activators is functional promiscuity, the ability of many viral activators to regulate not only viral but also a wide range of cellular genes. This functional promiscuity is often related to promoter targeting, and can be attributed to the ability of some of the viral activators, particularly those lacking a DNA-binding domain, to function through multiple cellular activators.

In this article, we will review some of the most extensively studied examples to illustrate how viral activators regulate transcription. Our emphasis will be how these activators target the promoter for transcriptional response, which represents a very important and probably the most versatile aspect in the function of viral activators. For comparison, we have placed the activators in three basic groups: the DNA-binding activators, non-DNA binding activators, and RNA binding activators. Another important aspect in the function of viral activators that we will discuss is how the activators, after being directed to the promoter, effect

Transcription Factors in Eukaryotes, edited by Athanasios G. Papavassiliou.

transcriptional activation; both activation at the level of initiation and elongation will be addressed.

VIRAL ACTIVATORS WITH DNA-BINDING DOMAINS

Papillomavirus E2 Proteins

The E2 translational open reading frame of papillomaviruses encodes three transcription regulatory factors (for a review see refs. 2, 3). The full-length product, E2TA, is a transcription activator with an activation domain located at the amino-terminus and a DNA-binding domain at the carboxy-terminus. These two functional domains are conserved among different papillomaviruses, and are separated by a variable hinge region. The two smaller E2 products, E2TR and E8/E2TR, lack the N-terminal activation domain, and are able to inhibit the activity of E2TA, presumably by competitive DNA binding or by forming inactive heterodimers with E2TA.

The consensus sequence of E2 DNA-binding sites is 5' -ACCN6GGT- 3'. E2TA binds multiple E2 sites cooperatively, and this cooperativity is dependent on the amino-terminal activation domain.[4,5] It has been shown that transactivation by E2TA through a single E2 DNA-binding site is very inefficient. However, multiple E2 sites, located close to or at a distance from the promoter, can act in synergy to achieve efficient transactivation.[6] E2TA can also activate transcription efficiently in synergy with some cellular activators, such as USF[7] and Sp1.[8] Interestingly, direct DNA binding is not always required for transcriptional activation by E2TA. It has been shown that Sp1 can mediate E2TA transactivation in the absence of an E2 binding site, presumably through direct interaction between these two proteins.[8]

Besides being able to activate transcription, E2TA can also negatively regulate certain papillomavirus promoters. These promoters usually contain E2 DNA-binding sites overlapping an essential promoter element, so the repression may be due to competitive DNA binding. One such promoter is the bovine papillomavirus type 1 P_{7185} promoter, which contains an E2 binding site overlapping an essential promoter element around the transcription start site.[9] Another E2TA-repressed promoter, the human papillomavirus type 18 (HPV-18) P_{105} promoter, which is equivalent to the HPV-16 P_{97}, contains a E2 binding site 3 base pair upstream of the TATA box. In this case, it has been shown that E2TA competes with the general transcription factor TFIID for binding to this promoter.[10]

Epstein-Barr Virus (EBV) Zta and Rta

After infection of B-lymphocytes, EBV is usually maintained in a latent form. Disruption of latency and the subsequent lytic cycle of EBV is associated with the expression of two transcriptional activators: the BZLF1 gene product Zta [also referred to as Epstein-Barr 1 (EB1), and ZEBRA] and the BRLF1 gene product Rta. Both Zta and Rta are DNA-binding factors (for a review see ref. 11).

The Zta DNA-binding domain is related to the basic region-leucine zipper (bZIP) DNA-binding domain of the AP1/cJun family of cellular transcription factors.[12,13] The highest homology is found in the basic region, which determines the DNA-binding specificity of such DNA-binding domains. Interestingly, in the region corresponding to the leucine zipper dimerization subdomain of a canonical bZIP, Zta does not have a leucine repeat, but instead it contains hydrophobic residues that can nevertheless support dimerization through coiled-coil structure.[14]

Both Zta and AP-1 can bind to the consensus AP-1 site, or the TPA response element (TRE), 5'-TGAGTCA-3'. However, these two factors differ in affinity to the variant sites.[13] Specifically, Zta, but not AP-1, can bind to the Zta response element (ZRE) found in the EBV promoters, and conversely some TRE variant sites are only recognized by AP-1. The Zta DNA-binding domain contains a serine residue that can be phosphorylated, and this phosphorylation abolishes its DNA-binding activity.[15]

Zta also appears to be able to activate transcription through a mechanism independent of direct DNA binding. For example, it has been shown that the promoter of Zta can be activated by a mutant Zta incapable of DNA binding. This transactivation has been found to be mediated by a DNA element distinct from ZRE.[16]

The other EBV activator, Rta, can activate many EBV genes by binding to responsive DNA elements (RRE). The consensus binding sequence of Rta, derived by a selection from random DNA sequences is 5' -GG/TCC(N7)GTGGTG- 3'.[17] Rta can activate transcription alone or cooperatively with Zta in a cell type-dependent manner.[11,18]

Herpes Simplex Virus (HSV) ICP4

HSV encodes several protein factors that can regulate gene expression. Among them, the immediate-early (IE) gene product ICP4 (also termed VmW175, IE175, and VP175) is the only DNA-binding factor. ICP4 is a 175 kDa phosphoprotein with a helix-turn-helix DNA-binding motif. The consensus sequence for ICP4 binding sites is 5' - RTCGTCNNYNYSG- 3' (where R is purine, S is C or G, and N is any base).[19] ICP4 can bind some DNA sites that deviate significantly from this sequence, albeit with decreased affinity. The DNA-binding activity of ICP4 can be regulated by phosphorylation. It has been shown that phosphorylation can strongly decrease the binding of ICP4 to low affinity sites while having a relatively small effect on its binding to the consensus site.[20]

ICP4 can both activate and repress transcription. While the HSV delayed-early (DE) and late genes are activated, the IE genes, including the gene encoding ICP4 itself, are downregulated by ICP4. Some of the activated genes do not contain obvious ICP4 binding sites, suggesting that binding of ICP4 to low affinity sites or nonspecific binding is sufficient for transactivation.[21] In contrast, repression by ICP4 requires a high-affinity DNA-binding site near the transcription initiation site.[22] The mechanism for repression by ICP4 will be discussed below.

The other HSV factors involved in gene expression include the viron activator VP16 (see below), the IE gene products ICP0, ICP22, and ICP27. ICP0 (also termed Vmw110), can activate transcription of HSV viral and cellular genes in general, sometimes in synergy with ICP4 (for a review see ref. 1). It contains a RING finger, a motif shared by a large family of factors with diverse cellular localization and functions.[23] ICP0 does not appear to bind DNA, and its mechanism in transactivation is not clear. Many studies have focused on its cellular localization, which changes during HSV infection cycle. In addition, ICP0 can also affect the cellular localization of certain cellular proteins.[24] This later finding suggests that ICP0 may affect transcription through indirect mechanisms or it may have multiple functions. Similar to ICP0, ICP22 also has a global effect on transcription, which seems to result from a novel phosphorylation of the RNA polymerase II large subunit.[25] Finally, ICP27, which can up and downregulate HSV as well as cellular genes, is able to exerts its regulatory effect at the level of cellular RNA metabolism including splicing[26] and RNA stability.[27]

NON-DNA BINDING VIRAL ACTIVATORS

Adenovirus Transactivator 13S E1a

Adenovirus E1a gene transcript undergoes alternative splicing to produce two messenger RNAs, the 12S and the 13S, which encode two E1a proteins of 243 and 298 amino acids respectively. The E1a gene products of adenoviruses have three conserved regions: CR1, CR2, and CR3 (for a review see ref. 28). CR1 and CR2 are common to both E1a proteins while CR3, which contains a zinc-binding structure essential for function, is unique to the 13S product. E1a can associate with a number of cellular factors in the cell. A related fact is that expression of E1a in the cell has multiple effects including changes in cell growth, induction of cell apoptosis, and extensive alteration in gene expression. The sequences

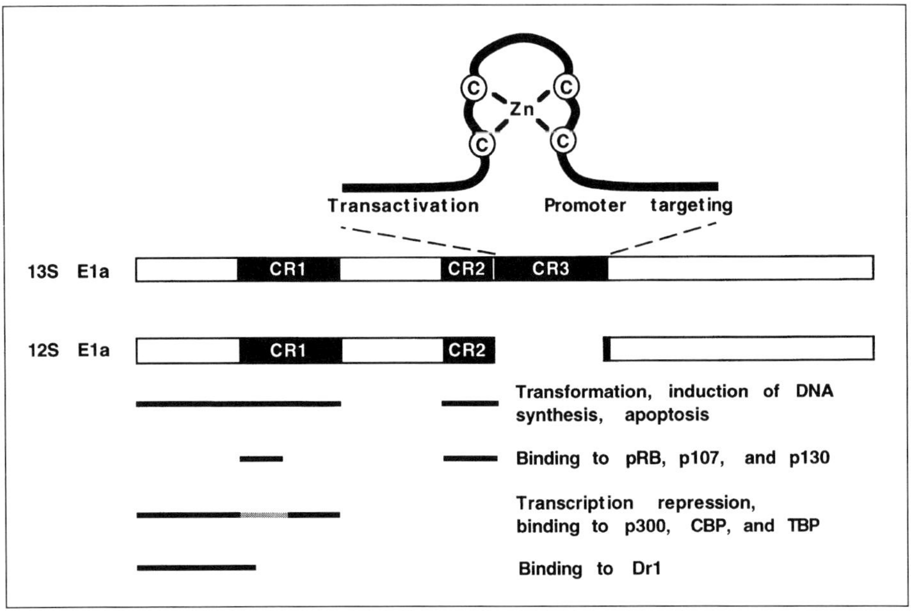

Fig. 17.1. Functional domains of the adenoviral E1a gene products. E1a proteins contain three conserved regions CR1, CR2, and CR3. CR3, which is unique to the 13S product and contains a zinc binding structure, is important for transcriptional activation, and consists of two overlapping subdomains required for promoter targeting and transactivation respectively. The other functional regions are explained in the text.

in E1a proteins required for each of these functions are relatively well-defined (Fig. 17.1). In this article, we will focus on their transcriptional effect only.

E1a can regulate transcription through multiple mechanisms. The most conspicuous mechanism that was the focus of most of the earlier studies involves CR3, and thus is specific to the 13S E1a. However, the 12S E1a can also regulate transcription, which mainly involves indirect mechanisms. The transcription regulatory effect of the 12S E1a, in contrast to that of the 13S product, is largely linked to induction of cell transformation and apoptosis. In this section, we will focus on the transcriptional induction by the 13S E1a, which will be referred to simply as E1a, and in a later section we will discuss the transcriptional effect of the 12S E1a.

The 13S E1a can induce the expression of adenoviral as well as many cellular genes. Multiple, often unrelated DNA elements recognized by different cellular activators have been shown to mediate this transcriptional induction (for a review see ref. 29). Because E1a lacks specific DNA-binding ability it has been suggested that E1a might induce transcription by interacting with these cellular activators. In support of this, it has been found that E1a contains, in CR3, a promoter targeting region, separable from the activation domain (Fig. 17.1).[30] This region is required for the transcriptional induction of a natural E1a-responsive promoter even though it is dispensable when the E1a activation domain is directed to the promoter through a heterologous DNA-binding domain. In addition, it has been shown that as a fusion protein, this region can target the HSV VP16 (see below) activation domain to the E1a responsive promoter for transcription activation.

One extensively characterized E1a responsive element is the activating transcription factor (ATF) binding site, which has been identified in all adenovirus promoters.[31] The ATF site and the cAMP-responsive element (CRE) share a consensus sequence, 5' -TGACGTCA- 3'. This sequence is recognized by a large group of cellular transcription factors that belong to the ATF/CREB family and contain a conserved bZIP DNA-binding domain, which is also shared by transcription factors of several other families, such as the AP-1/cJun family mentioned above.[32] It has been found that one of these factors, ATF-2, and its homologue ATF-a, can specifically mediate E1a response.[33,34] Consistent with the hypothesis that E1a functions by directly interacting with cellular activators, both in vivo and in vitro evidence suggest that ATF-2 and E1a can directly interact.[35] This interaction requires the bZIP of ATF-2 and is abolished by a deletion in the promoter targeting region of E1a. Interaction of E1a with the bZIPs of other ATF/CREB factors has also been observed. However, these factors can not mediate E1a response as efficiently as ATF-2 probably due to different nature of the activation domains.[35] Some evidence suggests that ATF-2 is inactive in most cell types due to direct intramolecular interaction between the activation domain and the bZIP;[36] the binding of E1a may disrupt such interaction leading to transcriptional activation.

Another activator that has been shown to mediate E1a response through direct interaction is Oct4,[37] a POU-homeo DNA-binding domain containing factor. The formation of a E1a-Oct4-DNA ternary complex has been detected using extract prepared from cells expressing E1a or with in vitro translated proteins. This complex formation depends on the CR3 region of E1a and is abolished by a point mutation in the promoter targeting region.

While E1a probably activates transcription primarily through direct interaction with cellular activators, some other mechanisms have also been suggested. For example, some studies have suggested that E1a might stimulate transcription by increasing the activity of general transcription factor TFIID,[38] even though further characterization failed to show a change in the amount or the phosphorylation state of this factor in the E1a expressing cells.

Besides class II genes, adenovirus infection also leads to induction of transcription by RNA polymerase III. This transcriptional induction requires E1a expression even though it is not known whether the CR3 region is important. This transcriptional effect involves the alteration of the general transcription factor TFIIIC. (ref. 39 and references therein) TFIIIC2, one of the two biochemically separable TFIIIC activities, can be isolated in two forms, recently named TFIIIC2a and TFIIIC2b. It has been found that in the cells expressing E1a or grown at high serum concentration, the ratio of the active form TFIIIC2a to the inactive form TFIIIC2b is higher than in the cells grown at low serum concentration. Recent studies have shown that this alteration is due to increased expression of the 110 kDa subunit (TFIIICβ) that is present in TFIIIC2a but not TFIIIC2b.[40] How the expression of TFIIICβ is regulated and whether the CR3 region is required await further studies.

HSV Viron Protein VP16

VP16 (also termed Vmw65, α-TIF, VF65, and ICP25) is probably the best understood among viral activators (for a review see refs. 41, 42). Like E1a, VP16 does not have a DNA-binding domain but instead contains a promoter targeting region.[43] However, VP16 activates transcription of only a very restricted set of genes.

Induction of transcription by VP16 involves the recruitment of VP16 to the promoter by Oct-1 to form a protein-DNA complex. This complex also contains a cellular accessory factor HCF (see below). Oct-1 contains a POU-homeo DNA-binding domain that by itself is sufficient for assembly of the VP16 containing complex. Despite the fact that the related protein Oct-2 contains a homologous POU-homeo domain and belongs to the same family of transcription factors as Oct-1, it has a much lower ability to form a VP-16 containing complex.[44,45] Mutational analysis has revealed that a single amino acid residue difference in the DNA-binding domains is responsible for selective recognition of Oct-1 by VP16.[46,47]

The DNA sequence recognized by Oct-1 as well as other POU-homeo domain factors is 5' -ATGCAAAT- 3', also referred to as the octamer binding site. Interestingly, this sequence, while sufficient for Oct-1 binding, is not sufficient for mediating VP16 response. Rather, the VP16 response elements contain an additional GARAT sequence at its 3' end. Based on footprinting studies, this sequence can induce a conformational change in the Oct1 - DNA complex, which presumably is important for efficient VP16 recognition.[48]

The VP16-Oct1-DNA ternary complex is stabilized by the cellular factor HCF,[49,50] which comprises a group of polypeptides derived through posttranslational processing of a single precussor.[51] The stimulatory effect of HCF also enables the assembly of the VP16 containing complex at VP16 responsive elements that do not contain the whole octamer sequence and consequently bind Oct-1 alone poorly.[52] Thus, multiple mechanisms govern VP16 selectivity in transcriptional activation.

Human T-Cell Leukemia Virus Type I (HTLV-I) Transactivator Tax

Tax is a nuclear phosphoprotein of about 40 kDa which is able to stimulate transcription activation of the HTLV-I LTR as well as several cellular genes. Tax does not bind DNA. Several studies have shown that Tax can function through different unrelated cellular transcription factors such as ATF/CREB, NF-κB and SRF. Although it is still unclear how Tax can recognize so many different factors, mutation analysis has revealed that Tax possesses distinct domains, each of them capable of responding to a specific class of cellular proteins. For example, substitutions in amino acids 137 and 138 abolish Tax responsiveness to NF-κB without affecting Tax regulation of promoters responsive to ATF/CREB,[53,54] while the opposite is true for mutations in amino acids 3 and 319.[53,54]

Different mechanisms have been proposed to explain how Tax can activate transcription. For example Tax seems to stimulate transcription of CRE containing promoters, such as the HTLV-I LTR, by increasing the DNA-binding activity of ATF/CREB factors.[54-57] It has been shown to interact specifically with the bZIP domain of ATF/CREB factors[56] and in particular with the basic region.[58] The bZIP-Tax interaction results in an increased level of bZIP dimerization, which is an essential step for bZIP DNA binding (Fig. 17.2). In addition Tax can also influence the DNA-binding specificity of bZIP.[58,59]

A different mechanism is used to activate NF-κB responsive promoters. Several lines of evidence suggest that Tax can increase the import of NF-κB heterodimers from the cytoplasm to the nucleus. NF-κB activity is negatively regulated by IκBα, which functions by retaining NF-κB heterodimers in the cytoplasm. By inducing IkBa phosphorylation and degradation, Tax facilitates the release of NF-κB heterodimers from IkBα inhibitor and promotes their fast transport to the nucleus.[60] A recent study has proposed an additional mechanism. It has shown that Tax can increase the level of active NF-κB heterodimers by favoring the interaction between NF-κB precursors and the proteosome. This results in an acceleration of the proteolytic maturation of NF-κB precursors.[61]

Tax can also interact with the cellular factor $p67^{SRF}$, thereby activating the transcription of some cellular immediate early genes containing CArG boxes, such as *c-fos*.[62] Although it is not clear how this occurs it seems that Tax can relieve the requirement for a serum induced activation of $p67^{SRF}$.

Hepatitis B Virus (HBV) X Protein

pX is a protein of 154 amino acids and can transactivate both viral and cellular genes. Several reports have suggested that pX is functionally similar to Tax. In fact, pX can activate transcription through the same classes of factors required by Tax such as AP-1, AP-2, NF-κB and ATF/CREB (for a review see ref. 63), and can be substituted by Tax in the activation of the HBV enhancer.[64] Like Tax, pX can activate transcription through different mechanisms. For example pX seems to stimulate AP-1 and NF-κB re-

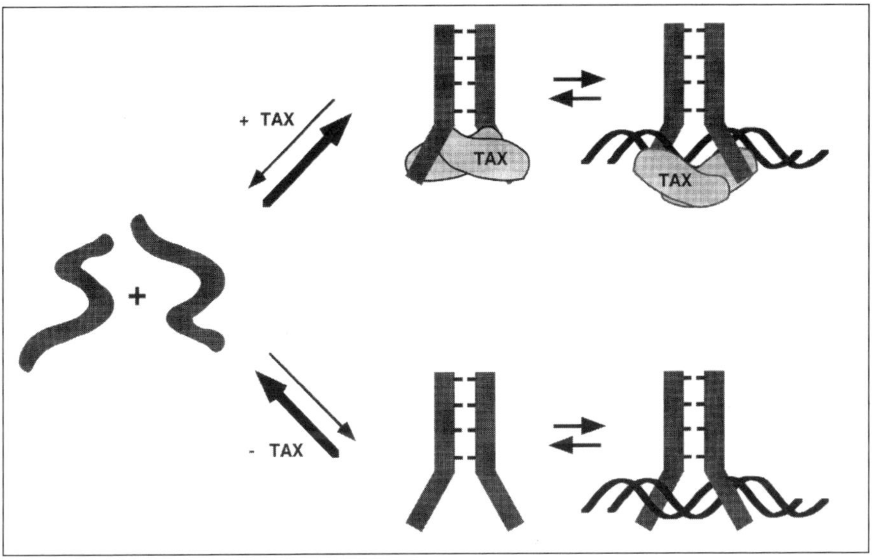

Fig. 17.2. HTLV transactivator Tax functions partly by increasing DNA binding by targeted cellular activators. Basic region-leucine zipper (bZIP) containing transactivators bind DNA as dimers. Tax increases DNA binding by such proteins by promoting dimer formation.

sponsive promoters by inducing specific signal transduction pathways.[65-67] On the contrary the activation of the HBV enhancer, which contains a CRE-like site, seems to occur through a direct recruitment of pX to the promoter. pX does not possess a DNA-binding domain and needs to interact with cellular transcription factors in order to target the promoter region. In vitro experiments have shown that pX can form stable complexes with CREB and ATF2.[68] The interaction with CREB results in an increase of CREB DNA-binding activity.[69]

HIV-1 TAT: AN RNA-BINDING ACTIVATOR

HIV-1 Tat is a strong activator of the HIV-1 LTR promoter (for a review see ref. 70). Unlike other known transcription factors, Tat contains an RNA binding domain and is directed to the promoter through an RNA element, Tar, located within the first 59 nucleotides of the nascent transcript (Fig. 17.3). The Tar element forms a stem-loop structure with a bulge in the stem region. The sequence within and around the bulge is important for the binding of Tat in vitro, while the sequence in the loop is not. However, alterations of the loop sequence have a strong effect on Tat function in vivo, suggesting that a cellular factor may bind to the loop region of the Tar element to promote Tat function (Fig. 17.3).

Tat contains two separable domains, one for transcriptional activation and the other for RNA binding. It has been shown that the Tat activation domain can efficiently activate transcription when it is directed to the promoter as a fusion to heterologous RNA binding domain, in transient transfection assays using a reporter whose transcript contains the corresponding RNA element in place of the Tar element.[71,72] This finding indicates that the Tar element is responsible solely for targeting Tat to the promoter but not for the transcriptional activation per se. For this reason, the putative Tar-loop binding factor may be required for efficient Tat binding to Tar in vivo.

That a Tar-loop binding factor is required for efficient Tat binding to Tar is also supported by studies on the species

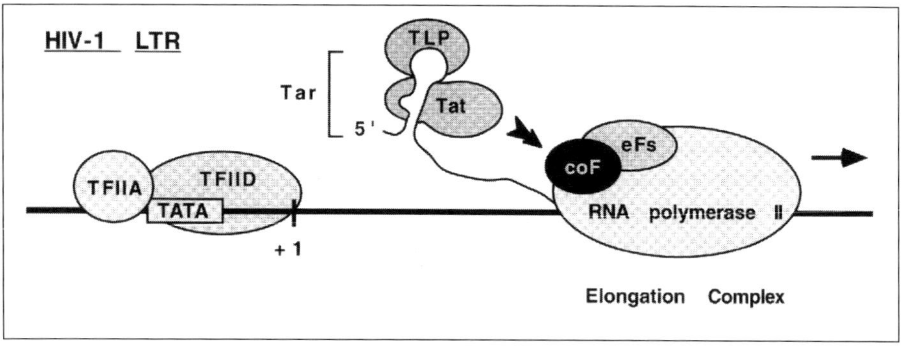

Fig. 17.3. HIV-1 transactivator Tat functions by binding to RNA to stimulate transcription elongation. Tar, the RNA element recognized by Tat located in the promoter proximal region of the HIV-1 LTR transcript; TLP, the putative Tar-loop binding protein; coF, cellular Tat-cofactor; eFs, elongation factors involved in Tat transactivation.

specificity in Tat function. The HIV-1 LTR promoter is activated by Tat more efficiently in primate cells than in rodent or CHO cells. However, it has been shown that when directed to the promoter as a fusion to a heterologous RNA binding domain Tat can activate transcription efficiently in CHO cells.[73] In addition, Tat can also function efficiently in CHO12 fusion cells, which contain human chromosome 12.[74] These findings suggest that the Tar-loop binding factor, required for efficient Tat binding, may be species specific and may be encoded by a gene on human chromosome 12. However, further studies have failed to detect a species specific Tar-loop binding factor.[75] One explanation for this is that the species specific factor, instead of binding to Tar, is probably required for proper modification of the Tar-loop binding factor. Several cellular factors that can bind the Tar element have been identified, but their significance in Tat function is unclear.[70]

Even though Tat normally functions through RNA, it can also activate transcription when directed to a promoter through a DNA-binding domain.[76,77] Complementary to this finding, one activator, VP16, which normally functions through DNA, can activate transcription when targeted to RNA.[78] Based on these studies, Tat may not be fundamentally different from other activators.

MECHANISMS OF TRANSACTIVATION BY VIRAL ACTIVATORS

It has been assumed that transcription is regulated mainly at the level of initiation. However, recent studies have suggested that the regulation at the level of elongation may just be as important in eukaryotic cells (for a review see ref. 79). In addition, it has been found that elongation can be regulated by conventional activators instead of by some special factors as earlier studies suggested. Interestingly, the ability to stimulate transcription elongation differs among activators and does not seem to correlate with the ability to stimulate transcription initiation. Specifically, while activators can usually stimulate transcription initiation, only some activators can also stimulate elongation.[80] In addition, HIV-1 Tat appears to stimulate transcription primarily at the level of elongation even though in some instances it has been shown to stimulate initiation as well.[80-82] Therefore, stimulation of transcription initiation and elongation may involve different mechanisms.

Stimulation of Transcription Initiation

During transcription initiation, the general transcription factors and RNA polymerase II assemble at the promoter to form a transcription preinitiation complex. Extensive studies have shown that an important aspect of transcriptional activation is interaction between transactivators and general transcription factors. Such interaction, which may be direct or mediated through a coactivator, allows activators to recruit general transcription factors to the promoter, or to overcome a rate limiting step during preinitiation complex formation and transcription initiation. Below, we will summarize some of the findings on the regulation of transcription initiation by viral activators, mainly focusing on the evidence of their interaction with the general transcription factors.

HSV VP16

The activation domain of VP16 and its mechanism of transactivation is probably the best characterized among all activators. The VP16 activation domain is rich in acidic residues, and is considered, together with activation domains of the yeast activators GAL4 and GCN4 to be the prototypic acidic activation domain.[83]

To determine the mechanism of transcriptional activation by VP16, the interaction between VP16 and general transcription factors TBP[84] and TFIIB[85,86] were detected using affinity column chromatography. The binding to both factors was abolished by the mutation (FP442) of a single critical residue of the VP16 activation domain. A recent study suggests that the interaction of VP16 with TBP may not be required for transcriptional activation.[87] In contrast, based on a variety of evidence, TFIIB is an important target for transactivation by VP16. It has been shown that during transcription preinitiation complex assembly, the binding of TFIIB is very inefficient and can be stimulated by VP16.[85,88] In addition, mutations of VP16 and TFIIB that disrupted the interaction between these two proteins also abolished activated transcription.[89] Interestingly, not only can VP16 recruit TFIIB to the preinitiation complex, one study showed that it can also change the conformation of TFIIB.[90] This conformational change is presumably important for recruiting other general transcription factors to the preinitiation complex. Besides TBP and TFIIB, VP16 has also been shown in a recent study to interact in vitro with TFIIH,[91] and one of the factors in the TFIID complex, TAF40.[92]

BPV Zta

It has been shown that Zta can stabilize the binding of TBP to TATA box. When TBP is substituted by TFIID in the binding reaction, Zta can stimulate the binding of TFIID to the promoter in the presence of TFIIA.[93] A recent study has shown that multiple Zta molecules bound to the promoter can stimulate TFIID/IIA/IIB-promoter complex formation in a synergistic manner.[94]

Adenovirus 13S E1a

The activation domain of 13S E1a is unique in containing a zinc-binding structure.[30,95] The E1a activation domain has been shown to interact with TBP in vitro. However, this interaction is not affected by several E1a activation domain mutations,[96] suggesting that either TBP is not the true target of E1a, or that an additional factor is required for E1a activation domain function. Indeed, one study suggests that the E1a activation domain may interact with an adapter protein.[95]

HSV ICP4

While little is known about the mechanism of transactivation by this protein, there are some very interesting findings regarding its transcriptional repression activity. As mentioned above, the promoters that are repressed by ICP4 contain ICP4 binding sites located at the transcription initiation site. In vitro studies have shown that ICP4 can inhibit the formation of transcription preinitiation complex but not transcription elongation. However, ICP4 does not affect the binding of TFIID and TFIIB, and in fact ICP4 binds cooperatively with TFIID and TFIIB to the promoter.[97] Therefore, the repression may result from a block of the preinitiation complex assembly at a step after TFIIB binding.

Stimulation of Transcription Elongation

The regulation of transcription elongation is poorly understood. The HIV-1 Tat remains the only eukaryotic activator that has been extensively studied in this regard, and will be the focus of this section.

That Tat can stimulate transcription elongation was shown by transient transfection assays soon after it was identified as the HIV-1 activator.[81,82] This in vivo finding has since been recapitulated in vitro, which has made it possible to investigate the underlying mechanisms.[98-100] It has been found that Tat can suppress transcription termination at heterologous termination sites placed downstream of the Tar element, which provided evidence that Tat can qualitatively alter transcription elongation complexes to increase transcription processivity.[101] Two lines of evidence suggest that Tat directly targets elongation complexes to stimulate elongation. First, it has been shown that Tat can stimulate transcription of preformed elongation complexes in vitro.[100] In addition, a recent study has shown that Tat can stably associate with the elongation complex.[102]

In many aspects, Tat is reminiscent of prokaryotic antitermination factor N encoded by bacteriophage λ (for a review see ref. 103). Like Tat, factor N, also an RNA binding factor, can associate with the elongation complex, and can suppress early transcription termination by increasing transcription processivity of the elongation complex. Transcription antitermination by factor N involves the association of multiple factors with the elongation complex. At the present, the factors required for Tat activated elongation have not been defined. However, two Tat-cofactors, required specifically for Tat activated transcription, have been identified.[104,105] One of these factors can bind Tat, while whether the other can also bind Tat or whether it corresponds to the Tar-binding factor implicated in molecular genetic studies (see above) is not clear. Our own studies suggest that an elongation factor can interact with Tat and is required for Tat transactivation (Li XY and Green MR, unpublished). Therefore, Tat-transactivation also seems to involve multiple factors. The finding of an elongation factor as the target of Tat is particularly interesting because it suggests that by analogy to the stimulation of transcription initiation by activators, Tat probably functions at least in part by recruiting elongation factors to the elongation complex to stimulate transcription elongation (Fig. 17.3).

Various evidence suggests that regulation of transcription elongation may involve phosphorylation of the CTD domain of RNA polymerase II large subunit. Even though CTD phosphorylation is generally associated with transcription initiation, which involves the TFIIH associated kinase, some evidence indicates that these two events can be separated. In particular, it has been found that nonprocessive elongation complexes associated with the *Drosophila* heat shock genes are underphosphorylated while actively transcribing elongation complexes are hyperphosphorylated in the cell.[106] At the present it is not clear how CTD phosphorylation is regulated during elongation even though some in vitro studies suggest that it may involve an elongation factor associated kinase sensitive to DRB inhibition.[107] It has been indicated that this kinase may be important for Tat activated transcription as well.[99]

As mentioned above, Tat can function through RNA as well as DNA. In addition, it has been found that when directed to the promoter through a heterologous DNA-binding domain, Tat can also stimulate transcription elongation. Thus, Tat seems to be functionally similar in both cases. It will be interesting to find out whether the same mechanisms are also used.

TRANSCRIPTIONAL REGULATION THROUGH INDIRECT MECHANISMS

Viruses often encode factors that do not fit the category of transactivators, but can regulate transcription. A very well studied example that we will discuss here is the adenovirus 12S E1a.

The 12S E1a can downregulate transcription from a variety of promoters under the control by different types of transactivators. (refs. 108, 109 and references therein) This general transcription repression function of the 12S E1a has been mapped to a region spanning the amino-terminal sequence and CR1, the same region also required for the cell transformation activity of E1a (Fig. 17.1). Paradoxically, even though we will refer to this region as a transcription repression region, it is actually homologous to the E2F activation domain and can activate transcription when expressed as a fusion to a heterologous DNA-binding domain.[110] The reason why this region in the 12S E1a is involved in transcription repression instead of transactivation may be due to the lack of DNA-binding ability or a promoter targeting function in the 12S E1a, as in the 13S product. Recent studies have suggested that transcription repression by the 12S E1a is correlated with its ability to interact with the cellular factors including CBP and p300 (see Fig. 17.1), two closely related proteins that function as the coactivators for a number of activators. It has been found that these interactions involve the same regions in p300 and CBP that are also required for binding by certain cellular activators, such as c-Fos, and the factors that are the functional targets of p300 and CBP, such as TFIIB and P/CAF, a histone-acetylating enzyme.[111,112] Therefore, E1a can apparently repress transcription by sequestering p300 and CBP and by inhibiting their activities. In addition, some other studies suggest that the 12S E1a may also repress transcription by interacting through the repression region, and consequently sequestering TFIID.[113]

While 12S E1a can repress transcription of many genes as discussed above, it can also upregulate some others. In contrast to the 13S E1a, which activates transcription by direct interaction with cellular activators, transcriptional activation by the 12S E1a involves indirect mechanisms. One of the mechanisms that is relatively well understood requires CR2 and a portion of the CR1 sequence, and correlates with binding of the 12S E1a with the retinoblastoma tumor suppressor gene product, pRB, and its homologues, p107 and p130 (Fig. 17.1). These proteins are known to associate with transcription factors of the E2F family at different phases of the cell cycle, leading to inhibition of their transcriptional activities. E1a, through its interaction with these cell cycle regulators, releases E2F from the complex, leading to transcriptional activation.[114] Similar transcription regulatory activity is also shared by several other viral oncogenic proteins including the papillomavirus E7, and the SV40 large T antigen.

Besides E2F, transcription induction by the 12S E1a can also be mediated by several other DNA elements and requires the E1a amino-terminal repression region described above. In the case of the *hsp70* promoter, the transcription induction is mediated by the TATA box, and involves binding of E1a to a transcription repressor Dr1 through the N-terminal sequence (Fig. 17.1).[115] For several other genes, the YY1 binding site has been shown to be responsible. YY1 can function as a transcription repressor. It has been suggested that the E1a proteins may relieve this repression by direct interaction even through a recent study suggests that direct interaction between YY1 and p300 may be important. (ref. 116 and references therein) A 12S E1a responsive element has also been found in the *c-jun* promoter, and this element is recognized by ATF2/cJun heterodimer.[117]

PERSPECTIVES

Viral activators comprise a large and diverse group of factors. Through extensive studies, many aspects concerning the structure and function of quite a few of these activators are relatively well understood, and many intriguing properties of the viral activators have been elucidated. For example, these studies have shown that many viral activators lack DNA-binding ability, and target to promoter through association with cellular activators. Most often, these activators can interact with multiple cellular activators, leading to transcription induction of multiple genes. In the cases of Tax and pX, the transcription stimulatory effect at least partly involves an increase of the DNA binding of the interacting cellular activators, which represent a novel mechanism for transcription regulation. In addition, some activators, such as HTLV-1 Tax and HBV pX, can regulate transcription through multiple mechanisms including indirect mechanisms, which also contributes to functional promiscuity for these factors. The HIV-1 Tat is a particularly fascinating factor; it not only acts by binding to RNA, it is a potent transcription elongation promoting factor even though cellular activators have also been implicated to be capable of stimulating elongation.

Despite the tremendous progress in our understanding about the functions of viral activators, many important questions remain. For example, what are the molecular bases for transcriptional promiscuity? How could a single factor interact with different types of factors, or even function through multiple mechanisms? Probably the most important and currently poorly understood aspect of viral activators is how, after binding to the promoter, the activators stimulate transcription, both at the level of initiation and elongation. This involves not only the question about which general transcription initiation/elongation factor(s) is the direct target, or whether a cofactor is involved, but also how such interaction leads to increased transcription efficiency. Undoubtedly, a lot more studies are needed before these and many other fundamental questions about transcription regulation can be understood.

References

1. Martin KJ, Green MR. Transcriptional activation by viral immediate-early proteins: variations on a common theme. In: McKnight SL, Yamamoto KR, eds. Transcriptional Regulation. Cold Spring Harbor: Cold Spring Harbor Laboratory Press, 1992: p695-725.
2. McBride AA, Romanczuk H, Howley PM. The papillomavirus E2 regulatory proteins. J Biol Chem 1991; 266: 18411-18414.
3. Ham J, Dostatni N, Gauthier JM et al. The papillomavirus E2 protein: A factor with many talents. Trends Biochem Sci 1991; 16:440-444.
4. Monini P, Grossman SR, Pepinsky B et al. Cooperative binding of the E2 protein of bovine papillomavirus to adjacent E2-responsive seqences. J Virol 1991; 65:2124-2130.
5. Knight JD, Li R, Botchan M. The activation domain of the bovine papillomavirus E2 protein mediates association of DNA-bound dimers to form DNA loops. Proc Natl Acad Sci USA 1991; 88:3204-3208
6. Thierry F, Dostatni N, Arnos F et al. Cooperative activation of transcription by bovine papillomavirus type 1 E2 can occur over a large distance. Mol Cell Biol 1990; 10:4431-4437.
7. Ham J, Dostatni N, Arnos F et al. Several different upstream promoter elements can potentiate transactivation by the BPV-1 E2 protein. EMBO J 1991; 10:2931-2940.
8. Li R, Knight JD, Jackson SP et al. Direct interaction between Sp1 and the BPV enhancer E2 protein mediates synergistic activation of transcription. Cell 1991; 65:493-505.
9. Stenlund A, Botchan M. The E2 transactivator can act as a repressor by interfering with the cellular transcription factor. Genes & Dev 1990; 4:123-136
10. Dostatni N, Lambert PF, Sousa R et al. The functional BPV-1 E2 trans-activating protein can act as a repressor by preventing formation of the initiation complex. Genes & Dev 1991; 5:1657-1671
11. Sinclair AJ, Farrell PJ. Epstein-Barr virus transcription factors. Cell Growth & Diff 1992; 3:557-563.
12. Farrell P, Rowe C, Rooney C et al. Epstein-Barr virus BZLF1 trans-activator specifically binds to consensus AP1 site and is related to c-fos. EMBO J 1989; 8:127-132
13. Chang Y, Dong D, Hayward G et al. Epstein-Barr virus Zta transactivator: a member of the bZIP family with unique DNA-binding specificity and a dimerization domain that lacks the characteristic heptad leucine sipper motif. J Virol 1990; 64:3358-3369

14. Flemington E, Speck SH. Evidence for coiled-coil dimer formation by Epstein-Barr virus trans-activator that lacks a heptad repeat of leucine residues. Proc Natl Acad Sci USA 1990; 87:9459-9463
15. Kolman JL, Taylor N, Marshak DR et al. Serine-173 of the Epstein-Barr virus ZEBRA protein is required for DNA binding and is a target for casein kinase II phosphorylation. Proc Natl Acad Sci USA. 1993; 90:10115-10119.
16. Flemington EK, Lytle JP, Cayrol C et al. DNA-binding-defective mutants of the Epstein-Barr virus lytic switch activator Zta transactivate with altered specificity. Mol Cell Biol 1994; 14:3041-3052
17. Gruffat H, Sergeant A. Characterization of the DNA-binding site repertoire for the Epstein-Barr virus transcription factor R. Nucl Acids Res 1994; 22: 1172-1178.
18. Quinlivan EB, Holley-Guthrie EA, Norris M et al. Direct BRLF1 binding is required for cooperative BZLF1/BRLF1 activation of the Epstein-Barr virus early promoter, BMRF1. Nucl Acids Res 1993; 21:1999-2007.
19. DiDonato JA, Spitzner JR, Muller MT. A predictive model for DNA recognition by the herpes simplex virus protein ICP4. J Mol Biol 1991; 219:451-470.
20. Papavassiliou AG, Wilcox KW, Silverstein SJ. The interaction of ICP4 with cell/infected-cell factors and its state of phosphorylation modulate differential recognition of leader sequence in herpes simplex virus DNA. EMBO J 1991; 10: 397-406.
21. Imbalzano AN, Shepard AA, DeLuca NA. Functional relevance of specific interactions between herpes simplex virus type 1 ICP4 and sequences from the promoter-regulatory domain of the viral thymidine kinase gene. J Virol 1990; 64: 2620-2631.
22. Roberts MS, Boundy A, O'Hare P et al. Direct correlation between a negative autoregulatory response element at the cap site of the herpes simplex virus type 1 IE175(a4) promoter and a specific binding site for the IE175(ICP4) protein. J Virol 1988; 62:4307-4320.
23. Saurin AJ, Borden KLB, Boddy MN et al. Does this have a familiar RING? Trends Biochem Sci 1996; 21:197-235.
24. Everett RD, Maul GG HSV-1 IE protein Vmw110 causes redistribution of PML. EMBO J 1994; 13:5062-5069.
25. Rice SA, Long MC, Lam V et al. Herpes simplex virus immediate-early protein ICP22 is required for viral modification of host RNA polymerase II and establishment of the normal viral transcription program. J Virol 1995; 69:5550-5559.
26. Hardy WR, Sandri-Goldin RM. Herpes simplex virus inhibits host cell splicing, and regulatory protein ICP27 is required for this effect. J Virol 1994; 68: 7790-7799.
27. Brown CR, Nakamura MS, Mosca JD et al. Herpes simplex virus trans-regulatory protein ICP27 stabilizes and binds to 3' ends of labile mRNA. J Virol 1995; 69:7187-7195.
28. Moran E, Mathews MB. Multiple functional domains in the adenovirus E1a gene. Cell 1987; 48:177-178.
29. Flint J, Shenk T. Adenovirus E1a protein paradigm viral transactivator. Annu Re Genet 1989; 23:141-161.
30. Lillie JW, Green MR. Transcriptional activation by the adenovirus E1a protein. Nature 1989; 338:39-44.
31. Green MR. Activities of adenovirus E1a proteins. In: Brugge et al, eds. Origins of human cancer: a comprehensive review. Cold Spring Harbor: Cold Spring Harbor Laboratory Press, 1991: p383-392.
32. Hai T, Liu F, Coukos WJ et al. Transcription factor ATF cDNA clones: An extensive family of leucine zipper proteins able to selectively form DNA-binding heterodimers. Genes & Dev 1989; 3:2083.
33. Liu F, Green MR. A specific member of the ATF transcription factor family can mediate transcription activation by the adenovirus E1a protein. Cell 1990; 61: 1217-1224.
34. Chatton B, Bocco JL, Gaire M et al. Transcriptional activation by the adenovirus larger E1a product is mediated by members of the cellular transcription factor ATF family which can directly associate with E1a. Mol Cell Biol 1993; 13:561-570.

35. Liu F, Green MR. Promoter targeting by adenovirus E1a through interaction with different cellular DNA-binding domains. Nature 1994; 368:520-525.
36. Li X-Y, Green MR. Intramolecular inhibition of activating transcription factor-2 function by its DNA-binding domain. Genes & Dev 1996; 10:517-527.
37. Scholer HR, Ciesiolka T, Gruss P. A nexus between Oct4 and E1a: implication for gene regulation in embryonic stem cells. Cell 1991; 66:291-304.
38. Berk AJ. Adenovirus E1a trans-activation: Understanding it will require learning how the general transcription factors function. In: McKnight SL, Yamamoto KR, eds. Transcriptional Regulation. Cold Spring Harbor: Cold Spring Harbor Laboratory Press, 1992: p695-725.
39. Hoeffler WK, Kovelman R, Roeder RG. Activation of transcription factor IIIC by adenovirus E1a protein. Cell 1988; 53: 907-920.
40. Sinn E, Wang Z, Kovelman R, Roeder RG. Cloning and characterization of a TFIIIC2 subunit (TFIIICb) whose presence correlates with activation of RNA polymerase-mediated transcription by adenovirus E1a expression and serum factors. Genes Dev 1995; 9:675-685.
41. Goding C, O'Hare P. Herpes simplex virus Vmw65-octamer-binding protein interaction: A paradigm for combinatorial control of transcription. Virol 1989; 173:363-367.
42. Thompson CC, McKnight CL. Anatomy of an enhancer. Trends Genet 1992; 8:232-236.
43. Triezenberg SJ, Kingsbury RC, McKnight SL. Functional dissection of VP16, the trans-activator of Herpes simplex virus immediate early gene expression. Genes & Dev 1988; 2:718-729.
44. Kristie TM, LeBowitz JH, Sharp PA. The octamer-binding proteins form multiprotein-DNA complexes with the HSV a-TIF regulatory protein. EMBO J 1989; 8:4229-4238.
45. Stern S, Herr W. The Oct1 homeodomain directs formation of a multiprotein-DNA complex with the HSVtransactivator VP16. Nature 1989; 341:624-630.
46. Pomerantz JL, Kristie TM, Sharp PA. Recognition of the surface of a homeo domain protein. Genes & Dev 1992; 6:2047-2057.
47. Lai JS, Cleary MA, Herr W. A single amino acid exchange transfers VP16-induced positive control from the Oct-1 to the Oct-2 homeo domain. Genes & Dev 1992; 6:2058-2065.
48. Walker S, Hayes S, O'Hare P. Site-specific conformational alteration of the Oct-1 POU domain-DNA complex as the basis for differential recognition by Vmw 65(VP16). Cell 1994; 79:841-852.
49. Kristie TM, Sharp PA. Interactions of the Oct-1 POU subdomains with specific DNA sequences and the HSV[a]-trans-activator protein. Genes & Dev 1990; 4:2383-2396.
50. Stern S, Herr W. The herpes simplex virus transactivator VP16 recognizes the Oct-1 homeo domain: Evidence for a homeo domain recognition subdomain. Genes & Dev 1991; 5:2555-2566.
51. Wilson AC, Lamarco K, Peterson MG et al. The VP16 accessory protein HCF is a family of polypeptides processed from a large precursor protein. Cell 1993; 74: 115-125.
52. Cleary MA, Stern S, Tanaka M et al. Differential positive control by Oct-1 and Oct-2: Activation of a transcriptionally silent motif through Oct-1 and VP16 corecruitment. Genes & Dev 1993; 7: 72-83.
53. Smith MR, Greene WC. Identification of the HTLV-I Tax transactivator mutants exhibiting novel transcriptional phenotypes. Genes Dev 1990; 4:1875-1885.
54. Smith MR, Greene WC. Type I human T-cell leukemia virus Tax protein transforms Rat fibroblasts through the cyclic adenosine monophosphate response element binding protein/activating transcription factor pathway. J Clin Invest 1991; 88:1038-1042.
55. Zhao LJ, Giam C-Z. Human T-cell lymphotrophic virus type I (HTLV-I) transcriptional activator, Tax, enhances CREB binding to the HTLV-I 21-basepair repeats by protein-protein interaction. Proc Natl Acad Sci USA 1992; 89:7070-7074.

56. Wagner S, Green MR. HTLV-I Tax protein stimulation of DNA binding of bZIP proteins by enhancing dimerization. Science 1993; 252:395-399.
57. Armstrong AP, Franklin AA, Uittenbogaard MN et al. Pleiotropic effect of the human T-cell leukemia virus Tax protein on the DNA binding activity of eukaryotic transcription factors. Proc Natl Acad Sci USA 1993; 90:7303-7307.
58. Perini G, Wagner S, Green MR. Recognition of bZIP proteins by the human T-cell leukemia virus transactivator Tax. Nature 1995; 376:602-605.
59. Baranger AM, Palmer RC, Hamm MK et al. Mechanism of DNA-binding enhancement by the human T-cell leukemia virus transactivator Tax. Nature 1995; 376:606-608.
60. Sun S-C, Elwood J, Beraud C et al. Human T-cell leukemia virus type I Tax activation of NF-κB/rel involves phosphorylation and degradation of the Iκ-Ba and relA (p65)-mediated induction of the c-rel gene. Mol Cell Biol 1994; 14:7377-7384.
61. Rousset R, Desbois C, Bantignies F et al. Effects on NF-κB1/p105 processing of the interaction between the HTLV-I transactivator Tax and the proteosome. Nature 1996; 381:328-331.
62. Fujii M, Tsuchiya H, Chuhjo T et al. Serum response factor has functional roles both in the indrect binding to the CArG box and in the transcriptional activation function of the human T-cell leukemia virus type I. J Virol 1994; 68:7275-7283.
63. Benedict Yen TS. Hepatitis B viral X protein: Review of recent progress. J Biomed Sci 1996; 3:20-25.
64. Faktor O, Shaul Y. The identification of the hepatitis B virus X gene responsive elements reveals functional similarity of X and HTLV-I Tax. Oncogene 1990; 5:867-872.
65. Kekule AS, Lauer U, Weiss L et al. Hepatitis B virus transactivator HBx uses a tumor promoter signaling pathway. Nature 1993; 361:742-745.
66. Mahe Y, Mukaida N, Kuno K et al. Hepatitis B virus X protein transactivates human interleukin-8 gene through acting on nuclear factor κB and CCAAT/enhancer binding protein-like cis-elements. J Biol Chem 1991; 266:17759-17763.
67. Natoli G, Avantaggiati ML, Chirillo P et al. Induction of the DNA-binding activity of the c-jun/c-fos heterodimers by the hepatitis B virus transactivator pX. Mol Cell Biol 1994; 14:989-998.
68. Maguire HF, Hoeffler JP, Siddiqui A. HBV X protein alters the DNA binding specificity of CREB and ATF-2 by protein-protein interactions. Science 1991; 252:842-844.
69. Williams JS, Andrisani OM. The hepatitis B virus X protein targets the basic region-leucine zipper domain of CREB. Proc Natl Acad Sci USA 1995; 92:3819-3823.
70. Jones KA, Peterlin BM. Control of RNA initiation and elongation at the HIV-1 promoter. Annu Rev Bochem 1994; 63:717-743.
71. Selby MJ, Peterlin BM. Trans-activation by HIV-1 Tat via a heterologous RNA binding protein. Cell 1990; 62:769-776.
72. Southgate C, Zapp ML, Green MR. Activation of transcription by HIV-1 Tat protein tethered to nascent RNA through another protein. Nature 1990; 345:640-642.
73. Alonso A, Derse D, Peterlin BM. Human chromosome 12 is required for optimal interactions between Tat and Tar of human immunodeficiency virus type 1 in rodent cells. J Virol 1992; 66:4617-4621.
74. Madore SJ, Cullen BR. Genetic analysis of the cofactor requirement for human immunodeficiency virus type 1 Tat function. J Virol 1993; 67:3703-3711.
75. Alonso A, Cujec TP, Peterlin BM. Effects of human chromosome 12 on interactions between Tat and Tar of human immunodeficiency virus type 1. J Virol 1994; 68:6505-6513
76. Berkout B, Gatignol A, Rabson AB et al. Tar-independent activation of the HIV-1 LTR: evidence that Tat requires specific regions of the promoter. Cell 1990; 62:757-767.
77. Southgate CD, Green MR. The HIV-1 Tat protein activates transcription from an upstream DNA-binding site: Implica-

tions for Tat function. Genes & Dev 1991; 5:2496-2507.
78. Tiley LS, Madore SJ, Malim MH et al. The VP16 transcription activation domain is functional when targeted to a promoter-proximal RNA sequence. Genes & Dev 1992; 6:2077-2087.
79. Bentley D. Regulation of transcriptional elongation by RNA polymerase II. Curr Op Genet Dev 1995; 5:210-216.
80. Blau J, Xiao H, McCracken S et al. Three functional classes of transcriptional activation domains. Mol Cell Biol 1996; 16:2044-2055.
81. Kao SY, Calman AF, Luciw PA et al. Antitermination of transcription within the long terminal repeat of HIV by the tat gene product. Nature 1987; 330: 489-493.
82. Laspia MF, Rice AP, Mathews MB. HIV-1 Tat protein increases transcriptional initiation and stablizes elongation. Cell 1989; 59:283-292.
83. Mitchell PJ, Tjian R. Transcriptional regulation in mammalian cells by sequence-specific DNA binding proteins. Science 1989; 245:371-378.
84. Stringer KF, Ingles CJ, Greenblatt J. Direct and selective binding of an acidic transcriptional activation domain to the TATA-box factor TFIID. Nature 1990; 345:783.
85. Lin YS, Green MR. Mechanism of action of an acidic transcriptional activator in vitro. Cell 1991; 971-981.
86. Lin YS, Ha I, Maldonado et al. Binding of general transcription factor TFIIB to an acidic activating region. Nature 1991; 353:569-571.
87. Tansey WP, Herr W. The ability to associate with activation domains in vitro is not required for the TATA box-binding protein to support activated transcription in vivo. Proc Natl Acad Sci USA 1995; 92:10550-10554.
88. Choy B, Green MR. Eukaryotic activators function during multiple steps of preinitiation complex assembly. Nature 1993; 366:5531-536.
89. Roberts SGE, Ha I, Maldonado E et al. Interaction between an acidic activator and transcription factor TFIIB is required for transcriptional activation. Nature 1993; 363:741-744.
90. Roberts SGE, Green MR. Activator-induced conformational change in general transcription factor TFIIB. Nature 1994; 371:717-720.
91. Xiao H, Pearson A, Coulombe B et al. Binding of basal transcription factor TFIIH to the acidic activation domains of VP16 and p53. Mol Cell Biol 1994; 14:7013-7024.
92. Goodrich JA, Hoey T, Thut CJ et al. *Drosophila* TAF(II)40 interacts with both a VP16 activation domain and the basal transcription factor TFIIB. Cell 1993; 75:519-530.
93. Lieberman PM, Berk AJ. A mechanism for TAFs in transcriptional activation: activation domain enhancement of TFIID-TFIIA-promoter DNA complex formation. Genes & Dev 1994; 8: 995-1006.
94. Chi T, Lieberman P, Ellwood K et al. A general mechanism for transcriptional synergy by eukaryotic activators. Nature 1995; 377:254-257.
95. Martin KJ, Lillie JW, Green MR. Evidence for interaction of different eukaryotic transcriptional activators with distinct cellular targets. Nature 1990; 346: 147-152.
96. Geisberg JV, Lee WS, Berk AJ et al. The zinc finger region of the adenovirus E1a transactivating domain complexes with the TATA box binding protein. Proc Natl Acad Sci USA 1994; 91:2488-2492.
97. Smith CA, Bates P, Rivera-Gonzalez R et al. ICP4, the major regulatory protein of herpes simplex virus type 1, forms a tripartite complex with TATA-binding protein and TFIIB. J Virol 1993; 67: 4676-4687.
98. Marciniak RA, Calnan BJ, Frankel AD et al. HIV-1 Tat protein transactivates transcription in vitro. Cell 1990; 63:791-802.
99. Marciniak RA, Sharp PA. HIV-1 Tat protein promotes formation of more processive elongation complexes. EMBO J 1991; 10: 4189-4196.
100. Kato H, Sumimoto H, Pognonec P et al. HIV-1 Tat acts as a processivity factor in vitro in conjunction with cellular elongation factor. Genes & Dev 1992; 6:655-666.

101. Graeble MA, Churcher MJ, Lowe AD et al. Human immunodeficiency virus type 1 transactivator protein, tat, stimulates transcriptional read-through of distal terminator sequences in vitro. Proc Natl Acad Sci USA 1993; 90:6148-6188.
102. Keen NJ, Gait MJ, Karn J. Human immunodeficiency virus type-1 Tat is an integral component of the activated transcription-elongation complex. Proc Natl Acad Sci USA 1996; 93:2505-2510.
103. Greenblatt J, Nodwell JR, Mason SW. Transcriptional antitermination. Nature 1993; 364:410-406.
104. Zhou Q, Sharp PA. Novel mechanism and factor for regulation by HIV-1 Tat. EMBO J 1995; 14:321-328.
105. Suñé C, García-Blanco MA. Transcriptional trans activation by human immunodeficiency virus type 1 Tat requires specific coactivators that are not basal factors. J Virol 1995; 69:3098-3107.
106. Marshall NF, Price DH. Control of formation of two distinct classes of RNA polymerase II elongation complexes. Mol Cell Biol 1992; 12:2078-2090.
107. O'Brien T, Hardin S, Greenleaf A et al. Phosphorylation of RNA polymerase II C-terminal domain and transcriptional elongation. Nature 1994; 370:75-77.
108. Rochette-Egly C, Fromental C, Chambon P. General repression of enhanson activity by the adenovirus-2 E1a proteins. Genes Dev 1990; 4:137-150.
109. Braun T, Bober E, Arnold HH. Inhibition of muscle differentiation by the adenovirus E1a protein: repression of the transcriptional activating function of the HLH protein Myf-5. Genes Dev 1992; 6:888-902.
110. Trouche D, Kouzarides T. E2F1 and $E1a_{12S}$ have a homologous activation domain regulated by RB and CBP. Proc Natl Acad Sci USA 1996; 93:1439-1442.
111. Janknecht R, Hunter T. Versatile molecular glue. Curr Biol 1996; 6:951-954.
112. Yang X-J, Ogryzko VV, Nishikawa J-I et al. A p300/CBP-associated factor that competes with the adenoviral oncoprotein E1a. Nature 1996; 382: 319-324.
113. Song CZ, Loewenstein PM, Toth K et al. Transcription factor TFIID is a direct functional target of the adenovirus E1a transcription-repression domain. Proc Natl Acad Sci USA 1995; 92: 10330-10333.
114. Cress WD, Nevins JR. Use of the E2F transcription factor by DNA tumor virus regulatory proteins. Curr Topics Microbiol Immunol 1996; 208:63-78
115. Kraus VB, Inostroza JA, Yeung K et al. Interaction of the Dr1 inhibitory factor with the TATA binding protein is disrupted by adenovirus E1a. Proc Natl Acad Sci USA 1994; 91:6279-6282.
116. Lee JS, Galvin KM, See RH et al. Relief of YY1 transcriptional repression by adenovirus E1a is mediated by E1a-associated protein p300. Genes & Dev 1995; 9:1188-1198.
117. Hagmeyer BM, König H, Herr I et al. Adenovirus E1a negatively and positively modulates transcription of AP-1 dependent genes by dimer-specific regulation of the DNA binding and transactivation activities of Jun. EMBO J 1993; 12: 3559-3572.

CHAPTER 18

DNA-Transcription Factor Interactions in the Context of Chromatin

Peter B. Becker

INTRODUCTION

The packaging of DNA into chromatin in the eukaryote nucleus establishes a repressed "ground state" of gene expression, above which selective activation of genes occurs according to elaborate regulatory programs. In general, active genes reside in less condensed, more accessible chromatin domains. Whether a particular domain is repressive or permissive is presumably largely dictated by the higher order folding of the nucleosomal fiber and the compartmentalization of chromatin domains in the nucleus, two aspects of nuclear architecture that are poorly understood to date. Correlations between gene activity and the presence of specific nonhistone proteins and histone modifications reflect structural differences between these functional domains yet to be unraveled.[1-3] Within a permissive domain, regulatory sites such as promoters or enhancers containing binding sites for sequence-specific DNA-binding transcription factors can be identified as discontinuities in the regular nucleosomal array. Their accessibility to nonspecific nucleases such as DNase I is strikingly increased, indicating that in general, the interaction of transcription factors with their target sequences correlates with serious perturbations of chromatin structure.[4]

Transcription control elements in chromatin have been classified as "preset" or requiring "remodeling", depending on whether they are constitutively accessible in chromatin or whether chromatin is opened up quickly in response to an inducer of transcription.[5] According to an attractive hypothesis the "presetting" of a promoter, i.e., the constitutive binding of transcription factors which prevents the formation of a closed chromatin configuration, occurs directly after the passage of the replication fork when chromatin is transiently disrupted. A delay between DNA replication and nucleosome assembly could provide a window of opportunity within which transcription factors might compete successfully for their binding sites. Evidence for this "pre-emptive" establishment of accessibility comes mainly

Transcription Factors in Eukaryotes, edited by Athanasios G. Papavassiliou.

from in vitro experiments in which the binding of transcription factors prior to chromatin assembly keeps a promoter active.[6] Whether this mechanism operates in vivo is unclear since there are some indications that regular nucleosomal arrays are formed immediately after replication prior to the disruption of particular nucleosomes covering regulatory sites.[7,8] "Remodeling" of nucleosomes by dedicated transcription factors may therefore be the prevailing mechanism to establish accessible promoters.

Transcription factor access to chromatin is an early step in gene expression, which provides a unique opportunity for regulation. Current ideas suggest that some types of factors are responsible for the initial opening up of chromatin to establish strategic patches of accessibility, while others rely on these pioneering actions to approach their target sequences. The action of the glucocorticoid receptor (GR) serves as a paradigm for transcription activation-associated chromatin perturbations which are likely to reflect profound alterations in nucleosome structure and/or positioning.[9] The fact that fast and reversible cases of "chromatin remodeling" in the absence of DNA replication have been described[10,11] point to a dynamic equilibrium of nucleosomes with less restrictive structures. The nature of the structural alterations and the mechanisms at work are difficult to derive from descriptive in vivo experiments. Recently, biochemical analyses of the phenomenon, some of which are described below, have led to important insights.

BINDING OF TRANSCRIPTION FACTORS TO NUCLEOSOMES

The mechanisms envisioned by which transcription factors may gain access to their binding sites in chromatin can be divided into two principal categories.[12] The first category subsumes models assuming that transcription factors never have to actually contact their binding sites on the surface of a nucleosome, due to dynamic properties of nucleosomes which lead to occasional and transient exposure of DNA to incoming factors. The second kind of model requires that a transcription factor recognizes its target site on a nucleosome at least transiently in order to initiate a "remodeling" reaction. A priori, DNA bent over the surface of a nucleosome cannot easily be bound by many transcription factors: much of the DNA is sterically not accessible due to its association with the histone octamer or the close proximity of the two gyres of DNA.[13] Due to its tight bend, nucleosomal DNA deviates significantly from B-form DNA and this may also explain the experimental observation that in general transcription factors bind nucleosomal DNA in vitro only with reduced affinity as compared to free DNA.[14] There are, however, notable exceptions to this rule, the most prominent being the above-mentioned glucocorticoid receptor (GR) which binds nucleosomal DNA in stable triple complexes in vitro.[15,16] In vitro analyses have identified a number of parameters that affect the affinity with which a nucleosomal binding site can be contacted by a transcription factor. Whether some transcription factors bind DNA on the surface of classical nucleosomes in vivo is not known because it is difficult to distinguish transcription factor-bound nucleosomes from DNA wrapped around protein complexes that consist entirely of transcription factors.[17-19]

Obviously the binding of a factor to a nucleosome will be influenced by the orientation of the binding site on the histone octamer, i.e., whether the recognition sequence will be facing in or out (rotational positioning, Fig. 18.1A).[20] Therefore, the efficient targeting of a nucleosome by a transcription factor in vivo may require that its recognition sequence is exposed on the surface in the appropriate rotational configuration. This could be achieved by either of two strategies: by tight and precise positioning of the nucleosome with respect to the underlying DNA sequence or, quite to the contrary, by flexibility in positioning and nucleosome movements that assure that the binding site is constantly moved over the nucleosome surface. In the latter scenario the "optimal" display of any sequence element will occur at least transiently without strict

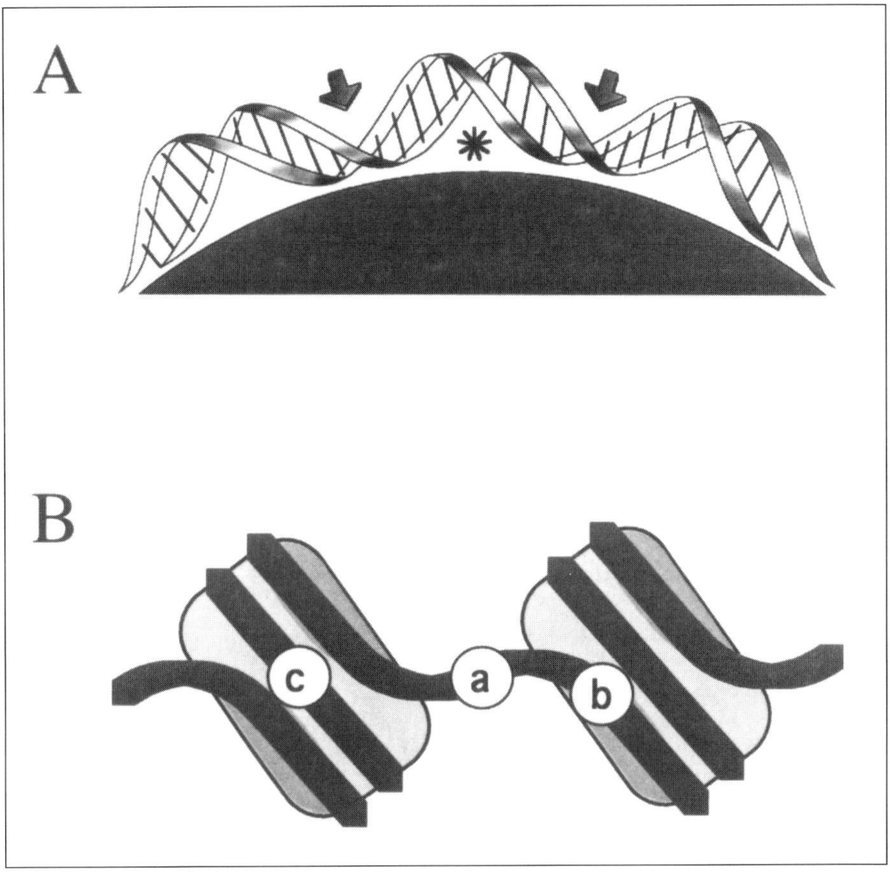

Fig. 18.1. The accessibility of nucleosomal DNA depends on the positioning of the histone octamer with respect to the underlying DNA sequence.
A. Rotational positioning. The arrows point to major grooves that are accessible from outside, the asterisk indicates inaccessible major groove sequences pointing towards the histones.
B. Translational positioning. a: Linker DNA between two neighboring nucleosomes is accessible. b: DNA at the entry/exit points is less restricted. c: DNA at the pseudodyad axis of the nucleosome is least accessible.

requirements for positioning, and transcription factors would be able to select the appropriate configuration from a series of positions. The concept of tight positioning of nucleosomes in vivo has recently been challenged by the observation of multiple, sequence-biased positions of nucleosome B in the mouse mammary tumor virus LTR, which had long served as a paradigm for tight positioning.[21] On the other hand short distance "diffusion" of DNA over the surface of a nucleosome has been well documented in vitro[22] and requires only little energy.[23] In addition, activities that increase the mobility of nucleosomes at the expense of energy have been found in chromatin reconstituted under physiological conditions in extracts from *Drosophila* embryos.[24]

The translational positioning, i.e., how far into the nucleosome a binding site lies, is also important.[25,26] As a rule of thumb, DNA at the nucleosomal entry and exit points (b in Fig. 18.1B) is much less tightly associated with histones than DNA at the nucleosomal dyad (c in Fig. 18.1B) and can therefore be more easily recognized by factors. The accessibility of sequences at the entry and exit points and in the linker DNA between two adjacent nucleosomes (a in Fig. 18.1B) will dependent on the presence or absence of linker binding proteins such as histone H1.[27] Linker histones enhance the wrapping of the DNA at the nucleosomal entry and exit points and interact with the linker DNA via their C-terminal tails.[28] The association of histone H1 prevents the interaction of transcription factors with their binding sites on nucleosomes in model experiments.[29,30] There are indications that histone H1 may be absent from DNase I-hypersensitive promoters[31]. Once a protein has bound nucleosomal DNA at a loose end it may facilitate the binding of other proteins closer to the dyad axis, effectively "invading" a nucleosome.[26] The binding of unrelated transcription factors to nucleosomal DNA is inherently cooperative: the simultaneous approach of two or three factors to the same nucleosome leads to a significant occupancy by both proteins even if each protein individually cannot interact significantly.[32] Based on theoretical considerations Polach and Widom have proposed a model that explains why two or more transcription factors that do not interact directly can nevertheless synergize by facilitating each other's binding to nucleosomal DNA.[33] The model assumes that nucleosomes are intrinsically dynamic, "breathing" structures in which patches of DNA detach transiently from the histone octamer in low salt.[34] The proposed accessibility of mononucleosomal DNA can be monitored by cleavage with restriction enzymes. Restriction enzymes have also been used to probe for general accessibility of more complex chromatin reconstituted under physiological conditions in *Drosophila* embryo extracts.[24] Interestingly, we observed that in this system a state of heightened accessibility is maintained at the expense of ATP-hydrolysis by an assembly of proteins termed "chromatin accessibility complex" (CHRAC; P.D. Varga-Weisz and P.B. Becker, unpublished observations). It is unclear at present whether the increased access and the energy-dependent nucleosomal mobility mentioned above are mechanistically related and whether the restriction sites cleaved in the presence of CHRAC are located on nucleosomal DNA or in nucleosomal linkers.[24] In any case, these results testify to the dynamic properties of nucleosomal arrays and indicate that the general accessibility of chromatin can be modulated by factors at the expense of energy.

The accessibility of nucleosomal DNA is also significantly restricted by interactions with the positively charged N-terminal histone domains (the histone "tails") that protrude from the otherwise compact histone octamer particle. The neutralization of these charges through acetylation of the tail lysines will weaken these interactions and lead to increased access by transcription factors.[35,26] The acetylation of the N-terminus of histone H4 appears to stimulate the binding of transcription factors particularly well when compared to corresponding modifications of H3.[36] Since H4 acetylation enhances the binding of factors at multiple translational positions on the nucleosome Workman and coworker suggest that the modification not only affects the accessibility of sequences at the specific contact sites but also more generally by changing the conformation of the nucleosome.[36] In vivo, increased acetylation characterizes active chromatin domains[64] and histone hyperacetylation relieves the nucleosome-mediated repression of a class of promoters that are well represented by the HIV-1 promoter.[37]

NUCLEOSOME DISASSEMBLY TRIGGERED BY TRANSCRIPTION FACTORS

Once transcription factors have contacted their nucleosomal binding site, the resulting ternary complex may be stable under the particular conditions of an experiment as in the case of the glucocorticoid receptor, or rather meta-stable, dissociating either into a nucleosome and free transcription factor, or into factor-bound DNA and displaced histones. The above-mentioned synergy of transcription factors for nucleosome binding also increases the likelihood for nucleosome disruption or distortions which lead to characteristic DNase I-hypersensitivity in nuclei.[70] Up to now, not a single case has been described where a transcription factor was able to disassemble a nucleosome on its own. Rather, a variety of accessory factors have been shown to facilitate the stepwise disassembly of nucleosomes. The removal of histones from DNA is promoted by free nucleic acids such as RNA[38] or competitor DNA that serve as a histone acceptor or "sink" during the reaction.[39] Beyond this nonspecific effect of charge neutralization, specific histone-binding proteins also encourage nucleosome disassembly by serving as acceptors for histones.[40,41] In the presence of a H2A/H2B binding protein, nucleoplasmin, a transcription factor, was able to trigger the transfer of H2A/H2B dimers from the histone octamer to the acceptor protein.[40] The remaining H3/H4 tetramer organizes only 80-100 bp of DNA and appears intrinsically less restricting.[42,35] It is possible that in vivo nucleosomes at regulatory sites are not always completely disassembled, but that H3/H4 tetramers or other subnucleosomal particles represent a "remodeled" or open structure of increased accessibility. In different experimental approaches evidence for a complete absence of nucleosomes at a promoter[43] or the persistence of a somehow altered nucleosome[18] have been obtained. In the biochemical assay of Workman and colleagues, however, H3/H4 tetramers were readily displaced if free competitor DNA was present[40] and hence the concerted action of a transcription factor with some ability to recognize binding sites on the nucleosomal surface, H2A/H2B binding proteins and a negatively charged acceptor molecule resulted in the complete disassembly of the targeted nucleosome (summarized in Fig. 18.2). Nucleosome disassembly under those conditions is stepwise, since it is prevented by cross-linking of the histones in the octamer.[41]

The fact that proteins that interact with H2A/H2B with some specificity were more potent initiators of nucleosome disassembly than simple polyanions such as polyglutamic acid[40] indicates that there is more to their action than just charge neutralization. H2A/H2B binding proteins such as nucleoplasmin and NAP1 that function in this assay have originally been described as nucleosome assembly factors that promote the coordinated assembly into nucleosomes by keeping histones in solution.[44,45] It is, therefore, tempting to speculate that the pathway that leads to nucleosome disassembly in vivo may simply be a reversal of the nucleosome assembly process. Nucleosome assembly may be in equilibrium with nucleosome disassembly, and dedicated transcription factors may shift this balance towards disassembly. Subnucleosomal particles with reduced content of H2A/H2B in vivo have been observed not only after the replication fork as an intermediate of octamer assembly,[42] but also in actively transcribed chromatin, presumably resulting from a disassembly reaction.[46]

DIRECT INTERACTION OF TRANSCRIPTION FACTORS WITH HISTONES

It is likely that transcription factors which contact DNA on the surface of a nucleosome also get in contact with the histones themselves. Such interactions could either promote the formation of a stable ternary complex or contribute to the destabilization of the nucleosome. Indirect evidence for interactions between transcription regulators and the histone N-termini have been obtained in the Grunstein laboratory. Removal of the H3 "tails" by genetic trickery in yeast derepresses the UAS-dependent transcription of genes involved in galactose metabolism, as expected under the assumption that the N-termini restrict the accessibility of incoming proteins to chromatin.[47] On the other hand, deletion of N-terminal amino acids from histone H4 results in a reduction of transcription by an order of magnitude[48] and in changes in accessibility around the TATA element.[49] These results foster the hypothesis that histone tails may contain information required for the targeted activation of promoters, perhaps via

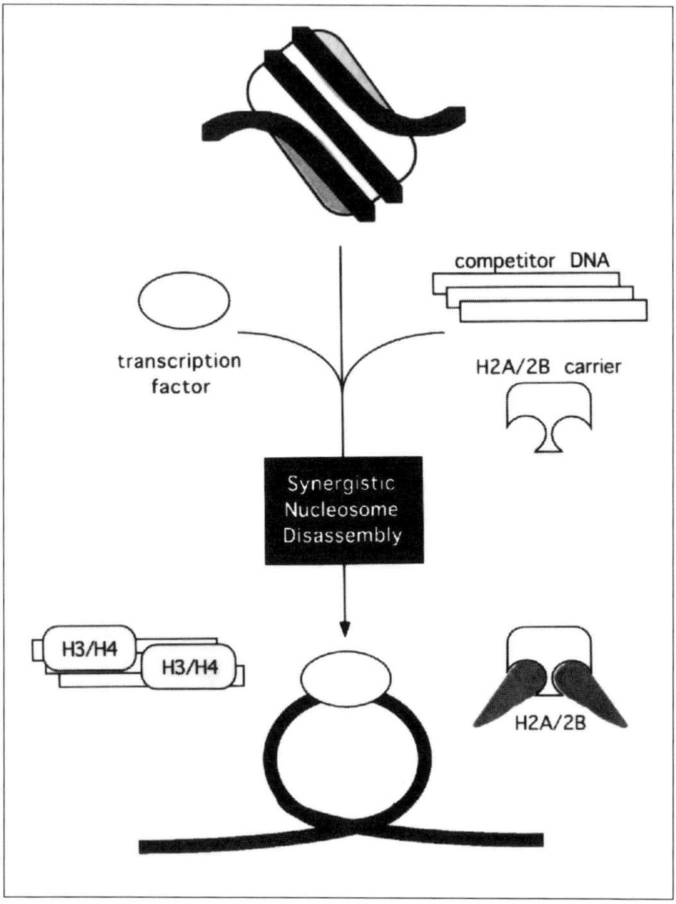

Fig. 18.2. A nucleosome can be disassembled through the concerted action of a sequence-specific transcription factor, a histone H2A/2B-binding protein ("carrier") and free competitor DNA. Details of the underlying mechanism are unclear, hence the black box. The experiment is summarized in the text and described in detail in references 40 and 41.

interaction with dedicated activator proteins. Interactions between transcription factors and the globular part of a histone in an octamer have recently been demonstrated between CTF-1/NF-1 and histone H3.[50] The tight correlation between the ability of the CTF-1 trans-activation domain to mediate induction by TGF-β and the interactions of this domain with histone H3 suggest that the mechanism of transcriptional activation by CTF-1/NF-1 may involve a direct contact with a repressive nucleosome.

NUCLEOSOME REMODELING MACHINES

As discussed above, the movement of nucleosomes, their disassembly or the modification of their histones may all increase the accessibility of DNA, thereby facilitating the interaction of a transcription factor with chromatin. It is likely that these principles contribute to the changes that lead to promoter activation which, in the absence of precise knowledge about the underlying processes, are often summarized as "nucleosome remodeling". The finding of Workman and coworkers[40] that multiple factors have to cooperate in a concerted action to disassemble a nucleosome prompts the question of how this coordination is achieved in vivo. Obviously, all the components required have to be concentrated at the site of action at the right time. A possible answer to this question may be the recent description of large multimeric protein complexes, so called "remodeling machines", that apparently contain all the components required to change the accessibility of nucleosomal DNA to transcription factors. The paradigm of such a molecular machine is the SWI/SNF ("Switch/Snif") complex, a protein assembly 20 times larger than a nucleosome, which was initially identified in yeast but later also in higher eukaryotes.[51] In vivo, SWI/SNF proteins are required as general coactivators for a number of transcription factors. Their action at the *SUC2* promoter, which was studied in some detail, correlates with an altered, accessible chromatin structure.[52] An important component of the SWI/SNF complex is the SWI2/SNF2 protein, a DNA-stimulated ATPase required for transcriptional activation.[53] Purified SWI/SNF complex from either yeast or human cells is able to facilitate the binding of a variety of transcription factors to nucleosomal DNA in a reaction that, remarkably, requires ATP hydrolysis by SWI2/SNF2.[54-56] While the mechanism underlying this action is unclear at present there is evidence for contacts of SWI/SNF complex with both, histones and DNA. Interaction of the SWI/SNF complex with histones is suggested by genetic analyses in yeast showing that *swi/snf* mutants can be suppressed by mutations in the structured parts of histones H3 and H4.[57] The DNA-binding properties of the complex were again revealed by biochemical studies on the purified complex.[58] The complex essentially binds DNA like an HMG-box protein, with preference for A/T-rich sequences and structured DNA, exemplified by four-way-junction DNA. Whether the observed changes in DNA topology that correlate with binding of the complex to DNA[58] are crucial to its action remains to be seen.

The SWI/SNF complex is not the only known chromatin remodeling machine and chances are that a whole battery of them exist that may differ in target specificity and perhaps mechanism of action. Recently, Wu and coworkers identified a new activity in a physiological chromatin reconstitution system derived from *Drosophila* embryos.[59] This nucleosome remodeling factor (NURF) utilizes energy in the form of ATP to alter nucleosomes such that transcription factors can efficiently bind within chromatin.[59,60] The patterns of nuclease digestion that indicate "remodeling" are still compatible with either the disruption of nucleosomes over the target site or simply their moving away from it.[60,24] NURF is, again, a molecular machine by the criteria that it is made of several proteins and that it consumes energy in order to achieve its goals.[61] The ATPase driving NURF is ISWI, a protein previously isolated on the basis of its sequence similarity to the SWI2 protein.[62,63] In interesting contrast to the SWI2 ATPase, the ISWI protein in NURF is not stimulated by free DNA nor isolated histones, but rather by DNA wrapped around a histone octamer.[63]

TARGETING OF REMODELING MACHINES

Changes in chromatin accessibility in a nucleus are likely to occur at different levels, presumably governed by different principles and mechanisms. The DNase I sensitivity that characterizes active genes or those that can potentially be activated appears to be a domain-wide phenomenon,[64] which may require the action of global factors. Within these "unfolded" domains, nucleosomes covering regulatory elements containing binding sites for specific transcription factors are remodeled, presumably by dedicated machines. How is this selectivity in chromatin remodeling achieved? The easy answer is, of course, that the specificity is determined by the sequence recognition of the respective transcription factors involved. Transcription factors may simply take advantage of ongoing nucleosome dynamics to trap their transiently accessible binding site (Fig. 18.3 B). The dynamic nature of

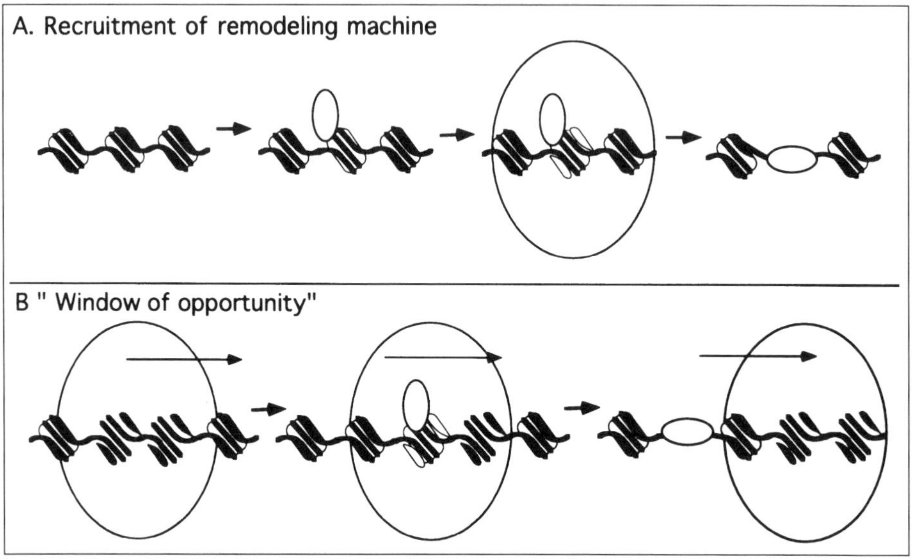

Fig. 18.3. Two models for chromatin remodeling.
A. Recruitment of a remodeling machine. The events are initiated by the interaction of a transcription factor (small ellipsoid) with its binding site which renders the nucleosome susceptible for "remodeling" by an assembly of proteins constituting a "remodeling machine" (large ellipsoid). The remodeling machine could be either ubiquitously present throughout chromatin or targeted by the sequence-specific factor to its site of action.
B. "Window of opportunity". This model assumes that chromatin is maintained in a dynamic state by a ubiquitous activity that disrupts nucleosomes transiently. A transcription factor can take advantage of this nonspecific disruption of nucleosomes and bind to its target sequence preventing the reformation of the nucleosome.

nucleosomes is best illustrated by the demonstrated mobility of nucleosomes under physiological conditions,[24] but it is also possible that the nucleosome octamer is in equilibrium with subnucleosomal particles through assembly/disassembly reactions under native conditions. Alternatively, transcription factors may bind to their sites in chromatin and synergize with either targeted or ubiquitous remodeling machines to trigger nucleosome disassembly (Fig. 18.3A). This scenario would benefit from specific interactions between transcription factors and remodeling machines.

Other mechanisms by which chromatin reorganizing activities may be targeted to a promoter are suggested by two recent findings: first, a coactivator of transcription has been identified as a histone acetylase, pointing to a close association of chromatin remodeling activities with transcription factors.[65] Secondly, it was found that SWI/SNF proteins can associate with the RNA polymerase II to form a huge "holoenzyme complex".[66] This again suggests that chromatin remodeling entities may be closely associated with or even part of a productive transcription machinery, but does not provide a convincing explanation of how accessibility at promoters is created in the first place. According to the current models it is the binding of the sequence-specific activators and TFIID, which is not part of a holoenzyme complex, to chromatin that recruits the polymerase complex. Other activities, not associated with the polymerase, should therefore be necessary to initiate the series of events that leads to the establishment of active promoters in chromatin.

CONCLUSION

It appears that transcription factors can be subdivided into categories according to their ability to induce nucleosome remodeling and to further affect the formation of productive preinitiation complexes.[5,67,70] Their potential will depend on their competence to interact with nucleosomal DNA and their ability to synergize with nucleosome remodeling machines. Key to our understanding of how important regulators function in the context of chromatin is the appreciation of the dynamic, flexible nature of chromatin, down to the ubiquitous nucleosome unit. The basic principles that govern chromatin accessibility will undoubtedly be unraveled in the years to come. They are of relevance not only for our complete understanding of transcriptional regulation in nuclei, but also for other processes that require a chromatin substrate, such as replication[68] and recombination.[69]

ACKNOWLEDGMENTS

I thank Karl Nightingale and Patrick Varga-Weisz for valuable comments on the manuscript.

REFERENCES

1. Orlando V, Paro R. Mapping *Polycomb*-repressed domains in the *bithorax* complex using in vivo formaldehyde cross-linked chromatin. Cell 1993; 75:1-20.
2. Lavender JS, Birley AJ, Palmer MJ et al. Histone H4 acetylated at lysine 16 and proteins of the *Drosophila* dosage compensation pathway co-localize on the male X chromosome through mitosis. Chromosome Research 1994; 2:398-404.
3. Turner BM, O'Neill LP. Histone acetylation in chromatin and chromosomes. Sem Cell Biol 1995; 6:229-236.
4. Elgin SCR. The formation and function of DNase I hypersensitive sites in the process of gene activation. J Biol Chem 1988; 263:19259-19262.
5. Wallrath LL, Quinn L, Granok H et al. Architectural variations of inducible eukaryotic promoters: Preset and remodeling chromatin structures. BioEssays 1994; 16:165-170.
6. Felsenfeld G. Chromatin as an essential part of the transcription mechanism. Nature 1992; 355:219-224.
7. Solomon MJ, Varshavsky A. A nuclease-hypersensitive region forms de novo after chromosome replication. Mol Cell Biol 1987; 7:3822-3825.
8. Lucchini R, Sogo JM. Replication of transcriptionally active chromatin. Nature 1995; 374:276-280.
9. Beato M, Herrlich P, Schütz G. Many actors in search of a plot. Cell 1995; 83:851-857.
10. Reik A, Schütz G, Stewart F. Glucocorticoids are required for establishment and maintenance of an alteration in chromatin structure: Induction leads to a reversible disruption of nucleosomes over an enhancer. EMBO J 1991; 10:2569-2576.
11. Schmid A, Fascher K-D, Hörz W. Nucleosome disruption at the yeast PHO5 promoter upon PHO5 induction occurs in the absence of DNA replication. Cell 1992; 71:853-864.
12. Becker PB. The establishment of active promoters in chromatin. BioEssays 1994; 16:541-547.
13. Wolffe AP. New insights into chromatin function in transcriptional control. FASEB J 1992; 6:354-3361.
14. Owen-Hughes T, Workman JL. Experimental analysis of chromatin function in transcriptional control. Crit Rev Euk Gene Expr 1994; 8:1-39.
15. Perlmann T, Wrange Ö. Specific glucocorticoid receptor binding to DNA reconstituted in a nucleosome. EMBO J 1988; 7:3073-3079.
16. Pina B, Bruggemeier U, Beato M. Nucleosome positioning modulates accessibility of regulatory proteins to the mouse mammary tumor virus promoter. Cell 1990; 60:719-31.
17. McPherson CE, Shim E, Friedman DS et al. An active tissue-specific enhancer and bound transcription factors existing in a precisely positioned nucleosomal array. Cell 1993; 75:387-398.
18. Truss M, Bartsch J, Schelbert A et al. Hormone induces binding of receptors and transcription factors to a rearranged nucleosome on the MMTV promoter in vivo. EMBO J 1995; 14:1737-1751.

19. Thanos D, Maniatis T. Virus induction of human IFNβ gene expression requires the assembly of an enhanceosome. Cell 1995; 83:1091-1100.
20. Li Q, Wrange Ö. Accessibility of a glucocorticoid response element in a nucleosome depends on its rotational positioning. Mol Cell Biol 1995; 15: 4375-4384.
21. Fragoso G, John S, Roberts MS et al. Nucleosome positioning on the MMTV LTR results from frequency-biased occupancy of multiple frames. Genes Dev 1995; 9:1933-1947.
22. Meersseman G, Pennings S, Bradbury EM. Mobile nucleosomes—a general behaviour. EMBO J 1992; 11:2951-2959.
23. Van Holde KE, Yager TD. Nucleosome motion: Evidence and Models. In: Structure anf Function of the Genetic Apparatus. Nicolini C, Ts'o POP, eds. Plenium Publishing Co New York 1985.
24. Varga-Weisz PD, Blank TA, Becker PB. Energy-dependent chromatin accessibility and nucleosome mobility in a cell-free system. EMBO J 1995; 14:2209-2216.
25. Li Q, Wrange Ö. Translational positioning of a nucleosomal glucocorticoid response element modulates glucocorticoid receptor affinity. Gene Dev 1993; 7: 2471-2482.
26. Vettese-Dadey M, Walter P, Chen H et al. Role of the histone amino termini in facilitated binding of a transcription factor, GAL4-AH, to nucleosome cores. Mol Cell Biol 1994; 14:970-981.
27. Pruss D, Hayes JJ, Wolffe AP. Nucleosomal anatomy - where are the histones? BioEssays 1995; 17:161-170.
28. Hamiche A, Schultz P, Ramakrishnan V et al. Linker histone-dependent DNA structure in linear mononucleosomes. J Mol Biol 1996; 257:30-42.
29. Juan L, Utley RT, Adams CC et al. Differential repression of transcription factor binding by histone H1 is regulated by the core histone amino termini. EMBO J 1994; 13:6031-6040.
30. Ura K, Hayes JJ, Wolffe AP. A positive role for nucleosome mobility in the transcriptional activity of chromatin templates: Restriction by linker histones. EMBO J 1995; 14:3752-3765.
31. Bresnick EH, Bustin M, Marsaud V et al. The transcriptionally active MMTV promoter is depleted of histone H1. Nucl Acids Res 1991; 20:273-278.
32. Adams CC, Workman JL. Binding of disparate transcriptional activators to nucleosomal DNA is inherently cooperative. Mol Cell Biol 1995; 15:1405-1421.
33. Polach KJ, Widom J. A model for cooperative binding of eukaryotic regulatory proteins to nucleosomal target sites. J Mol Biol 1996; in press
34. Polach KJ, Widom J. Mechanism of protein access to specific DNA sequences in chromatin: A dynamic equilibrium model for gene regulation. J Mol Biol 1995; 254:130-149.
35. Lee DY, Hayes JJ, Pruss D et al. A positive role for histone acetylation in transcription factor access to nucleosomal DNA. Cell 1993; 72:73-84.
36. Vettese-Dadey M, Grant PA, Hebbes TR et al. Acetylation of histone H4 plays a primary role in enhancing transcription factor binding to nucleosomal DNA in vitro. EMBO J 1996; 15:2508-2518.
37. Van Lint C, Emiliani S, Ott M et al. Transcriptional activation and chromatin remodeling of the HIV-1 promoter in response to histone acetylation. EMBO J 1996; 15:1112-1120.
38. Ilyin YV, Varshavsky AY, Mickelsaar UN et al. Redistribution of proteins in mixtures of deoxyribonucleoproteins, DNA and RNA. Eur J Biochem 1971; 22: 235-245.
39. Workman JL, Kingston RE. Nucleosome core displacement in vitro via a metastable transcription factor-nucleosome complex. Science 1992; 258:1780-1784.
40. Chen H, Li B, Workman JL. A histone-binding protein, nucleoplasmin, stimulates transcription factor binding to nucleosomes and factor-induced nucleosome disassembly. EMBO J 1994; 13: 380-390.
41. Walter PP, Owen-Hughes TA, Coté J et al. Stimulation of transcription factor binding and histone displacement by nucleosome assembly protein 1 and nucleoplasmin requires disruption of the histone octamer. Mol Cell Biol 1995; 15:6178-6187.

42. Gruss C, Wu J, Koller T et al. Disruption of nucleosomes at the replication fork. EMBO J. 1993; 12:4533-4545.
43. Nacheva GA, Guschin DY, Preobrazhenskaya OV et al. Change in pattern of histone binding to DNA upon transcriptional activation. Cell 1989; 58:27-36.
44. Earnshaw WC, Honda BM, Laskey RA. Assembly of nucleosomes: The reaction involving X. laevis nucleoplasmin. Cell 1980; 21:373-383.
45. Ishimi Y, Yasuda H, Hirosumi J et al. A protein which facilitates assembly of nucleosome-like structures in vitro in mammalian cells. J Biochem 1983; 94:735-744.
46. Baer BW, Rhodes D. Eukaryotic RNA polymerase II binds to nucleosome cores from transcribed genes. Nature 1983; 301:482-488.
47. Mann RK, Grunstein M. Histone H3 N-terminal mutations allow hyperactivation of the yeast GAL 1 gene in vivo. EMBO J 1992; 11:3297-3306.
48. Wan JS, Mann RK, Grunstein M. Yeast histone H3 and H4 N-termini function through different Gal 1 regulatory elements to repress and activate transcription. Proc Natl Acad Sci USA 1995; 92:5664-5668.
49. Fisher-Adams G, Grunstein M. Yeast histone H4 and H3 N-termini have different effects on the chromatin structure of the GAL1 promoter. EMBO J 1995; 14:1468-1477.
50. Alevizopoulos A, Dusserre Y, Tsai-Pflugfelder M et al. A proline-rich TGF-β-responsive transcriptional activator interacts with histone H3. Genes Dev 1995; 9:3051-3066.
51. Peterson CL, Tamkun JW. The SWI-SNF complex: A chromatin remodeling machine? TIBS 1995; 20:143-146.
52. Hirschhorn JN, Brown SA, Clark CD et al. Evidence that SNF2/SWI2 and SNF5 activate transcription in yeast by altering chromatin structure. Genes Dev 1992; 6:2288-2298.
53. Laurent BC, Treich I, Carlson M. The yeast SNF2/SWI2 protein has DNA-stimulated ATPase activity required for transcriptional activation. Genes Dev 1993; 7:583-591.
54. Côté J, Quinn J, Workman JL et al. Stimulation of GAL4 derivative binding to nucleosomal DNA by the yeast SWI/SNF complex. Science 1994; 265:53-60.
55. Kwon H, Imbalzano AN, Khavari PA et al. Nucleosome disruption and enhancement of activator binding by a human SWI/SNF complex. Nature 1994; 370: 477-481.
56. Imbalzano AN, Kwon H, Green MR et al. Facilitated binding of TATA-binding protein to nucleosomal DNA. Nature 1994; 370:481-485.
57. Kruger W, Peterson CL, Sil A et al. Amino acid substitutions in the structured domains of histones H3 and H4 partially relieve the requirement of the yeast SWI/SNF complex for transcription. Genes Dev 1995; 9:2770-2779.
58. Quinn J, Fyrberg AM, Ganster RW et al. DNA-binding properties of the yeast SWI/SNF complex. Nature 1996; 379: 844-847.
59. Tsukiyama T, Becker PB, Wu C. ATP-dependent nucleosome disruption at a heat-shock promoter mediated by binding of GAGA transcription factor. Nature 1994; 367:525-532.
60. Wall G, Varga-Weisz PD, Sandaltzopoulos R et al. Chromatin remodeling by GAGA factor and heat shock factor at the hypersensitive *Drosophila* hsp26 promoter in vitro. EMBO J 1995; 1727-1736.
61. Tsukiyama T, Wu C. Purification and properties of an ATP-dependent nucleosome remodeling factor. Cell 1995; 83: 1011-1020.
62. Elfring LK, Deuring R, McCallum CM et al. Identification and characterization of *Drosophila* relatives of the yeast transcriptional activator SNF2/SWI2. Mol Cell Biol 1994; 14:2225-2234.
63. Tsukiyama T, Wu C. ISWI, a member of the SWI2/SNF2 ATPase family, encodes the 140 kD subunit of the nucleosome remodeling factor. Cell 1995; 83: 1021-1026.
64. Hebbes TR, Clayton AL, Thorne AW et al. Core histone acetylation co-maps with generalized DNase I sensitivity in the chicken β-globin chromosomal domain. EMBO J 1994; 13:1823-1830.
65. Brownell JE, Zhou T, Kobayashi R et al. Tetrahymena histone acetyltransferase A: A homolog to yeast Gcn5p linking histone acetylation to gene activation. Cell 1996; 84:843-851.

66. Wilson CJ, Chao DM, Imbalzano AN et al. RNA polymerase II holoenzyme contains SWI/SNF regulators involved in chromatin remodeling. Cell 1996; 84:235-244.
67. Pazin MJ, Sheridan PL, Cannon K et al. NF-κB-mediated chromatin reconfiguration and transcriptional activation of the HIV-1 enhancer in vitro. Genes Dev 1996; 10:37-49.
68. Ramsperger U, Stahl H. Unwinding of chromatin by the SV40 large T antigen helicase. EMBO J 1995; 14:3215-3225.
69. Ohta K, Shibata T, Nicolas A. Changes in chromatin structure at recombination initiation sites during meiosis. EMBO J 1994; 13:57554-5763.
70. Boyes J, Felsenfeld G. Tissue-specific factors additively increase the probability of the all-or-none formation of a hypersensitive site. EMBO J 1996; 15:2496-2507.

INDEX

F

G

H

I

J

K

L

P

R

S

T

U

V

W

X

Y

Z